Complex Dynamic System Modelling, Identification and Control

Complex Dynamic System Modelling, Identification and Control

Editors

Quanmin Zhu
Giuseppe Fusco
Jing Na
Weicun Zhang
Ahmad Taher Azar

MDPI • Basel • Beijing • Wuhan • Barcelona • Belgrade • Manchester • Tokyo • Cluj • Tianjin

MDPI

Editors

Quanmin Zhu
School of Engineering
University of the West of England
Bristol
United Kingdom

Giuseppe Fusco
Department of Electrical and Information Engineering
University of Cassino and Southern Lazio
Cassino
Italy

Jing Na
Electrical Engineering
Kunming University of Science and Technology
Yunnan
China

Weicun Zhang
School of Automation and Electrical Engineering
University of Science and Technology Beijing
Beijing
China

Ahmad Taher Azar
College of Computer & Information Sciences (CCIS)
Prince Sultan University
Riyadh
Saudi Arabia

Editorial Office
MDPI
St. Alban-Anlage 66
4052 Basel, Switzerland

This is a reprint of articles from the Special Issue published online in the open access journal *Entropy* (ISSN 1099-4300) (available at: www.mdpi.com/journal/entropy/special_issues/Complex_Dyn_Syst).

For citation purposes, cite each article independently as indicated on the article page online and as indicated below:

LastName, A.A.; LastName, B.B.; LastName, C.C. Article Title. *Journal Name* **Year**, *Volume Number*, Page Range.

ISBN 978-3-0365-7661-9 (Hbk)
ISBN 978-3-0365-7660-2 (PDF)

Contents

About the Editors

Quanmin Zhu

Quanmin Zhu is a Professor in control systems at the School of Engineering, University of the West of England, Bristol, UK. He obtained his MSc from Harbin Institute of Technology, China, in 1983, and his PhD from the Faculty of Engineering, University of Warwick, UK, in 1989, and worked as a postdoctoral research associate at the Department of automatic control and systems engineering, University of Sheffield, UK, between 1989-2004. His main research interest is in nonlinear system modeling, identification, and control. He has published over 300 papers on these topics, edited various books with Springer, Elsevier, and other publishers, and provided consultancy to various industries. Currently, Professor Zhu is acting as Editor of Elsevier's book series Emerging Methodologies and Applications in Modelling, Identification and Control.

Giuseppe Fusco

In 1988, Giuseppe Fusco received the Laurea in Electrical Engineering at the University of Naples Federico II. In 1990 he won a grant entitled "Ottimizzazione delle reti elettriche di trazione urbana e suburbana con metodi non lineari" by the Ansaldo Trasporti. In 1990, he started working as a scientific assistant at the Laboratory of Industrial Engineering at the University of Cassino. In 1995, he served as an assistant professor, and since 2003, he has been an associate professor in control engineering. He has been a member of several Italian and European projects such as ARCAS, IGREENGrid, AEROARMS, WiMUST, DeXRoV, ROBUST, C4E, Robilaut, and Canopies. He is a patentee concerning an innovative relay that detects unintentional islanding operation of a part of a distribution grid. Since 2009, he has been the European regional editor for the "International Journal of Modelling, Identification and Control". He is a co-author of about 110 international journal and conference papers. He is the co-author of the Springer book titled "Adaptive Voltage Control in Power Systems: Modeling Design and Applications". His main research interest is in the control of electric power systems and smart grids.

Jing Na

Jing Na is currently a Professor at the Faculty of Electrical & Mechanical Engineering at the Kunming University of Science & Technology. He received his B.S. and Ph.D. degrees from the School of Automation, Beijing Institute of Technology, China, in 2004 and 2010, respectively. From 2011 to 2012, he was a Monaco/ITER Postdoctoral Fellow with the ITER Organization, France. From 2015 to 2016, he was a Marie Curie Fellow with the University of Bristol, UK. He has also held visiting positions with the Universitat Politecnica de Catalunya, Spain, and with University of Bristol, UK.

His current research interests include parameter estimation, adaptive optimal control, and nonlinear control with application to vehicle systems, servo mechanisms, and energy conversion plants (e.g., engine, wave energy converters, etc.). He has authored/co-authored more than 100 peer-reviewed journal and conference papers. Dr Na has been awarded the Best Application Paper Award of the 3rd IFAC International Conference on Intelligent Control and Automation Science (IFAC ICONS 2013), and the 2017 Hsue-shen Tsien Paper Award. He is currently an Associate Editor of the IEEE Transactions on Industrial Engineering, Neurocomputing, and International Journal of Modelling, Identification and Control. He has served as an international program committee Chair of ICMIC 2017 and DDCLS 2019 and as an IPC member of many international conferences. He has organized/co-organized several special issues in international journals and invited sessions at several prestigious conferences.

Weicun Zhang

Weicun Zhang received his Ph.D. degree in Control Theory and Applications from Tsinghua University in 1993. His research interests include self-tuning control, multiple model adaptive control, and estimation for both linear and nonlinear dynamic systems. As representative research work, he established a Virtual Equivalent System theory for a unified analysis of a general class of self-tuning control systems and multiple model adaptive control systems, which is independent of specific control strategy and parameter estimation algorithms.

Ahmad Taher Azar

Prof. Ahmad Azar is a Research Professor at Prince Sultan University, Riyadh, the Kingdom of Saudi Arabia. He is also a professor at the Faculty of Computers and Artificial intelligence, Benha University, Egypt. He is a leader of the Automated Systems and Soft Computing Lab (ASSCL), Prince Sultan University, Saudi Arabia.

Prof. Azar is the Editor in Chief of several renowned international journals, including the *International Journal of Service Science, Management, Engineering, and Technology* (IJSSMET), the *International Journal of System Dynamics Applications* (IJSDA), and the *International Journal of Sociotechnology and Knowledge Development* (IJSKD), published by IGI Global, USA. He is also the Editor in Chief of the *International Journal of Intelligent Engineering Informatics* (IJIEI) published by Inderscience Publishers, UK.

With over 430 research papers published in prestigious peer-reviewed journals, book chapters, and conference proceedings, Prof. Azar has made significant contributions to areas such as control systems, robotics, artificial intelligence, and machine learning.

Prof. Azar is actively involved in the academic community as a reviewer for international journals and a committee member for international conferences. He has received several awards and honors, including the Egyptian State Encouragement Award in Engineering Sciences and the Abdul Hameed Shoman Arab Researchers Award in Machine Learning and Big Data Analytics. He has also been recognized as one of the top computer scientists in Saudi Arabia and as one of the top 2% of scientists in the world in artificial intelligence by Stanford University.

Additionally, Prof. Azar holds various leadership positions in professional organizations, such as the Vice Chair of the International Federation of Automatic Control (IFAC) Technical Committee of Control Design, and the Vice President (North) of the System Dynamics Africa Regional Chapter. He is actively involved in IEEE committees and task forces related to emerging technologies, robotics, fuzzy logic, and computational collective intelligence.

Preface to "Complex Dynamic System Modelling, Identification and Control"

Systems are naturally or purposely formed with functional components and connection structures. Their complicities are induced by nonlinearity, dynamics, time delay, uncertainties, disturbances, irreversible processes, and those characteristics that are generally explained in other literature. Modeling represents the innate tendency of humans to find rules or mechanisms that govern phenomena (a process/plant in a human-made system or a natural system, such as the Earth's global climate, organisms, and the human brain). This is generally consistent with the journal titled *Entropy*, as the idea of entropy provides a mathematical way to encode/model the intuitive notion of which processes are obviously complex due to their irreversible characteristics, even though they would not violate the fundamental law of the conservation of energy. There are two predominant approaches to establishing models: principle-based (e.g., information theory, statistic physics, statistical mechanics, etc.) analytical equations and data (measured and simulated) driven input/output fitted sets of regression numerical polynomials (most commonly called identification). Control is a way to improve a system's behavior/performance by adding additional functional components and revising the system structure to form a closed-loop framework with adaptation and robustness against uncertainties. Accordingly, modeling, identification, and control (MIC) is a cross-discipline from all engineering (human-made) systems to all natural scientific research discoveries. This reprint encourages those emerging insights and approaches to provide concise/effective solutions in complex dynamic system modeling, identification, and control. The philosophy embedded in the SI is to seek simplicity (solutions) from complicity (problems). This reprint is a forum for presenting new and improved insight, methodologies, and techniques of MIC for complex systems that are challenging for research and (potentially) significant for a wide range of applications in real-world natural and engineering domains. Fundamentally, the papers should justify why the works have not been undertaken by other colleagues and what the bottleneck issues have been for such research progression and applications.

As the editors, we hope that the chapters in this reprint will stimulate further research in complex system modeling and utilize them in real-world applications. We hope that this reprint, covering so many different aspects, will be of value to all readers.

We would like to also thank the reviewers for their diligence in reviewing the chapters.

Quanmin Zhu, Giuseppe Fusco, Jing Na, Weicun Zhang, and Ahmad Taher Azar
Editors

entropy

MDPI

Editorial

Special Issue "Complex Dynamic System Modelling, Identification and Control"

Quanmin Zhu [1,*], Giuseppe Fusco [2], Jing Na [3], Weicun Zhang [4] and Ahmad Taher Azar [5,6]

1 Department of Engineering Design and Mathematics, University of the West of England, Frenchy Campus Coldharbour Lane, Bristol BS16 1QY, UK
2 Department of Electrical and Information Engineering, Universita degli Studi di Cassino e del Lazio Meridionale, 03043 Cassino, FR, Italy; fusco@unicas.it
3 Faculty of Mechanical & Electrical Engineering, Kunming University of Science and Technology, No. 727 Jingming South Road, Chenggong, Kunming 650500, China; najing25@kust.edu.cn
4 School of Automation and Electrical Engineering, University of Science and Technology Beijing, Beijing 100083, China; weicunzhang@ustb.edu.cn
5 College of Computer & Information Sciences (CCIS), Prince Sultan University, Riyadh 11586, Saudi Arabia; aazar@psu.edu.sa or ahmad.azar@fci.bu.edu.eg or ahmad_t_azar@ieee.org
6 Faculty of Computers and Artificial Intelligence, Benha University, Benha 13518, Egypt
* Correspondence: quan.zhu@uwe.ac.uk

Citation: Zhu, Q.; Fusco, G.; Na, J.; Zhang, W.; Azar, A.T. Special Issue "Complex Dynamic System Modelling, Identification and Control". *Entropy* **2022**, *24*, 380. https://doi.org/10.3390/e24030380

Received: 16 February 2022
Accepted: 25 February 2022
Published: 8 March 2022

1. Introduction

Systems are naturally or purposely formed with functional components and connection structures. The complicities could be induced from nonlinearity, dynamics, time delay, uncertainties, disturbances, irreversible processes, and those characteristics generally explained in other literature.

Modeling represents an innate intuition for humans to find rules or mechanisms that govern phenomena (a process/plant in a human-made system or a natural system, such as earth's global climate, organisms, and the human brain). This is generally consistent with the journal titled Entropy, as the idea of entropy provides a mathematical way to encode/model the intuitive notion of which processes are obviously complex due to their irreversible characteristics, even though they would not violate the fundamental law of the conservation of energy. There are two predominant approaches to establish models, principle-based (e.g., information theory, statistic physics, statistical mechanics, etc.) analytical equations and data (measured and simulated) driven input/output fitted sets of regression numerical polynomials (most commonly called identification). Control is a way to improve a system behavior/performance by adding additional functional components and revising the system structure to form a closed-loop framework with adaptation and robustness to the uncertainties. Accordingly, modeling, identification, and control (MIC) is a cross-discipline from all engineering (human-made) systems to all natural scientific research discoveries.

This Special Issue (SI) encourages those emerging insights and approaches to provide concise/effective solutions in complex dynamic system modeling, identification, and control. The philosophy embedded in the SI is to seek simplicity (solutions) from complicity (problems). This Special Issue is a forum for presenting new and improved insight, methodologies, and techniques of MIC for complex systems that are challenging for research and (potentially) significant for a wide range of applications in real-world natural and engineering domains. Fundamentally, the papers should justify why the works have not been undertaken by the other colleagues and what the bottleneck issues have been for such research progression and applications.

2. The Expanded SI Publication List

This SI accepted/published 18 papers. Here, a brief summary is presented as an expanded content list for quick view of the SI.

Li et al. [1] propose a U-model-based two-degree-of-freedom internal model control (UTDF-IMC) structure with strength in nonlinear dynamic inversion and the separation of tracking design and robustness design. This approach can effectively accommodate modeling error and disturbance while removing those widely used linearization techniques for nonlinear plants/processes. To assure the expansion and applications, it analyses the key properties associated with the UTDF-IMC. For initial benchmark testing, computational experiments are conducted using MATLAB/Simulink for two mismatched linear and nonlinear plants. Further tests consider an industrial system, in which the IMC of a permanent magnet synchronous motor (PMSM) is simulated to demonstrate the effectiveness of the design procedure for potential industrial applications.

Kang et al. [2] consider the situations: Android devices are currently widely used in many fields, such as automatic control, embedded systems, the Internet of Things, and so on. At the same time, Android applications (apps) always use multiple permissions, and permissions can be abused by malicious apps that disclose users' privacy or breach the secure storage of information. FlowDroid has been extensively studied as a novel and highly precise static taint analysis for Android applications. Aiming to resolve the problem of complex detection and false alarms in FlowDroid, an improved static detection method based on feature permission and risk rating is proposed. Firstly, the Chi-square test is used to extract correlated permissions related to malicious apps, and mutual information is used to cluster the permissions to generate feature permission clusters. Secondly, risk calculation method based on permissions and combinations of permissions are proposed to identify dangerous data flows. Experiments show that this method can significantly improve detection efficiency while maintaining the accuracy of dangerous data flow detection.

Jin et al. [3] consider trend prediction based on sensor data in a multi-sensor system as an important topic. As the number of sensors increases, we can measure and store more and more data. However, the increase in data has not effectively improved prediction performance. This paper focuses on this problem and presents a distributed predictor that can overcome unrelated data and sensor noise: First, the causality entropy is defined to calculate the measurement's causality. Then, the series causality coefficient (SCC) is proposed to select the high causal measurement as the input data. To overcome the traditional deep learning network's over-fitting to the sensor noise, the Bayesian method is used to obtain the weight distribution characteristics of the sub-predictor network. A multi-layer perceptron (MLP) is constructed as the fusion layer to fuse the results from different sub-predictors. The experiments were implemented to verify the effectiveness of the proposed method by meteorological data from Beijing. The results show that the proposed predictor can effectively model the multi-sensor system's big measurement data to improve prediction performance.

Liang et al. [4] proposed a mechanistic kinetic model of cobalt–hydrogen electrochemical competition for the cobalt removal process in zinc hydrometallurgy. In addition, to overcome the parameter estimation difficulties arising from the model nonlinearities and the lack of information on the possible value ranges of parameters to be estimated, a constrained guided parameter estimation scheme was derived based on model equations and experimental data. The proposed model and the parameter estimation scheme have two advantages: (i) The model reflected for the first time the mechanism of the electrochemical competition between cobalt and hydrogen ions in the process of cobalt removal in zinc hydrometallurgy. (ii) The proposed constrained parameter estimation scheme did not depend on the information of the possible value ranges of parameters to be estimated. (iii) The constraint conditions provided in that scheme directly linked the experimental phenomenon metrics to the model parameters, thereby providing deeper insights into the model parameters for model users. Numerical experiments showed that the proposed constrained parameter estimation algorithm significantly improved the estimation efficiency. Meanwhile, the proposed cobalt–hydrogen electrochemical competition model allowed for accurate simulation of the impact of hydrogen ions on cobalt removal rate as well as simulation of the trend of hydrogen ion concentration, which would be helpful for the actual cobalt removal process in zinc hydrometallurgy.

Luan et al. [5] consider that an industrial process, namely the estimation of feeding composition, is important for analyzing production status and making control decisions. However, random errors or even gross ones inevitably contaminate the actual measurements. Feeding composition is conventionally obtained via discrete and low-rate artificial testing. To address these problems, a feeding composition estimation approach based on a data reconciliation procedure is developed. To improve the variable accuracy, a novel robust M-estimator is first proposed. Then, an iterative robust hierarchical data reconciliation and estimation strategy is applied to estimate the feeding composition. The feasibility and effectiveness of the estimation approach are verified on a fluidized bed roaster. The proposed M-estimator showed better overall performance.

Pan et al. [6] propose a data-driven fault diagnosis method using the deep convolutional neural network (DCNN). The DCNN is used to deal with sensor and actuator faults of robot joints, such as gain error, offset error, and malfunction for both sensors and actuators, and different fault types are diagnosed using the trained neural network. In order to achieve the above goal, the fused data of sensors and actuators are used, where both types of fault are described in one formulation. Then, the deep convolutional neural network is applied to learn characteristic features from the merged data to try to find discriminative information for each kind of fault. After that, the fully connected layer performs prediction work based on learned features. In order to verify the effectiveness of the proposed deep convolutional neural network model, different fault diagnosis methods, including support vector machine (SVM), artificial neural network (ANN), conventional neural network (CNN) using the LeNet-5 method, and long-term memory network (LTMN), are investigated and compared with the DCNN method. The results show that the DCNN fault diagnosis method can realize high fault recognition accuracy while needing less model training time.

Li et al. [7] investigates a critical hazard identification method for railway accident prevention. A new accident causation network is proposed to model the interaction between hazards and accidents. To realize consistency between the most likely and shortest causation paths in terms of hazards for accidents, a method for measuring the length between adjacent nodes is proposed, and the most likely causation path problem is first transformed to the shortest causation path problem. To identify critical hazard factors that should be alleviated for accident prevention, a novel critical hazard identification model is proposed based on a controllability analysis of hazards. Five critical hazard identification methods are proposed to select critical hazard nodes in an accident causality network. A comparison of results shows that the combination of an integer programming-based critical hazard identification method and the proposed weighted direction accident causality network considering length has the best performance in terms of accident prevention.

Han et al. [8] propose a new active fault tolerant control scheme based on active fault diagnosis to address the component/actuator faults for systems with state and input constraints. Firstly, the active fault diagnosis is composed of diagnostic observers, constant auxiliary signals, and separation hyperplanes, all of which are designed offline. In online applications, only a single diagnostic observer is activated to achieve fault detection and isolation. Compared with the traditional multi-observer parallel diagnosis methods, such a design is beneficial to improve the diagnostic efficiency. Secondly, the active fault tolerant control is composed of outer fault tolerant control, inner fault tolerant control, and a linear-programming-based interpolation control algorithm. The inner fault tolerant control is determined offline and satisfies the prescribed optimal control performance requirement. The outer fault tolerant control is used to enlarge the feasible region, and it needs to be determined online together with the interpolation optimization. In online applications, the updated state estimates trigger the adjustment of the interpolation algorithm, which in turn enables control reconfiguration by implicitly optimizing the dynamic convex combination of outer fault tolerant control and inner fault tolerant control. This control scheme contributes to further reducing the computational effort of traditional constrained predictive fault tolerant control methods. In addition, each pair of inner fault tolerant control and diagnostic observer is designed in an integrated manner to suppress the robust interaction of influences between estimation error and control error. The soft constraint method is further integrated to handle some cases that lead to constraint violations. The effectiveness of these designs is finally validated by a case study of a wastewater treatment plant model.

Zhang et al. [9] consider that spacecraft with large flexible appendages are characterized by multiple system modes. They suffer from inherent low-frequency disturbances in the operating environment that consequently result in considerable interference in the operational performance of the system. It is required that the control design ensures the system's high pointing precision, and it is also necessary to suppress low-frequency resonant interference as well as take into account multiple performance criteria, such as attitude stability and bandwidth constraints. Aiming to address the comprehensive control problem of this kind of flexible spacecraft, a control strategy is proposed using a structured H-infinity controller with low complexity that was designed to meet multiple performance requirements, so as to reduce the project cost and implementation difficulty. According to the specific resonant mode of the system, the design strategy of adding an internal mode controller, a trap filter, and a series PID controller to the structured controller is proposed, so as to achieve the comprehensive control goals through cooperative control of multiple control modules. A spacecraft with flexible appendages (solar array) is presented as an illustrative example in which a weighted function was designed for each performance requirement of the system (namely robustness, stability, bandwidth limit, etc.), and a structured comprehensive performance matrix with multiple performance weights and decoupled outputs was constructed. A structured H-infinity controller meeting the comprehensive performance requirements is given, which provides a structured integrated control method with low complexity for large flexible systems that is convenient for engineering practice and provides a theoretical basis and reference examples for structured H-infinity control. The simulation results show that the proposed controller gives better control performance compared with the traditional H-infinity one and can successfully suppress the vibration of large flexible appendages at 0.12 Hz and 0.66 Hz.

Li et al. [10] examine the adaptive control of high-order nonlinear systems with strict-feedback form. An adaptive fixed-time control scheme is designed for nonlinear systems with unknown uncertainties. In the design process of a backstepping controller, the Lyapunov function, an effective controller, and adaptive law are constructed. Combined with the fixed-time Lyapunov stability criterion, it is proven that the proposed control scheme can ensure the stability of the error system in finite time, and the convergence time is independent of the initial condition. Finally, simulation results verify the effectiveness of the proposed control strategy.

Luo et al. [11] investigate the cluster-delay mean square consensus problem of a class of first-order nonlinear stochastic multi-agent systems with impulse time windows. Specifically, on the one hand, a discrete control mechanism (i.e., impulsive control) was applied in the system instead of a continuous one, which has the advantages of low control cost and high convergence speed. On the other hand, the existence of impulse time windows was considered when modeling the system, i.e., a single impulse appears randomly within a time window rather than an ideal fixed position. In addition, this paper also considers the influence of stochastic disturbances caused by fluctuations in the external environment. Then, based on algebraic graph theory and Lyapunov stability theory, some sufficiency conditions that the system must meet to reach the consensus state are given. Finally, we designed a simulation example to verify the feasibility of the obtained results.

Pang et al. [12] consider that the continuous development of spacecraft with large flexible structures has resulted in an increase in the mass and aspect ratio of launch vehicles, while the wide application of lightweight materials in the aerospace field has increased the flexible modes of launch vehicles. In order to solve the problem of deviation from the nominal control or even destabilization of the system caused by uncertainties, such as unknown or unmodelled dynamics, frequency perturbation of the flexible mode, changes in its own parameters, and external environmental disturbances during the flight of such large-scale flexible launch vehicles with simultaneous structural deformation, rigid-elastic coupling and multimodal vibrations, an improved adaptive augmentation control method based on model reference adaption, and spectral damping are proposed in this paper, including a basic PD controller, a reference model, and an adaptive gain adjustment based on spectral damping. The baseline PD controller was used for flight attitude control in the nominal state. In the non-nominal state, the spectral dampers in the adaptive gain adjustment law extracted and processed the high-frequency signal from the tracking error and control-command error between the reference model and the actual system to generate the adaptive gain. The adjustment gain was multiplied by the baseline controller gain to increase/decrease the overall gain of the system to improve the system's performance and robust stability, so that the system had the ability to return to the nominal state when it was affected by various uncertainties and deviated from the nominal state or even destabilized.

Zhang et al. [13] address the problem of local fault (unknown input) reconstruction for interconnected systems. This contribution consists of a geometric method which solves the fault reconstruction (FR) problem based on observations and a differential algebraic concept. The fault diagnosis (FD) problem is tackled using the concept of the differential transcendence degree of a differential field extension and the algebraic observability. The goal is to examine whether the fault occurring in the low-level subsystem can be reconstructed correctly by the output at the high-level subsystem under given initial states. By introducing the fault as an additional state of the low subsystem, an observer based approached is proposed to estimate this new state. Particularly, the output of the lower subsystem is assumed unknown, and is considered as auxiliary output. Then, the auxiliary outputs are estimated by a sliding mode observer, which is generated by using global outputs and inverse techniques. After this, the estimated auxiliary outputs are employed as virtual sensors of the system to generate a reduced-order observer, which is capable of estimating the fault variable asymptotically. Thus, the purpose of multi-level fault reconstruction is achieved. Numerical simulations on an intensified heat exchanger are presented to illustrate the effectiveness of the proposed approach.

Azar et al. [14], an all-guest editor involved work, presents the robust stabilization and synchronization of a novel chaotic system. First, a novel chaotic system is presented, which is realized by implementing a sigmoidal function to generate the chaotic behavior. A bifurcation analysis is provided in which by varying three parameters of this chaotic system, the respective bifurcations plots are generated and evinced to analyze and verify when this system is in the stability region or in a chaotic regimen. Then, a robust controller is designed to drive the system variables from the chaotic regimen to stability so that these variables reach the equilibrium point in finite time. The robust controller is obtained by selecting an appropriate robust control Lyapunov function to obtain the resulting control law. For synchronization purposes, the novel chaotic system designed in this study is used as a drive and response system, considering that the error variable is implemented in a robust control Lyapunov function to drive this error variable to zero in finite time. In the control law design for stabilization and synchronization purposes, an extra state is provided to ensure that the saturated input sector condition must be mathematically tractable. A numerical experiment and simulation results are evinced, along with the respective discussion and conclusion.

Wang et al. [15] propose comprehensive fault diagnosis method of rolling bearing about noise interference, fault feature extraction, and identification. Based on complete ensemble empirical mode decomposition with adaptive noise (CEEMDAN), detrended fluctuation analysis (DFA), and improved wavelet thresholding, a denoising method of CEEMDAN-DFA-improved wavelet threshold function was presented to reduce the distortion of the noised signal. Based on quantum-behaved particle swarm optimization (QPSO), multiscale permutation entropy (MPE), and support vector machine (SVM), the QPSO-MPE-SVM method was presented to construct the fault-features sets and realize fault identification. Simulation and experimental platform verification showed that the proposed comprehensive diagnosis method can not only better remove the noise interference and maintain the original characteristics of the signal by CEEMDAN-DFA-improved wavelet threshold function, but also overcome overlapping MPE values by the QPSO-optimizing MPE parameters to separate the features of different fault types. The experimental results showed that the fault identification accuracy of the fault diagnosis can reach 95%, which is a great improvement compared with the existing methods.

Li et al. [16] propose a novel, adaptive, fixed-time neural network tracking control scheme for nonlinear interconnected systems. An adaptive backstepping technique is used to address unknown system uncertainties in the fixed-time settings. Neural networks are used to identify the unknown uncertainties. The study shows that, under the proposed control scheme, all states in the system can converge into small regions near zero with fixed-time convergence time via Lyapunov stability analysis. Finally, the simulation example is presented to demonstrate the effectiveness of the proposed approach. A step-by-step procedure for engineers in industry process applications is proposed.

Chen et al. [17] consider that the coupling between variables in the multi-input multi-output (MIMO) systems brings difficulties to the design of the controller. Aiming to address this problem, this paper combines the particle swarm optimization (PSO) with the coefficient diagram method (CDM) and proposes a robust controller design strategy for the MIMO systems. The decoupling problem is transformed into a compensator parameter optimization problem, and PSO optimizes the compensator parameters to reduce the coupling effect in the MIMO systems. For the MIMO system with measurement noise, the effectiveness of CDM in processing measurement noise is analyzed. This paper gives the control design steps of the MIMO systems. Finally, simulation experiments of four typical MIMO systems demonstrate the effectiveness of the proposed method.

Liu et al. [18] consider that pulsars, especially X-ray pulsars detectable for small-size detectors, are highly accurate natural clocks, suggesting potential applications, such as interplanetary navigation control. Due to various complex cosmic background noise, the original pulsar signals, namely photon sequences, observed by detectors have low signal-to-noise ratios (SNRs) that obstruct practical use. This paper presents a pulsar denoising strategy developed based on the variational mode decomposition (VMD) approach. This is in fact the initial work of the authors' interplanetary navigation control research. The original pulsar signals are decomposed into intrinsic mode functions (IMFs) via VMD, by which the Gaussian noise contaminating the pulsar signals can be attenuated because of the filtering effect during signal decomposition and reconstruction. Comparison experiments based on both simulation and HEASARC-archived X-ray pulsar signals are carried out to validate the effectiveness of the proposed pulsar denoising strategy.

3. Conclusions

In conclusion, the SI has witnessed and promoted great interest in <Complex Dynamic System Modelling, Identification and Control>. As always, the research topics associated with this SI are quite widely demanded, from academic research to real applications, particularly in those manmade systems (e.g., engineering products). Therefore, the guest editors hope that the readers can benefit from these published articles, the meaningful concepts and insights presented, emerging techniques, and inspiration for their future research and publications.

Once again, the guest editors wish to show their heartfelt gratitude to all the support and passion devoted to the SI from the authors, the publisher, and many others.

Conflicts of Interest: The authors declare no conflict of interest.

References

1. Li, R.; Zhu, Q.; Narayan, P.; Yue, A.; Yao, Y.; Deng, M. U-Model-Based Two-Degree-of-Freedom Internal Model Control of Nonlinear Dynamic Systems. *Entropy* **2021**, *23*, 169. [CrossRef] [PubMed]

2. Kang, H.; Liu, G.; Wu, Z.; Tian, Y.; Zhang, L. A Modified FlowDroid Based on Chi-Square Test of Permissions. *Entropy* **2021**, *23*, 174. [CrossRef] [PubMed]
3. Jin, X.; Yu, X.; Su, T.; Yang, D.; Bai, Y.; Kong, J.; Wang, L. Distributed Deep Fusion Predictor for a Multi-Sensor System Based on Causality Entropy. *Entropy* **2021**, *23*, 219. [CrossRef]
4. Liang, Y.; Zhang, Y.; Li, Y. Constrained Parameter Estimation for a Mechanistic Kinetic Model of Cobalt–Hydrogen Electrochemical Competition during a Cobalt Removal Process. *Entropy* **2021**, *23*, 387. [CrossRef] [PubMed]
5. Luan, Y.; Jiang, M.; Feng, Z.; Sun, B. Estimation of Feeding Composition of Industrial Process Based on Data Reconciliation. *Entropy* **2021**, *23*, 473. [CrossRef] [PubMed]
6. Pan, J.; Qu, L.; Peng, K. Sensor and Actuator Fault Diagnosis for Robot Joint Based on Deep CNN. *Entropy* **2021**, *23*, 751. [CrossRef] [PubMed]
7. Li, Q.; Zhang, Z.; Peng, F. Causality-Network-Based Critical Hazard Identification for Railway Accident Prevention: Complex Network-Based Model Development and Comparison. *Entropy* **2021**, *23*, 864. [CrossRef] [PubMed]
8. Han, K.; Chen, C.; Chen, M.; Wang, Z. Constrained Active Fault Tolerant Control Based on Active Fault Diagnosis and Interpolation Optimization. *Entropy* **2021**, *23*, 924. [CrossRef] [PubMed]
9. Zhang, Y.; Pang, A.; Zhu, H.; Feng, H. Structured H∞ Control for Spacecraft with Flexible Appendages. *Entropy* **2021**, *23*, 930. [CrossRef] [PubMed]
10. Li, Y.; Zhang, J.; Ye, X.; Chin, C. Adaptive Fixed-Time Control of Strict-Feedback High-Order Nonlinear Systems. *Entropy* **2021**, *23*, 963. [CrossRef] [PubMed]
11. Luo, H.; Wang, Y.; Zhan, R.; Zhang, X.; Wen, H.; Yang, S. Cluster-Delay Mean Square Consensus of Stochastic Multi-Agent Systems with Impulse Time Windows. *Entropy* **2021**, *23*, 1033. [CrossRef] [PubMed]
12. Pang, A.; Zhou, H.; Cai, W.; Zhang, J. Improved Adaptive Augmentation Control for a Flexible Launch Vehicle with Elastic Vibration. *Entropy* **2021**, *23*, 1058. [CrossRef] [PubMed]
13. Zhang, M.; Dahhou, B.; Wu, Q.; Li, Z. Observer Based Multi-Level Fault Reconstruction for Interconnected Systems. *Entropy* **2021**, *23*, 1102. [CrossRef] [PubMed]
14. Azar, A.; Serrano, F.; Zhu, Q.; Bettayeb, M.; Fusco, G.; Na, J.; Zhang, W.; Kamal, N. Robust Stabilization and Synchronization of a Novel Chaotic System with Input Saturation Constraints. *Entropy* **2021**, *23*, 1110. [CrossRef] [PubMed]
15. Wang, Y.; Xu, C.; Wang, Y.; Cheng, X. A Comprehensive Diagnosis Method of Rolling Bearing Fault Based on CEEMDAN-DFA-Improved Wavelet Threshold Function and QPSO-MPE-SVM. *Entropy* **2021**, *23*, 1142. [CrossRef] [PubMed]
16. Li, Y.; Zhang, J.; Xu, X.; Chin, C. Adaptive Fixed-Time Neural Network Tracking Control of Nonlinear Interconnected Systems. *Entropy* **2021**, *23*, 1152. [CrossRef] [PubMed]
17. Liu, K.; Meng, F.; Meng, S.; Wang, C. Robust Controller Design for Multi-Input Multi-Output Systems Using Coefficient Diagram Method. *Entropy* **2021**, *23*, 1180. [CrossRef] [PubMed]
18. Chen, Q.; Zhao, Y.; Yan, L. X-ray Pulsar Signal Denoising Based on Variational Mode Decomposition. *Entropy* **2021**, *23*, 1181. [CrossRef] [PubMed]

entropy

MDPI

Article

Robust Stabilization and Synchronization of a Novel Chaotic System with Input Saturation Constraints

Ahmad Taher Azar [1,2,*], Fernando E. Serrano [3,4], Quanmin Zhu [5], Maamar Bettayeb [6], Giuseppe Fusco [7], Jing Na [8], Weicun Zhang [9] and Nashwa Ahmad Kamal [10]

1 College of Computer and Information Sciences, Prince Sultan University, Riyadh 11586, Saudi Arabia
2 Faculty of Computers and Artificial Intelligence, Benha University, Benha 13511, Egypt
3 Instituto de Investigaci6n en Energía, Universidad Nacional Autonoma de Honduras (UNAH), Tegucigalpa 11101, Honduras; serranofer@eclipso.eu or fernandoserr@gmail.com
4 Research Collaborator, Prince Sultan University, Riyadh 11586, Saudi Arabia
5 FET–Engineering, Design and Mathematics, University of the West of England, Bristol BS16 1QY, UK; Quan.Zhu@uwe.ac.uk
6 Electrical Engineering Department, University of Sharjah, Sharjah 27272, United Arab Emirates; maamar@sharjah.ac.ae
7 Department of Electrical and Information Engineering, Universita degli Studi di Cassino e del Lazio Meridionale, 03043 Cassino, Italy; fusco@unicas.it
8 Faculty of Mechanical & Electrical Engineering, Kunming University of Science and Technology, No. 727 Jingming South Road, Chenggong, Kunming 650500, China; najing25@kust.edu.cn
9 School of Automation and Electrical Engineering, University of Science and Technology Beijing, Beijing 100083, China; weicunzhang@ustb.edu.cn
10 Faculty of Engineering, Cairo University, Giza 12613, Egypt; nashwa.ahmad.kamal@gmail.com
* Correspondence: aazar@psu.edu.sa or ahmad.azar@fci.bu.edu.eg or ahmad_t_azar@ieee.org

Citation: Azar, A.T.; Serrano, F.E.; Zhu, Q.; Bettayeb, M.; Fusco, G.; Na, J.; Zhang, W.; Kamal, N.A. Robust Stabilization and Synchronization of a Novel Chaotic System with Input Saturation Constraints. *Entropy* **2021**, 23, 1110. https://doi.org/10.3390/e23091110

Academic Editor: José A. Tenreiro Machado

Received: 29 July 2021
Accepted: 23 August 2021
Published: 27 August 2021

Abstract: In this paper, the robust stabilization and synchronization of a novel chaotic system are presented. First, a novel chaotic system is presented in which this system is realized by implementing a sigmoidal function to generate the chaotic behavior of this analyzed system. A bifurcation analysis is provided in which by varying three parameters of this chaotic system, the respective bifurcations plots are generated and evinced to analyze and verify when this system is in the stability region or in a chaotic regimen. Then, a robust controller is designed to drive the system variables from the chaotic regimen to stability so that these variables reach the equilibrium point in finite time. The robust controller is obtained by selecting an appropriate robust control Lyapunov function to obtain the resulting control law. For synchronization purposes, the novel chaotic system designed in this study is used as a drive and response system, considering that the error variable is implemented in a robust control Lyapunov function to drive this error variable to zero in finite time. In the control law design for stabilization and synchronization purposes, an extra state is provided to ensure that the saturated input sector condition must be mathematically tractable. A numerical experiment and simulation results are evinced, along with the respective discussion and conclusion.

Keywords: chaos theory; bifurcation; stabilization; chaos synchronization; robust control

1. Introduction

Chaotic systems were studied several decades ago, as these types of systems are found in nature and other physical systems. Because of their wide range of applications, chaotic dynamical systems have received much attention over the last three decades. As the discovery of new physical systems in engineering and exact sciences is increasing, it is critical to suppress chaotic behavior because this phenomenon generates unwanted behavior when implementing this type of system. Furthermore, in recent years, the synchronization of chaotic systems has become critical in the synchronization of coupled chaotic systems, as found in applications such as like optics and other implementations in physics.

There are many results found in the literature related to different types of chaotic systems. This research study will mainly focus on self-excited attractors in their integer and fractional order. For example, the authors of [1] presented hidden and self-excited attractors for an oligopoly model by showing the respective bifurcation analysis. Another example is found in papers like that in [2], in which a describing function is implemented in a Chua circuit to find a hidden attractor. Then, in [3], a hidden chaotic attractor is found in a Lorenz system. In [4], the digital signal processing (DSP) implementation of fractional-order chaotic hidden and self-excited attractors is proposed.

Saturation is one of the constraints found in the input of many complex dynamic systems, something that can generate instability if a standard controller or synchronizer is implemented. There are a vast amount of research studies found in the literature which refer to this input nonlinearity. For example, in [5], a neural network control law is implemented for the stabilization and synchronization of a chaotic system with input saturation. Then, in [6], an adaptive controller is designed and presented for the control of a Lorenz chaotic system with input saturation. Other studies found in the literature are not necessarily related to the control and synchronization of chaotic systems where the design of appropriate controllers with input saturation is evinced. For example, in [7], the robust attitude control for a 3-DOF helicopter is shown when input saturation is presented in the system. Then, in [8], the saturation control of a switched system is given. In [9], the saturation control of a chain of integrators system is proposed. In [10], a robust controller for a class of nonlinear saturation system is studied.

Different control strategies for the stabilization and chaos suppression in this kind of system are essential considering that these control approaches provide a very efficient method when uncertainties or disturbances are found in a complex dynamic system [11]. The control of chaotic systems is crucial considering the physical applications in which this control strategy is needed. For example, in [12], a sliding mode controller and synchronization strategy for a chaotic system is designed based on a cubic reaching law. Then, in [13], a Duffing oscillator is controlled and stabilized by an active controller. Other examples are found in [14] in which an internal model principle does the robust control of a Chua circuit. Another example can be found in [15], where a robust controller is designed for a Lorenz system subjected to mismatched uncertainties. There are many control strategies found in the literature related to the control of many physical systems, such as in [16], in which a H-infinity controller is designed for a piezo electric actuator. Then, in [17], a robust controller for a permanent magnet synchronous motor is presented in which uncertainties are found in the system. Then, in [18], the suppression of chaos in a permanent magnet synchronous motor is achieved by a robust adaptive dynamic surface control.

The synchronization of a chaotic system is essential when two chaotic systems must be synchronized. Chaotic synchronization can be categorized as identical and non-identical [19–25]. There are plenty of papers in the literature in which this kind of synchronization controller strategy is evinced. Thus, for example, in [26], the inverse synchronization of two chaotic coupled systems with time delays are presented. Then in [27], the synchronization of a Chua system is achieved for a master–slave piecewise linear system. Finally, in [28], a Lorenz system is synchronized with an application to image encryption.

Other interesting results related to chaos stabilization are found in the litertaure shch as [29–33]. The studies like that in [34], in which a controller and stabilizer of magneto-elastic chaos found in a beam system, represented by the Duffing equation, is achieved by a delayed feedback control method. Then, in [35], the stabilization and synchronization of fractional-order discrete-time system is evinced. Another research study that reports interesting results is found in [36], in which a nonlinear dynamic inversion stabilizer and controller is obtained for a chaotic system. Then other results are found in studies like [37,38], in which in the first paper, period-doubling bifurcations are stabilized implementing smooth feedback, and in the second paper, a feedback controller to stabilize chaos synchronization is presented.

Moreover, there are significant research studies found in the literature related to chaos synchronization [39–42]. Therefore, for example, in [43], the synchronization and circuit design of an antimonotonic hyperjerk system is proposed in which reverse period-doubling bifurcation is used as a control parameter. Then, in [44], the synchronization of two Josephson junctions in series is presented. Another study related to synchronization is found in [45], where a complete synchronization is implemented for a system with non-inertial coupling. Finally, in [46,47], an output feedback controller for the chaos synchronization of stochastic reaction-diffusion time-delayed neural network and the chaos synchronization by optical feedback are presented, respectively.

In this paper, the stabilization and synchronization of a novel chaotic system are achieved by a robust controller. First, a novel chaotic system is presented to generate chaos by implementing a sigmoidal function. A dynamic analysis is done in which the equilibrium points, bifurcation analysis, and phase portraits are evinced to analyze the system's chaotic behavior to show the domain of attraction, the system orbits, and the bifurcations of this novel chaotic system. The robust stabilization is done by selecting an appropriate robust control Lyapunov functional by adding an auxiliary system to find two control laws for chaos suppression purposes. The synchronization control law is obtained similarly by defining the system's error variable. In this case, drive and response systems are defined, and then by selecting an appropriate robust control Lyapunov function and adding an additional system, two synchronization control laws are found. This research study presents two numerical experiments to validate the obtained theoretical results along with the respective discussion and conclusion. The self-excited attractor obtained in this research study provides a novel chaotic system that posses stability and chaotic region, as evinced in the bifurcation diagram, by estimating the parameters the chaotic regime is reached to generate an unstable equilibrium point.

One of the novelties of this new chaotic system is the use of a sigmoidal function which allows finding the chaotic regimen and behavior by approaching the equilibrium points as the domain of attraction. The sigmoid function nonlinearity can be switched such that the unstable limit cycle reaches the equilibrium points. It is important to mention that this novel's chaotic system is probably extended to multi-wing, multi scroll, or even to a complex variable chaotic dynamic system, which could attract many researchers' attention in chaos theory. Besides, the stabilization and synchronization in engineering-related physical systems can be achieved appropriately using several control techniques such as robust control, sliding mode control, and backstepping control. The control techniques for this kind of novel chaotic system can be implemented where the chaos is suppressed, or two identical systems are synchronized successfully even in the presence of nonlinearities.

As explained before, the contributions of this research study are that a novel chaotic system is provided for research purposes and modeling of many kinds of physical phenomena and other types of engineering-related systems, taking into consideration the dynamic behavior of this novel chaotic system. The input saturation in chaotic dynamic systems yields unwanted performance or even instability in this kind of chaotic system, so that a robust control is implemented for stabilization and synchronization of the novel chaotic system when saturation is found in the system inputs, so by a Lyapunov approach and by the use of the appropriate theoretical background, a robust and reliable control synchronization strategy is found.

The paper starts by evincing first the novel chaotic system. A sigmoidal function is used as a nonlinearity to drive the domain of attraction to the equilibrium points. Then, a bifurcation analysis reveals that this novel chaotic system becomes stable to chaotic by varying some system parameters, something essential to observe that a period-doubling occurs when the system enters a chaotic regime. Then, the robust stabilization of this novel chaotic system with input saturation is achieved by deriving a robust control Lyapunov function to find the appropriate control law that annihilates chaos and drives the system variables of this novel chaotic attractor their equilibrium points. A novel robust control synchronization law is found by obtaining the error of the system dynamic variables to obtain the synchronization law by designing the robust control Lyapunov function necessary to stabilize the error dynamics of the proposed novel chaotic system.

This research study aims to provide a novel fundamental theoretical background for chaos theory and its respective synchronization, anti-synchronization, and stabilization of any chaotic system. This work can be extended to other types of variations of chaotic systems, such as hyperchaotic systems, complex variable hyperchaotic systems, multi-wing, and multi-scroll chaotic system stabilization, synchronization, and anti-synchronization. The robust control strategy for synchronization and stabilization purposes provides a new theoretical framework useful for input saturated nonlinear dynamic systems. The selected robust Lyapunov function considers the system variables for stabilization purposes to reach the equilibrium points in finite time. The sector condition considers the saturated input to deal with this nonlinearity found in this chaotic system. An essential contribution of this research study is that synchronization of the two identical proposed systems is achieved by considering that the system input of the response system is saturated.

This paper is divided into the following sections. In Section 2, the related work to this study is presented, then in Section 3, the novel chaotic attractor is evinced. In Section 4, the main results of this study are shown. In Section 5, two numerical experiments are presented. Finally, in Sections 6 and 7, the respective discussion and conclusion of this research study are presented.

2. Related Work

The dissipation properties of chaotic systems are one of the most important issues to consider in the dynamic analysis of chaotic systems. As shown in [48], the dissipation properties of a chaotic system's convergence region or domain of attraction are contracted. As is well known, the divergence of a chaotic system's vector field is given by [48]

$$\nabla . F = \frac{\partial F}{\partial x_1} + \frac{\partial F}{\partial x_2} + \ldots + \frac{\partial F}{\partial x_n} \tag{1}$$

Meanwhile, for a chaotic system described by the following dynamic model $\frac{dX}{dt} = f(X,t)$, it is critical that this dynamic system meets the Lipschitz continuity property, which has been demonstrated in numerous papers in the literature, including in [49]:

$$\|f(X_s,t) - f(X_m,t)\| \leq \|L(X_s - X_m)\| \tag{2}$$

for a positive constant L. After considering the chaotic systems' characteristics and properties, it is necessary to mention some research studies on chaos stabilization. As stated in the preceding section, it is critical to remember that the stabilization and control of various types of chaotic systems are critical, especially given that chaos suppression is one of the primary strategies that must be implemented to avoid this undesirable phenomenon. There are different research studies related to chaos stabilization, such as that in [50], in which a simple polynomial function of the system states is done by implementing adaptive feedback control for a chaotic system. Then, in [51], the control and anti-synchronization of a novel fractional-order chaotic system and its analysis are proposed. Another example is found in [52], where the chaos control of a fractional-order neural network with electromagnetic radiation is presented. Then, in [53], the chaos control of a piezoelectric auto parametric vibration system is shown.

The synchronization problem consists in following the drive system trajectory in time by the response system in order to drive the error variable to zero in finite time as appears in [28]:

$$\epsilon = \sqrt{\sum_{i=1}^{n} |Y_{Di} - Y_{Ri}|^2} \tag{3}$$

where $Y_D \in \mathbb{R}^{\kappa}$ represents the discretized drive response in time and $Y_R \in \mathbb{R}^{\kappa}$ represents the discretized response system evolution in time. This root mean square error (RMSE) describes the accuracy of the control synchronizing strategy in order to make sure that the control synchronization will be the minimum and zero in finite time. Other results in the literature related to synchronization control of chaotic systems can be found in papers such as that in [54], which achieves chaos synchronization in the frequency domain. Then, in [55], a novel control approach does the synchronization of two non-identical 4-D hyperchaotic systems. Finally, in [56,57], the synchronization of two novel hyperchaotic system with unknown parameters and the circuit realization, control and synchronization of a novel hyperchaotic system are presented, respectively.

This study considers the discovery of a novel chaotic system, as well as the homoclinic and heteroclinic chaotic properties of this type of system. For example, in papers such as that in [58], it is demonstrated how some co-dimensions two bifurcations originate regions of chaotic and simple dynamics, and bifurcation structures such as Bykov T-points spirals are evinced. In papers such as that in [59], it is demonstrated how multiple chaos can arise from single parametric perturbations of a degenerated homoclinic orbit. The system under consideration in this paper is a periodical perturbed differential equation in which a Former's perturbation is used to find the homoclinic orbits. Apart from the paper in [60], homoclinic chaos in piecewise smooth oscillators is demonstrated by using small parametric perturbations in this type of oscillator. In this paper, the Melnikov approach is used first to find that homoclinic chaos exists in a system without small perturbations, and then the same approach is used to find sufficient conditions to find the control parameters. Then, in [61], the homoclinic bifurcation and control of chaotic MEMS are presented. The Melnikov function is used as an analytical methodology for homoclinic chaos, which is expressed as an inequality in terms of system parameters. Another intriguing study can be found in [62], which shows novel bifurcation diagrams for piecewise smooth systems in which the transversability of homoclinic points does not imply chaos. Four scenarios are presented in this research study in which the systems are assumed to be subject to small non-autonomous perturbations, yielding four new bifurcation diagrams. Then, in [63], minimal topological chaos is discovered with respect to finite sets of homoclinic and periodic orbits.

Another important topic to mention is heteroclinic chaos, which is demonstrated in this paper through the design of a new chaotic attractor. There are numerous research studies in the literature that deserve to be mentioned in this research paper. In papers such as that in [64], for example, it is demonstrated that heteroclinic cycles connecting repellers and saddles in locally compact metric space induce chaos. The results presented in this paper are critical because they are based on a topological analysis in which the maps in the criteria are shown to have positive topological entropy and to be chaotic in the Li–Yorke sense. Then, in [65], another topological analysis is presented, in which chaos generated by heteroclinic curves connects repellers in complete metric spaces. This paper includes two classifications of heteroclinic cycles: regular and singular, as well as degenerated and nondegenerated. Other findings can be found in papers such as that in [66], which demonstrates a design methodology and algorithm for implementing geometric features with focus-saddle and center node equilibrium points.

Because entropy issues in novel chaotic systems are important in the theoretical development of this research study, it is worth mentioning some articles found in the literature on the entropy of novel chaotic systems. As an example, in papers such as that in [67], the enhancing of chaos complexity of a plasma model is demonstrated. It is demonstrated in this paper that by increasing the power input, the system can change from monostable to multistable without the addition of any input terms. It is also demonstrated that there is a transition, in terms of system complexity, from transient chaos to steady periodic behavior. Then, in [68], a novel measure inspired by Lyapunov exponents is demonstrated. In this paper, a network measure for characterizing state-transition networks is implemented. Another interesting study about chaotic system entropy found in the literature is that in [69], which shows evidence of strange attractors found in a C-Class amplifier with a bipolar transistor. The nonlinear bilateral behavior is shown in this paper to be a necessary but not sufficient condition for finding a complex behavior when the transistor is modeled as a two port admittance parameter. Meanwhile, the study in [70] demonstrates an important research study related to entropy in which the origin and fundamentals of entropy are presented, explaining how the entropy concept is used in physics, information theory, chaos theory, and data mining, and providing researchers with important issues that can be selected to choose the right variant of entropy for their research study. In addition, the study in [71] shows a chaotic time delay signature suppression by frequency band extracting is investigated the time delay signature and entropy grow enhancement in a chaotic optical feedback semiconductor laser. Then, in [72], a two-parameter bifurcation diagram for the computational analysis of Ca^{2+} oscillatory biosignals is presented. This paper explains how those types of diagrams provide different types of information about the analyzed autonomous system and how they complement one another. Aside from robust control, which is the strategy for stabilization and synchronization used in this research study, some research studies found in the literature related to the stability and chaos synchronization of various types of chaotic systems are worth mentioning. For example, in papers such as that in [73], the synchronization patterns in Kuramoto oscillators are demonstrated, in which phase locked states with constant phase shifts between these oscillators are studied. The synchronization estimation for complex time series using cross sample entropy measure is then demonstrated in papers such as that in [74]. Other papers with interesting results include that in [75], in which a modified Chua's circuit is used with a five segment piecewise linear Chua's diode. This paper demonstrates that the attractors have small basins of attraction. Finally, in [76], it is shown that a tuned pendulum absorber can reduce vibration and, at the same time to harvest energy. Then, in [77], it is shown how Lagrangian descriptors can be implemented to characterize invariant tori of generic systems. Finally, in [78], the magnetic confinement of a neutral atom is presented. In this paper, a neutral atom inside a double-wire waveguide in the presence of two uniform bias fields is presented.

It is important to remember that, in contrast to other research studies found in the literature, such as those in [72–74,76], the stabilization is done using techniques that do not guarantee stability and performance when saturation input is found in the respective chaotic systems shown in these papers. Note that although these control strategies do not take saturation into account, the high nonlinearity and complexity exhibited in these research papers make them ideal candidates for implementing robust control and synchronization strategies, but it is worth noting that there is sometimes a trade-off between the complexity of the designed controllers exhibited in these papers, so the authors consider. As previously stated, the obtained novel chaotic system shown in this paper can be used in engineering implementations as well as other exact science implementations such as metheorology, astrophysics, and astrochemistry.

Because of the high performance of the robust control synchronization law provided in this research study, the synchronization control law, as explained in previous sections, is achieved faster. Unfortunately, the literature on the implementation of robust control or other controller techniques such as sliding mode control and backstepping control for controller for chaotic system with input saturation is limited. Therefore, given the high nonlinearity of this novel chaotic system as a sigmoidal function, a roust control strategy is proposed to minimize the convergence error between the drive and response system variables until it reaches the origin and, as previously stated, faster than other types of control strategies that take into account the novel chaotic system's high complexity.

Unfortunately, due to the lack of similar research studies found in the literature, a comparative analysis was not possible in the numerical experiment section, but as explained later in this paper, it is not necessary to do a comparative analysis taking into account that the stabilized and synchronized systems are novel chaotic systems, as observed in the numerical experiments.

3. Definition of the Novel Chaotic System

This section presents the definition of a novel chaotic attractor as well as a bifurcation analysis. The new chaotic attractor is made up of a sigmoidal function that causes the system to enter a chaotic state. The bifurcation analysis demonstrates how the system transitions to a chaotic regime and the period-doubling exhibited by the respective bifurcations. The applications of this novel chaotic system is very wide an can be implemented for the mathematical modeling of phenomenons found in meteorology [79], astrophysics [80], fluid dynamics [81], and in other engineering related systems such as electrical and power, mechanical, mechatronics, and chemical systems. The reason of the applicability of this novel chaotic system is that by estimating the right parameters and considering some topological aspects such as the limit set related to the unstable limit cycle and the domain of attraction this chaotic system can be implemented in the mathematical modeling of different types of phenomena.

3.1. Definition of the Novel Chaotic System

Consider the following novel chaotic attractor with sigmoidal functions:

$$\begin{aligned} \dot{x}_1 &= asigmoid(z_1)x_1 - z_1^3 \\ \dot{y}_1 &= bz_1 + x_1y_1 \\ \dot{z}_1 &= csigmoid(x_1) + dx_1y_1 \end{aligned} \tag{4}$$

in which *sigmoid* is the sigmoidal function $sigmoid(x) = 1/(1+e^{-x})$. The equilibrium point of this system is $[x_1, y_1, z_1]^T = [-2.3, 0.78, 1.21]^T$ for a system with initial conditions $[x_1(0), y_1(0), z_1(0)]^T = [-2.4, 0.8, 1.2]^T$. The equilibrium points were obtained by the numerical nonlinear algebraic equation solver of GNU Octave *fsolve* of GNU Octave version 4.2.2. The set of nonlinear algebraic equations is done by considering the vector field $f(x_1, y_1, z_1) = 0$ so by solving this equation the equilibrium points are found efficiently. The values of the parameters used for this novel chaotic system are as follows: $a = -1$, $b = 4,9$, $c = 13$, and $d = 1$ in order that the system reach the chaotic regimen using this parameters in all the numerical experiments of this paper. The Lyapunov exponents of this chaotic system are $L_1 = -0.00195243$, $L_2 = 0.00852359$, and $L_3 = -0.00394307$ verifying that only the Lyapunov exponent L_2 has a positive real component so the system is considered as chaotic. It is important to remark that by changing the parameters of this novel chaotic system the system solution can be varied from a quasiperiodic orbit to a chaotic orbit. The divergence of the system (4) is obtained in the following way:

$$\nabla.F(x_1, y_1, z_1) = \frac{a}{1+e^{-z_1}} + x_1 \tag{5}$$

The range of a in which the proposed novel chaotic system is dissipative is shown as follows:

$$-max(x_1(1+e^{-z_1})) < a < -min(x_1(1+e^{-z_1})) \tag{6}$$

In this range of the constant a, the energy of the system dissipates until the equilibrium point reach the equilibrium points, so in this way, the periodic orbits of the novel chaotic attractor reach the equilibrium points. By this dissipation proof, it is verified and validated that the the dissipation characteristics for the novel chaotic system are found by establishing the range of the constant a in which this property is found.

The phase portraits in different planes are evinced in Figure 1 which corroborates the chaotic behavior of the novel chaotic system. It is possible to observe the orbit of this chaotic system, which follows a helicoidal trajectory around the domain of attraction, which is the equilibrium point. A vortex is formed around the domain of attraction, which is desirable in order to model certain natural systems. It is important to notice that the sigmoidal nonlinearities found in this novel chaotic systems yields the trajectory of the orbits of this proposed system to the equilibrium point in finite time. As verified later, in the bifurcation diagrams, the unstable limit cycle generated by chaos is obtained by selecting the appropriate parameters values in order to obtain the chaotic orbit. Due to the appropriate selection of the parameter values, the novel chaotic system enters in chaotic regimen considering that the initial conditions are crucial to find other domains of attractions in other equilibrium points. Because the domain of attraction in different equilibrium points of the novel chaotic systems is sensitive to the initial conditions, the initial conditions were chosen as long as they are close to the first encountered equilibrium point by determining that the inner product of two vector fields over the inner product of the initial condition and the equilibrium points is equal to infinity [82].

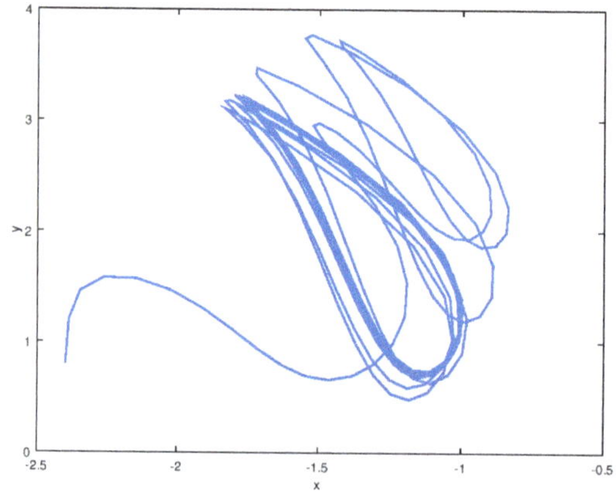

(**a**) Phase portrait in the $x - y$ plane

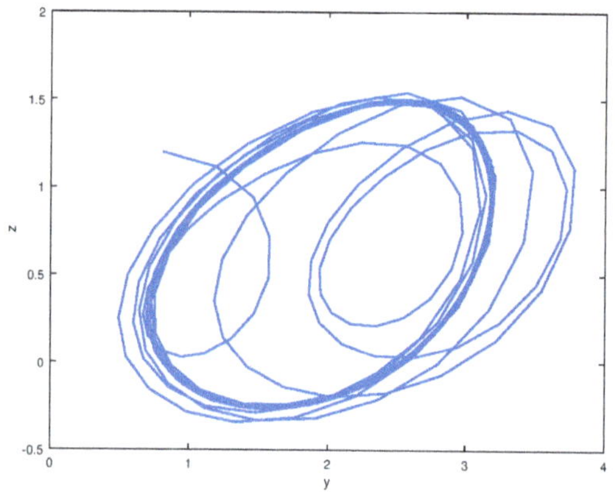

(**b**) Phase portrait in the $y - z$ plane

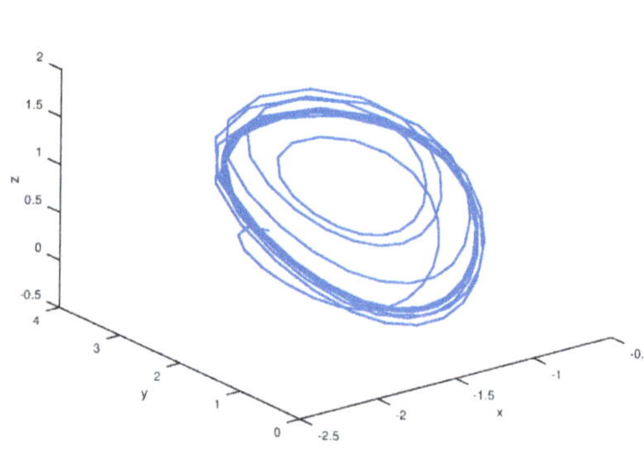

(**c**) Phase portraits in x, y and z

Figure 1. Phase portrait of the novel chaotic attractor (4).

3.2. *Bifurcation Analysis*

The bifurcation diagrams while varying the constants a, b, and d in (4) are shown in Figure 2 in which it can be shown how the system states are driven from the stable region to the chaotic regimen by varying the parameters a, b, and d, as shown in the three figures. It is also demonstrated that, despite the fact that this is a novel chaotic system, the system exhibits chaotic behavior because period doubling occurs in the bifurcation diagram until the limit cycles appear in the novel chaotic system proposed in this paper.

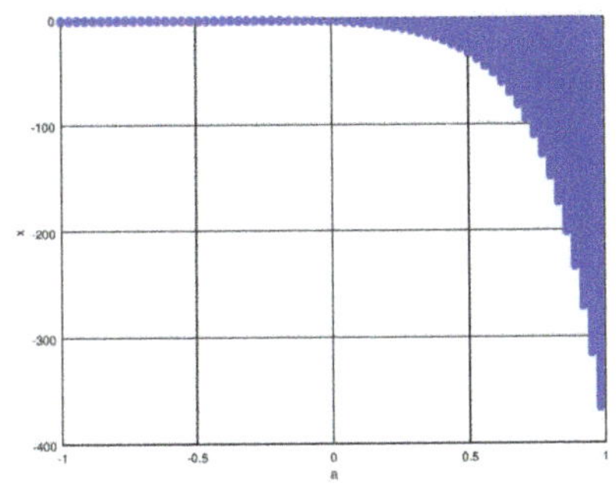

(**a**) Bifurcation diagram of system (4) while varying a

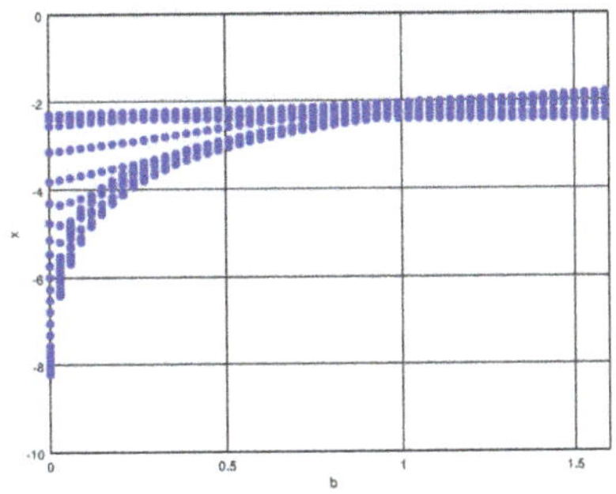

(**b**) Bifurcation diagram of system (4) while varying b

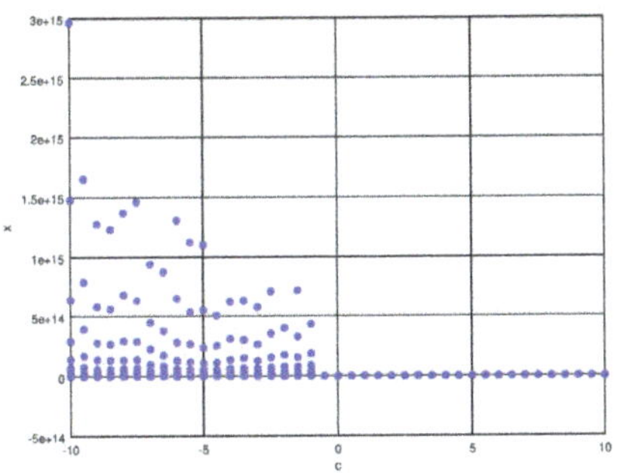

(**c**) Bifurcation diagram of system (4) while varying c

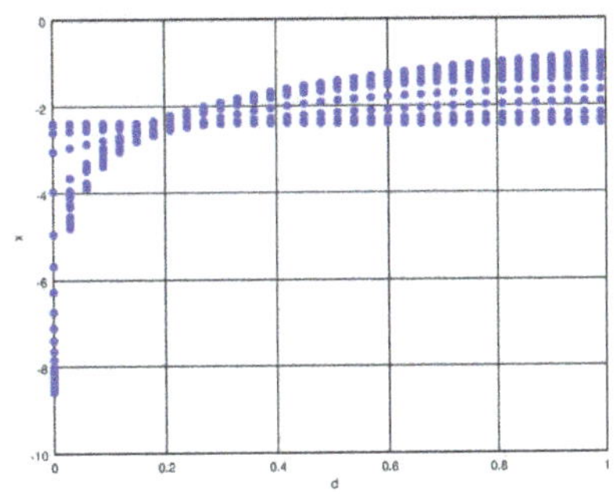

(**d**) Bifurcation diagram of system (4) while varying d

Figure 2. Bifurcation diagrams of the novel chaotic system (4).

The resulting diagrams demonstrate that the parameters a, b, and d can influence chaotic behavior; however, as explained later, there is one parameter that influences the system variables even more than the others in order to drive the system variables from the stable region to the chaotic regime. It is important to remark that while varying the parameter a, specifically when this parameter increases, the tendency of the variable x is to increase. This occurs because the parameter a is the only parameter that has to do with the divergence of this novel chaotic system as explained in (5) with the range (6), so the variable increases when it is outside the limits of the parameter a.

In Figure 2a, it can be seen how the bifurcation diagram is stable until it reaches the chaotic regimen in approximately $a = 0.3$, and it can be seen that this parameter affects the transition between the stable region and the chaotic regimen the most. For the bifurcation diagrams shown in Figure 2b,d the bifurcations in chaotic regimen can be seen until stability is reached around $b = 1.3$ and $d = 0.8$. Meanwhile in Figure 2c it can be noticed the transition from the chaotic regimen to stability in the interval $-10 \leq c \leq 0$ while stability is found in the rest of the regions as noticed in the bifurcation diagram.

It can be seen in these bifurcation diagrams that the density of these diagrams evinces how the chaotic regimen is reached as well as the effects of sigmoidal nonlinearity. One of the most important conclusions is that the novel chaotic system generates chaos, which is significant when considering not only the phase portrait diagrams, but also the bifurcation diagrams when the parameters of the novel chaotic system are varied.

As a result of this section, it is possible to confirm the existence of a novel chaotic system. It is worth noting that the presented nonlinear dynamic system employs sigmoidal nonlinearity to generate chaos. This designed novel chaotic system is very important because new phenomena are discovered in nature every day, so it is critical to provide new chaotic systems in order to model novel complex physical systems discovered in exact sciences and engineering.

Another significant contribution of the novel chaotic systems is that it can be extended to other types of chaotic-related systems, such as hyperchaotic, multi-wing, or multi-scroll chaotic systems, in the latest kinds of chaotic system is important because the sigmoidal nonlinearity can extend this results to the kind of desired chaotic system by following an approach similar to that seen in the chaotic oscillator hidden attractor found in [82].

The design methodology of the robust control and synchronization laws that are used to synchronize this novel chaotic attractor is explained in the following section. It is critical to consider the saturation input nonlinearity as explained in this section in order to perform the stabilization and synchronization so as to provide a theoretical framework for applied sciences such as physics, chemistry, and biology, as well as engineering applications that require the stabilization, chaos suppression, and synchronization of chaotic systems.

This paper demonstrates that, regardless of the complexity of the input nonlinearity found in chaotic systems, it can be efficiently addressed by selecting the appropriate mathematical framework to stabilize and synchronize the proposed chaotic systems. Finally, it is validated how the designed controller and synchronizer stabilizes this novel chaotic system in the equilibrium points, as discovered and explained in this section, by providing the respective numerical experiment and simulation framework.

4. Main Results

The robust control laws for the stabilization and synchronization of the novel chaotic system presented in this study are deduced in this section. The control laws are deduced into two theorems, and the following drive (7) and response systems (8) are taken into account for stabilization and synchronization. In this section, it is demonstrated how the control strategy for stabilization purposes is designed while taking into account the saturation input nonlinearities, which can cause instability and deteriorated performance. When this occurs in a chaotic system, the consequences can be significant.

Note that the robust control strategy was chosen with the complexity of the sigmoidal function nonlinearity in mind, which is part of the novel chaotic system explained and demonstrated in this research study. The stabilization of this novel chaotic system is an important contribution because the findings can be applied to hyperchaotic, multi-wing, and multi-scroll chaotic systems. Note that in the synchronization case, only the identical case is considered for the purposes of this paper, and the results obtained and presented in this paper can later be extended to the non-identical case.

A robust control Lyapunov function is designed for the stabilization case in order to find the control law that stabilizes the system at the equilibrium points. Note that if the chaotic system presented in this paper needs to be stabilized in a different domain of attraction or at a different equilibrium point, a regulator is required to drive the system variables to the desired final value. Note that in the case of the synchronization of the novel two identical chaotic systems, these results can be extended to non-identical systems and other types of coupled chaotic systems.

$$\begin{aligned} \dot{x}_1 &= asigmoid(z_1)x_1 - z_1^3 \\ \dot{y}_1 &= bz_1 + x_1y_1 \\ \dot{z}_1 &= csigmoid(x_1) + dx_1y_1 \end{aligned} \tag{7}$$

$$\begin{aligned} \dot{x}_2 &= asigmoid(z_2)x_2 - z_2^3 + \phi(u_1) + \alpha_1 \\ \dot{y}_2 &= bz_2 + x_2y_2 + \phi(u_2) + \alpha_2 \\ \dot{z}_2 &= csigmoid(x_2) + dx_2y_2 + \phi(u_3) + \alpha_3 \end{aligned} \tag{8}$$

in which the function $\phi(.)$ is the saturation nonlinearity. For the derivation of the controller and the stabilization and synchronization purposes, the following extra system is needed:

$$\begin{aligned} \dot{\alpha}_1 &= v_1 \\ \dot{\alpha}_2 &= v_2 \\ \dot{\alpha}_3 &= v_3 \end{aligned} \tag{9}$$

so the following vectors are considered $X_1 = [x_1, y_1, z_1]^T, X_2 = [x_2, y_2, z_2]^T, U = [u_1, u_2, u_3]^T$ and $\alpha = [\alpha_1, \alpha_2, \alpha_3]^T$. Rearranging (7) and (8) yields

$$\dot{X}_1 = F(X_1) = \begin{bmatrix} asigmoid(z_1)x_1 - z_1^3 \\ bz_1 + x_1y_1 \\ csigmoid(x_1) + dx_1y_1 \end{bmatrix} \tag{10}$$

$$\begin{aligned} \dot{X}_2 &= G_1(X_2) + \phi(U) + \alpha \qquad (11) \\ \dot{X}_2 &= \underbrace{\begin{bmatrix} asigmoid(z_2)x_2 - z_2^3 \\ bz_2 + x_2y_2 \\ csigmoid(x_2) + dx_2y_2 \end{bmatrix}}_{G_1(X_2)} + \underbrace{\begin{bmatrix} \phi(u_1) \\ \phi(u_2) \\ \phi(u_3) \end{bmatrix}}_{\phi(U)} + \underbrace{\begin{bmatrix} \alpha_1 \\ \alpha_2 \\ \alpha_3 \end{bmatrix}}_{\alpha} \end{aligned}$$

For the saturation nonlinearity, consider the following sector condition property:

Property 1. *Considering a saturation nonlinearity $\phi(U)$ and the deadzone nonlinearity $D_z(U)$ with system input U, the following sector condition is met:*

$$U^T[U - D_z(U)] \geq 0 \tag{12}$$

with:

$$D_z(U) = U - \phi(U) \tag{13}$$

with these definitions the robust control laws for stabilization and synchronization purposes can be defined.

In the two theorems of this paper, it does not matter the values of the saturation because the sector condition ensures that for any value of the saturation the stability of the stabilizer and synchronizer in ensured in the presence of saturation input, as explained in this paper. The saturation function used in the whole paper is given by

$$\phi(x) = \begin{cases} -d_1 & \textbf{IF} \quad x < -l_2 \\ x & \textbf{IF} \quad -l_1 \leq x < l_2 \\ d_2 & \textbf{IF} \quad x \geq l_2 \end{cases} \tag{14}$$

4.1. Robust Stabilization of the Novel Chaotic System

The closed loop stabilization system's block diagram is shown in Figure 3 where the robust controller and stabilizator are located in the feedback loop, connected to the saturation nonlinearity and receiving the vector field G_1. It is necessary to design the appropriate robust control Lyapunov function in order to obtain the respective control law that drives the system states to the equilibrium points in finite time in order to design the stabilization control law for this novel chaotic system. The robust control law obtained in this section includes a switching component, but no chattering is observed as a result of this. Consider the following theorem for the derivation of the robust control law for stabilization purposes.

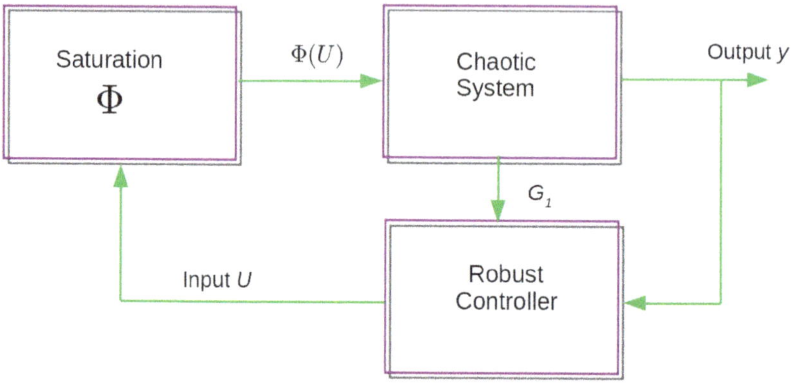

Figure 3. Block diagram of the closed loop stabilization controller.

Theorem 1. *Consider the controlled system (11) and the extra system (9), so the following control laws stabilizes the novel chaotic system:*

$$\begin{aligned} U &= -X_2 \\ v &= \frac{-\alpha}{\|\alpha\|^2} X_2^T G_1(X_2) - X_2 \end{aligned} \tag{15}$$

By selecting an appropriate robust control Lyapunov function.

Proof. Consider the following robust control Lyapunov function:

$$V = \frac{1}{2} X_2^T X_2 + \frac{1}{2} \alpha^T \alpha \tag{16}$$

Now, by taking the time derivative of (16), the following results are obtained:

$$\begin{aligned} \dot{V} &= X_2^T \dot{X}_2 + \alpha^T \dot{\alpha} \\ \dot{V} &= X_2^T G_1(X_2) + X_2^T \phi(U) + X_2^T \alpha + \alpha^T v \\ \dot{V} &= X_2^T G_1(X_2) - U^T [U - D_z(U)] + X_2^T \alpha + \alpha^T v \end{aligned} \tag{17}$$

Then, by substituting the control laws (15) in (17) and by using the sector condition which appears in Property 1 the following result is obtained:

$$\dot{V} = -U^T [U - D_z(U)] \le 0 \tag{18}$$

Therefore, with this conclusion the novel chaotic system is stabilized and the proof of the theorem is completed. □

As demonstrated in the previous theorem, it is possible to design a robust control law for the stabilization of the novel chaotic system.

4.2. Robust Synchronization of the Novel Chaotic System

The closed-loop synchronization controller for the novel chaotic drive and response system is shown in Figure 4. As can be seen, the system error, which is the difference between the state variables of the drive and response system, is used as input for the robust controller, and the additional state variable is also used for the synchronizer to drive the state error to zero in finite time. This input is applied to the input saturation function along with the input U to prevent saturation effects from causing instability or other undesirable effects. The robust control law for synchronization is then obtained by selecting an appropriate Lyapunov robust control law and then obtaining the first derivative of this Lyapunov function to obtain the synchronization control law. When the saturation nonlinearity is found in the response system input, the sector condition, as shown in Property 1 is critical in order to find the required stability conditions. As stated in the introduction section, the theoretical results presented in this paper are based on the identical chaotic system synchronization of the novel chaotic system described in Section 3. The most important results are evinced in the following theorem in this subsection, in order to provide the theoretical framework for these types of systems. Consider the following theorem for the Robust synchronization of this identical novel chaotic system.

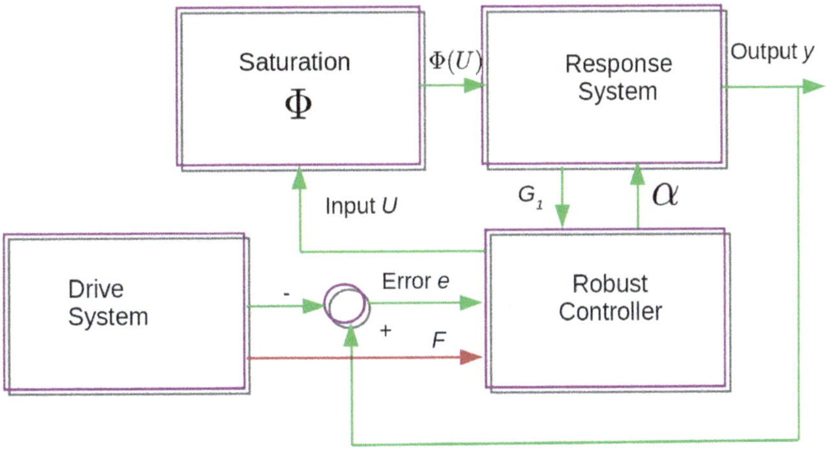

Figure 4. Block diagram of the control synchronization closed loop system.

Theorem 2. *Considering the following error variable and its derivative:*

$$\begin{aligned} e &= X_2 - X_1 \\ \dot{e} &= G_1(X_2) + \phi(U) + \alpha - F(X_1) \end{aligned} \tag{19}$$

so the following robust synchronization laws are obtained for systems (10) and (11):

$$\begin{aligned} U &= -e \\ v &= \frac{-\alpha}{\|\alpha\|^2} e^T [G_1(X_2) - F(X_1)] - e \end{aligned} \tag{20}$$

By selecting an appropriate robust control Lyapunov function.

Proof. Consider the following robust control Lyapunov function:

$$V = \frac{1}{2} e^T e + \frac{1}{2} \alpha^T \alpha \tag{21}$$

Now, by taking the time derivative of (21) yields

$$\begin{aligned} \dot{V} &= e^T \dot{e} + \alpha^T \dot{\alpha} \\ \dot{V} &= e^T [G_1(X_2) - F(X_1)] + e^T [U - D_z(U)] + e^T \alpha + \alpha^T v \end{aligned} \tag{22}$$

so by substituting (20) into (22) and by using Property 1 the following result is obtained:

$$\dot{V} = -U^T [U - D_z(U)] \leq 0 \tag{23}$$

so the robust synchronization stability is achieved and the proof is completed. □

This proof demonstrates how the closed-loop stability of the error variable dynamics is ensured in order for the state variables of the response system to reach the state variables of the drive system. A very accurate and fast robust control synchronization is achieved with the obtained control law and by the implementation of an extra variable that aids in the closed loop stability. As a result, it is important to remember that, similar to the stabilization case, the error system dynamics reach zero in finite time faster despite the presence of an input saturation nonlinearity in the system. Before concluding this section, note that the theoretical proofs evinced in two theorems allow for faster and more accurate stability of the closed-loop system in the stabilization and synchronization case. In the following section, two numerical experiments demonstrate how both of these control strategies, for stabilization and synchronization purposes, provide accurate results as demonstrated theoretically in this section.

5. Numerical Experiments

Two numerical experiments are presented in this section: the first for the stabilization of the novel chaotic system and the second for the control synchronization of two identical novel controllers. The novel chaotic system is robustly stabilized in the first experiment by selecting specific initial conditions with saturation input nonlinearity. In this manner, the robust controller used for stabilization is validated. In the case of robust synchronization, two novel chaotic systems are synchronized by a robust control law so that the response system variables follow the evolution of the drive system variables in time. This goal must be met despite the fact that the response system's input is saturated nonlinear. In these numerical experiments, it is verified and validated that the robust control synchronization law successfully synchronizes both systems in order to reduce the synchronization error to zero.

5.1. Experiment 1: Robust Stabilization of the Novel Chaotic System

This experiment involves the robust stabilization of a novel chaotic system. Because the saturation nonlinearity is in the system's input, the robust controller is used to overcome or outperform the performance of the closed loop system in the presence of this nonlinearity. This numerical example effectively validates the proposed control strategy. The response of the novel chaotic system variable is stabilized while eliminating or suppressing the chaotic behavior in this novel chaotic system, allowing the state variables to be driven to the equilibrium points in finite time without exhibiting chattering or other undesirable phenomena in this type of robust controller.

Note that if the initial conditions are close to another domain of attraction, the system variables will be stabilized in different equilibrium points. As shown in the block diagram in Figure 3, the vector field G_1 must be calculated in order to obtain the required control input action, implying that some kind of compensation is required to deal with the saturation input, as shown in the theoretical results obtained in this research study. The results of this numerical experiment are as follows: first, the evolution of the state variables over time, then the evolution of the input variables over time, and finally the respective phase portraits.

For the robust stabilization of the novel chaotic controller with input saturation (8), the following initial condition is considered $X_2 = [-2.4, 0.8, 1.2]^T$ with the saturation function:

$$\phi(x) = \begin{cases} -10 & \textbf{IF} \quad x < -15 \\ x & \textbf{IF} \quad -15 \leq x < 15 \\ 10 & \textbf{IF} \quad x \geq 15 \end{cases} \tag{24}$$

This saturation nonlinearity is found in the system input, as shown in Figure 3. It can be demonstrated that the theoretical derivation of the stabilization controller for this novel chaotic system ensures the overall system's stability, given that due to a rigorous theoretical derivation, independent of the saturation parameters, the stabilization of the variables of this novel chaotic system is achieved in order for these state variables to reach the eigenvalues. The robust control law suppresses the instability by assuming that the overall closed-loop system is stabilized by an efficient Lyapunov design. In this numerical experiment section, it is demonstrated that the saturation nonlinearity found in the input of this novel chaotic system has no effect on the overall performance of the closed loop chaotic system, regardless of the saturation parameters. Because of an efficient controller design, the robust controller effectively overcomes saturation nonlinearity in order to stabilize the novel chaotic system in the equilibrium points. Note that in this numerical experiment, the stabilization occurs quickly and precisely, with no oscillations or chattering.

The stabilized state variables x, y, and z are shown in Figure 5, and it is noted that these variables reach equilibrium in finite time without the presence of unwanted oscillation. It has been demonstrated that the robust stabilization controller completely suppresses the chaos behavior, demonstrating that the main purpose of this proposed controller has been met. As previously stated, the state variable evolution reaches the equilibrium points; however, by varying the initial conditions close to other domains of attraction, other equilibrium points can be reached using the same robust control law; this is due to the robust control law's robustness. The Lyapunov functional design is independent of the equilibrium points, because the controller is designed in such a way that the domain of attraction related to the equilibrium point is close to the robust controller performance. The stabilized phase portrait is shown in Figure 6, and as can be seen, chaos suppression is used to eliminate all unwanted oscillations. The limit cycle vanishes when equilibrium is reached. In comparison to the phase portrait shown in Figure 1, it is demonstrated that this variable is stabilized along a smooth trajectory rather than the chaotic trajectory shown in the previous figure. As a result, the robust controller action eliminates the chaotic behavior satisfactorily, and the system variables shown in this phase portrait are efficiently driven to the equilibrium point.

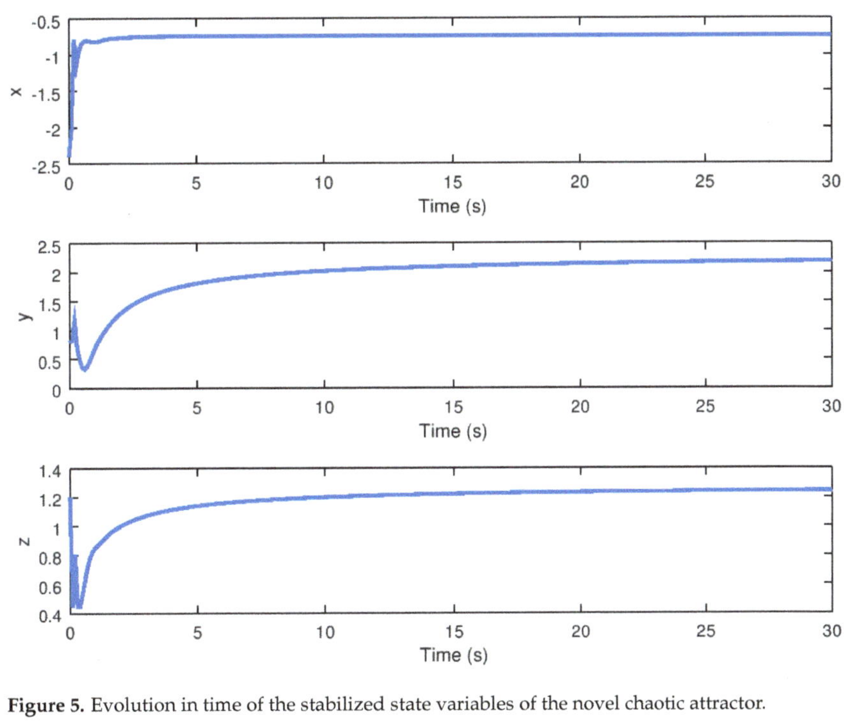

Figure 5. Evolution in time of the stabilized state variables of the novel chaotic attractor.

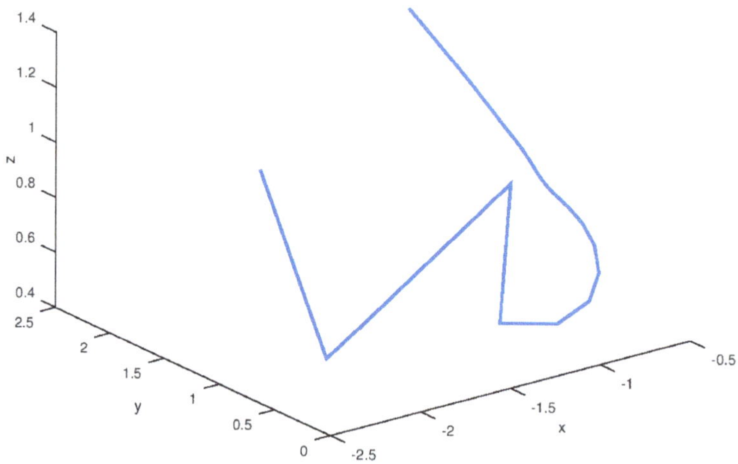

Figure 6. Stabilized phase portrait of the novel chaotic system.

The input variables U and v are depicted in Figures 7 and 8, respectively. It is possible to see how these variables arrive at a final value until the trajectory or orbit of the novel chaotic system depicted in this paper reaches equilibrium. Note that these input variables provide a low control effort and chattering ausence, which demonstrates the high performance of the proposed robust controller for the stabilization of the novel chaotic system. Another important consideration is that the controller action would be significantly small control effort in the case that this robust control strategy is implemented in a real physical system, given that the saturation input nonlinearity yields instability and poor performance that can avoid chaos suppression in an efficient manner.

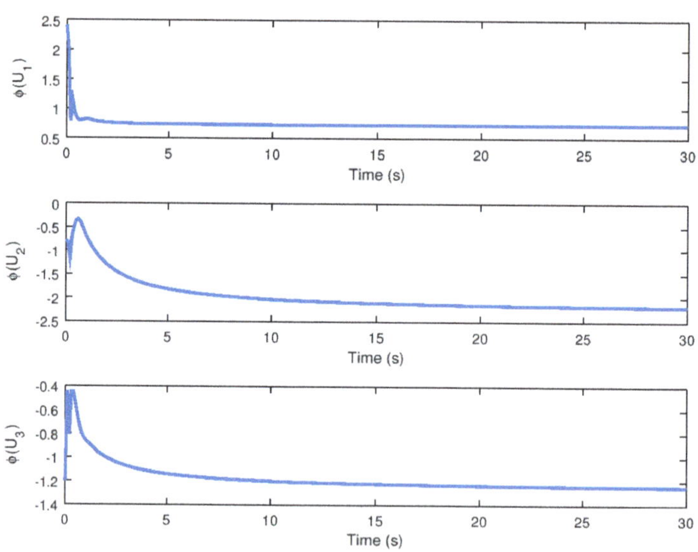

Figure 7. Input variable U of the novel chaotic system by the implementation of the proposed robust controller.

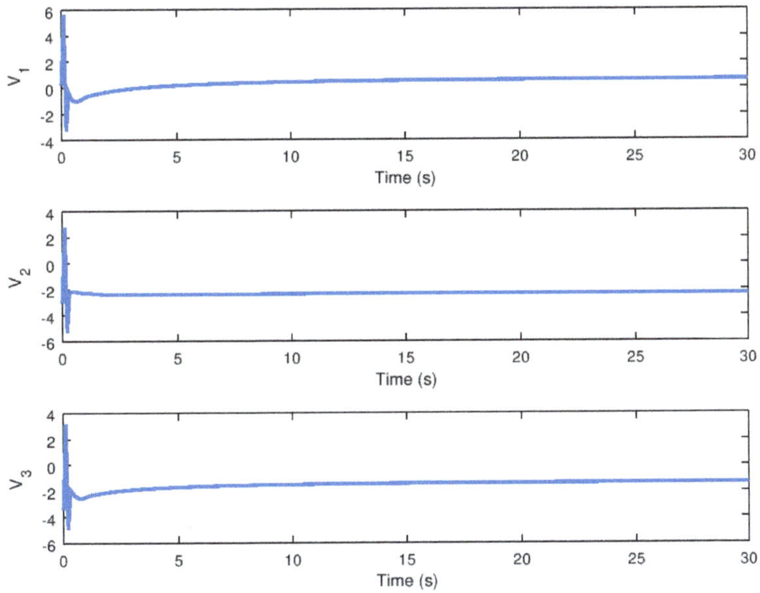

Figure 8. Input variable v of the novel chaotic system with the proposed robust controller.

Finally, as shown in Figure 9, the evolution in time of the auxiliary variable α for stabilization purposes is evinced in which is the extra state variable with input v as shown in Figure 8. When saturation is detected in this novel chaotic system, this provides adequate input for stabilization.

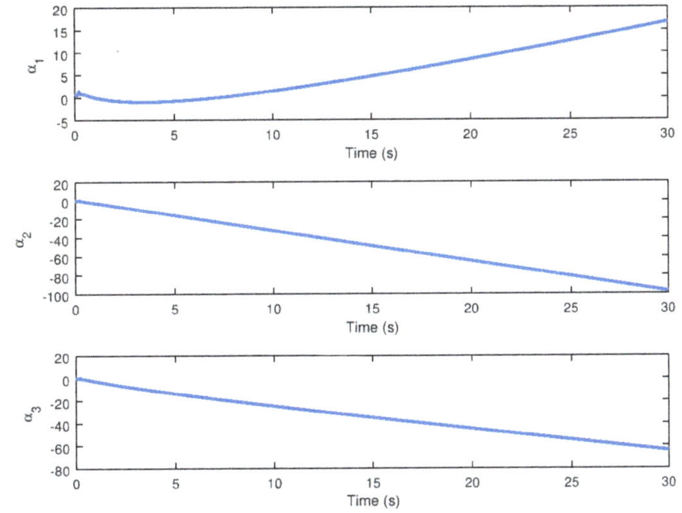

Figure 9. Evolution in time of the variable α for stabilization purposes.

5.2. Experiment 2: Robust Synchronization of the Novel Chaotic System

The synchronization of two identical chaotic systems is demonstrated in this numerical experiment. This experiment's chaotic system is a novel chaotic system shown in this paper. One system serves as a drive system, while the other serves as a response system.

The saturation nonlinearity is found in the input of the response, as explained in the block diagram shown in Figure 4, so it is important to note that this kind of nonlinearity in the input of the novel chaotic system is not commonly found in the literature, but in real physical systems this phenomenon is commonly found considering the type of application. It is also worth noting that the vector fields F and G_1 must be calculated because, similar to the stabilization case, they act as a compensator to overcome the saturation nonlinearity regardless of its characteristics. The following initial conditions are implemented in this numerical experiment for the synchronization of two identical novel chaotic systems, where (10) is the drive system and (11) is the response system: $X_1 = [-2.4, 0.8, 1.2]^T$ and $X_2 = [-2.9, 0.4, 1.7]^T$ with the saturation input shown in (24).

The evolution in time of the synchronized state variables, as well as their respective error variables, is depicted in Figures 10 and 11. As can be seen, the response variables x, y, and z reach the drive variable's trajectory in finite time, whereas the error reaches the origin or zero value faster. Note in these figures that the robust controller performs optimally by rapidly and precisely driving the error variables to zero. Note that the proposed robust control synchronization strategy provides a fast and accurate system response even in the presence of saturation nonlinearity, regardless of the characteristics of this phenomenon found in the drive system's input. It should be noted that this robust controller is only suitable for the synchronization of identical chaotic systems; for the synchronization of non-identical chaotic systems, this synchronization robust control strategy must be modified in order to achieve the required closed loop system performance.

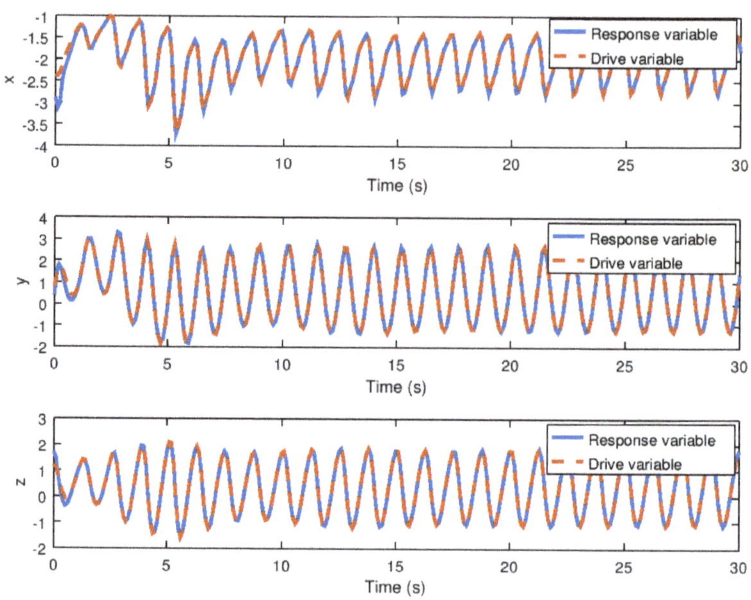

Figure 10. Evolution in time of the synchronized state variables of the drive and response novel chaotic system.

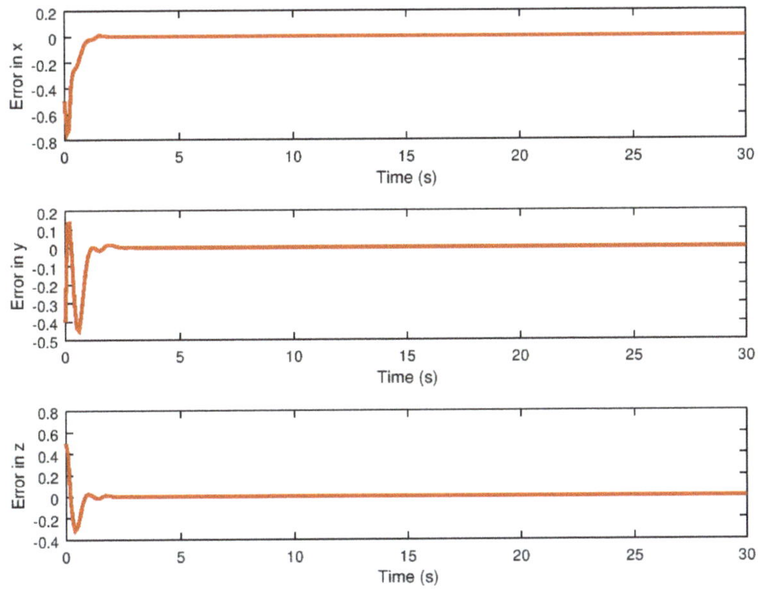

Figure 11. Evolution in time of the synchronization error variables.

The derivative of the error variable is shown in Figure 12. The velocity of error convergence is shown in this figure, indicating that the error variable reaches the origin in finite time faster due to the action of the robust control strategy even in the presence of saturation input nonlinearity. The peaks of the error variable derivative are thought to be small, which is related to the acceleration with which the error dynamics reach the origin in finite time, which is desirable when the proposed robust control synchronization strategy is implemented in real physical systems.

The evolution of the synchronization input variables U and v in time is depicted in Figures 13 and 14. As with the stabilization case, these input variables reach their final values in finite time until the drive and response variables are synchronized. Note that, despite the discontinuity in the robust controller part, the proposed control synchronization law does not produce chattering.

In these figures, the control variable U provides a non-oscillatory control action and a low control effort, which is important when this synchronization strategy is used in a variety of physical systems such as those found in physics, chemistry, biology, and engineering. Meanwhile, the time evolution of the virtual input variable v is shown, demonstrating that this error variable contributes only a minor control effort, implying that the overall control action must be as small as possible even in the presence of saturation input nonlinearity. Note that the control law design can be implemented in specific cases where the novel chaotic system must be synchronized in a real physical system or implemented in the form of a circuit or mechanical chaotic system.

Finally, in Figure 15 the evolution in time of the variable α for stabilization purposes is depicted, with the variable reaching equilibrium when synchronization is completed. This variable, like the extra state variable in the control technique for stabilization purposes, is basically the extra state variable with input v, as shown in Figure 14. Note that this auxiliary variable aids in the stability of the novel chaotic system in the presence of input saturation nonlinearity, which is significant given that this variable is appropriate for implementation in real physical systems in order to synchronize two identical systems, which in this case are essentially the novel chaotic system presented in this paper.

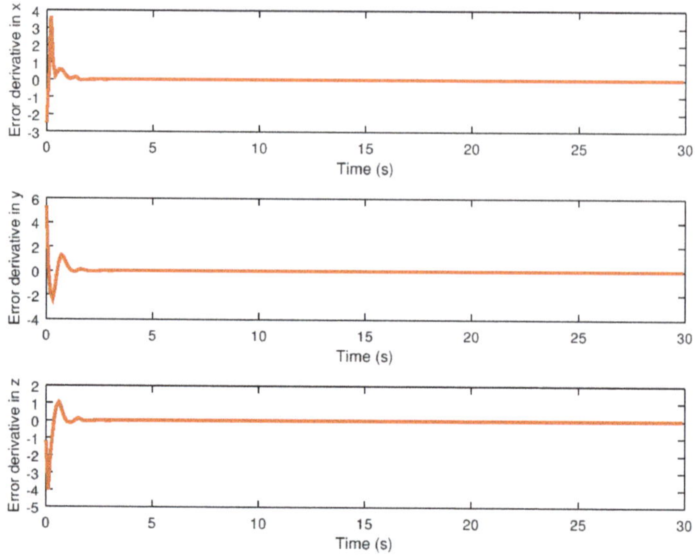

Figure 12. Derivative of the synchronization error variables.

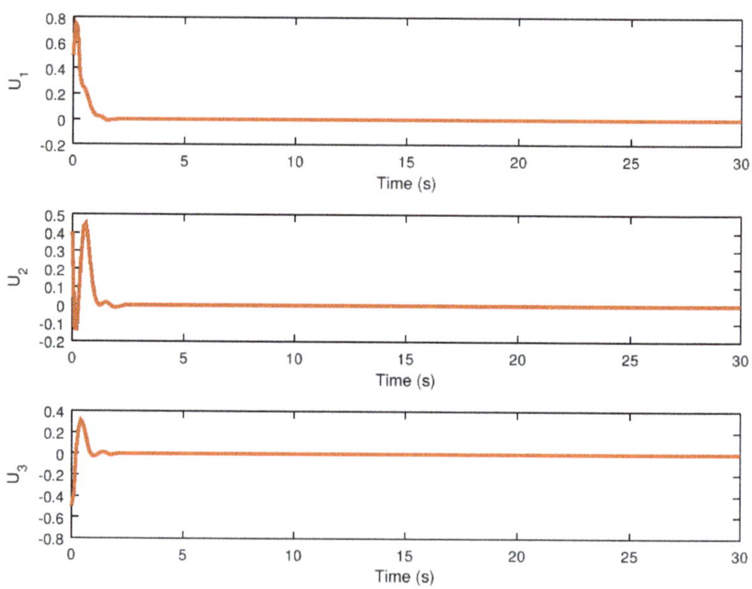

Figure 13. Input variable U for the response novel chaotic system.

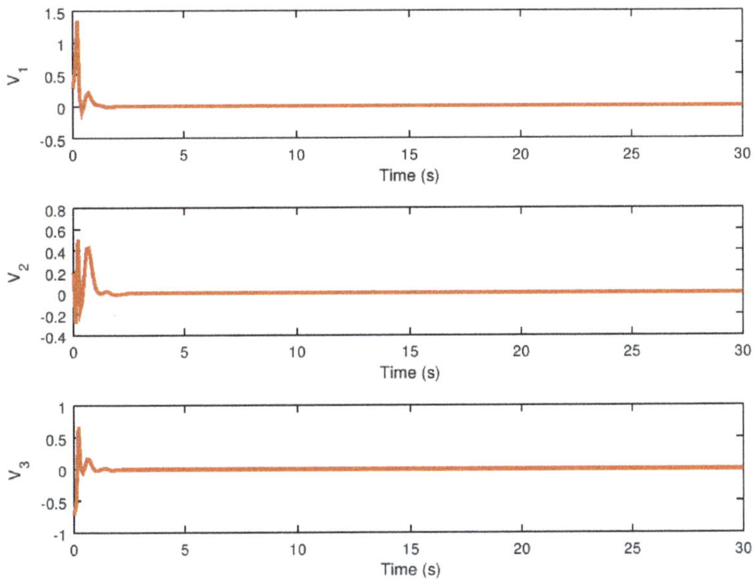

Figure 14. Input variable v for the novel chaotic response system.

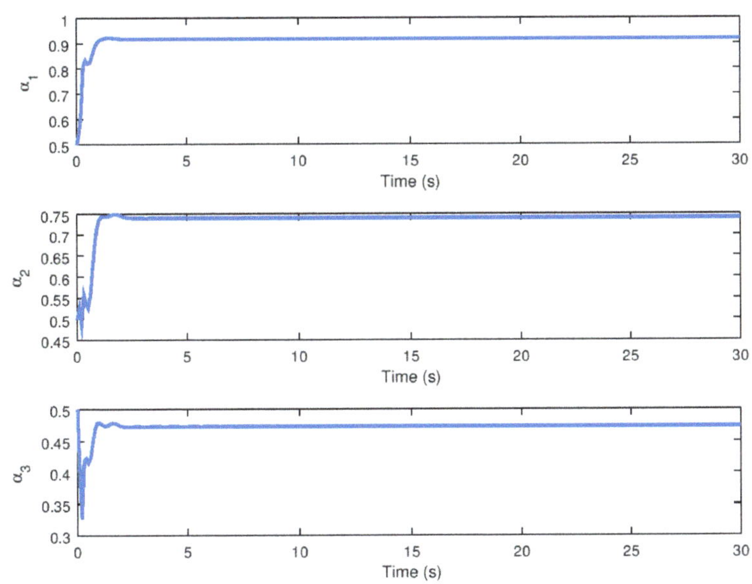

Figure 15. Evolution in time of the α variable for synchronization purposes.

5.3. Experimental Results Analysis

The experimental results for stabilization and control purposes, as explained in this section, numerically corroborate the results obtained in this study. It is possible to see how the variables reach equilibrium while suppressing chaotic behavior in the stabilization experiment. The three variables x, y, and z of the stabilized chaotic system are shown in Figure 5 how these variables reach the equilibrium point in a fast and accurate way in approximately 1 s despite the input saturation that is found in the chaotic system input providing a stable response in finite time. This is an important consideration, especially if this controller is used in a real-world physical system. It is demonstrated in Figure 1 how the proposed controller drives the state variable to a stable equilibrium rather than, as shown in Figure 1, to a chaotic region of attraction suppressing the limit cycle. The time evolution of the input variable U, as shown in Figure 7 depicts how a small control effort is generated in order to avoid input saturation of the stabilized chaotic system, which is important because input saturation suppresses some phenomena such as instability or chattering.

Meanwhile, the evolution in time of the extra additional input variable v and the state variable α can be seen in Figures 8 and 9 demonstrating that these variables reach their respective equilibrium until the chaotic system variables are stabilized. In the case of synchronization, it can be seen how the response system follows the drive variable in finite time, reducing the error to zero, as shown in Figures 10 and 11, respectively, given that it is difficult to drive the synchronization error variable in the presence of saturation while avoiding chattering or instability. It is also important to note in Figure 13 that the synchronization input variable provides an anti chattering response and a small control effort in a very short time until the proposed control synchronization law achieves ideal synchronization. Finally, as in the stabilization case, the variables v and α, as shown in Figures 14 and 15 depict the time evolution of the auxiliary variables, indicating how these variables reach the equilibrium.

6. Discussion

Taking into account the theoretical results, note that a novel chaotic system is created by combining two sigmoidal functions. It can be seen that a chaotic attractor is designed in which, as confirmed, the domain of attraction of this chaotic system is the equilibrium point, resulting in this chaotic system being a self-excited chaotic system. The proposed novel chaotic system is well suited to modeling many chaotic behaviors observed in nature and physical systems. The bifurcation diagrams demonstrate the chaotic behavior of this system as it transitions from a stable to a chaotic regime, and vice versa.

The robust controller is implemented appropriately for stabilization purposes by selecting an appropriate Lyapunov functional. The addition of an additional system provides the required inputs, which are considered dynamic inputs and include the discontinuity as part of the overall system input with saturation. The saturation is modeled by implementing the sector condition while accounting for the deadzone nonlinearity and taking into account the dynamic characteristics of this input nonlinearity. The conditions for obtaining an appropriate control law are provided by an anti-chattering stabilization control law and a state feedback input. A similar axiomatic methodology is used to find the robust control synchronization law by first defining the errors in the system dynamics and then selecting the appropriate robust control Lyapunov function.

In the case of the novel chaotic system's robust stabilization, the state feedback controller part moves the eigenvalues of the linearized novel chaotic system to a point where the state variables reach equilibrium in finite time, while the robust controller part switches this input variable to obtain a fast, accurate, and reliable response. Something similar happens with this novel chaotic system's robust control synchronization law. The linearized error system's eigenvalues are moved to an appropriate position in order to drive this error variable to zero in finite time. A state feedback controller is implemented to reach the zero value of the state variable while avoiding the input system chattering.

Meanwhile, in the experimental results section, it is possible to see how the robust controller drives the state variables of the stabilized system to the equilibrium point in finite time, with small peaks, and without chattering. When input saturation is detected in the system input, the small control effort generated by the variables U and v is crucial. Despite the fact that the input saturates at very low values, the system variables reach the equilibrium point in finite time, avoiding instability and, more importantly, poor performance. As seen in the synchronization case of this drive and response novel chaotic system, the drive and response variables are synchronized more quickly, accurately reaching the zero error between these variables. In the case of synchronization, saturation has no effect on synchronization performance, avoiding instability or poor performance, which is desirable given that saturation is almost unavoidable in most physical systems.

Given the vast number of physical systems in which the chaos phenomenon can be found, the results presented in this study can be implemented without difficulty in any of these physical systems represented or modeled as dynamic systems. One of the main advantages is that the chaos suppression robust control strategy for stabilization effectively eliminates nonlinear saturation effects on this novel chaotic system, such as instability of degraded performance, while also efficiently stabilizing the novel chaotic system's equilibrium points.

Apart from whether this control strategy is implemented in any type of physical system, the results provided in this research study can be extended to various types of physical systems in which the chaotic phenomenon is found, such as the physical systems mentioned previously in this research study. These results can be modified so that the robust controller can be implemented in other systems where hyperchaos or other phenomena are present. Note that the results obtained in this study can be extended for chaos synchronization in various types of systems in which not only two systems are synchronized, because it can be extended to chaos synchronization for multi-coupled chaotic systems. Furthermore, the results obtained in this study can be easily extended to complex chaotic networks, where it is important to note that robust control is always necessary given the vast number of physical systems in which chaotic synchronization is required.

7. Conclusions

This paper presents a novel chaotic attractor, as well as its robust control stabilization and synchronization with input saturation. First, the novel attractor is presented, in which two sigmoidal functions are implemented to generate chaotic dynamics, followed by a bifurcation analysis in which the stability and chaotic regimen regions for the novel chaotic attractor's parameters a, b, and d are presented. The phase portraits plot confirms that this chaotic attractor reaches the domain of attraction at the equilibrium point, despite the fact that it is self-excited. The control laws are obtained by using an extra system after a robust controller for stabilization purposes is derived by considering the sector condition property of the saturation input and implementing a robust control Lyapunov function. To obtain the synchronization control law, a similar methodology is used, but this time the synchronization error variable is established. Finally, two numerical experiments are carried out in conjunction with the research study's respective discussions and conclusions. The results presented in this numerical experiment section demonstrate that, in the stabilization case, the novel chaotic system can be driven to the equilibrium points in finite time with very little control effort and no oscillation or chattering. Note that the resulting control stabilization law can be efficiently implemented even in many different types of physical systems, such as electrical mechanical or other systems found in exact sciences. As in the stabilization case, it is demonstrated that in the synchronization case, an efficient robust control synchronization law is used, even in the presence of an input saturation nonlinearity in the response system, and that synchronization between the drive and response systems is achieved faster and more accurately. A gain matrix will be included in the stabilization and control laws in the future to be tuned by linear matrix inequalities.

Author Contributions: Conceptualization, A.T.A., F.E.S., N.A.K.; Formal analysis, A.T.A., F.E.S., N.A.K., Q.Z., M.B., G.F., J.N., W.Z.; Supervision, Q.Z., A.T.A.; Funding acquisition, A.T.A.; Methodology, A.T.A., F.E.S., N.A.K., Q.Z., M.B., G.F., J.N., W.Z.; Resources, Q.Z., M.B., G.F., J.N., W.Z.; Software, F.E.S., N.A.K.; Investigation, A.T.A., F.E.S., N.A.K., Q.Z., M.B., G.F., J.N., W.Z.; Validation, A.T.A., F.E.S., N.A.K., Q.Z., M.B., G.F., J.N., W.Z.; Visualization, A.T.A., F.E.S., N.A.K., Q.Z., M.B., G.F., J.N., W.Z.; Writing—original draft, A.T.A., F.E.S., N.A.K., Q.Z., M.B., G.F., J.N., W.Z.; Writing—review &

editing, A.T.A., F.E.S., N.A.K., Q.Z., M.B., G.F., J.N., W.Z. All authors have read and agreed to the published version of the manuscript.

Funding: This research is funded by Prince Sultan University, Riyadh, Saudi Arabia.

Data Availability Statement: Not applicable.

Acknowledgments: The authors would like to thank Prince Sultan University, Riyadh, Saudi Arabia for supporting this work. Also, the authors wish to acknowledge the editor and anonymous reviewers for their insightful comments, which have improved the quality of this publication.

Conflicts of Interest: The authors declare no conflict of interest.

References

1. Danca, M.F.; Lampart, M. Hidden and self-excited attractors in a heterogeneous Cournot oligopoly model. *Chaos Solitons Fractals* **2021**, *142*, 110371. [CrossRef]
2. Kuznetsov, N.; Kuznetsova, O.; Leonov, G.; Mokaev, T.; Stankevich, N. Hidden attractors localization in Chua circuit via the describing function method. *IFAC-PapersOnLine* **2017**, *50*, 2651–2656. [CrossRef]
3. Munmuangsaen, B.; Srisuchinwong, B. A hidden chaotic attractor in the classical Lorenz system. *Chaos Solitons Fractals* **2018**, *107*, 61–66. [CrossRef]
4. Liu, T.; Yan, H.; Banerjee, S.; Mou, J. A fractional-order chaotic system with hidden attractor and self-excited attractor and its DSP implementation. *Chaos Solitons Fractals* **2021**, *145*, 110791. [CrossRef]
5. Sanchez, E.N.; Ricalde, L.J. Chaos control and synchronization, with input saturation, via recurrent neural networks. *Neural Netw.* **2003**, *16*, 711–717. [CrossRef]
6. Yau, H.T.; Chen, C.L. Chaos control of Lorenz systems using adaptive controller with input saturation. *Chaos Solitons Fractals* **2007**, *34*, 1567–1574. [CrossRef]
7. Zhu, X.; Li, D. Robust attitude control of a 3-DOF helicopter considering actuator saturation. *Mech. Syst. Signal Process.* **2021**, *149*, 107209. [CrossRef]
8. Zhang, J.; Raïssi, T. Saturation control of switched nonlinear systems. *Nonlinear Anal. Hybrid Syst.* **2019**, *32*, 320–336. [CrossRef]
9. Niu, X.; Lin, W.; Gao, X. Static output feedback control of a chain of integrators with input constraints using multiple saturations and delays. *Automatica* **2021**, *125*, 109457. [CrossRef]
10. Yong, K.; Chen, M.; Shi, Y.; Wu, Q. Flexible performance-based robust control for a class of nonlinear systems with input saturation. *Automatica* **2020**, *122*, 109268. [CrossRef]
11. Azar, A.T.; Serrano, F.; Kamal, N.A.; Koubaa, A. Robust Kinematic Control of Unmanned Aerial Vehicles with Non-holonomic Constraints. *Proc. Int. Conf. Adv. Intell. Syst. Inform.* **2020**, *1261*, 839–850.
12. Kocamaz, U.E.; Cevher, B.; Uyaroglu, Y. Control and synchronization of chaos with sliding mode control based on cubic reaching rule. *Chaos Solitons Fractals* **2017**, *105*, 92–98. [CrossRef]
13. Aguilar-López, R.; Martínez-Guerra, R. Chaos suppression via observer based active control scheme: Application to Duffing's oscillator. *Chaos Solitons Fractals* **2007**, *32*, 1887–1897. [CrossRef]
14. Lee, K.W.; Singh, S.N. Robust control of chaos in Chua's circuit based on internal model principle. *Chaos Solitons Fractals* **2007**, *31*, 1095–1107. [CrossRef]
15. Lin, J.S.; Yan, J.J.; Liao, T.L. Robust control of chaos in Lorenz systems subject to mismatch uncertainties. *Chaos Solitons Fractals* **2006**, *27*, 501–510. [CrossRef]
16. E, J.; Qian, C.; Liu, H.; Liu, G. Design of the H-Infinity robust control for the piezoelectric actuator based on chaos optimization algorithm. *Aerosp. Sci. Technol.* **2015**, *47*, 238–246. [CrossRef]
17. Hu, J.; Qiu, Y.; Lu, H. Adaptive robust nonlinear feedback control of chaos in PMSM system with modeling uncertainty. *Appl. Math. Model.* **2016**, *40*, 8265–8275. [CrossRef]
18. Wei, D.Q.; Luo, X.S.; Wang, B.H.; Fang, J.Q. Robust adaptive dynamic surface control of chaos in permanent magnet synchronous motor. *Phys. Lett. A* **2007**, *363*, 71–77. [CrossRef]
19. Hamiche, H.; Takhi, H.; Messadi, M.; Kemih, K.; Megherbi, O.; Bettayeb, M. New synchronization results for a class of nonlinear discrete-time chaotic systems based on synergetic observer and their implementation. *Math. Comput. Simul.* **2021**, *185*, 194–217. [CrossRef]
20. Hamoudi, A.; Djeghali, N.; Bettayeb, M. Speech encryption based on synchronisation of chaotic Takagi-Sugeno systems using PI observer. In Proceedings of the 2021 18th International Multi-Conference on Systems, Signals Devices (SSD), Monastir, Tunisia, 22–25 March 2021; pp. 1285–1290. [CrossRef]
21. Djennoune, S.; Bettayeb, M.; Saggaf, U.M.A. Impulsive observer with predetermined finite convergence time for synchronization of fractional-order chaotic systems based on Takagi–Sugeno fuzzy model. *Nonlinear Dyn.* **2019**, *98*, 331–1354. [CrossRef]
22. Megherbi, O.; Hamiche, H.; Djennoune, S.; Bettayeb, M. Secure Color Image Transmission Based on the Impulsive Synchronization of Fractional-Order Chaotic Maps Over a Single Channel. In *Proceedings of the 4th International Conference on Electrical Engineering and Control Applications*; Bououden, S., Chadli, M., Ziani, S., Zelinka, I., Eds.; Springer: Singapore, 2021; pp. 1081–1095.

23. Djennoune, S.; Bettayeb, M.; Al-Saggaf, U.M. Synchronization of fractional–order discrete–time chaotic systems by an exact delayed state reconstructor: Application to secure communication. *Int. J. Appl. Math. Comput. Sci.* **2019**, *29*, 179–194. [CrossRef]
24. Wang, Z.; Volos, C.; Kingni, S.T.; Azar, A.T.; Pham, V.T. Four-wing attractors in a novel chaotic system with hyperbolic sine nonlinearity. *Opt. Int. J. Light Electron Opt.* **2017**, *131*, 1071–1078. [CrossRef]
25. Bettayeb, M.; Al-Saggaf, U.M.; Djennoune, S. High gain observer design for fractional-order non-linear systems with delayed measurements: Application to synchronisation of fractional-order chaotic systems. *IET Control Theory Appl.* **2017**, *11*, 3171–3178. [CrossRef]
26. Shahverdiev, E.; Bayramov, P.; Qocayeva, M. Inverse chaos synchronization between bidirectionally coupled variable multiple time delay systems. *Opt. Int. J. Light Electron Opt.* **2013**, *124*, 3427–3429. [CrossRef]
27. Mkaouar, H.; Boubaker, O. Chaos synchronization for master slave piecewise linear systems: Application to Chua's circuit. *Commun. Nonlinear Sci. Numer. Simul.* **2012**, *17*, 1292–1302. [CrossRef]
28. Moon, S.; Baik, J.J.; Seo, J.M. Chaos synchronization in generalized Lorenz systems and an application to image encryption. *Commun. Nonlinear Sci. Numer. Simul.* **2021**, *96*, 105708. [CrossRef]
29. Vaidyanathan, S.; Azar, A.T. Dynamic analysis, adaptive feedback control and synchronization of an eight-term 3-D novel chaotic system with three quadratic nonlinearities. In *Advances in Chaos Theory and Intelligent Control*; Springer: Berlin, Germany, 2016; pp. 155–178.
30. Vaidyanathan, S.; Azar, A.T. A novel 4-D four-wing chaotic system with four quadratic nonlinearities and its synchronization via adaptive control method. In *Advances in Chaos Theory and Intelligent Control*; Springer: Berlin, Germany, 2016; pp. 203–224.
31. Vaidyanathan, S.; Azar, A.T. Adaptive control and synchronization of Halvorsen circulant chaotic systems. In *Advances in Chaos Theory and Intelligent Control*; Springer: Berlin, Germany, 2016; pp. 225–247.
32. Vaidyanathan, S.; Azar, A.T. Adaptive backstepping control and synchronization of a novel 3-D jerk system with an exponential nonlinearity. In *Advances in Chaos Theory and Intelligent Control*; Springer: Berlin, Germany, 2016; pp. 249–274.
33. Vaidyanathan, S.; Azar, A.T. Generalized projective synchronization of a novel hyperchaotic four-wing system via adaptive control method. In *Advances in Chaos Theory and Intelligent Control*; Springer: Berlin, Germany, 2016; pp. 275–290.
34. Hikihara, T.; Kawagoshi, T. An experimental study on stabilization of unstable periodic motion in magneto-elastic chaos. *Phys. Lett. A* **1996**, *211*, 29–36. [CrossRef]
35. Khennaoui, A.A.; Ouannas, A.; Bendoukha, S.; Grassi, G.; Lozi, R.P.; Pham, V.T. On fractional–order discrete–time systems: Chaos, stabilization and synchronization. *Chaos Solitons Fractals* **2019**, *119*, 150–162. [CrossRef]
36. Mukherjee, M.; Halder, S. Stabilization and Control of Chaos Based on Nonlinear Dynamic Inversion. *Energy Procedia* **2017**, *117*, 731–738. [CrossRef]
37. Abed, E.; Wang, H.; Chen, R. Stabilization of period doubling bifurcations and implications for control of chaos. *Phys. D Nonlinear Phenom.* **1994**, *70*, 154–164. [CrossRef]
38. Duan, C.; Yang, S.; Min, W.; Xia, S. Feedback-control to realize and stabilize chaos synchronization. *Chaos Solitons Fractals* **1998**, *9*, 921–929. [CrossRef]
39. Ouannas, A.; Azar, A.T.; Abu-Saris, R. A new type of hybrid synchronization between arbitrary hyperchaotic maps. *Int. J. Mach. Learn. Cybern.* **2017**, *8*, 1887–1894. [CrossRef]
40. Ouannas, A.; Azar, A.T.; Vaidyanathan, S. New hybrid synchronisation schemes based on coexistence of various types of synchronisation between master-slave hyperchaotic systems. *Int. J. Comput. Appl. Technol.* **2017**, *55*, 112–120. [CrossRef]
41. Vaidyanathan, S.; Azar, A.T. Qualitative Study and Adaptive Control of a Novel 4-D Hyperchaotic System with Three Quadratic Nonlinearities. In *Advances in Chaos Theory and Intelligent Control*; Azar, A.T., Vaidyanathan, S., Eds.; Springer International Publishing: Cham, Switzerland, 2016; pp. 179–202.
42. Ouannas, A.; Azar, A.T.; Vaidyanathan, S. A robust method for new fractional hybrid chaos synchronization. *Math. Methods Appl. Sci.* **2017**, *40*, 1804–1812. [CrossRef]
43. Zambrano-Serrano, E.; Anzo-Hernández, A. A novel antimonotic hyperjerk system: Analysis, synchronization and circuit design. *Phys. D Nonlinear Phenom.* **2021**, *424*, 132927. [CrossRef]
44. Shahverdiev, E.; Bayramov, P.; Nuriev, R.; Qocayeva, M.; Hashimova, L. Chaos synchronization between Josephson junctions coupled in series and driven by a central junction. *Phys. C Supercond. Its Appl.* **2019**, *557*, 26–32. [CrossRef]
45. Korneev, I.; Semenov, V.; Slepnev, A.; Vadivasova, T. Complete synchronization of chaos in systems with nonlinear inertial coupling. *Chaos Solitons Fractals* **2021**, *142*, 110459. [CrossRef]
46. Tai, W.; Teng, Q.; Zhou, Y.; Zhou, J.; Wang, Z. Chaos synchronization of stochastic reaction-diffusion time-delay neural networks via non-fragile output-feedback control. *Appl. Math. Comput.* **2019**, *354*, 115–127. [CrossRef]
47. Al Bayati, B.M.; Ahmad, A.K.; Al Naimee, K.A. Effect of control parameters on chaos synchronization by means of optical feedback. *Opt. Commun.* **2020**, *472*, 126032. [CrossRef]
48. Xu, F.; Yu, P. Chaos control and chaos synchronization for multi-scroll chaotic attractors generated using hyperbolic functions. *J. Math. Anal. Appl.* **2010**, *362*, 252–274. [CrossRef]
49. Rehan, M. Synchronization and anti-synchronization of chaotic oscillators under input saturation. *Appl. Math. Model.* **2013**, *37*, 6829–6837. [CrossRef]
50. Wang, D.; Zheng, Y. Stabilizing a Class of Chaotic Systems by Using Adaptive Feedback Control. *Phys. Procedia* **2012**, *24*, 1922–1927. [CrossRef]

51. Mahmoud, E.E.; Trikha, P.; Jahanzaib, L.S.; Eshmawi, A.A.; Matoog, R. Chaos control and Penta-compound combination anti-synchronization on a novel fractional chaotic system with analysis and application. *Results Phys.* **2021**, *24*, 104130. [CrossRef]
52. Allehiany, F.; Mahmoud, E.E.; Jahanzaib, L.S.; Trikha, P.; Alotaibi, H. Chaos control and analysis of fractional order neural network under electromagnetic radiation. *Results Phys.* **2021**, *21*, 103786. [CrossRef]
53. Tan, T.; Wang, Z.; Zhang, L.; Liao, W.H.; Yan, Z. Piezoelectric autoparametric vibration energy harvesting with chaos control feature. *Mech. Syst. Signal Process.* **2021**, *161*, 107989. [CrossRef]
54. Liu, M.K.; Suh, C.S. Synchronization of chaos in simultaneous time-frequency domain. *Appl. Math. Model.* **2013**, *37*, 9524–9537. [CrossRef]
55. AL-Azzawi, S.F.; Aziz, M.M. Chaos synchronization of nonlinear dynamical systems via a novel analytical approach. *Alex. Eng. J.* **2018**, *57*, 3493–3500. [CrossRef]
56. Elsonbaty, A.; Elsaid, A.; Nour, H. Circuit realization, chaos synchronization and estimation of parameters of a hyperchaotic system with unknown parameters. *J. Egypt. Math. Soc.* **2014**, *22*, 550–557. [CrossRef]
57. El-Sayed, A.; Nour, H.; Elsaid, A.; Matouk, A.; Elsonbaty, A. Dynamical behaviors, circuit realization, chaos control, and synchronization of a new fractional order hyperchaotic system. *Appl. Math. Model.* **2016**, *40*, 3516–3534. [CrossRef]
58. Pusuluri, K.; Meijer, H.; Shilnikov, A. (INVITED) Homoclinic puzzles and chaos in a nonlinear laser model. *Commun. Nonlinear Sci. Numer. Simul.* **2021**, *93*, 105503. [CrossRef]
59. Zhu, C.; Zhang, W. Multiple chaos arising from single-parametric perturbation of a degenerate homoclinic orbit. *J. Differ. Equ.* **2020**, *268*, 5672–5703. [CrossRef]
60. Li, H.; Liao, X.; Huang, J.; Chen, G.; Dong, Z.; Huang, T. Diverting homoclinic chaos in a class of piecewise smooth oscillators to stable periodic orbits using small parametrical perturbations. *Neurocomputing* **2015**, *149*, 1587–1595. [CrossRef]
61. Siewe, M.S.; Hegazy, U.H. Homoclinic bifurcation and chaos control in MEMS resonators. *Appl. Math. Model.* **2011**, *35*, 5533–5552. [CrossRef]
62. Franca, M.; Pospísil, M. New global bifurcation diagrams for piecewise smooth systems: Transversality of homoclinic points does not imply chaos. *J. Differ. Equ.* **2019**, *266*, 1429–1461. [CrossRef]
63. Huaraca, W.; Mendoza, V. Minimal topological chaos coexisting with a finite set of homoclinic and periodic orbits. *Phys. D Nonlinear Phenom.* **2016**, *315*, 83–89. [CrossRef]
64. Li, Z. Chaos induced by heteroclinic cycles connecting repellers and saddles in locally compact metric spaces. *Nonlinear Anal. Theory Methods Appl.* **2009**, *71*, 1379–1388. [CrossRef]
65. Li, Z.; Shi, Y.; Zhang, C. Chaos induced by heteroclinic cycles connecting repellers in complete metric spaces. *Chaos Solitons Fractals* **2008**, *36*, 746–761. [CrossRef]
66. Barajas-Ramírez, J.G.; Franco-López, A.; González-Hernández, H.G. Generating Shilnikov chaos in 3D piecewise linear systems. *Appl. Math. Comput.* **2021**, *395*, 125877.
67. Natiq, H.; Kamel Ariffin, M.R.; Asbullah, M.A.; Mahad, Z.; Najah, M. Enhancing Chaos Complexity of a Plasma Model through Power Input with Desirable Random Features. *Entropy* **2021**, *23*, 48. [CrossRef]
68. Sándor, B.; Schneider, B.; Lázár, Z.I.; Ercsey-Ravasz, M. A Novel Measure Inspired by Lyapunov Exponents for the Characterization of Dynamics in State-Transition Networks. *Entropy* **2021**, *23*, 103. [CrossRef]
69. Petrzela, J. Evidence of Strange Attractors in Class C Amplifier with Single Bipolar Transistor: Polynomial and Piecewise-Linear Case. *Entropy* **2021**, *23*, 175. [CrossRef]
70. Ribeiro, M.; Henriques, T.; Castro, L.; Souto, A.; Antunes, L.; Costa-Santos, C.; Teixeira, A. The Entropy Universe. *Entropy* **2021**, *23*, 222 . [CrossRef]
71. Guo, Y.; Liu, T.; Zhao, T.; Zhang, H.; Guo, X. Chaotic Time-Delay Signature Suppression and Entropy Growth Enhancement Using Frequency-Band Extractor. *Entropy* **2021**, *23*, 516. [CrossRef]
72. Marszalek, W.; Sadecki, J.; Walczak, M. Computational Analysis of Ca2+ Oscillatory Bio-Signals: Two-Parameter Bifurcation Diagrams. *Entropy* **2021**, *23*, 876. [CrossRef] [PubMed]
73. Dénes, K.; Sándor, B.; Néda, Z. Synchronization patterns in rings of time-delayed Kuramoto oscillators. *Commun. Nonlinear Sci. Numer. Simul.* **2021**, *93*, 105505. [CrossRef]
74. Shang, D.; Shang, P.; Zhang, Z. Efficient synchronization estimation for complex time series using refined cross-sample entropy measure. *Commun. Nonlinear Sci. Numer. Simul.* **2021**, *94*, 105556. [CrossRef]
75. Wang, N.; Zhang, G.; Kuznetsov, N.; Bao, H. Hidden attractors and multistability in a modified Chua's circuit. *Commun. Nonlinear Sci. Numer. Simul.* **2021**, *92*, 105494. [CrossRef]
76. Kecik, K. Simultaneous vibration mitigation and energy harvesting from a pendulum-type absorber. *Commun. Nonlinear Sci. Numer. Simul.* **2021**, *92*, 105479. [CrossRef]
77. Montes, J.; Revuelta, F.; Borondo, F. Lagrangian descriptors and regular motion. *Commun. Nonlinear Sci. Numer. Simul.* **2021**, *102*, 105860. [CrossRef]
78. Salas, J.P.; Iñarrea, M.; Lanchares, V.; Palacián, J.; Yanguas, P. Magnetic confinement of a neutral atom in a double-wire waveguide: A nonlinear dynamics approach. *Commun. Nonlinear Sci. Numer. Simul.* **2021**, *101*, 105662. [CrossRef]
79. Adewole, A.; Falayi, E.; Roy-Layinde, T.; Adelaja, A. Chaotic time series analysis of meteorological parameters in some selected stations in Nigeria. *Sci. Afr.* **2020**, *10*, e00617. [CrossRef]

80. Chabab, M.; El Moumni, H.; Iraoui, S.; Masmar, K.; Zhizeh, S. Chaos in charged AdS black hole extended phase space. *Phys. Lett. B* **2018**, *781*, 316–321. [CrossRef]
81. Shirmohammadi, F.; Tohidi, A. Mixing enhancement using chaos theory in fluid dynamics: Experimental and numerical study. *Chem. Eng. Res. Des.* **2019**, *141*, 350–360. [CrossRef]
82. Azar, A.T.; Serrano, F.E. Stabilization of Port Hamiltonian Chaotic Systems with Hidden Attractors by Adaptive Terminal Sliding Mode Control. *Entropy* **2020**, *22*, 122. [CrossRef] [PubMed]

entropy

Article

U-Model-Based Two-Degree-of-Freedom Internal Model Control of Nonlinear Dynamic Systems

Ruobing Li [1,*], Quanmin Zhu [1], Pritesh Narayan [1], Alex Yue [1], Yufeng Yao [1] and Mingcong Deng [2]

1 Department of Engineering Design and Mathematics, University of the West of England, Frenchay Campus, Coldharbour Lane, Bristol BS16 1QY, UK; quan.zhu@uwe.ac.uk (Q.Z.); Pritesh.Narayan@uwe.ac.uk (P.N.); Alex.Yue@uwe.ac.uk (A.Y.); Yufeng.Yao@uwe.ac.uk (Y.Y.)

2 Department of Electrical and Electronic Engineering, Tokyo University of Agriculture and Technology, 2-24-16 Nakacho, Koganei, Tokyo 183-8538, Japan; deng@cc.tuat.ac.jp

* Correspondence: Ruobing2.Li@live.uwe.ac.uk

Abstract: This paper proposes a U-Model-Based Two-Degree-of-Freedom Internal Model Control (UTDF-IMC) structure with strength in nonlinear dynamic inversion, and separation of tracking design and robustness design. This approach can effectively accommodate modeling error and disturbance while removing those widely used linearization techniques for nonlinear plants/processes. To assure the expansion and applications, it analyses the key properties associated with the UTDF-IMC. For initial benchmark testing, computational experiments are conducted using MATLAB/Simulink for two mismatched linear and nonlinear plants. Further tests consider an industrial system, in which the IMC of a Permanent Magnet Synchronous Motor (PMSM) is simulated to demonstrate the effectiveness of the design procedure for potential industrial applications.

Keywords: Internal Model Control (IMC); U-model; U-model-based control (U-control); Two-Degree-of-Freedom IMC (TDF-IMC); dynamic inversion; invariance entropy

Citation: Li, R.; Zhu, Q.; Narayan, P.; Yue, A.; Yao, Y.; Deng, M. U-Model-Based Two-Degree-of-Freedom Internal Model Control of Nonlinear Dynamic Systems. *Entropy* **2021**, *23*, 169. https://doi.org/10.3390/e23020169

Received: 8 January 2021
Accepted: 24 January 2021
Published: 29 January 2021

Publisher's Note: MDPI stays neutral with regard to jurisdictional claims in published maps and institutional affiliations.

1. Introduction

With the development of science and technology, the scale of industrial production in almost all fields, such as petrochemical, metallurgy, electric power, machinery, aerospace, etc. continues to expand, and the corresponding operational systems have had a demand for high quality and better quantity [1], which is inevitably at the cost of bringing complexity to the control system design. These challenges have motivated academic research and industrial development.

The classical Proportional Integral Derivative (PID) control and its integrations with various control strategies such as fuzzy PID [2], and neural PID [3] have been widely used in industrial systems. Although these control strategies can cope with complexities such as uncertainty, nonlinear dynamics, and large time delays, it is still worthwhile seeking other effective control system design methodologies to further upgrade control system performance while improving the design effectiveness. For example, a commonly observed practical situation, is that the success of tuning PID controller parameters often depends on a combination of the applicant's engineering experience and tedious effort on trial and error. Although this is workable, this unsystematically experienced approach often causes largely inefficient use of human resource and equipment to obtain satisfactory tuning results.

For this paper's interest, model validity is a fundamental basis for model-based control system design. A better model makes control system design and tuning easier/more efficient. However, for most engineering systems, there can be difficulties in obtaining accurate plant/process models, primarily due to equipment diversity and environment complexity; such as internal uncertainties and external disturbances. Even though a mathematical model can be established from physical principles (such as energy conservation law) and/or data driven (identification), it is usually taken as nominal reference (nominal model is an approximate description to accurate model). There are two main streams in dealing with such model uncertainties, i.e., adaptive control and robust control. This paper will follow the route of robust control.

The other practical topic is digital control, which has been used in almost all modern control engineering systems. To deal with a digital channel between the sensor and the controller, entropy [4], the concepts, terminologies, and techniques, has been adopted in digital control systems. The invariance entropy [5] has been used to determine the smallest average rate of information transmission to guarantee a considered subset of the state-space invariant to achieve the integrated control system performance. Accordingly control system design in connection to online applications, with potential contribution to this field, is probably making the controllers more efficient in online computation, so that reduces the burden to the communication capacity or at least does not increase significantly the information processing complexity.

Internal Model Control (IMC) [6,7] has been widely accepted as an efficient robust control approach. IMC selects the model inverse as the controller and integrates a robust filter to control an explicit plant/process model. The IMC structure is characterized with (1) capable robustness to overcome model uncertainties and system disturbances, (2) effective procedures for designing and tuning, (3) successful application across different industries [8–10]. However, the control performance of classical IMC is not desirable, because the adjustable parameters only exist in the filter. At the same time, higher robustness demand could degrade tracking performance [11], which must compromise with some of the other performances. Although a Two-Degree-of-Freedom IMC (TDF-IMC) structure can solve the aforementioned problems with the classical IMC structure, its control performance still cannot be separately designed [12–14].

When a linear model is completely reversible, the design of linear IMC is straightforward to take the controller as the inverse of the model and select a suitable filter. Even though when the model is not completely reversible, the model can be decomposed into reversible parts and irreversible parts, in which the inverse of the reversible part is taken as the controller. Appropriate filter selection can then also ensure that the control system has the smallest output variance for both stabilization and tracking control. However, for controlling nonlinear plants/processes, these approaches are not applicable, and effective algorithms for nonlinear dynamic inversion are very limited [15].

To deal with nonlinear control plants/processes, the approaches used by most of the IMC structures can be divided into (1) linearizing the controlled plants/processes and using linear method to invert [16]; (2) using PID [17], neural network [18,19] and fuzzy control-based [20,21] dynamic inversion; (3) using some numerical tools, such as the Newton–Raphson method [22]. However, the linearized and the other approximating modeling methods could lose accurate representation of input-output relationship and degrade the performance of the designed systems. Therefore, deriving the nonlinear model inversion and enabling the two performance indicators (i.e., tracking and robustness) of the IMC structure to be independently designed are the main challenges and focuses in this paper. Accordingly, this study proposes a framework of U-Model-Based Two-Degree-of-Freedom Internal Model Control (U-TDF-IMC) of nonlinear dynamic systems

U-model is a derived control-oriented model set to map almost all classical models into their U-model realization, and converts classical models into controller output u-based with time-varying parameters [23] expressions. U-model establishes a platform for solution of dynamic inversion by solving roots of polynomial equations, which is more generally attractive compared to the other ad hoc approaches/algorithms [24]. U-model-based control [25] (denoted as 'U-control' thereafter), takes advantages of U-model in dynamic inversion with the following characteristics:

1. Design control systems in a universal procedure, separate two dynamic inversions, invariant controller implementation by inversing specified system performance in a feedback configuration and plant utilization by plant inversion. These two designs are parallel and separable;

2. The difference in U-control between linear models and nonlinear models is the solution with the first-order or higher-order polynomial root-solving. The difference in U-control between polynomial models and state-space models is the one-layer or multi-layer polynomial root-solving;
3. U-control is seamlessly supplemented to the other exist control schemes, for example U-Pole Placement Control (UPPC) [26], U-General Predictive Control (UGPC) [27], U-Neuro-Control (UNC) [28], U-Total Nonlinear Control (UNLC) [23], and U-Internal Model Control (UIMC) [29].

This paper is aimed at using U-control, an enhanced tool supplemented to classical approaches, to integrate the strengths exhibited in U-control and IMC to provide an enhanced version of IMC with strength in system configuration and nonlinear dynamic inversion. To further improve TDF-IMC, this study expands the previous IMC work [29,30] by effectively introducing U-model-based dynamic inversion within a revised system structure configuration. By doing so, the new framework presents a new U-model-based Two-Degree-of-Freedom IMC (UTDF-IMC) structure to achieve the completely independent design in rejecting disturbance and tracking operational set-point. Compared with the classical IMC and TDF-IMC, this proposed structure has better control performance and more convenient tuning methods without introducing additional design work and maintaining the same hardware configuration. Accordingly, the major impacts of this paper are outlined below:

1. Propose a general U-model-based Two-Degree-of-Freedom IMC (UTDF-IMC) structure for controlling a class of open-loop stable polynomial/state-space modeled linear and nonlinear dynamic plants. The new control system structure accommodates both linear and nonlinear plants consistently and separate the tracking control filter design from robust control filter design.
2. Tailor the UM-dynamic inversion platform [31] in conjunction with IMC, which removes the necessity of either linearizing the nonlinear model, or converting it to a quasi-linear parameter-varying (quasi-LPV) model in advance. This UM-dynamic inversion platform directly provides algorithms dealing with all types of inversions in IMC structured systems.
3. Analyze the designed UTDF-IMC control system properties to provide a valid reference for future study expansions and applications.
4. Verify the control system performance through benchmark tests of simulated case studies and illustrate application procedure from an industrial case demonstration.

For the remainder of the paper, Section 2 presents the basis of using IMC and U-control for the next step development of the new UTDF-IMC system structure. Section 3 elaborates on the principle of TDF-IMC structure and establishes the U-model-based TDF-IMC (UTDF-IMC) framework; consequently, it analyses the control system properties. Section 4 showcases two computational investigations to benchmark test/demonstrate the proposed UTDF-IMC system performance. Then an industrial backgrounded permanent magnet synchronous motors (PMSM) system is added to demonstrate the application procedure and the comparative studies. Section 5 concludes this study with key findings and observations.

2. Preliminaries

2.1. Internal Model Control (IMC)

A classical IMC control scheme [7] is shown in Figure 1a, in which the plants/process P is approximated by model P_0 (specifically known as internal model) and the controller Q. Figure 1b shows the equivalently rearranged IMC structure, which the controller is expressed in the inner loop

$$C = \frac{Q}{1 - QP_0} \tag{1}$$

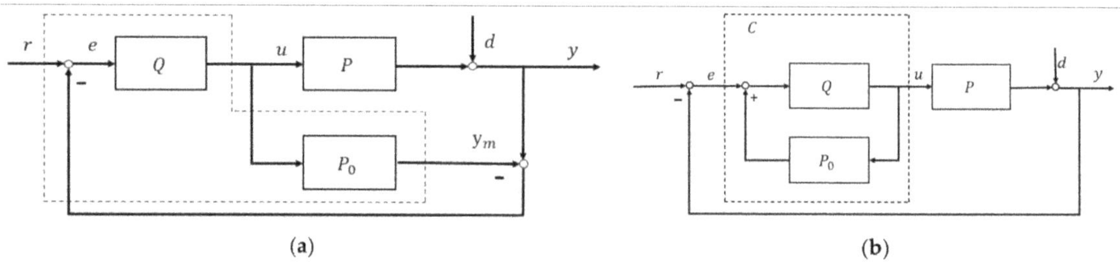

Figure 1. IMC method structure (**a**) Internal model control structure; (**b**) Equivalent IMC structure.

For a given set-point reference r, the control system is designed to keep the output y following a pre-specified output response y_m (Figure 1a) with the desired transient and steady-state performance. With reference to Figure 1a,

Plant output:

$$y = \frac{QP}{1+Q(P-P_0)}r + \frac{1-QP_0}{1+Q(P-P_0)}d \tag{2}$$

Error output:

$$e = \frac{1}{1+Q(P-P_0)}(r-d) \tag{3}$$

Controller output:

$$u = \frac{Q}{1+Q(P-P_0)}(r-d) \tag{4}$$

Remark 1. *(2) can also be rewritten as:*

$$y = \alpha r + \beta d \tag{5}$$

where $\alpha = \frac{QP}{1+Q(P-P_0)}$ *specifies tracking performance and* $\beta = \frac{1-QP_0}{1+Q(P-P_0)}$ *denotes the contribution to robustness. These two weights meet the condition of* $\alpha + \beta = 1$.

The main features of IMC [7] include:

1. Dual stability: For $P = P_0$ and $d = 0$, and $y = y_m$, the feedback error signal e is obviously zero. IMC system becomes an open-loop structure and both controller Q and plant P stable.
2. Perfect control: This requests plant $P = P_0$ minimum-phase and invertible and controller as the model inverse $Q = {P_0}^{-1}$. Accordingly (2) becomes:

$$y = \frac{{P_0}^{-1}P}{1+{P_0}^{-1}P - {P_0}^{-1}P_0}r + \frac{1-{P_0}^{-1}P_0}{1+{P_0}^{-1}(P-P_0)}d = r, \text{ and } \alpha = 1, \beta = 0 \tag{6}$$

3. Augmented robust IMC is shown in Figure 2. It decomposes model and dynamic inversion by factorizing P_0 into P_{0+} and P_{0-}, namely: $P_0 = P_{0+}P_{0-}$, where P_{0+} is the part containing pure delay and uncertain zero, and P_{0-} is the minimum-phase part. There are certain factorization techniques, such as simple factorization, all-pass factorization [32]. Hence, the controller is kept as the inverse of the plant/process model with invertible portion, i.e.,

$$Q_1 = \frac{1}{P_{0-}} \tag{7}$$

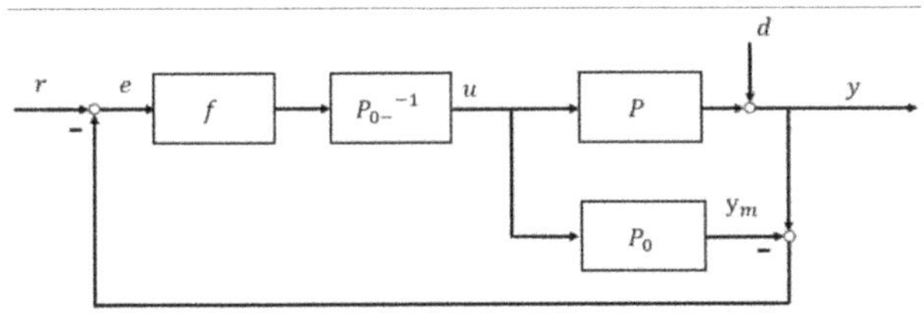

Figure 2. Robust IMC structure.

Filter: When designing the IMC controller, it should add a low-pass filter for the inverse of the factorized minimum-phase model to ensure the controller be proper and robust to against the model mismatching and disturbance. Define the IMC controller and the filter as:

$$Q = fQ_1 \tag{8}$$

$$f = \frac{1}{(1+\lambda s)^{\rho}} \tag{9}$$

where ρ is the order of the filter, normally assigned with a large value to ensure Q_1 be proper or semi-proper; λ is the time constant, the sole design parameter of the controller and is inversely proportional to the closed-loop response speed. Therefore, λ is a trade-off between the performances.

Remark 2. *Substituting (7) and (8) into (2) obtains the plant output:*

$$y = \frac{\frac{f}{P_{0-}}P}{1+\frac{f}{P_{0-}}P-\frac{f}{P_{0-}}P_0}r + \frac{1-\frac{f}{P_{0-}}P_0}{1+\frac{f}{P_{0-}}P-\frac{f}{P_{0-}}P_0}d = \frac{\frac{f}{P_{0-}}P}{(1-fP_{0+})+\frac{f}{P_{0-}}P}r + \frac{1-fP_{0+}}{(1-fP_{0+})+\frac{f}{P_{0-}}P}d \tag{10}$$

To track the reference signal with a faster speed and effectively reject the modeling errors and system disturbance, it requires output (10) satisfying $fP_{0+} = 1$ necessarily, which is achieved by selecting λ in the filter.

2.2. *U-Model-Based Control (U-Control)*

2.2.1. U-Models

A general U-polynomial-model of G_P [33], with a triplet of $(y(t), u(t), \alpha(t))$, $y(t) \in \mathbb{R}$, $u(t) \in \mathbb{R}$ $\alpha(t) \in \mathbb{R}^J$ for the output, input, and time-varying parameter vector respectively at time $t \in \mathbb{R}^+$, is defined for Single-Input and Single-Output (SISO) dynamic processes as

$$\overset{(M)}{y} = \mathcal{A}^T\mathcal{U} = \sum_{j=0}^{J} \alpha_j f_j\left(\overset{(N)}{u}{}^{j}\right), \; M \geq N \tag{11}$$

where $\overset{(M)}{y}$ and $\overset{(N)}{u}$ are the *Mth* and *Nth* order derivatives of the plant output y and the plant input u respectively. The time-varying parameter $\alpha_j \in \mathbb{R}^+$ absorbing all other output terms such as $\left[\begin{array}{ccc}\overset{(m-1)}{y} & ,\ldots, & y\end{array}\right] \in \mathbb{R}^M$ and input terms $\left[\begin{array}{ccc}\overset{(n-1)}{u} & ,\ldots, & u\end{array}\right] \in \mathbb{R}^N$. Function $f_j(*)$ is associated with the input $\left(\overset{(N)}{u}\right)^j$. $\mathcal{A}^T = \left[\begin{array}{ccc}\alpha_0 & ,\ldots, & \alpha_J\end{array}\right]$ and $\mathcal{U} = \left[\begin{array}{ccc}f_0 & ,\ldots, & f_J\end{array}\right]$ are the operators mapping the underlying input, output, and parameters into the condensed vector expressions. To illustrate the U-representation of classical models, consider a general polynomial model:

$$\ddot{y} = \left(1-e^{-|y|}\right)\dot{y} + \left(1+y^2\right)\sin(u) + \left(1+\dot{y}^2\right)u^2 + y + \left(y+\dot{y}^2\right)u^3 \tag{12}$$

Its U-model is transformed with the U-mappings of $\mathcal{A}$ and $\mathcal{U}$

$$\begin{cases} \ddot{y} = \alpha_0 f_0(u) + \alpha_1 f_1(u) + \alpha_2 f_2(u)^2 + \alpha_3 f_3(u)^3 \\ \alpha_0 = \left(1 - e^{-|y|}\right)\dot{y} + y,\ f_0(u^0) = 1 \\ \alpha_1 = 1 + y^2,\ f_1(u) = \sin(u) \\ \alpha_2 = 1 + \dot{y}^2,\ f_2(u^2) = u^2 \\ \alpha_3 = y + \dot{y}^2,\ f_3(u^3) = u^3 \end{cases} \tag{13}$$

Remark 3. *For representation to classical linear models, assign degree $J = 1$ and function $f_0(u) = 1,\ f_1(u) = u$, then the linear U-model is expressed with*

$$\overset{(M)}{y} = \alpha_0 + \alpha_1 \overset{(N)}{u},\ M \geq N \tag{14}$$

Remark 4. *For U-stats space-models, expand the single layer U-polynomial model (11) into multi-layer systems of polynomials [31].*

2.2.2. UM-Dynamic Inversion

For determining the output u of G_p^{-1}, a general UM-dynamic inversion algorithm is developed [19] to determine one of the solutions of $\overset{(N)}{u}$ from solving the following general polynomial equation

$$G_P^{-1} \Leftrightarrow \overset{(N)}{u} \in \overset{(M)}{y} - \sum_{j=0}^{J} \alpha_j f_j\left(\overset{(N)^j}{u}\right) = 0,\ M \geq N \tag{15}$$

For the solution exist, the systems must be Bounded Input and Bounded Output (BIBO) stable and no unstable zero dynamic (nonminimum phase). The solution platform has been expanded including the root-solving algorithms for continuous/discrete time, linear/nonlinear, polynomial/state-space models [31]. For online root-solving, Zhu [34] has proposed Newton–Raphson iterative algorithm.

2.2.3. U-Control

Let G_p be a general representation of both polynomial and state-space-based linear and nonlinear models for dynamic plants. In assumption, the plant has most properties as those requested in the other classical works [35]. Consequently,

a. Model of G_p is invertible, i.e., G_P^{-1} exists
b. Meet the continuity of Lipschitz, G_p and its inverse G_P^{-1} are globally unified and diffeomorphic in $\mathbb{R}^n$, i.e.,

$$\|G_p(x_1) - P(x_2)\| \leq \gamma_1 G_p \|x_1 - x_2\|,\ \forall x_1, x_2 \in \mathbb{R}^n \|G_P^{-1}(x_1) - G_P^{-1}(x_2)\| \leq \gamma_2\, G_P^{-1} \|x_1 - x_2\|,\ \forall x_1, x_2 \in \mathbb{R}^n \tag{16}$$

where x_1, x_2 are the states while G_p in the expression of state-space equation, γ_1 and γ_2 are the Lipschitz coefficients. This study takes SISO (input $u \in \mathbb{R}^1$ and output $y \in \mathbb{R}^1$) prototype in consideration. U-control system framework [25] is shown in Figure 3, in which $\mathcal{F}$ is for U-control system structure, G_{c1} is a linear invariant controller to be designed, and G_P^{-1} is the inverter of the controlled plant G_P to be designed as well. It is noted that U-control framework is applicable to various plants/processes when the dynamic inverse G_P^{-1} exist.

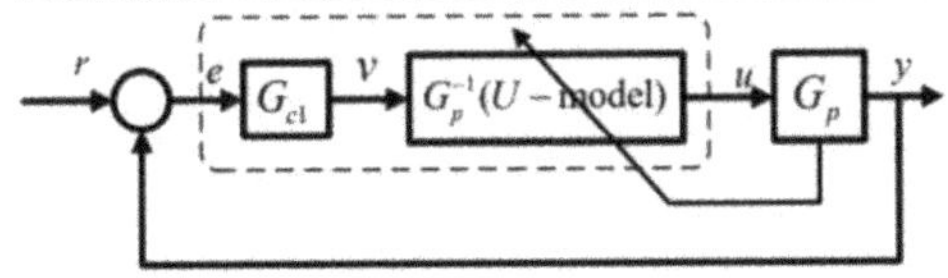

Figure 3. U-control system framework.

The U-control system is structured

$$\sum = \left(\mathcal{F}, \mathcal{C}\left(G_{c1}, G_P^{-1}\right), G_p\right) \Leftrightarrow \sum = \left(\mathcal{F}, G_{c1}, G_{ip}\right) \tag{17}$$

where $\mathcal{C}(*)$ is a set to be designed and $G_{ip} = G_P^{-1} G_P$.

In general, the design of U-control system can be divided into two separate designs:

1. Designed dynamic inverter G_P^{-1} to achieve $G_P^{-1} G_P = 1$. This gives $\sum = (F, G_{c1})$
2. Design invariant controller G_{c1}, which is a typically linear controller. Let the specified closed-loop performance in transfer function G, in form of $G = \frac{G_{c1}}{1+G_{c1}}$, which can be comfortably assigned using damping ratio ζ and undamped natural frequency ω_n for linear system dynamic/steady-state response.

3. U-Model-Based Two-Degree-of-Freedom IMC (UTDF-IMC)

3.1. Classical Two-Degree-of-Freedom IMC (TDF-IMC) Structure

Figure 4 shows a TDF-IMC structure to be incorporated with U-control, which comprises feedback controller F added in the external loop within the classical IMC structure. Clearly, if the feedback filter F is a unit constant, this structure is the same as that in Figure 1a.

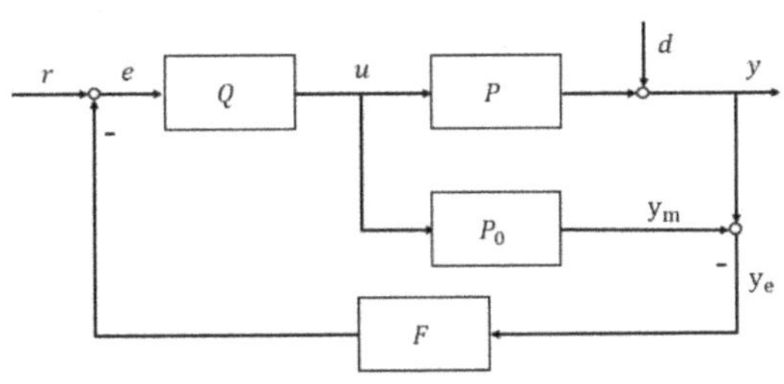

Figure 4. IMC structure with the feedback filter.

From Figure 4, the system output $y = y_m + y_e$. Therefore,

$$y = (r - y_e F) Q P_0 + y_e = r Q P_0 + y_e (1 - F Q P_0) \tag{18}$$

In the TDF-IMC system, if the controlled plant is a minimum-phase system, then the controller $Q(s) = f(s)/P_0(s)$. The output of (18) can be re-organised as:

$$y = r f + y_e (1 - F f) \tag{19}$$

The explicit input/output relationships from Figure 4 can be written as follows:

$$u = \frac{Q}{1 + (P - P_0) F Q} (r - dF) \tag{20}$$

$$y = \frac{QP}{1 + (P - P_0) F Q} r + \frac{1 - Q F P_0}{1 + (P - P_0) F Q} d \tag{21}$$

If the controlled system does not contain uncertain parameters or control disturbance, then $y_e = 0$, otherwise, $|y_e| > 0$. From (19), rf determines the system tracking performance, while $y_e(1-F)f$ determines the system robustness.

To achieve desired control performance, a condition must hold true below:

$$\lim_{t\to\infty} f(t) = 1,\ \lim_{t\to\infty} \mathcal{L}^{-1}(F(s)f(s)) = 1 \tag{22}$$

where $\mathcal{L}^{-1}(*)$ is the inverse Laplace transform operator, $F(s)$ and $f(s)$ are the Laplace functions of filters F and f respectively. Thus, output y equals to the reference r eventually, and the system disturbance and modeling errors will be eliminated. The performance of the IMC control system will depend on these two filters F and f. The setting time and rise time of these two filters should be as short as possible. However, response speed which are too fast will cause the amplitude of the controller output signal to increase sharply.

From Figure 4, the controller $Q(s)$ output u is:

$$u = (r - y_e F)Q \tag{23}$$

From (20), when controller Q is determined, the faster the response of the filter F, the larger value the initial controller output u. In general, this can be observed from (19) that the tracking ability and robustness of IMC system cannot be separately designed, as well as its design flexibility is relatively limited. Therefore, this is one of paper aims, to separate IMC's designing of tracking ability control and robustness and improve its design effectiveness without affecting its desired control performance.

3.2. U-Model-Based Two-Degree-of-Freedom IMC (UTDF-IMC) Structure

Based on the IMC problem stated in introduction and TDF-IMC analysis in Section 3.1, this paper changes the classical TDF-IMC structure in Figure 4 to a UTDF-IMC structure as shown in Figure 5.

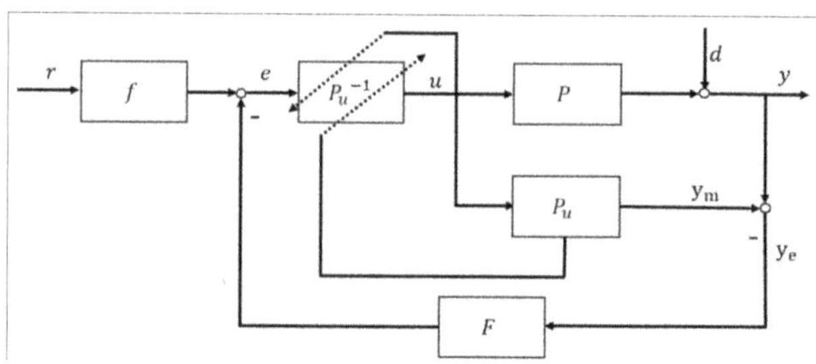

Figure 5. U-model-based Two-Degree-of-Freedom IMC structure.

In Figure 5, the original controller Q in classical TDF-IMC shown in Figure 4 has been split into two parts: the feedforward filter f and the inversion $P_u{}^{-1}$ of the U-realization controlled plant model P_u, where the original IMC's controller $Q = fP_0{}^{-1}$. In contrast to the classical IMC structure, feedforward filter f appears outside the system feedback loop. However, generally the plant model inversion $P_0{}^{-1}$ cannot exist alone because of its irrationality and unrealizable property. For polynomial-based modeling of the controlled plant expressed by Laplace transfer function, its inversion will make the order of the numerator higher than the order of the denominator, which cannot be achieved in the actual control system. Therefore, this paper introduces UM-dynamic inversion algorithm to design the plant's inversion part in UTDF-IMC structure.

From Figure 5, the system output $y = y_m + y_e$. Therefore,

$$y = (rf - y_e F)P_u{}^{-1}P_u + y_e = rf + y_e(1 - Ff) \tag{24}$$

3.3. UTDF-IMC Design Procedures

Figure 5 presents the U-model-based Two-Degree-of-Freedom IMC system framework, where f and F are the designed feedforward and feedback filters, respectively. P is the controlled plant or process, which is allowed to be linear or nonlinear. P_u is U-model-based approximation to the controlled plant P. ${P_u}^{-1}$ is the inverter designed by the U-model-based root-solving algorithm. From (15), the parameters absorbed by α_j can be obtained from the output signal y_m of the plant model $P_u(s)$ and controller output u. In general, similar to the classical IMC design, UTDF-IMC system design has the following two steps:

1. Assume the controlled plant or process P is stable and bounded, and its inverse P^{-1} exists. Use U-model to describe P or convert the plant model P_0 into its U-realization P_u. The specific U-modeling process can refer to Section 2.1. In contrast to the classical IMC or classical TDF-IMC, U-realization of the original model P_0 can comfortably cover nonlinear dynamics, therefore, remove linearization restrictions.
2. Design filters f and F according to system control performance requirements, then re-optimize the parameters of the filters according to the controller output limit. The feedforward filter determines the system's set-point tracking ability (response time) while the feedback filter determines the system's robustness. Because the control system performance is completely designed according to the two filters independently, designers can select the appropriate filters according to performance requirements, hardware limitations, controller output limitations, etc.

3.4. Property Analysis

1. Property 1 (Dual stability): Assume the plant model is perfectly matched ($P_u = P$) and system disturbance is absent $d = 0$, then from Table 1, the closed-loop stability is characterized by the stability of the plant $P(P^{-1})$ and the feedforward filter f. In this case, the system output signal will be: $y = rf$.
2. Property 2 (Perfect control): Assume that the dynamic inverter ${P_u}^{-1}$ is satisfied with $P_u = P$ and P stable, then the closed-loop system is stable and perfectly controlled. In this case, the system output is $y = rf + (1 - F)d$. The faster the response speed of feedback filter F, the better the system robustness.
3. Property 3 (Zero offset): Assume that the steady-state gain of the controller equals to steady-state gain of the inverse model, and this closed-loop system is input-output stable with this controller, then offset free control is obtained asymptotically to step or ramp type inputs and disturbances.
4. Property 4: Separability of designing the tracking filter and the robust filter: This is shown in the tables, which UTDF-IMC has no product of the two filters Ff.

Table 1. Input/output comparison of IMC, TDF-IMC, and UTDF-IMC against disturbance.

	Controller Output u	System Output y
IMC	$u = \frac{1}{P_0+(P-P_0)f}(rf - df)$	$y = \frac{fP}{P_0+(P-P_0)f}r + \frac{P_0(1-f)}{P_0+f(P-P_0)}d$
TDF-IMC	$u = \frac{1}{P_0+(P-P_0)Ff}(rf - dFf)$	$y = \frac{fP}{P_0+(P-P_0)Ff}r + \frac{P_0(1-fF)}{P_0+(P-P_0)Ff}d$
UTDF-IMC	$u = \frac{1}{P_u+(P-P_u)F}(rf - dF)$ where P_u is the U-realization of P_0	$y = \frac{fP}{P_u+(P-P_u)F}r + \frac{P_u(1-F)}{P_u+(P-P_u)F}d$

Comparison with IMC and TDF-IMC, Tables 1 and 2 list the three IMC types of control system configurations against disturbance and model mismatching, respectively. For UTDF-IMC the typical properties are analyzed below.

Table 2. Output comparison of IMC, TDF-IMC, and UTDF-IMC against model mismatching.

	System Output y
IMC	$y = rf - (1 - f)y_e$
TDF-IMC	$y = rf - (1 - Ff)y_e$
UTDF-IMC	$y = rf + (1 - F)y_e$

From Table 1, the factor associated with d is called the disturbance rejection designed. It is clear that this rejection part only depends on the feedforward filter f in IMC, depends on two filters F and f in TDF-IMC but only depends on the feedback filter F in UTDF-IMC structure. In case of model mismatch, it can also use the output error signal y_e to analyze the system performance in Table 2:

From Table 2, regarding UTDF-IMC, the function associated with y_e is robustness designed, where y_e absorbs all whole modeling error and system disturbance; the function associated with signal r is for tracking designed. Obviously, when the controller equals to plant model inversion, all the tracking design only depends on the feedforward filter f and robustness designed is the same as previously discussed. In summary, compared with the classical IMC and TDF-IMC structure, the main differences of UTDF-IMC structure are as follows:

1. Classical TDF-IMC structure can make tracking ability and robustness be designed separately but not wholly independent due to the product of Ff in robustness specification. The UTDF-IMC overcomes this shortcoming without resorting to a more complex structure. Therefore, when the robustness performance of the system is determined, UTDF-IMC structure will have a faster response speed than the classical TDF-IMC structure.
2. U-model is used to facilitate control system design, which can be easily to form an inversion of the plants to cancel both dynamic and nonlinearities. Accordingly, it converts the nonlinear control system into a linear model-based control with a nonlinear dynamic inverter.
3. UM-dynamic inversion algorithm is used to design the inversion part in UTDF-IMC structure, which has a faster convergence speed and allows the inversion part exists alone properly without the feedforward filter.
4. This structure where feedforward filter f from outside the control loop allows the tracking ability and robustness performance to be completely independently designed.
5. The improved control performance is not complicating the system structure and/or increasing the additional computation burden throughout the design process.

4. Simulation Demonstrations

This simulation demonstration selects three plants to test the proposed U-model-based TDF-IMC structure. Both plants will be controlled by IMC, TDF-IMC, and UTDF-IMC structure.

4.1. Linear Internal Model (Also Called Nominal Model in the Study)

$$P_0(s) = \frac{{\omega_n}^2}{s^2 + 2\zeta\omega_n s + {\omega_n}^2} = \frac{1}{s^2 + 3s + 1} \tag{25}$$

This is characterized with the damping ratio $\zeta = 1.5$ and the undamped natural frequency $\omega_n = 1$.

For designing the UTDF-IMC system:

1. Convert plant model (25) into its corresponding U-model:

$$\begin{cases} P_u(s): \ddot{y} = u - 3\dot{y} - y = \alpha_0 f_0(u) + \alpha_1 f_1(u) \\ \alpha_0 = -3\ddot{y}\frac{1}{s} - \ddot{y}\frac{1}{s^2},\ f_0(u^0) = 1 \\ \alpha_1 = 1,\ f_1(u) = u \end{cases} \tag{26}$$

2. Design the inverter of the plant model $P_u(s)$:

$$u = \ddot{y} + 3\ddot{y}\frac{1}{s} + \ddot{y}\frac{1}{s^2} \tag{27}$$

3. Design feedforward filter $f(s)$ and feedback filter $F(s)$

In this paper, based on the UTDF-IMC system design procedure in Section 3.2, to make the system achieve a fast response speed and no overshoot, $f(s) = \frac{1}{(0.2s+1)^2}$ and $F(s) = \frac{1}{(0.1s+1)^2}$. To compare control performance fairly, TDF-IMC system uses the same filters as UTDF-IMC. To ensure the same robustness, the classical IMC system uses $f'(s) = \frac{1}{(0.1s+1)^2}$.

To test the performance of the designed control system, assume the plant a 2nd order dynamic with $\zeta = 1$ and $\omega_n = 0.5$, and an external disturbance added at the system output, i.e.,

$$P(s) = \frac{1}{4s^2 + 1s + 1} + D(s) \tag{28}$$

The system disturbance is a band-limit white noise with changing rate of 1hz, system signal-noise ratio (SNR) of 26.9db. The noise sequence is shown in Figure 6.

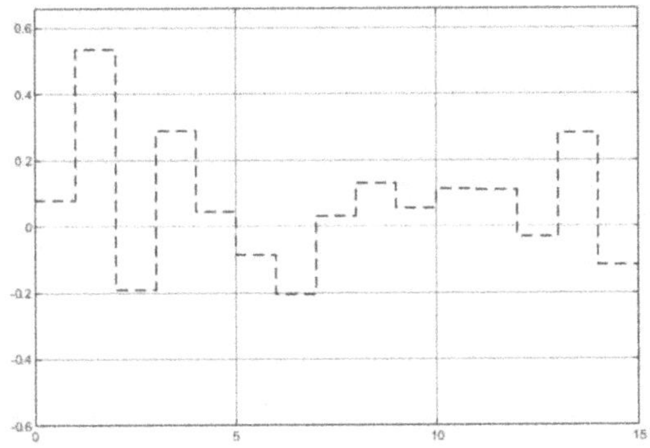

Figure 6. System disturbance noise.

Figure 7 shows the simulation results under the three different IMC schemes. From Figure 7a,b, UTDF-IMC and IMC have better robustness performance in rejection of system disturbance and modeling error. IMC system has a faster tracking speed because of its fast respond-speed filter; however due to modeling errors, stronger tracking ability brings larger overshoot. The simulation results also demonstrate the analysis in Section 3.4. From Figure 7c, UTDF-IMC structure does not increase the maximum peak output of the controller compared with TDF-IMC structure. However, fast tracking speed also brings a large controller output peak in the IMC system, which may cause the controller to overload in real-time applications. Consider the control performance and controller load, in case of selecting the same filters (control parameters), UTDF-IMC system shows better control performance.

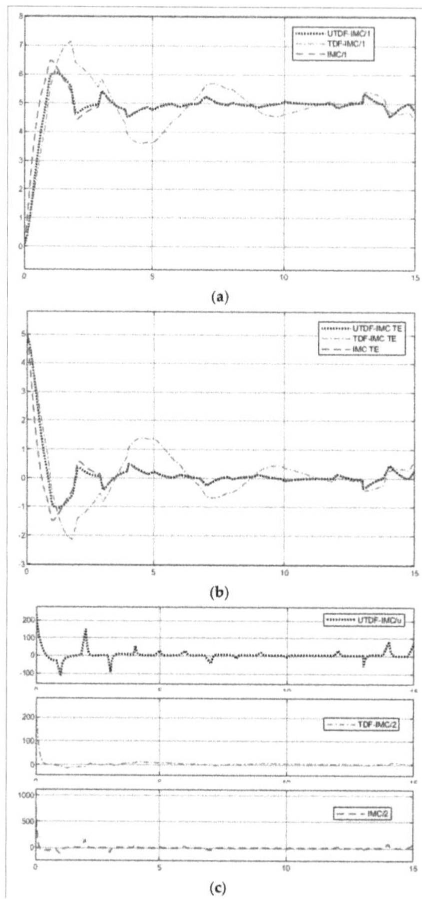

Figure 7. Simulation results of plant1 (**a**) System outputs; (**b**) Tracking errors; (**c**) Controller outputs.

4.2. Nonlinear Internal Model

$$P_0: \dot{y} = a\dot{u}^3 + b\dot{u}^2 - c\dot{u} - ky + e^u \tag{29}$$

where the coefficients $a = b = c = 1$, $k = 0.5$, then $\mathrm{P}_0(\mathrm{s}) = \dot{u}^3 + \dot{u}^2 - \dot{u} - 0.5y + e^u$.

For designing the UTDF-IMC system,

1. Convert plant model (29) into its corresponding U-model:

$$\begin{cases} \mathrm{P}_u(\mathrm{s}): \dot{y} = \alpha_0 f_0(\dot{u}) + \alpha_1 f_1(\dot{u}) + \alpha_2 f_2(\dot{u}) + \alpha_3 f_3(\dot{u}) \\ \alpha_0 = -0.5\dot{y}\frac{1}{s} + e^u,\ f_0\left(\dot{u}^0\right) = 1 \\ \alpha_1 = 1,\ f_1(\dot{u}) = -\dot{u} \\ \alpha_2 = 1,\ f_1\left(\dot{u}^2\right) = \dot{u}^2 \\ \alpha_3 = 1,\ f_1\left(\dot{u}^3\right) = \dot{u}^3 \end{cases} \tag{30}$$

2. Design the inverter of the plant model $\mathrm{P}_u(\mathrm{s})$:

$$u = root\left(\alpha_0 f_0(\dot{u}) + \alpha_1 f_1(\dot{u}) + \alpha_2 f_2(\dot{u}) + \alpha_3 f_3(\dot{u}) - \dot{y} = 0\right) \tag{31}$$

It should be noted that because Equation (32) is a cubic equation of one variable about $\dot{u}$, to ensure that the controller output is rational, the real root of Equation (32) is selected as the output of the controller.

3. Design feedforward filter $f(s)$ and feedback filter $F(s)$

Same as previous work, to make the system achieve a fast response speed and no overshoot, this paper chooses $f(s) = \frac{1}{(0.1s+1)^2}$ and $F(s) = \frac{1}{(0.2s+1)^2}$ for the plant 2. To compare control performance fairly, TDF-IMC system uses the same filters as UTDF-IMC. To ensure the same tracking speed, the classical IMC system uses $f'(s) = \frac{1}{(0.2s+1)^2}$.

To demonstrate the performance of the designed control system, assume plant with the same structure as the IM, but $c = 1.4$ and $k = 0.8$, and an external noise added at the system output, i.e.,

$$P(s) : \dot{y} = \dot{u}^3 + 1.4\dot{u}^2 - \dot{u} - 0.8y + e^u + d \quad (32)$$

The system noise is a band-limit white noise with changing rate of 1 hz and SNR of 20.9 db. The noise sequence is shown in Figure 8.

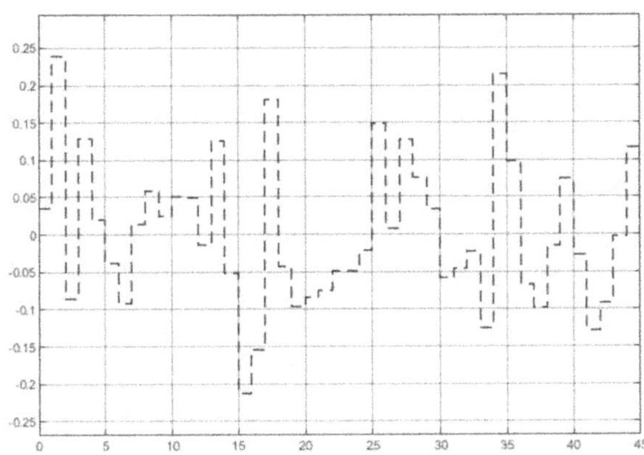

Figure 8. System noise.

Figure 9a–c show the simulation results under the three IMC schemes, Figure 9d shows the tracking reference signal. From Figure 9a,b, UTDF-IMC both has a better robustness performance in rejection of system disturbance and modeling error and faster tracking speed. When the reference signal suddenly jumps sharply, the response of TDF-IMC system also shakes sharply although it has the same filters as UTDF-IMC's. These simulation results demonstrate the analysis in Section 3.4. From Figure 9c, UTDF-IMC structure does not increase the burden on the controller, although it has a better control performance. The outputs of controller show that the UTDF-IMC is not overloaded. Once again, consider the control performance and controller load, in case of selecting the same filters (control parameters), UTDF-IMC system shows better control performance.

4.3. Control of PMSM System

In the past few decades, Permanent Magnet Synchronous Motors (PMSM) have been widely used in industry because of their high-power density, high efficiency, and large torque inertia ratio. PMSM is essentially a nonlinear Multiple-Input-Multiple-Output (MIMO) system, so parameter uncertainty and interference acting on torque will make it difficult for PMSM control systems to obtain higher control performance [36]. Most advanced control strategies [37–39] for PMSM servo system position control ignore the nonlinear term in the speed equation, assuming that A=B and load torque disturbance does not change. Therefore, it is still a challenge to provide an efficient set-point value tracking control strategy for a general PMSM system affected by time-varying system disturbance and uncertain parameters. Therefore, this section applies the proposed UTDF-IMC structure combined with the U-modeling of the PMSM system to achieve high-precision set-point robust tracking control of the PMSM operation.

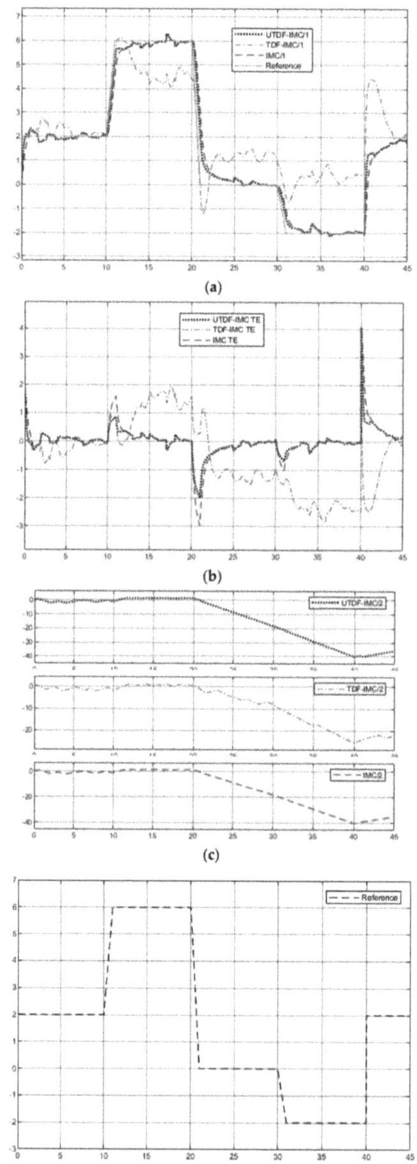

Figure 9. Simulation results of plant2 (**a**) System outputs; (**b**) Tracking errors; (**c**) Controller outputs; (**d**) Tracking reference.

4.3.1. Modeling of PMSM System

It should be noted that the permanent magnets used in the PMSM are a type of modern rare-earth varieties with high resistivity, so the induced current in the rotor can be negligible. The model of the PMSM is based on s number of equations in the d-q reference frame [40].

The electric torque of the PMSM is:

$$T_e = 3p\left[\Phi_v i_q + \left(L_d - L_q\right)i_d i_q\right]/2 \tag{33}$$

And its motor dynamics can be modeling as:

$$T_e = T_L + B\omega_r + J\Delta\omega_r \tag{34}$$

The relationship between voltages and currents in motor are:

$$\begin{bmatrix} V_d \\ V_q \end{bmatrix} = \begin{bmatrix} R_s + L_d\Delta & -p\omega_r L_q \\ p\omega_r L_d & R_s + L_q\Delta \end{bmatrix}\begin{bmatrix} i_d \\ i_q \end{bmatrix} + \begin{bmatrix} 0 \\ p\omega_r\Phi_v \end{bmatrix} \tag{35}$$

The rotor flux rotates at rotor speed ω_r and is positioned by the rotor angular position:

$$\theta_r = \int \omega_r dt \tag{36}$$

Therefore, the PMSM in the rotating *d-q* reference frame can be modeled in the following state-space equation [41],

$$\left\{\begin{array}{c} \frac{d\theta_r}{dt} = \omega_r \\ \frac{d\omega_r}{dt} = \frac{3p\Phi_v}{2J}i_q + \frac{3p}{2J}\left(L_d - L_q\right)i_d i_q - \frac{B}{J}\omega_r - \frac{1}{J}T_L \\ \frac{di_d}{dt} = -\frac{R_s}{L_d}i_d + \frac{pL_q}{L_d}i_q\omega_r + \frac{1}{L_d}V_d \\ \frac{di_q}{dt} = -\frac{R_s}{L_d}i_q - \frac{pL_d}{L_q}i_d\omega_r - \frac{p\Phi_v}{L_q}\omega_r + \frac{1}{L_q}V_q \end{array}\right. \tag{37}$$

where

Δ: differential operator $(\frac{d*}{dt})$

θ_r and ω_r: the rotor angular position and rotor speed

i_d, i_q and V_d, V_q: stator currents and voltages in *d-q* reference frame

L_d and L_q : axes inductances in *d-q* reference frame

T_L: load torque, Φ_v:rotor flux, J: inertia, R_s: stator resistance, B: viscous friction coefficient and p: number of pole pairs.

The design aim is controlling voltages V_d and V_q in (37) to make rotor position θ_r track a desired constant reference position θ_d and the current i_d is regulated to zero asymptotically, concretely, this PMSM control system is two-input two-output with $u = [u_1\ u_2] = [V_d\ V_q]$ and $y = [y_1\ y_2] = [\theta_r\ i_d]$. The same as used [41], the commonly used nonlinear load torque disturbance to test the system performance is generated by the following disturbance dynamic model:

$$\left\{\begin{array}{c} \dot{v}_1 = v_2 \\ \dot{v}_2 = -av_1 + b\left(1 - {v_1}^2\right)v_2 \end{array}\right. \tag{38}$$

where $v_1 = T_L$ is the solution of this Van der Pol oscillator.

Let

$$\left\{\begin{array}{c} x_1 = \theta_r, x_2 = \omega_r, x_3 = i_d, x_4 = i_q \\ a_1 = \frac{3p\Phi_v}{2J}, a_2 = \frac{3p}{2J}\left(L_d - L_q\right), a_3 = \frac{B}{J}, a_4 = \frac{1}{J} \\ b_1 = \frac{R_s}{L_d}i_d, b_2 = \frac{pL_q}{L_d}, b_3 = \frac{1}{L_d} \\ c_1 = \frac{R_s}{L_d}, c_2 = \frac{pL_d}{L_q}, c_3 = \frac{p\Phi_v}{L_q}, c_4 = \frac{1}{L_q} \end{array}\right. \tag{39}$$

Then system (37) can be rewritten into standard state-space equation of:

$$\begin{cases} \dot{x}_1 = x_2 \\ \dot{x}_2 = a_1x_4 + a_2x_3x_4 - a_3x_2 - a_4v_1 \\ \dot{x}_3 = -b_1x_3 + b_2x_3x_2 + b_3u_1 \\ \dot{x}_4 = -c_1x_4 - c_2x_3x_2 - c_3x_2 + c_4u_2 \end{cases} \text{and} \begin{cases} y_1 = x_1 + d \\ y_2 = x_3 \end{cases} \tag{40}$$

where d is the system disturbance. Linearize system (40) gives:

$$\dot{\overline{x}} = A\overline{x} + B\overline{u},\ \overline{y} = C\overline{x} + d \tag{41}$$

where $A = \begin{bmatrix} 0 & 1 & 0 & 0 \\ 0 & -a_3 & a_1 & 0 \\ 0 & -b_3 & -b_1 & 0 \\ 0 & 0 & 0 & -c_1 \end{bmatrix}$, $B = \begin{bmatrix} 0 & 0 \\ 0 & 0 \\ b_4 & 0 \\ 0 & c_3 \end{bmatrix}$ and $C = \begin{bmatrix} 1 & 0 & 0 & 0 \\ 0 & 0 & 1 & 0 \end{bmatrix}$.

4.3.2. Simulation Test

In this section, the following three controllers are compared with simulation tests.

a. IMC: The filter time parameter shown in equation (9) is chosen as $\lambda = 0.01$, use linearization to approximate the inverse of PMSM.

b. TDF-IMC: Based on the structure in Figure 4, the feedforward filter and feedback filter are chosen as $f = \frac{1}{(1+0.1s)^{\gamma}}$, $F = \frac{1}{(1+0.01s)^{\gamma}}$, use UM-dynamic inversion to design the inverse of PMSM.

c. UTPF-IMC: To test the performance of UTDF-IMC fairly, based on the structure in Figure 5, the feed forward filter and feedback filter are chosen as $f = \frac{1}{(1+0.1s)^{\gamma}}$, $F = \frac{1}{(1+0.01s)^{\gamma}}$, use UM-dynamic inversion to design the inverse of PMSM.

Comparison test of controller a and controller c is to demonstrate the superiority of UM-dynamic inversion algorithm for modeling nonlinear controlled plants/processed and inversion calculation, and comparison test of controller b and controller c is to show the efficiency of the proposed UTPF-IMC structure under the same modeling and calculation accuracy. The nominal values of PMSM parameters [41] for the simulations are $p = 3$, $R_s = 1.2\ \Omega$, $\Phi_v = 0.18$ Vs/rad, $\overline{L_d} = 0.011$ H, $\overline{L_q} = 0.015$ H, $\overline{B} = 0.0001$ Nms/rad $\overline{J} = 0.006$ kgm^2. Choose $a = 9$, $b = 1$. The initial values are chosen as follows: $\theta_r(0) = 0$ rad, $\omega_r(0) = 0$ rad/s, $i_d(0) = 0$ A, $i_q(0) = 0$ A.

4.3.3. Matched Model with System Disturbance

To test the property 2 in Section 3.4 while the process model is perfectly matched, i.e., $P_0 = P$, assign the step reference signal with tracking positions $\theta_d = \pi$ rad and current $i_d = 0$, plus a squared disturbance shown in Figure 10 is added.

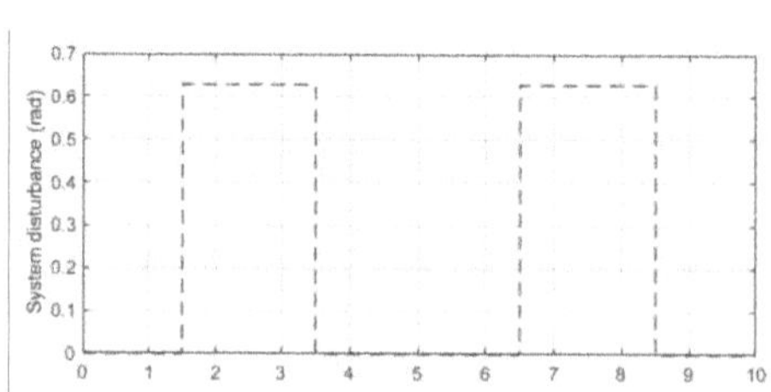

Figure 10. System disturbance.

Figure 11 shows the simulation results. Clearly, all the controllers can track the desired set-point and reject the system disturbance but the robustness of TDF-IMC system is worse than others. IMC system has a faster response speed; however, it has overshoot due to linearization error. From Figure 11b, when θ_r reaches the designated angular position, rotor speed ω_r is stabilized at zero. From Figure 11c, all control systems current i_d can stay at 0, but its peak value in IMC system is much larger than the others obviously. These simulation results demonstrate properties justified in Section 3.4. From Figure 11d,e, the controller outputs have large peak values at initial phase in the IMC system, especially output voltage V_d.

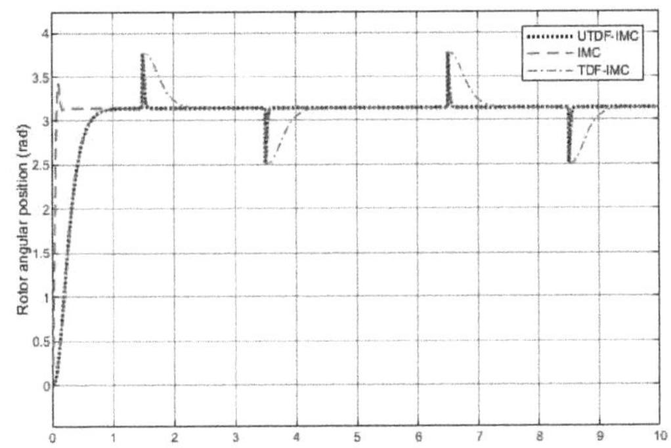

(a)

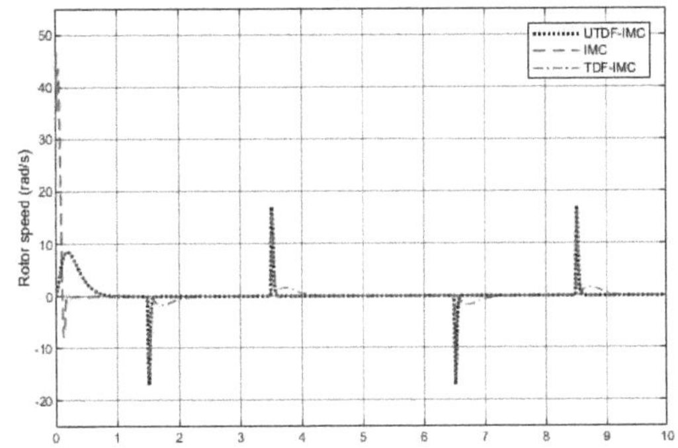

(b)

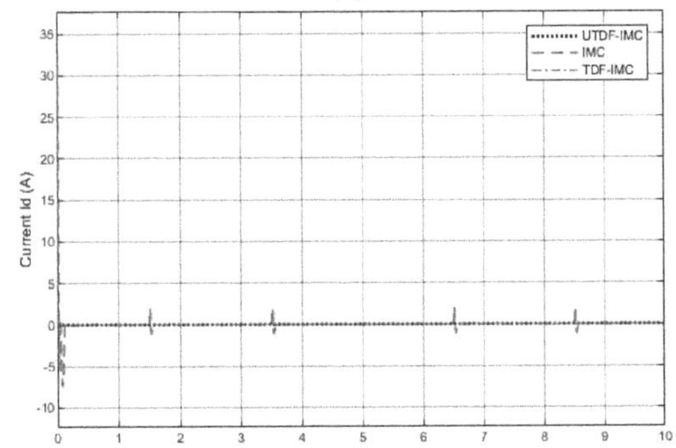

(c)

Figure 11. *Cont.*

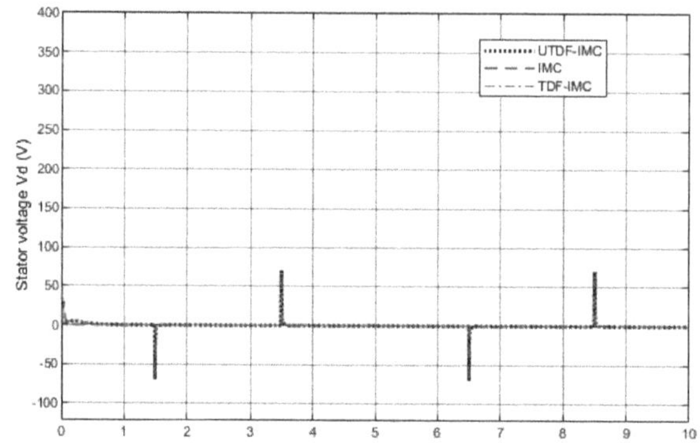

(d)

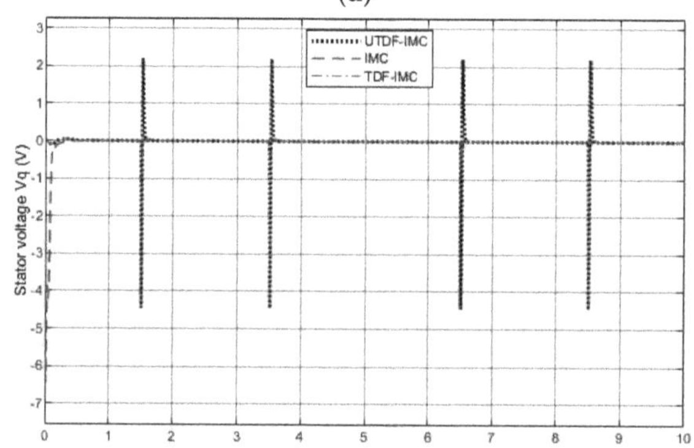

(e)

Figure 11. Simulation results with only system disturbance (**a**) Output angular position θ_r; (**b**) Output rotor speed ω_r; (**c**) Output current i_d; (**d**) Controller output voltage V_d; (**e**) Controller output voltage V_q.

4.3.4. Mismatched Model with System Disturbance

In this part, three controllers under a more actual situation (with modeling error) will be tested to investigate property 3 in Section 3.4. The parameters of PMSM become: $L_d = 0.5\overline{L_d}$, $L_q = 1.3\overline{L_q}$, $B = 1.45\overline{B}$, $J = 0.75\overline{J}$. The load torque disturbance generated by (38) with initial values of $v_1(0) = 0$ and $v_2(0) = 0.1$ is also added in PMSM system, which is shown in Figure 12. System disturbance is the same as previous experiment shown in Figures 10a and 13 shows the comparative simulation results.

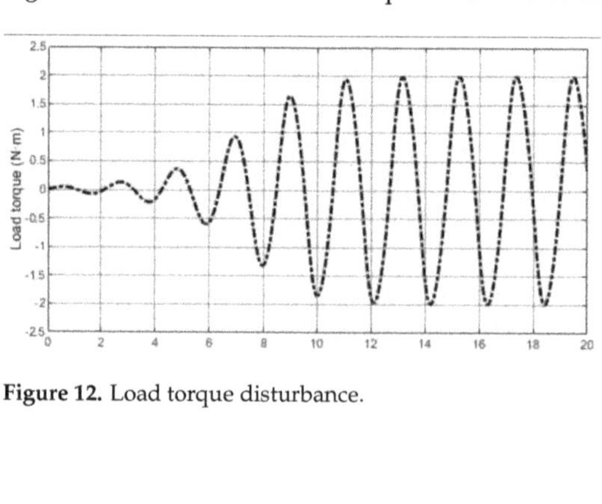

Figure 12. Load torque disturbance.

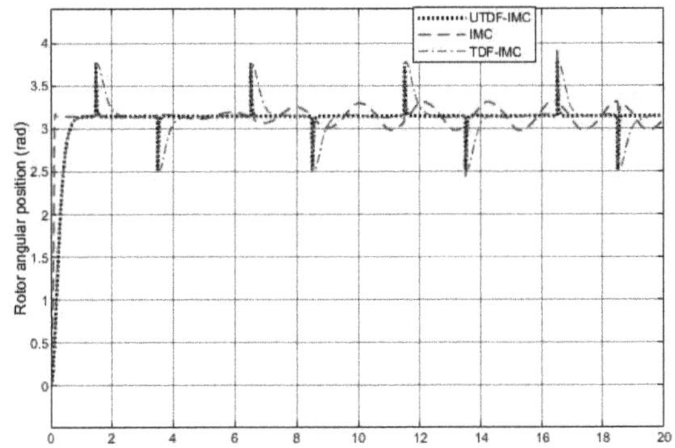

(a)

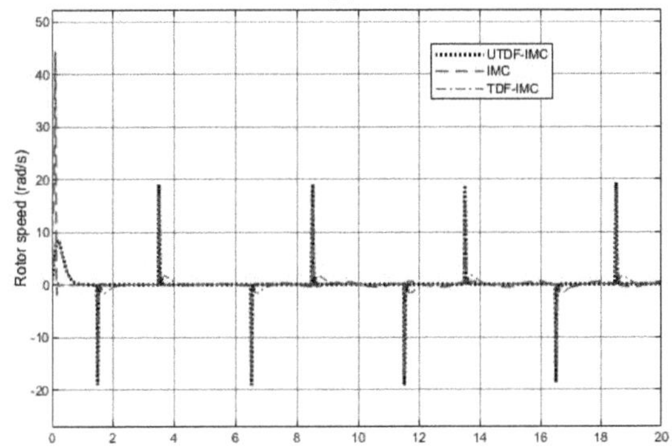

(b)

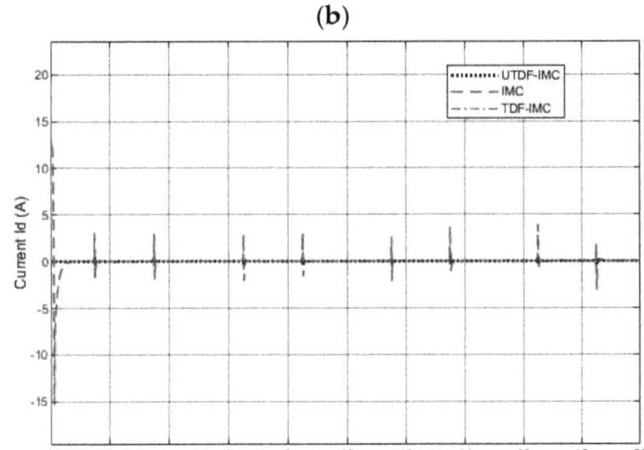

(c)

Figure 13. *Cont.*

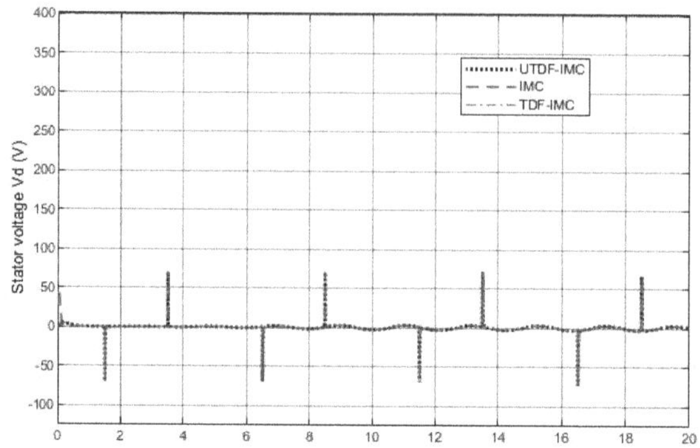

(d)

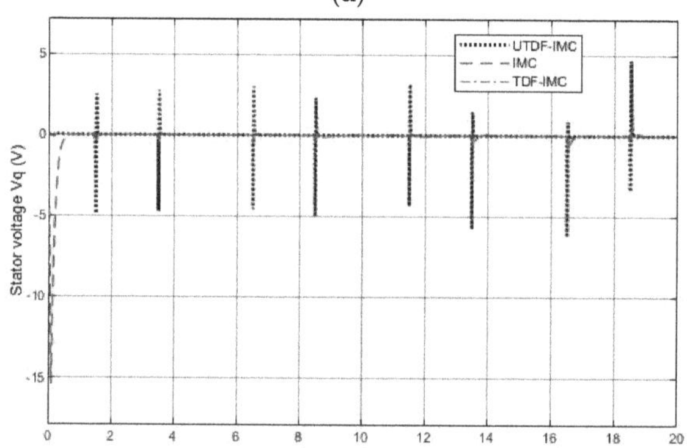

(e)

Figure 13. Simulation results with modeling error and disturbance (**a**) Output angular position θ_r; (**b**) Output rotor speed ω_r; (**c**) Output current i_d; (**d**) Controller output voltage V_d; (**e**) Controller output voltage V_q.

From Figure 13a, IMC system has tracking error due to the accuracy limitation of linearization, which makes IMC unable to reject strong nonlinear load torque disturbance. Both TDF-IMC and UTDF-IMC systems can achieve the prescribed set-point tracking performance because UM-dynamic inversion does not lose any nonlinear features. UTDF-IMC system has better robustness than TDF-IMC system due to the difference in their structures. From Figure 13b, when θ_r reaches the designated angular position, rotor speed ω_r in UTDF-IMC and TDF-IMC systems is stabilized at zero; however, the rotor revolves slightly in IMC system. From Figure 13c, all current i_d staying at zero but its peak value with IMC is larger than the others, this is because of the cost of faster response speed in IMC system. From Figure 13d,e, the controller outputs also have large peak values at initial phase in the IMC system, especially output voltage V_d.

In summary, from all simulation results, the control system using the linearization method does degrade the control performance while there is a strong nonlinear disturbance. Additionally, by using UM-dynamic inversion, UTDF-IMC and TDF-IMC systems can achieve reasonably good set-point tracking performance, and UTDF-IMC system has better robustness than classical TDF-IMC system with the same parameters chosen in the filters.

5. Conclusions

This paper introduces an effective U-model-based Two-Degree-of-Freedom IMC framework. Consistently with the simulation test results of linear and nonlinear controlled plants, the proposed UTDF-IMC framework shows its strong robustness and effectiveness in control system design compared with the classical IMC and TDF-IMC approaches. It is believed that UTDF-IMC, enhanced with nonlinear dynamic inverter, could be applied more effectively to a wide range of industrial control system design. Therefore, this study has established a platform for possible further expansion, for example controlling Multi-Input and Multi-Output (MIMO) systems, which involves solution challenges with nonlinear set equation in case of under, full, and over actuated control system design. Another research direction is to expand the UTDF-IMC to deal with nonminimum phase/unstable zero dynamic systems.

Author Contributions: R.L.: Writing—original draft, Software, Methodology, Formal analysis, Conceptualization; Q.Z.: Methodology, Formal analysis, Writing—review & editing, Supervision, Project administration, Funding acquisition; P.N.: Methodology, Writing—review & editing, Supervision, Project administration, Funding acquisition; A.Y.: Methodology, Writing—review & editing, Supervision, Project administration, Funding acquisition; Y.Y.: Writing—review & editing, Project administration, Funding acquisition; M.D.: Writing—review & editing. All authors have read and agreed to the published version of the manuscript.

Funding: This research received no external funding.

Institutional Review Board Statement: Not applicable.

Informed Consent Statement: Not applicable.

Data Availability Statement: The data used to support the findings of this study are available from the corresponding author upon request.

Acknowledgments: The first author is grateful to the partial PhD studentship from the Engineering Modeling and Simulation Research Group of the University of the West of England, UK. The authors are grateful to the editors and the anonymous reviewers for their constructive comments and suggestions regarding the revision of the paper.

Conflicts of Interest: The authors declare no conflict of interest.

References

1. ElMaraghy, W.; ElMaraghy, H.; Tomiyama, T.; Monostori, L. Complexity in engineering design and manufacturing. *CIRP Ann.* **2012**, *61*, 793–814.
2. Sari, N.N.; Jahanshahi, H.; Fakoor, M. Adaptive fuzzy PID control strategy for spacecraft attitude control. *Int. J. Fuzzy Syst.* **2019**, *21*, 769–781.
3. Rosales, C.; Soria, C.M.; Rossomando, F.G. Identification and adaptive PID control of a hexacopter UAV based on neural networks. *Int. J. Adapt. Control Signal Process.* **2019**, *33*, 74–91.
4. Kawan, C. *Invariance Entropy for Deterministic Control Systems*; Springer: Berlin/Heidelberg, Germany, 2013.
5. Kawan, C. Special Issue, Entropy in Networked Control, Entropy. 2019. Available online: https://www.mdpi.com/journal/entropy/specialissues/control (accessed on 11 April 2019).
6. Morari, M.; Zafiriou, E. *Robust Process Control*; PTR Prentice Hall: Upper Saddle River, NJ, USA, 2002.
7. Garcia, C.E.; Morari, M. Internal model control. A unifying review and some new results. *Ind. Eng. Chem. Process Des. Dev.* **1982**, *21*, 308–323.
8. Li, P.; Zhu, G.; Zhang, M. Linear Active Disturbance Rejection Control for Servo Motor Systems with Input Delay via Internal Model Control Rules. *IEEE Trans. Ind. Electron.* **2020**, *68*, 1077–1086.
9. Liu, J.; Yin, Z.; Bai, C.; Du, N. Internal model control of induction motors based on extended state observer. *J. Power Electron.* **2020**, *20*, 163–175.
10. Reddy, P.S.; Patwardhan, S.C.; Rani, K.Y. Robust trajectory tracking in a reactive batch distillation process using multirate nonlinear internal model control. *Ind. Eng. Chem. Res.* **2019**, *58*, 11364–11381.
11. Saxena, S.; Hote, Y.V. Advances in internal model control technique: A review and future prospects. *IETE Tech. Rev.* **2012**, *29*, 461–472.
12. Sonker, B.; Kumar, D.; Samuel, P. Design of two-degree-of-freedom-internal model control configuration for load frequency control using model approximation. *Int. J. Model. Simul.* **2019**, *39*, 27–37.
13. Singh, J.; Chattterjee, K.; Vishwakarma, C.B. Two-degree-of-freedom internal model control-PID design for LFC of power systems via logarithmic approximations. *ISA Trans.* **2018**, *72*, 185–196.
14. Zhu, Q.; Yin, Z.; Zhang, Y.; Niu, J.; Li, Y.; Zhong, Y. Research on two-degree-of-freedom internal model control strategy for induction motor based on immune algorithm. *IEEE Trans. Ind. Electron.* **2015**, *63*, 1981–1992.

15. Qiu, Z.; Sun, J.; Jankovic, M.; Santillo, M. Nonlinear internal model controller design for wastegate control of a turbocharged gasoline engine. *Control Eng. Pract.* **2016**, *46*, 105–114.
16. Economou, C.G.; Morari, M.; Palsson, B.O. Internal model control: Extension to nonlinear system. *Ind. Eng. Chem. Process Des. Dev.* **1986**, *25*, 403–411.
17. Cirtoaje, V. A practical unified algorithm of P-IMC type. *Processes* **2020**, *8*, 165.
18. Sun, X.; Chen, L.; Jiang, H.; Yang, Z.; Chen, J.; Zhang, W. High-performance control for a bearingless permanent-magnet synchronous motor using neural network inverse scheme plus internal model controllers. *IEEE Trans. Ind. Electron.* **2016**, *63*, 3479–3488.
19. Rivals, I.; Personnaz, L. Nonlinear internal model control using neural networks: Application to processes with delay and design issues. *IEEE Trans. Neural Netw.* **2000**, *11*, 80–90.
20. Zhang, X.; Hu, J.; Liu, Z. Fuzzy adaptive internal model control for a pneumatic muscle actuator. In Proceedings of the 2019 11th International Conference on Machine Learning and Computing, Zhuhai, China, 22–24 February 2019; pp. 546–550.
21. Ahmed, I.B.C.; Othman, C.; Soudani, D. On the fuzzy adaptative internal model control of the discrete-time nonlinear systems. In Proceedings of the 2017 International Conference on Advanced Systems and Electric Technologies (IC_ASET), Hammamet, Tunisia, 14–17 January 2017; IEEE: Hammamet, Tunisia, 2017; pp. 395–400.
22. Abbasbandy, S. Improving Newton–Raphson method for nonlinear equations by modified Adomian decomposition method. *Appl. Math. Comput.* **2003**, *145*, 887–893.
23. Zhang, W.; Zhu, Q.; Mobayen, S.; Yan, H.; Qiu, J.; Narayan, P. U-Model and U-control methodology for nonlinear dynamic systems. *Complexity* **2020**. [CrossRef]
24. Zhu, Q.; Wang, Y.; Zhao, D.; Li, S.; Billings, S.A. Review of rational (total) nonlinear dynamic system modelling, identification, and control. *Int. J. Syst. Sci.* **2015**, *46*, 2122–2133.
25. Zhu, Q.; Li, S.; Zhao, D. A universal U-model based control system design. In Proceedings of the 33rd Chinese Control Conference, Nanjing, China, 28–30 July 2014; IEEE: Nanjing, China, 2014; pp. 1839–1844.
26. Zhu, Q.M.; Guo, L.Z. A pole placement controller for non-linear dynamic plants. *Proc. Inst. Mech. Eng. Part I J. Syst. Control Eng.* **2002**, *216*, 467–476.
27. Geng, X.; Zhu, Q.; Liu, T.; Na, J. U-model based predictive control for nonlinear processes with input delay. *J. Process Control* **2019**, *75*, 156–170.
28. Zhu, Q.; Zhang, W.; Zhang, J.; Sun, B. U-neural network-enhanced control of nonlinear dynamic systems. *Neurocomputing* **2019**, *352*, 12–21.
29. Hussain, N.A.A.; Ali, S.S.A.; Ovinis, M.; Arshad, M.R.; Al-Saggaf, U.M. Underactuated coupled nonlinear adaptive control synthesis using u-model for multivariable unmanned marine robotics. *IEEE Access* **2019**, *8*, 1851–1865.
30. Shafiq, M.; Haseebuddin, M. U-model-based internal model control for non-linear dynamic plants. *Proc. Inst. Mech. Eng. Part I J. Syst. Control Eng.* **2005**, *219*, 449–458.
31. Li, R.; Zhu, Q.; Kiely, J.; Zhang, W. Algorithms for U-model-based dynamic inversion (UM-dynamic inversion) for continuous time control systems. *Complexity* **2020**. [CrossRef]
32. Hu, M.; Chen, S. One-Pass Incomplete Multi-View Clustering. In Proceedings of the AAAI Conference on Artificial Intelligence, Honolulu, HI, USA, 27 January–1 February 2019; Volume 33, pp. 3838–3845.
33. Zhu, Q.; Zhang, W.; Na, J.; Sun, B. U-model based control design framework for continuous-time systems. In Proceedings of the 2019 Chinese Control Conference (CCC), Guangzhou, China, 27–30 July 2019; IEEE: Guangzhou, China, 2019; pp. 106–111.
34. Zhu, Q.M.; Warwick, K.; Douce, J.L. Adaptive general predictive controller for nonlinear systems. *IEE Proc. D (Control Theory Appl.)* **1991**, *138*, 33–40.
35. Isidori, A. *Nonlinear Control Systems*; Springer Science Business Media: London, UK, 2000.
36. Wang, H.; Li, S.; Yang, J.; Zhou, X. Continuous sliding mode control for permanent magnet synchronous motor speed regulation systems under time-varying disturbances. *J. Power Electron.* **2016**, *16*, 1324–1335.
37. Sun, X.; Shi, Z.; Lei, G.; Guo, Y.; Zhu, J. Analysis and design optimization of a permanent magnet synchronous motor for a campus patrol electric vehicle. *IEEE Trans. Veh. Technol.* **2019**, *68*, 10535–10544.
38. Zaihidee, F.M.; Mekhilef, S.; Mubin, M. Application of fractional order sliding mode control for speed control of permanent magnet synchronous motor. *IEEE Access* **2019**, *7*, 101765–101774.
39. Fei, Q.; Deng, Y.; Li, H.; Liu, J.; Shao, M. Speed ripple minimization of permanent magnet synchronous motor based on model predictive and iterative learning controls. *IEEE Access* **2019**, *7*, 31791–31800.
40. Wang, T.; Ping, Z.; Huang, Y.; Lu, J.G. Nonlinear internal model based two-step controller design for PMSM position servo system. In Proceedings of the 2019 IEEE 15th International Conference on Control and Automation (ICCA), Edinburgh, UK, 16–19 July 2019; IEEE: Edinburgh, UK, 2019; pp. 1264–1269.
41. Pillay, P.; Krishnan, R. Modeling, simulation, and analysis of permanent-magnet motor drives. I. The permanent-magnet synchronous motor drive. *IEEE Trans. Ind. Appl.* **1989**, *25*, 265–273.

entropy

MDPI

Article

Adaptive Fixed-Time Control of Strict-Feedback High-Order Nonlinear Systems

Yang Li [1,*], Jianhua Zhang [1], Xiaoyun Ye [1] and Cheng Siong Chin [2]

[1] School of Information and Control Engineering, Qingdao University of Technology, Qingdao 266525, China; jianhuazhang@qut.edu.cn (J.Z.); yexiaoyun@qut.edu.cn (X.Y.)
[2] Faculty of Science, Agriculture, and Engineering, Newcastle University Singapore, Singapore 599493, Singapore; cheng.chin@ncl.ac.uk
* Correspondence: yang_li@qut.edu.cn

Abstract: This paper examines the adaptive control of high-order nonlinear systems with strict-feedback form. An adaptive fixed-time control scheme is designed for nonlinear systems with unknown uncertainties. In the design process of a backstepping controller, the Lyapunov function, an effective controller, and adaptive law are constructed. Combined with the fixed-time Lyapunov stability criterion, it is proved that the proposed control scheme can ensure the stability of the error system in finite time, and the convergence time is independent of the initial condition. Finally, simulation results verify the effectiveness of the proposed control strategy.

Keywords: adaptive fixed-time control; neural network control; strict-feedback high-order nonlinear systems

Citation: Li, Y.; Zhang, J.; Ye, X.; Chin, C.S. Adaptive Fixed-Time Control of Strict-Feedback High-Order Nonlinear Systems. *Entropy* **2021**, *23*, 963. https://doi.org/10.3390/e23080963

Academic Editors: Quanmin Zhu, Giuseppe Fusco, Jing Na, Weicun Zhang and Ahmad Taher Azar

Received: 2 July 2021
Accepted: 26 July 2021
Published: 27 July 2021

1. Introduction

Recently, the adaptive trajectory tracking control of uncertain nonlinear systems has made a significant breakthrough [1–3]. In addition, neural network adaptive control has become a popular method in the past decades [4–6]. Many remarkable results have extended to strict-feedback systems, pure-feedback systems, and Brunovsky systems, and neural networks are combined with various techniques, such as the backstepping technique, the adaptive technique, and the sliding mode control method [7–9]. The neural network is used to identify the nonlinear term of the uncertain system, which combines the advantages of adaptive control. Many excellent articles and monographs have been published. In the design of these control systems, the neural network is used as a general approximator to the uncertain nonlinear term of the systems [10–12]. In these systems, the unknown nonlinear systems are approximate by neural networks, which are valid only within a compact set, and the neural network controller is designed. Based on Lyapunov uniformly bounded (UUB) theory, the closed-loop error systems are bounded [13–15]. In order to overcome the problem of uncertainty or disturbance that does not meet the specific matching conditions, the adaptive controller is usually constructed by combining backstepping control technology with the adaptive neural network. The high-order system is divided into multiple subsystems. The virtual controller of the low-order subsystem is designed first. Then, the recursive design is used until the final design of the neural network adaptive controller to achieve stability of the system, allowing it to possess the desired performance indicators.

In practical engineering applications, the research of high-order nonlinear systems has attracted much attention, and their application is also extensive, for example, as financial systems, communication systems, biological systems, and machine systems [16–18]. Some results regarding high-order system control have been obtained following the development of adding a power integrator [19]. The problems studied in recent years involve robust control [20,21], adaptive global stabilization [17], global asymptotic stabilization [22], output feedback stabilization [23], and state feedback output tracking [16]. Many methods have been proposed, such as backstepping technology, adaptive technology, sliding mode control, neural network control, and fuzzy control. However, the above results need to be precise with some unknown coefficients in the system model. In [20], the unknown function in the system is described by the mathematical model of an online neural network. In addition to this pioneering result, high-order system control based on neural networks has been widely developed and applied [24–26].

In the actual industrial process, such as in missile systems, aircraft attitude control systems, robot control systems and other industrial control systems, the purpose of controller design is to achieve stability of the controlled system and maintain it for a limited time. However, the control method without considering the convergence time cannot achieve this objective. Compared with the traditional Lyapunov stability theory, the finite-time Lyapunov stability theory has attracted the attention of many researchers because it can make the controlled system stable near the equilibrium state in finite time [27–29].

Many researchers combine finite-time control with neural network adaptive control for nonlinear systems with nonlinear functions and dynamic uncertainties based on backstepping and propose many related adaptive finite-time control schemes [30–32]. However, there are still many problems to be solved in these existing control strategies. For finite-time control, the convergence time is dependent on the initial condition. However, the ideal weights of NNs are unknown, and it is difficult to obtain a convergence time. Therefore, to solve this issue, fixed-time neural network control is an appropriate selection of the control method.

The high-order systems' neural network control problem is discussed in the articles [33–35]. The fixed-time neural network adaptive controller is present for nonlinear high-order systems. Based on the fixed-time adaptive technology, the strict-feedback high-order system has fixed-time Lyapunov stability based on Lyapunov stability theory [36–38]. The convergence time of the system can be accurately calculated, and the settling time does not rely on the initial situation. The main contributions of this article are as follows:

(1) The combination of the neural network adaptive control with fixed-time Lyapunov stability theory for high-order nonlinear system control problems.
(2) The design of the fixed-time adaptive law of the error systems for neural networks. The parameters of neural networks are iteratively in fixed time based on the Lyapunov fixed-time stability theorem.
(3) The convergence time set by control parameters and adaptive law gain parameters without initial conditions to ensure the control performance.

This article consists of the following sections: in Section 2, a strict-feedback high-order nonlinear mathematical description of the problem is presented; in Section 3, the adaptive fixed-time neural network control scheme for the strict-feedback high-order nonlinear system is designed; in Section 4, simulation results show the effectiveness of the proposed control strategy; in Section 5, the conclusion of the article is presented.

2. Problem Formation and Preliminaries

Consider the following strict-feedback high-order nonlinear system:

$$\begin{aligned} \dot{x}_i &= g_i x_{i+1}^{\eta_i} + f_i(\overline{x}_i) \\ \dot{x}_n &= g_n u^{\eta_n} + f_n(\overline{x}_n) \\ y &= x_1 \end{aligned} \tag{1}$$

where $x_i \in R$ is the state of the system; $\overline{x}_i = [x_1, \ldots, x_i]^T \in R^i$ is the state vector of the system; $f_i(\overline{x}_i) : R^i \to R$ is the unknown smooth function; $y \in R$ is the output of the system; $u \in R$ is the corresponding control input of the system; η_i is the order of the system; g_i is the unknown control gain parameter and satisfies $0 < \underline{g}_i \leq g_i \leq \overline{g}_i$, where $\underline{g}_i$, $\overline{g}_i$ are known parameters; and the desired trajectory y_d and its derivative are continuous and bounded.

Lemma 1. *For positive real numbers $p, q, p \in (0,1), q \in (1,\infty)$ with a denominator and numerator, both are odd numbers and positive real numbers $\rho, \sigma, \rho_1, \rho_2, \sigma_1, \sigma_2$; then, the following inequalities hold:*

$$\begin{aligned} -\rho\widetilde{\theta}\hat{\theta}^p &\leq -\rho_1\widetilde{\theta}^{p+1} + \rho_2\theta^{p+1} \\ -\sigma\widetilde{\theta}\hat{\theta}^q &\leq -\sigma_1\widetilde{\theta}^{q+1} + \sigma_2\theta^{q+1} \end{aligned} \tag{2}$$

where $\rho_1, \rho_2, \sigma_1, \sigma_2$ are determined by p, q, ρ, σ [39].

Lemma 2. *For any constant where $x, y \in R$ and p, q are odd, the following inequality holds:*

$$x^{\xi} - y^{\xi} \leq \xi|x-y|\left(x^{\xi-1} + y^{\xi-1}\right) \leq \zeta|x-y|\left((x-y)^{\xi-1} + y^{\xi-1}\right) \tag{3}$$

where $\xi = \frac{q}{p}$ and $q > p > 1$, $\zeta = \xi(2^{\xi-2}+2)$.

Proof. Assuming $x \geq y$, for any constant, the following equation holds:

$$\frac{x^{\xi} - y^{\xi}}{x-y} = \xi c^{\xi-1} \tag{4}$$

where c is an existent constant and satisfies $y \leq c \leq x$; therefore,

$$\begin{aligned} x^{\xi} - y^{\xi} &= \xi c^{\xi-1}(x-y) \\ &\leq \xi c^{\xi-1}|x-y| \end{aligned} \tag{5}$$

because $y \leq c \leq x$, then $c^{\xi-1} \leq \max\{x^{\xi-1}, y^{\xi-1}\} \leq x^{\xi-1} + y^{\xi-1}$; therefore,

$$x^{\xi} - y^{\xi} \leq \xi|x-y|\left(x^{\xi-1} + y^{\xi-1}\right) \tag{6}$$

On the other hand, based on $\xi > 1$, for $x^{\xi-1}, (x-y)^{\xi-1}, y^{\xi-1}$, we have

$$x^{\xi-1} \leq \begin{cases} (x-y)^{\xi-1} + y^{\xi-1}, 1 < \xi < 2 \\ 2^{\xi-2}\left((x-y)^{\xi-1} + y^{\xi-1}\right), \xi \geq 2 \end{cases} \tag{7}$$

then, we choose

$$x^{\xi-1} \leq \left(2^{\xi-2}+1\right)\left((x-y)^{\xi-1} + y^{\xi-1}\right) \tag{8}$$

therefore,

$$x^{\xi-1} + y^{\xi-1} \leq \left(2^{\xi-2}+2\right)\left((x-y)^{\xi-1} + y^{\xi-1}\right) \tag{9}$$

Then,

$$\xi|x-y|\left(x^{\xi-1} + y^{\xi-1}\right) \leq \zeta|x-y|\left((x-y)^{\xi-1} + y^{\xi-1}\right) \tag{10}$$

where $\zeta = \xi(2^{\xi-2}+2)$. □

3. Main Results

In this section, for the strict-feedback high-order nonlinear system, the neural network is used to identify the nonlinear system, and an adaptive algorithm is used to adjust the weight coefficient of the neural network. Based on fixed-time Lyapunov stability theory, a neural network adaptive tracker based on backstepping control strategy is designed so that the system state can track the preset trajectory. Theoretical proof and a numerical simulation are given.

The design block diagram of the closed-loop system is shown in Figure 1. For high-order nonlinear systems with strict-feedback form, a neural network adaptive controller is designed to make the system track a given target signal in finite time. The convergence time is independent of the initial condition to achieve fixed-time Lyapunov stability of the closed-loop error system. The controller design can be divided into the following N steps:

Step 1: First, for the system, the following variables are selected:

$$z_1 = x_1 - y_d \tag{11}$$

the dynamics of z_1 can be obtained as

$$\dot{z}_1 = g_1 x_2^{\eta_1} + f_1(x_1) - \dot{y}_d \tag{12}$$

Moreover, we have

$$f_1(x_1) = W_1^T \Psi(Z_1) + \varepsilon_1(Z_1) \tag{13}$$

where $|\varepsilon_1(x_1)| \leq \varepsilon_1$, we have

$$z_1 f_1(x_1) \leq \theta_1 |z_1| \|\Psi(Z_1)\| + |z_1| \varepsilon_1 \tag{14}$$

$\theta_1 = \|W_1\|$ is defined, and the Lyapunov candidate functional is chosen as

$$V_1 = \frac{1}{2} z_1^2 + \frac{1}{2\mu_1} \widetilde{\theta}_1^2 \tag{15}$$

where $\mu_1 > 0$ is positive constant, and $\widetilde{\theta}_1 = \hat{\theta}_1 - \theta_1$. Differentiating V_1 with respect to time t yields

$$\dot{V}_1 \leq g_1 z_1 x_2^{\eta_1} + \theta_1 |z_1| \|\Psi(Z_1)\| + |z_1| \varepsilon_1 - z_1 \dot{y}_d + \frac{1}{\mu_1} \widetilde{\theta}_1 \dot{\hat{\theta}}_1 \tag{16}$$

The virtual control signal α_1 is selected as

$$\alpha_1 = -\underline{g}_1^{-\frac{1}{\eta_1}} \left(sign(z_1) \hat{\theta}_1 \|\Psi(Z_1)\| + \frac{z_1 \varepsilon_1^2}{|z_1| \varepsilon_1 + \eta_1} + sign(z_1) |\dot{y}_d| + \kappa_1 z_1^p + \iota_1 z_1^q \right)^{\frac{1}{\theta_1}} \tag{17}$$

Then, based on Lemma 2, we have

$$\dot{V}_1 \leq c_1 \overline{g}_1 |z_1| \left(\left| z_2^{\eta_1} \right| + |z_2| x_2^{\eta_1 - 1} \right) - |z_1| \widetilde{\theta}_1 \|\Psi(Z_1)\| + \delta_1 + \frac{1}{\mu_1} \widetilde{\theta}_1 \dot{\hat{\theta}}_1 - \kappa_1 z_1^{p+1} - \iota_1 z_1^{q+1} \tag{18}$$

where

$$z_2 = x_2 - \alpha_1 \tag{19}$$

then the adaptive law design as

$$\dot{\hat{\theta}}_1 = \mu_1 \left(|z_1| |\Psi_1| - \rho_1 \hat{\theta}_1^p - \sigma_1 \hat{\theta}_1^q \right) \tag{20}$$

based on Equation (18), we have

$$\begin{aligned} \dot{V}_1 \leq\ & c_1 \overline{g}_1 |z_1| \left(\left| z_2^{\eta_1} \right| + |z_2| x_2^{\eta_1 - 1} \right) - |z_1| \widetilde{\theta}_1 \|\Psi(Z_1)\| + \delta_1 + \widetilde{\theta}_1 |z_1| |\Psi_1| \\ & - \rho_1 \widetilde{\theta}_1 \hat{\theta}_1^p - \sigma_1 \widetilde{\theta}_1 \hat{\theta}_1^q - \kappa_1 z_1^{p+1} - \iota_1 z_1^{q+1} \end{aligned} \tag{21}$$

based on Lemma 1, we have

$$\begin{aligned} -\rho_1 \widetilde{\theta}_1 \hat{\theta}_1^p &\leq -\varsigma_1 \widetilde{\theta}_1^{p+1} + \upsilon_1 \theta_1^{p+1} \\ -\sigma_1 \widetilde{\theta}_1 \hat{\theta}_1^q &\leq -\omega_1 \widetilde{\theta}_1^{q+1} + \vartheta_1 \theta_1^{q+1} \end{aligned} \tag{22}$$

then

$$\dot{V}_1 \leq c_1\overline{g}_1|z_1|\left(\left|z_2^{\theta_1}\right| + |z_2|x_2^{\theta_1-1}\right) + \delta_1 - \varsigma_1\widetilde{\theta}_1^{p+1} + v_1\theta_1^{p+1} - \omega_1\widetilde{\theta}_1^{q+1} + \vartheta_1\theta_1^{q+1} - \kappa_1 z_1^{p+1} - \iota_1 z_1^{q+1} \tag{23}$$

Step i: the tracking error can be described as

$$z_i = x_i - \alpha_{i-1} \tag{24}$$

Based on dynamics and tracking error, the dynamics of z_i can be obtained as

$$\dot{z}_i = g_i x_{i+1}^{\theta_i} + f_i(x_i) - \dot{\alpha}_{i-1} \tag{25}$$

Moreover, we have

$$f_i(x_i) - \dot{\alpha}_{i-1} = W_i^T\Psi(Z_i) + \varepsilon_i(Z_i) \tag{26}$$

where $|\varepsilon_i(x_i)| \leq \varepsilon_i$; we have

$$z_i\left(f_i(\overline{x}_i) - \dot{\alpha}_{i-1}\right) \leq \theta_i|z_i|\|\Psi(Z_i)\| + |z_i|\varepsilon_i \tag{27}$$

$\theta_i = \|W_i\|$ is defined, and the Lyapunov candidate functional is chosen as

$$V_i = \frac{1}{2}z_i^2 + \frac{1}{2\mu_i}\widetilde{\theta}_i^2 \tag{28}$$

where $\mu_i > 0$ is positive constant and $\widetilde{\theta}_i = \hat{\theta}_i - \theta_i$. Differentiating V_i with respect to time t, yields

$$\dot{V}_i \leq g_i z_i x_{i+1}^{\theta_i} + \theta_i|z_i|\|\Psi(Z_i)\| + |z_i|\varepsilon_i + \frac{1}{\mu_i}\widetilde{\theta}_i\dot{\hat{\theta}}_i \tag{29}$$

The virtual control signal α_i is designed as

$$\alpha_i = -\underline{g}_i^{-\frac{1}{\eta_i}}\begin{pmatrix} c_{i-1}sign(z_i)\overline{g}_{i-1}|z_{i-1}|\left(z_i^{\eta_{i-1}-1} + x_i^{\eta_{i-1}-1}\right) + sign(z_i)\hat{\theta}_i\|\Psi(Z_i)\| \\ +\frac{z_i\varepsilon_i^2}{|z_i|\varepsilon_i+\delta_i} + \kappa_i z_i^p + \iota_i z_i^q \end{pmatrix}^{\frac{1}{\theta_i}} \tag{30}$$

then, based on Lemma 2, we have

$$\begin{aligned} \dot{V}_i \leq & -c_{i-1}\overline{g}_{i-1}|z_{i-1}|\left(\left|z_i^{\eta_{i-1}}\right| + |z_i|x_i^{\eta_{i-1}-1}\right) + c_i\overline{g}_i|z_i|\left(\left|z_{i+1}^{\eta_i}\right| + |z_{i+1}|x_{i+1}^{\eta_i-1}\right) \\ & -|z_i|\widetilde{\theta}_i\|\Psi(Z_i)\| + \eta_i + \tfrac{1}{\mu_i}\widetilde{\theta}_i\dot{\hat{\theta}}_i - \kappa_i z_i^{p+1} - \iota_i z_i^{q+1} \end{aligned} \tag{31}$$

where

$$z_{i+1} = x_{i+1} - \alpha_i \tag{32}$$

Then, the adaptive law design as

$$\dot{\hat{\theta}}_i = \mu_i\left(|z_i||\Psi_i| - \rho_i\hat{\theta}_i^p - \sigma_i\hat{\theta}_i^q\right) \tag{33}$$

based on Equation (31), we have

$$\begin{aligned} \dot{V}_i \leq & -c_{i-1}\overline{g}_{i-1}|z_{i-1}|\left(\left|z_i^{\eta_{i-1}}\right| + |z_i|x_i^{\eta_{i-1}-1}\right) + c_i\overline{g}_i|z_i|\left(\left|z_{i+1}^{\eta_i}\right| + |z_{i+1}|x_{i+1}^{\eta_i-1}\right) - |z_i|\widetilde{\theta}_i\|\Psi(Z_i)\| + \eta_i \\ & +\widetilde{\theta}_i|z_i||\Psi_i| - \rho_i\widetilde{\theta}_i\hat{\theta}_i^p - \sigma_i\widetilde{\theta}_i\hat{\theta}_i^q - \kappa_i z_i^{p+1} - \iota_i z_i^{q+1} \end{aligned} \tag{34}$$

based on Lemma 1, we have

$$\begin{aligned} -\rho_i\widetilde{\theta}_i\hat{\theta}_i^p &\leq -\varsigma_i\widetilde{\theta}_i^{p+1} + v_i\theta_i^{p+1} \\ -\sigma_i\widetilde{\theta}_i\hat{\theta}_i^q &\leq -\omega_i\widetilde{\theta}_i^{q+1} + \vartheta_i\theta_i^{q+1} \end{aligned} \tag{35}$$

then

$$\dot{V}_i \le -c_{i-1}\overline{g}_{i-1}|z_{i-1}|\left(\left|z_i^{\eta_{i-1}}\right| + |z_i|x_i^{\eta_{i-1}-1}\right) + c_i\overline{g}_i|z_i|\left(\left|z_{i+1}^{\eta_i}\right| + |z_{i+1}|x_{i+1}^{\eta_i-1}\right) + \delta_i - \varsigma_i\widetilde{\theta}_i^{p+1} + \upsilon_i\theta_i^{p+1} \\ -\omega_i\widetilde{\theta}_i^{q+1} + \vartheta_i\theta_i^{q+1} - \kappa_i z_i^{p+1} - \iota_i z_i^{q+1} \quad (36)$$

Step n: the time derivative of z_n can be described as

$$z_n = x_n - \alpha_{n-1} \quad (37)$$

Based on dynamics and tracking error, the dynamics of z_n can be obtained as

$$\dot{z}_n = g_n u^{\eta_n} + f_n(\overline{x}_n) - \dot{\alpha}_{n-1} \quad (38)$$

Moreover, we have

$$f_n(\overline{x}_n) - \dot{\alpha}_{n-1} = W_n^T\Psi(Z_n) + \varepsilon_n(Z_n) \quad (39)$$

where $|\varepsilon_n(x_n)| \le \varepsilon_n$, we have

$$z_n(f_n(\overline{x}_n) - \dot{\alpha}_{n-1}) \le \theta_n|z_n|\|\Psi(Z_n)\| + |z_n|\varepsilon_n \quad (40)$$

$\theta_n = \|W_n\|$ is defined, and the Lyapunov candidate functional is chosen as

$$V_n = \frac{1}{2}z_n^2 + \frac{1}{2\mu_n}\widetilde{\theta}_n^2 \quad (41)$$

where $\mu_n > 0$ is positive constant and $\widetilde{\theta}_n = \hat{\theta}_n - \theta_n$. Differentiating V_n with respect to time t yields

$$\dot{V}_n \le g_n z_n u^{\theta_n} + \theta_n|z_n|\|\Psi(Z_n)\| + |z_n|\varepsilon_n + \frac{1}{\mu_n}\widetilde{\theta}_n\dot{\hat{\theta}}_n \quad (42)$$

The actual control is designed as

$$u = -\underline{g}_n^{-\frac{1}{\eta_n}}\begin{pmatrix} c_{n-1}sign(z_n)\overline{g}_{n-1}|z_{n-1}|\left(z_n^{\theta_{n-1}-1} + x_n^{\theta_{n-1}-1}\right) + sign(z_n)\hat{\theta}_n\|\Psi(Z_n)\| \\ +\frac{z_n\varepsilon_n^2}{|z_n|\varepsilon_n+\delta_n} + \kappa_n z_n^p + \iota_n z_n^q \end{pmatrix}^{\frac{1}{\eta_n}} \quad (43)$$

then, based on Lemma 2, we have

$$\dot{V}_n \le -c_{n-1}\overline{g}_{n-1}|z_{n-1}|\left(\left(\left|z_n^{\eta_{n-1}}\right| + |z_n|x_n^{\eta_{n-1}-1}\right)\right) - |z_n|\widetilde{\theta}_n\|\Psi(Z_n)\| + \eta_n \\ +\frac{1}{\mu_n}\widetilde{\theta}_n\dot{\hat{\theta}}_n - \kappa_n z_n^{p+1} - \iota_n z_n^{q+1} \quad (44)$$

then, the adaptive law design as

$$\dot{\hat{\theta}}_n = \mu_n\left(|z_n||\Psi_n| - \rho_n\hat{\theta}_n^p - \sigma_n\hat{\theta}_n^q\right) \quad (45)$$

based on Equation (20), we have

$$\dot{V}_n \le -c_{n-1}\overline{g}_{n-1}|z_{n-1}|\left(\left(\left|z_n^{\eta_{n-1}}\right| + |z_n|x_n^{\eta_{n-1}-1}\right)\right) + \eta_n - \kappa_n z_n^{p+1} - \iota_n z_n^{q+1} - \rho_n\widetilde{\theta}_n\hat{\theta}_n^p - \sigma_n\widetilde{\theta}_n\hat{\theta}_n^q \quad (46)$$

based on Lemma 1, we have

$$-\rho_n\widetilde{\theta}_n\hat{\theta}_n^p \le -\varsigma_n\widetilde{\theta}_n^{p+1} + \upsilon_n\theta_n^{p+1} \\ -\sigma_n\widetilde{\theta}_n\hat{\theta}_n^q \le -\omega_n\widetilde{\theta}_n^{q+1} + \vartheta_n\theta_n^{q+1} \quad (47)$$

then

$$\dot{V}_n \leq -c_{n-1}\bar{g}_{n-1}|z_{n-1}|\left|z_n^{\eta_n-1}\right| + \delta_n - \kappa_n z_n^{p+1} - \iota_n z_n^{q+1} \\ -\varsigma_n \tilde{\theta}_n^{p+1} + \upsilon_n \theta_n^{p+1} - \omega_n \tilde{\theta}_n^{q+1} + \vartheta_n \theta_n^{q+1} \tag{48}$$

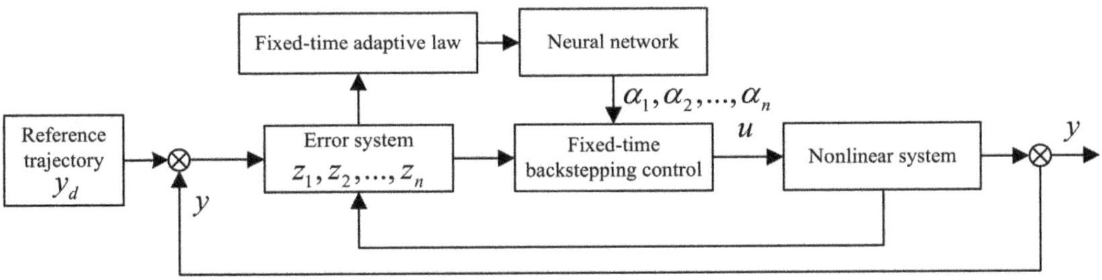

Figure 1. Block diagram of the closed-loop system.

Theorem 1. *For the strict-feedback high-order nonlinear system with unknown nonlinearity (1), based on the feasible virtual control signal (17), (30), actual controller (43), and adaptive function (20) (33), the error system is fixed-time Lyapunov stable, and the convergence time is independent of the initial condition.*

Proof. Based on Lyapunov candidate functional (15), (28), (41), the Lyapunov candidate functional is chosen.

$$V = \sum_{j=1}^{n} V_i \tag{49}$$

The virtual control signal is chosen as (17), (30), and the fixed-time adaptive function is chosen as (20), (33); the controller is designed as (43) according to the fixed-time Lyapunov stability theory. Based on fixed-time adaptive neural network control and backstepping technology, and taking the trajectory along the system, we have

$$\begin{aligned} \dot{V} &\leq -\sum_{j=1}^{n} \kappa_j z_j^{p+1} - \sum_{j=1}^{n} \iota_j z_j^{q+1} - \sum_{j=1}^{n} \varsigma_j \tilde{\theta}_j^{p+1} - \sum_{j=1}^{n} \omega_j \tilde{\theta}_j^{q+1} \\ &+ \sum_{j=1}^{n} \eta_j + \sum_{j=1}^{n} \upsilon_j \theta_j^{p+1} + \sum_{j=1}^{n} \vartheta_j \theta_j^{q+1} \\ &\leq -aV^{\frac{p+1}{2}} - bV^{\frac{q+1}{2}} + c \end{aligned} \tag{50}$$

where

$$\begin{aligned} a &= \frac{\min\{\kappa_{j\in N}, \varsigma_{j\in N}\}}{\left(\max\left\{\frac{1}{2}, \frac{1}{2\mu_{j\in N}}\right\}\right)^{\frac{p+1}{2}}}, b = \frac{(2n)^{\frac{1-q}{2}} \min\{\iota_{j\in N}, \omega_{j\in N}\}}{\left(\max\left\{\frac{1}{2}, \frac{1}{2\mu_{j\in N}}\right\}\right)^{\frac{q+1}{2}}} \\ c &= \sum_{j=1}^{n} \delta_j + \sum_{j=1}^{n} \upsilon_j \theta_j^{p+1} + \sum_{j=1}^{n} \vartheta_j \theta_j^{q+1} \end{aligned} \tag{51}$$

Therefore, according to the lemma in [39], all closed-loop signals possess fixed-time Lyapunov stability. □

The design details are summarized in Figure 2 to show the procedure of the control process.

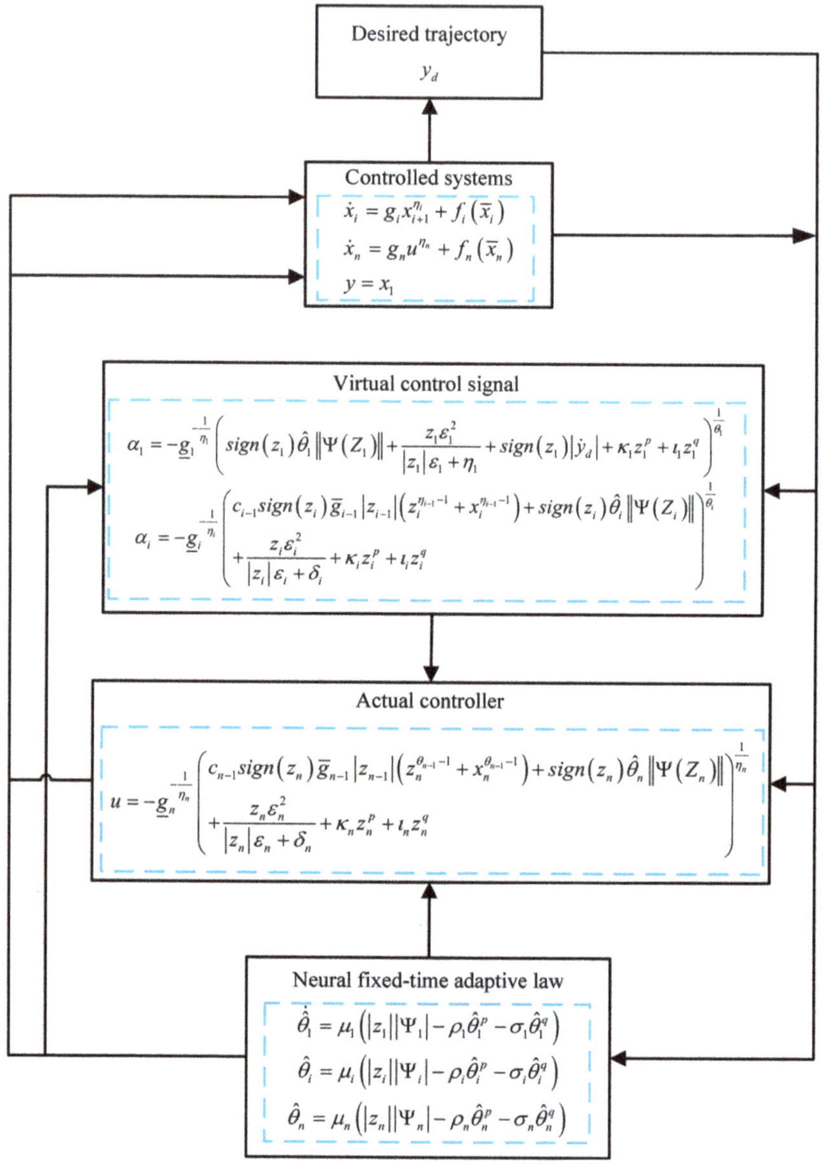

Figure 2. Design procedure.

4. Numerical Examples

In this paper, the feasibility and effectiveness of the algorithm are verified by numerical simulation. A strict-feedback high-order system is considered as follows:

$$\begin{cases} \dot{x}_1 = g_1 x_2^{\eta_1} + f_1(x_1) \\ \dot{x}_2 = g_2 u^{\eta_2} + f_2(x_1, x_2) \\ y = x_1 \end{cases} \tag{52}$$

where the function $f_1(x_1) = x_1(t) + \sin(0.1x_1(t))$, $f_2(x_1, x_2) = x_2(t)$, $g_1 = 1$, $g_2 = 1$, $\eta_1 = \frac{5}{3}$, $\eta_2 = \frac{7}{5}$, and the control input under the adaptive law is designed

$$\dot{\hat{\theta}}_1 = 0.01\left(|z_1||\Psi_1| - 0.1\hat{\theta}_1^{\frac{5}{3}} - 0.1\hat{\theta}_1^{\frac{1}{3}}\right) \tag{53}$$

$$\dot{\hat{\theta}}_2 = 0.01\left(|z_2||\Psi_2| - 0.1\hat{\theta}_2^{\frac{5}{3}} - 0.1\hat{\theta}_2^{\frac{1}{3}}\right) \tag{54}$$

the control input is designed as

$$u(t) = \begin{pmatrix} -5sign(z_2(t)) * |z_1(t)| * \left(|z_2(t)| + |x_2(t)|^{\frac{2}{5}}\right) + sign(z_2(t))\hat{\theta}_2\Psi_2(Z_2) \\ + \frac{0.01z_2}{0.1|z_2(t)|+0.1} + 5z_2(t)^{\frac{1}{3}} + 5z_2(t)^{\frac{5}{3}} \end{pmatrix}^{\frac{5}{7}} \tag{55}$$

The desired reference signal is $y_d = \sin(t)$. The initial condition is selected as $x_1(0) = 1$, $x_2(0) = 1, \hat{\theta}_1(0) = 1, \hat{\theta}_2(0) = 1$. The neural network consists of seven nodes, centers $c = [-3, -2, -1, 0, 1, 2, 3]$, and widths $b = 1$.

Figure 3 shows that under the action of the neural network adaptive controller, the state of the controlled system state can track the preset trajectory in finite time. Figure 4 shows the state trajectory of the error system. It can be seen from the figure that under the action of the controller, the error system achieves fixed-time Lyapunov stability. The adaptive function curve is shown in Figure 5. For fixed-time control, α_1 is designed by $f_1(x_1)$, which is approximated by NNs, but its derivative is not easy to approximate; therefore, $f_2(\overline{x}_2) - \dot{\alpha}_1$ is not easy to approximate, and the amplitude is inevitable. Figure 6 shows that the system's controllers are bounded. It can be seen from the figures that the designed method is effective.

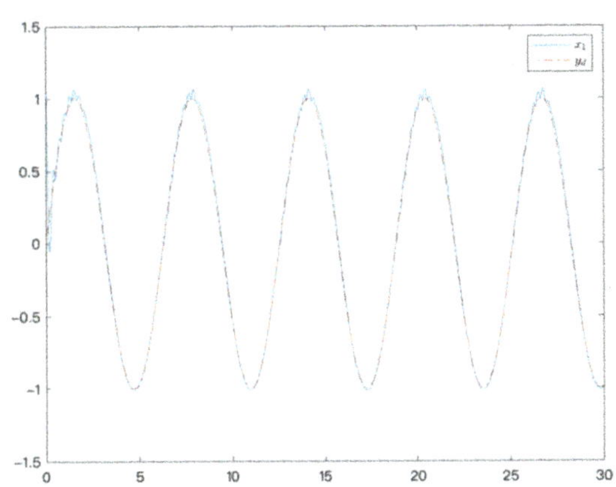

Figure 3. Trajectories of x_1 and y_d of a strict-feedback high-order nonlinear system.

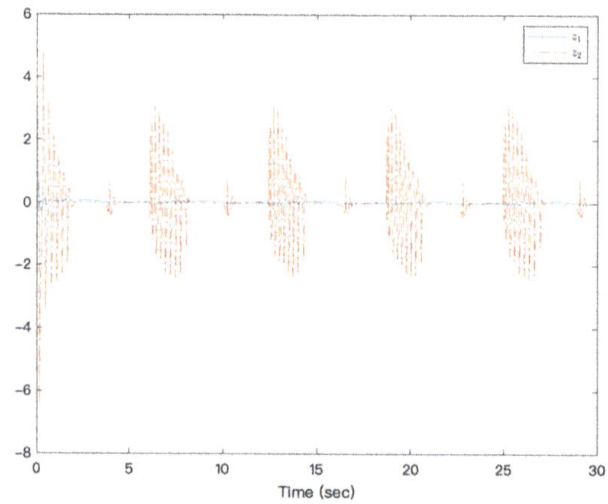

Figure 4. Trajectories of error states of a strict-feedback high-order nonlinear system.

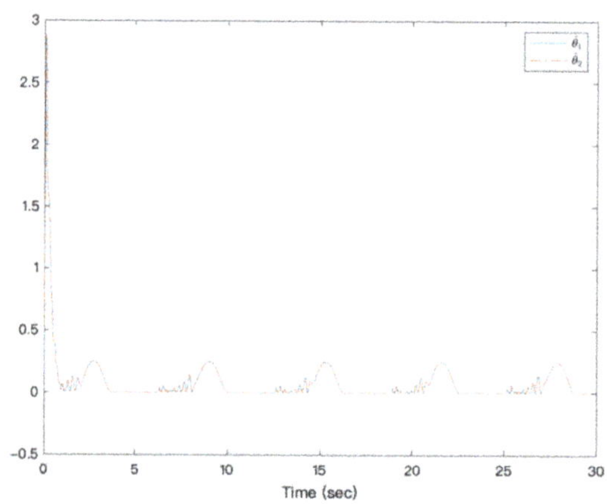

Figure 5. Trajectories of adaptive functions of a strict-feedback high-order nonlinear system.

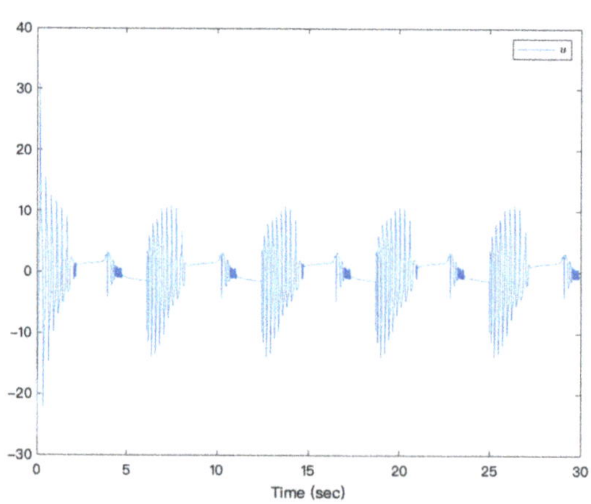

Figure 6. Trajectories of system input of a strict-feedback high-order nonlinear system.

5. Conclusions

In this paper, based on backstepping adaptive control technology, a neural network is used to approximate some unknown signals in a system. Combined with Lyapunov stability theorem and fixed time stability, an effective adaptive control scheme is designed. A class of strict-feedback high-order systems is further studied. The main contributions of this paper are as follows: the fixed-time control problem of strict-feedback high-order nonlinear systems is solved; the Lyapunov function is designed for each subsystem; at the same time, combined with adaptive backstepping technology, an adaptive neural network fixed-time controller is designed. The tracking error converges in finite time through stability analysis, and the convergence time does not relay on the initial condition. The most popular controller is designed in a linear control strategy, which controls the state's exponential stability. At present, the adaptive neural network control method based on backstepping has some limitations, and many problems need to be further studied and solved. In the finite-time adaptive control method for the multi-agent system, the finite time obtained by most finite-time control strategies often depends on the initial conditions of the system. Therefore, a finite-time control scheme independent of the initial value for a multi-agent system must be designed.

Author Contributions: Conceptualization, Y.L. and J.Z.; methodology, Y.L.; software, X.Y.; validation, Y.L., J.Z. and X.Y.; formal analysis, J.Z.; investigation, Y.L.; resources, C.S.C.; data curation, C.S.C.; writing—original draft preparation, Y.L.; writing—review and editing, Y.L.; visualization, Y.L.; supervision, Y.L.; project administration, Y.L.; funding acquisition, Y.L. All authors have read and agreed to the published version of the manuscript.

Funding: This research received no external funding.

Data Availability Statement: Not applicable.

Conflicts of Interest: The authors declare no conflict of interest.

References

1. Utkin, V.I. Sliding Mode Control Design Principles and Applications to Electric Drives. *IEEE Trans. Ind. Electron.* **2002**, *40*, 23–36. [CrossRef]
2. Li, H.; Wang, Y.; Yao, D.; Lu, R. A Sliding Mode Approach to Stabilization of Nonlinear Markovian Jump Singularly Perturbed Systems—ScienceDirect. *Automatica* **2018**, *97*, 404–413. [CrossRef]
3. Zhou, J.; Wen, C. Robust adaptive control of uncertain nonlinear systems in the presence of input saturation. *IFAC Proc. Vol.* **2006**, *39*, 149–154. [CrossRef]
4. Niu, B.; Duan, P.; Li, J.; Li, X. Adaptive Neural Tracking Control Scheme of Switched Stochastic Nonlinear Pure-Feedback Nonlower Triangular Systems. *IEEE Trans. Syst. Man Cybern. Syst.* **2021**, *51*, 975–986. [CrossRef]
5. Li, X.; Cheah, C.C. Adaptive Neural Network Control of Robot Based on a Unified Objective Bound. *IEEE Trans. Control Syst. Technol.* **2014**, *22*, 1032–1043. [CrossRef]
6. Dai, S.L.; Wang, C.; Wang, M. Dynamic learning from adaptive neural network control of a class of nonaffine nonlinear systems. *IEEE Trans. Neural Netw. Learn. Syst.* **2013**, *25*, 111–123.
7. Park, J.; Kim, S.; Park, T. Output-Feedback Adaptive Neural Controller for Uncertain Pure-Feedback Nonlinear Systems Using a High-Order Sliding Mode Observer. *IEEE Trans. Neural Netw. Learn. Syst.* **2019**, *30*, 1596–1601. [CrossRef]
8. Ge, S.S.; Wang, C. Adaptive NN control of uncertain nonlinear pure-feedback systems. *Automatica* **2002**, *38*, 671–682. [CrossRef]
9. Zhao, Q.; Lin, Y. Adaptive dynamic surface control for pure-feedback systems. *Int. J. Robust Nonlinear Control* **2012**, *22*, 1647–1660. [CrossRef]
10. Na, J.; Ren, X.; Zheng, D. Adaptive control. for nonlinear pure-feedback systems with high-order sliding mode observer. *IEEE Trans. Neural Netw. Learn. Syst.* **2013**, *24*, 370–382.
11. Sun, G.; Wang, D.; Peng, Z. Adaptive control based on single neural network approximation for non-linear pure-feedback systems. *IET Control Theory Appl.* **2012**, *6*, 2387–2396. [CrossRef]
12. Wang, H.; Chen, B.; Lin, C.; Sun, Y. Observer-based adaptive neural control for a class of nonlinear pure-feedback systems. *Neurocomputing* **2015**, *171*, 1517–1523. [CrossRef]
13. Wang, H.; Liu, P.X.; Bao, J.; Xie, X.J.; Li, S. Adaptive neural output-feedback decentralized control for large-scale nonlinear systems with stochastic disturbances. *IEEE Trans. Neural Netw. Learn. Syst.* **2020**, *31*, 972–983. [CrossRef]
14. Zhou, Q.; Zhao, S.; Li, H.; Lu, R.; Wu, C. Adaptive neural network tracking control. for robotic manipulators with dead zone. *IEEE Trans. Neural Netw. Learn. Syst.* **2019**, *30*, 3611–3620. [CrossRef] [PubMed]

15. Zhang, X.; Wang, Y.; Chen, X.; Su, C.Y.; Li, Z.; Wang, C.; Peng, Y. Decentralized adaptive neural approximated inverse control for a class of large-scale nonlinear hysteretic systems with time delays. *IEEE Trans. Syst. Man Cybern. Syst.* **2019**, *49*, 2424–2437. [CrossRef]
16. Qian, C.; Lin, W. Practical output tracking of nonlinear systems with uncontrollable unstable linearization. *IEEE Trans. Autom. Control* **2002**, *47*, 21–36. [CrossRef]
17. Lin, W.; Pongvuthithum, R. Nonsmooth adaptive stabilization of cascade systems with nonlinear parameterization via partial-state feedback. *IEEE Trans. Autom. Control* **2003**, *48*, 1809–1816. [CrossRef]
18. Zhang, Y.; Li, S.; Liao, L. Consensus of High-Order Discrete-Time Multiagent Systems with Switching Topology. *IEEE Trans. Syst. Man Cybern. Syst.* **2021**, *51*, 721–730. [CrossRef]
19. Lin, W.; Qian, C. Robust regulation of a chain of power integrators perturbed by a lower-triangular vector field. *Int. J. Robust Nonlinear Control* **2000**, *10*. [CrossRef]
20. Zhao, X.; Shi, P.; Zheng, X.; Zhang, J. Intelligent tracking control for a class of uncertain high-order nonlinear systems. *IEEE Trans. Neural Netw. Learn. Syst.* **2016**, *27*, 1976–1982. [CrossRef]
21. Khandekar, A.A.; Malwatkar, G.M.; Patre, B.M. Discrete sliding mode control for robust tracking of higher order delay time systems with experimental application. *ISA Trans.* **2013**, *52*, 36–44. [CrossRef]
22. Gao, F.; Wu, Y. Further results on global state feedback stabilization of high-order nonlinear systems with time-varying delays. *ISA Trans.* **2015**, *55*, 41–48. [CrossRef]
23. Sun, Z.Y.; Zhang, X.H.; Xie, X.J. Global continuous output-feedback stabilization for a class of high-order nonlinear systems with multiple time delays. *J. Frankl. Inst.* **2014**, *351*, 4334–4356. [CrossRef]
24. Feng, S.S.; Sun, Z.Y.; Zhou, C.Q.; Chen, C.C.; Meng, Q. Output tracking control via neural networks for high-order stochastic nonlinear systems with dynamic uncertainties. *Int. J. Fuzzy Syst.* **2021**, *23*, 716–726. [CrossRef]
25. Shahriari-Kahkeshi, M.; Afrush, A.; Pham, V.T. Adaptive consensus control of high-order uncertain nonlinear multi-agent systems with fuzzy dead-zone. *Int. J. Fuzzy Syst.* **2021**, *23*, 743–754. [CrossRef]
26. Aghababa, M.P.; Moradi, S. Robust adaptive dynamic surface back-stepping tracking control of high-order strict-feedback nonlinear systems via disturbance observer approach. *Int. J. Control* **2020**, *17*. [CrossRef]
27. Bhat, S.P.; Bernstein, D.S. Continuous finite-time stabilization of the translational and rotational double integrators. *IEEE Trans. Autom. Control* **1998**, *43*, 678–682. [CrossRef]
28. Bhat, S.P.; Bernstein, D.S. Finite-time stability of continuous autonomous systems. *SIAM J. Control Optim.* **2000**, *38*, 751–766. [CrossRef]
29. Qian, C.; Li, J. Global finite-time stabilization by output feedback for planar systems without observable linearization. *IEEE Trans. Autom. Control* **2005**, *50*, 885–890. [CrossRef]
30. Ma, L.; Zong, G.; Zhao, X.; Huo, X. Observed-based adaptive finite-time tracking control for a class of nonstrict-feedback nonlinear systems with input saturation. *J. Frankl. Inst. Eng. Appl. Math.* **2020**, *357*, 11518–11544. [CrossRef]
31. Li, Y.; Li, K.; Tong, S. Adaptive Neural Network finite-time control for multi-input and multi-output nonlinear systems with positive powers of odd rational numbers. *IEEE Trans. Neural Netw. Learn. Syst.* **2020**, *31*, 2532–2543. [CrossRef] [PubMed]
32. Na, J.; Wang, S.; Liu, Y.J.; Huang, Y.; Ren, X. Finite-time convergence adaptive neural network control for nonlinear servo systems. *IEEE Trans. Cybern.* **2020**, *50*, 2568–2579. [CrossRef] [PubMed]
33. Wang, M.; Wang, Z.; Chen, Y.; Sheng, W. Adaptive neural event-triggered control for discrete-time strict-feedback nonlinear systems. *IEEE Trans. Cybern.* **2020**, *50*, 2946–2958. [CrossRef] [PubMed]
34. Huang, J.; Zhang, M.; Ri, S.; Xiong, C.; Li, Z.; Kang, Y. High-order disturbance-observer-based sliding mode control for mobile wheeled inverted pendulum systems. *IEEE Trans. Ind. Electron.* **2020**, *67*, 2030–2041. [CrossRef]
35. Fang, L.; Ma, L.; Ding, S.; Zhao, D. Finite-time stabilization for a class of high-order stochastic nonlinear systems with an output constraint. *Appl. Math. Comput.* **2019**, *358*, 63–79. [CrossRef]
36. Xin, B.; Liu, L.; Hou, G.; Ma, Y. Chaos synchronization of nonlinear fractional discrete dynamical systems via linear control. *Entropy* **2017**, *19*, 351. [CrossRef]
37. Li, R.; Zhu, Q.; Narayan, P.; Yue, A.; Yao, Y.; Deng, M. U-model-based two-degree-of-freedom internal model control of nonlinear dynamic systems. *Entropy* **2021**, *23*, 169. [CrossRef]
38. Olvera-Guerrero, O.A.; Prieto-Guerrero, A.; Espinosa-Paredes, G. Non-linear stability analysis of real signals from nuclear power plants (boiling water reactors) based on noise assisted empirical mode decomposition variants and the shannon entropy. *Entropy* **2017**, *19*, 359. [CrossRef]
39. Zhang, J.; Li, Y.; Fei, W. Neural network-based nonlinear fixed-time adaptive practical tracking control for quadrotor unmanned aerial vehicles. *Complexity* **2020**, *2020*, 13. [CrossRef]

entropy

MDPI

Article

Structured H∞ Control for Spacecraft with Flexible Appendages

Yuntian Zhang [1], Aiping Pang [1,2,*], Hui Zhu [1] and Huan Feng [1]

[1] College of Electrical Engineering, Guizhou University, Guiyang 550025, China; ee.ytzhang18@gzu.edu.cn (Y.Z.); gs.huizhu19@gzu.edu.cn (H.Z.); gs.hfeng20@gzu.edu.cn (H.F.)
[2] Guizhou Provincial Key Laboratory of Internet + Intelligent Manufacturing, Guiyang 550025, China
* Correspondence: appang@gzu.edu.cn

Abstract: Spacecraft with large flexible appendages are characterized by multiple system modes. They suffer from inherent low-frequency disturbances in the operating environment that consequently result in considerable interference in the operational performance of the system. It is required that the control design ensures the system's high pointing precision, and it is also necessary to suppress low-frequency resonant interference as well as take into account multiple performance criteria such as attitude stability and bandwidth constraints. Aiming at the comprehensive control problem of this kind of flexible spacecraft, we propose a control strategy using a structured H-infinity controller with low complexity that was designed to meet the multiple performance requirements, so as to reduce the project cost and implementation difficulty. According to the specific resonant mode of the system, the design strategy of adding an internal mode controller, a trap filter, and a series PID controller to the structured controller is proposed, so as to achieve the comprehensive control goals through cooperative control of multiple control modules. A spacecraft with flexible appendages (solar array) is presented as an illustrative example in which a weighted function was designed for each performance requirement of the system (namely robustness, stability, bandwidth limit, etc.), and a structured comprehensive performance matrix with multiple performance weights and decoupled outputs was constructed. A structured H-infinity controller meeting the comprehensive performance requirements is given, which provides a structured integrated control method with low complexity for large flexible systems that is convenient for engineering practice, and provides a theoretical basis and reference examples for structured H-infinity control. The simulation results show that the proposed controller gives better control performance compared with the traditional H-infinity one, and can successfully suppress the vibration of large flexible appendages at 0.12 Hz and 0.66 Hz.

Keywords: structured control; flexible spacecraft; prevent oscillations

Citation: Zhang, Y.; Pang, A.; Zhu, H.; Feng, H. Structured H∞ Control for Spacecraft with Flexible Appendages. *Entropy* **2021**, *23*, 930. https://doi.org/10.3390/e23080930

Academic Editors: Quanmin Zhu, Giuseppe Fusco, Jing Na, Weicun Zhang and Ahmad Taher Azar

Received: 22 June 2021
Accepted: 19 July 2021
Published: 22 July 2021

1. Introduction

With the rapid development of the aerospace industry and of composite material technology, along with its broad application in aerospace, the structure of spacecrafts is becoming larger and more flexible, featuring multi-system modalities. The resonant mode of a flexible system of this type leads to tremendous changes in amplitude features. Meanwhile, the inherent low-frequency interference caused by the complex launch environment and the high-altitude environment during orbit operation, as well as the flexible mode of the system, greatly limit the choice of bandwidths, i.e., robust stability. The flexible modes and low-frequency disturbances inherent in the high-altitude environment impair the stability and performance of the spacecraft, which will cause performance degradation and failure to meet mission requirements, or result in unstable control or even failure of the spacecraft.

With the increasing diversification of spacecraft missions, the requirements for pointing accuracy of large spacecraft have become increasingly stringent, which makes control research more complicated and difficult to delve into. The difficulties of the control design of such large flexible systems are as follows: to suppress the external interference caused by the complex space environment and the inherent low-frequency resonance interference of flexible spacecraft; to meet "high-precision" performance requirements; and to ensure attitude stability and bandwidth amount (robustness requirement).

Previous studies show that in terms of a synthesis control over these spacecraft with multi-performance requirements, it is difficult to apply classic analysis methods to balance the requirements. Traditional control design schemes, in general, fail to simultaneously meet the requirements of pointing accuracy and robustness [1–5]. All of them [6–8] used some fuzzy/neural control scheme to deal with at least two of these undesirable aspects: presence of inertia uncertainties, misalignment, unknown or external disturbance, vibration, actuator saturation, and faults, to ensure spacecraft stability. H∞ control theory is a comprehensive control theory that can take multiple performance requirements into consideration in the design and is suitable for such comprehensive control problems with multiple performance requirements. Currently, robust adaptive control, robust H∞ control, and μ synthesis control are mainly adopted to realize vibration control during the stable operation of flexible spacecraft [9–13].

Despite its synthesis advantages, H∞ design has engineering application limitations [14] mainly due to its high-order and complicated controller. Apart from high costs, it is a tremendous challenge to decompose a high-order and complicated controller into multiple low-complexity control structures based on experience in engineering practices [15,16], thus leading to a low feasibility of the practical use of traditional H∞ controllers. A new H∞ control method combining the advantages of the traditional H∞ control [15–17] was proposed by Apkarian in recent years, which factors into system performances in all aspects and overcomes the infeasibility of engineering applications of traditional H∞ controllers due to non-transparency and high complexity. The new method has attracted much attention and been widely applied since it was proposed [18–20]. The logic of the new structured H∞ control method is as follows: the structure of a controller is first designed according to actual needs and control objectives; on this basis, appropriate weighting functions are selected as per the specific performance requirements of the control object in order to form an H∞ performance matrix with multi-dimensional performance output; and finally, the structured H∞ controller with optimal parameters can be obtained through parameter optimization of the controller with a fixed structure [21–23].

In view of problems in integrated control of spacecraft with large flexible solar panels, based on the structured H∞ control design strategy and the specific resonance mode of the system, this paper proposes incorporating into the controller structure an internal mode controller, a notch filter, and a serial PID controller, which can achieve integrated control through multi-control module collaboration. A control solution that satisfies the comprehensive performance requirements is provided, thus reducing the project cost and implementation difficulty. Apart from the introduction, this paper deals with system modeling in the second part, structured H∞ controller design in the third part, performance simulation and analysis in the fourth part, and conclusions in the last part.

2. System Model Case and Control Analysis

The spacecraft with large flexible appendages is shown in Figure 1. It mainly comprises solar arrays that provide energy, velocity gyroscopes, precision guidance sensors, and star trackers that provide spacecraft attitude data, as well as a reaction wheel and an electromagnetic torque device for momentum management, and a digital computer.

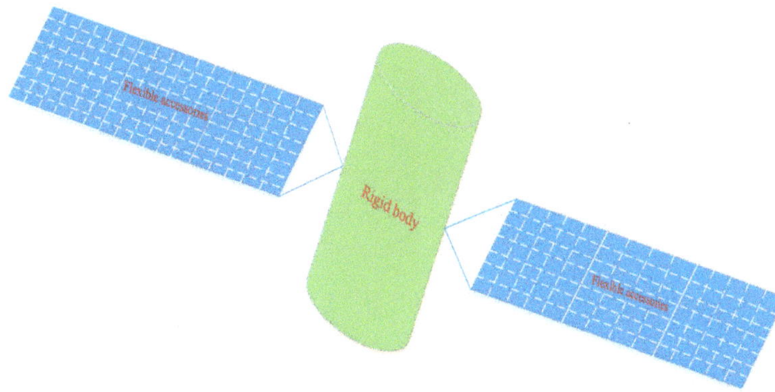

Figure 1. Spacecraft with large flexible appendages.

Only the pitch axis model that is most jitter-prone and the most important in the entire system is considered when constructing the simulation model. Official data show that the flexible spacecraft's pitch axis model is composed of a rigid body model and several flexible modules, as shown in the following formula:

$$\frac{\theta(s)}{u(s)} = \frac{1}{IS^2} + \sum_{i=1} \frac{K_i/I}{s^2 + 2\zeta w_i s + w_i^2} \tag{1}$$

In the formula, θ is the angular error of the pitch axis affected by the jitter of solar arrays; u is the given input of the pitch axis torque; s is the Laplace operator; I is the spacecraft pitch inertia constant with the value of 77,076 kg·m^2; ξ = 0.005 is the passive damping ratio constant of the system; Ki is the flexibility gain of the flexible module; and ω_i is the flexible frequency of the flexible module. The data are shown in Table 1.

Table 1. Flexible spacecraft system model parameters.

K_i/kg×m^2	ω_i/Hz
0.018	0.110
0.012	0.432
0.057	0.912
0.024	10.834
0.155	12.133
−1.341	13.201
−1.387	14.068
−0.806	14.285
−0.134	15.264

The Bode plot of the system without a controller is shown in Figure 2. The Bode plot shows that the cut-off frequency ω_c is only 0.16 Hz and the bandwidths of the system are very small, indicating poor interference suppression.

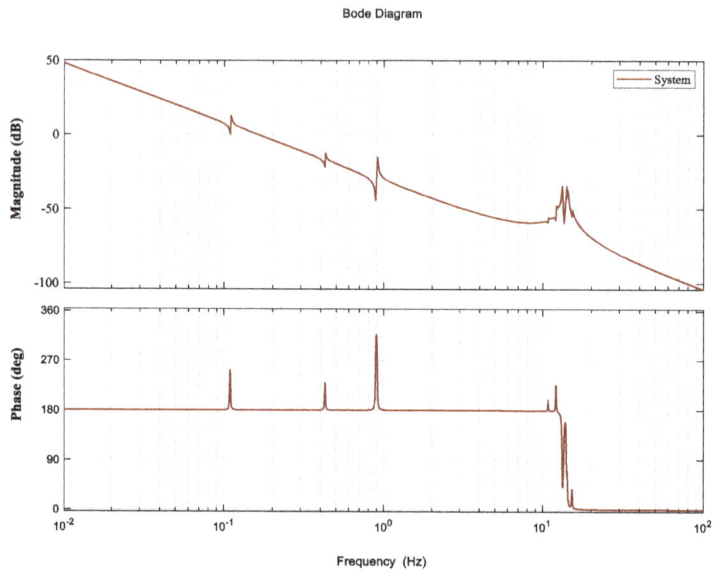

Figure 2. Bode plot without a control system.

Since they were put into operation, flexible spacecraft, as high-precision spacecraft, have never produced an output error of pitch axis exceeding 0.007 arcsec, which requires good control performance of the system. An effective controller should be good at disturbance suppression of solar panels and inherent flexibility suppression of the system, with a certain number of bandwidths.

The control redesign requirements can be stated as follows [24]:

1. Maintain at least 5 dB gain margin and 20 deg phase margin.
2. Provide at least 6 dB gain suppression (roll off) of the high-frequency spacecraft structural modes at 14 Hz.
3. Provide at least 20 dB additional disturbance attenuation at both 0.12 Hz and 0.66 Hz with respect to the original design.
4. Maintain the bandwidth (the open-loop gain crossover frequency) close to 1.5 Hz.

3. Design of Structured H∞ Control

In general, the complete design of structured control is divided into three steps: firstly, to design a structured controller according to design requirements and control objectives; secondly, to select and design proper weighting functions based on control objectives and performance requirements; and finally, to obtain a desired structured controller according to performance requirements and selected weighting functions.

In Section 3.1, the author briefly analyzes the control objectives and performance requirements in the control design of the large flexible spacecraft in this case and presents controller structure design in terms of the flexible modes and disturbance model of the system.

3.1. Structured Controller Setting

The main problems confronted in the control of large flexible spacecraft are as follows: first, the multiple low-frequency resonance modes of the system result in huge changes in amplitude characteristics; second, the inherent low-frequency interference caused by the complex launch environment and the high-altitude environment during orbit operation, as well as the flexible mode of the system, greatly limit the choice of bandwidths. The controller is designed to suppress the inherent low-frequency resonance disturbance of the system and achieve high pointing accuracy, while ensuring the bandwidth amount of the flexible system (robustness) and stability.

To meet the above control target, based on the flexible mode and inherent disturbance frequency of the system in this case, the authors design a structured integrated controller as shown in Figure 3.

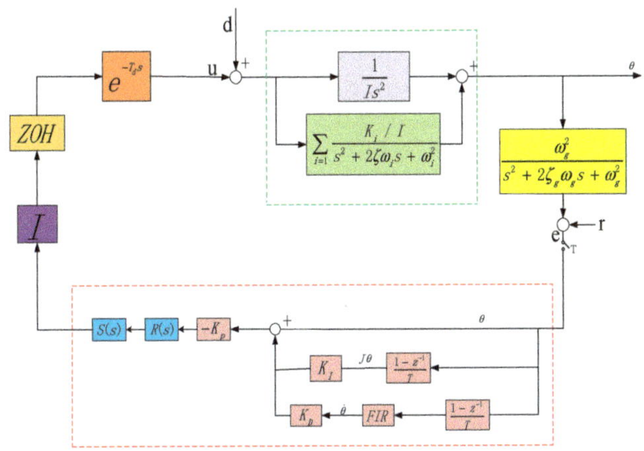

Figure 3. Control system block diagram.

The dotted line in Figure 3 shows the structured controller of the system. The red modules are for adjustable parameters, and the blue modules are fixed parameters.

The structured H∞ controller consists of three parts. The first part is an internal mode controller $S(s)$ designed for the system's resonance modes to suppress resonance disturbance.

$$S(s) = \frac{\left[\left(\frac{s}{87}\right)^2 + \frac{0.002s}{87} + 1\right]}{\left[\left(\frac{s}{45}\right)^2 + \frac{1.4s}{45} + 1\right]} \tag{2}$$

Its frequency characteristics are shown in Figure 4 and its purpose is to reduce system vibration by increasing the damping of the system for system vibration at 14 Hz.

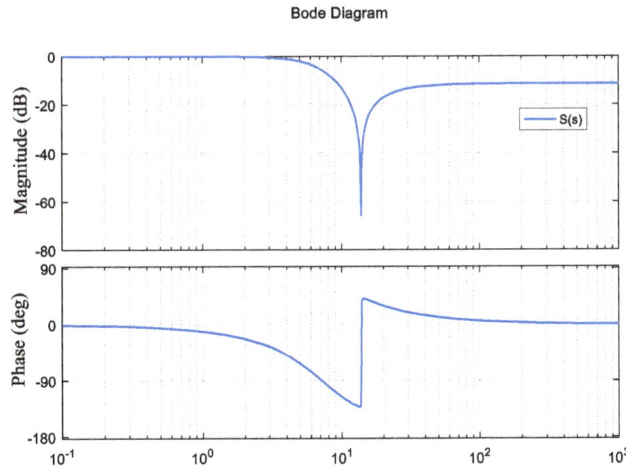

Figure 4. Internal mode controller $S(s)$.

The second part is a notch filter $R(s)$ set by the disturbance characteristics of the space environment to ensure sufficient suppression of the disturbance of solar panels at 0.12 Hz and 0.66 Hz on the premise of not affecting the stability of the frequency band in the system. The Bode plot is shown in Figure 5.

$$R(s) = \frac{\left[\left(\frac{s}{0.77872}\right)^2 + \frac{0.728s}{0.77872} + 1\right]}{\left[\left(\frac{s}{0.7536}\right)^2 + 1\right]} \times \frac{\left[\left(\frac{s}{3.84336}\right)^2 + \frac{0.254s}{3.84336} + 1\right]}{\left[\left(\frac{s}{4.1448}\right)^2 + 1\right]} \tag{3}$$

The third part is an adjustable parameter PID controller set to ensure the stability of the system, whose rate path is supplemented with an FIR filter to provide gain suppression.

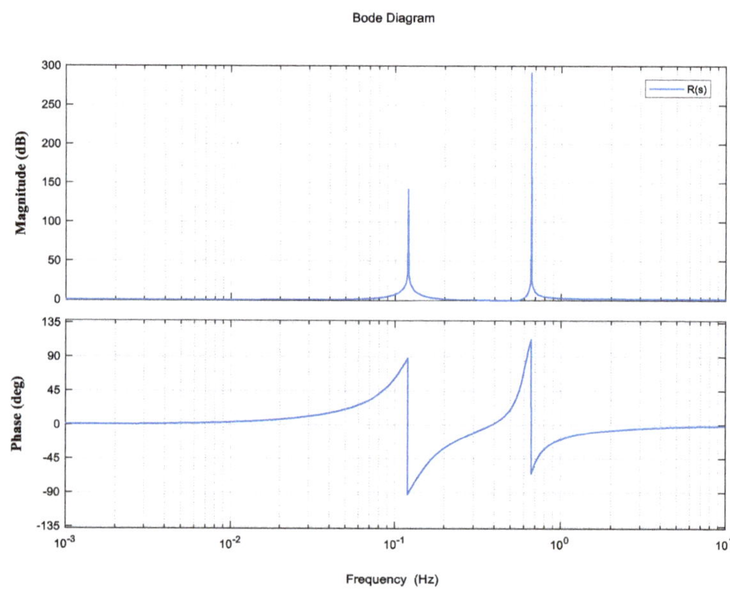

Figure 5. Notch filter $R(s)$.

3.2. Selection of Performance Weighting Functions

The following indicators need to be factored into the selection of weighting functions: the requirement of pointing accuracy; the stability and sensitivity of the system after the internal model controller of disturbance suppression is added; and the bandwidth limitations (robustness) of the flexible system.

By setting T_1 as the transfer function of $r - e$, the stability of the system is the distance from the transfer function T_1 to the critical operating point, which is also the upper limit of the gain of tracking performance, requiring:

$$\|W_1(s)T_1(s)\|_\infty \le \gamma \tag{4}$$

where γ is the norm index, and $W_1(s)$ is the weighting function. The upper limit of the design stability margin is 1, $W_1(s) = 1$.

T_2 is set as the transfer function from d to θ, which is the robust stability requirement of the system:

$$\|W_2(s)T_2(s)\|_\infty \le \gamma \tag{5}$$

where the weighting function $W_2(s) = 0.8$.

By setting T_3 as the transfer function from r to θ, the bandwidth requirement of the system is:

$$\|W_3(s)T_3(s)\|_\infty \le \gamma \tag{6}$$

In order to limit the bandwidth of the system, the weighting function $W_3(s)$ can be selected in the high-pass filter form as follows: The Bode diagram for $W_3(s)$ is shown in Figure 6.

$$W_3(s) = \frac{2s}{s+220} \tag{7}$$

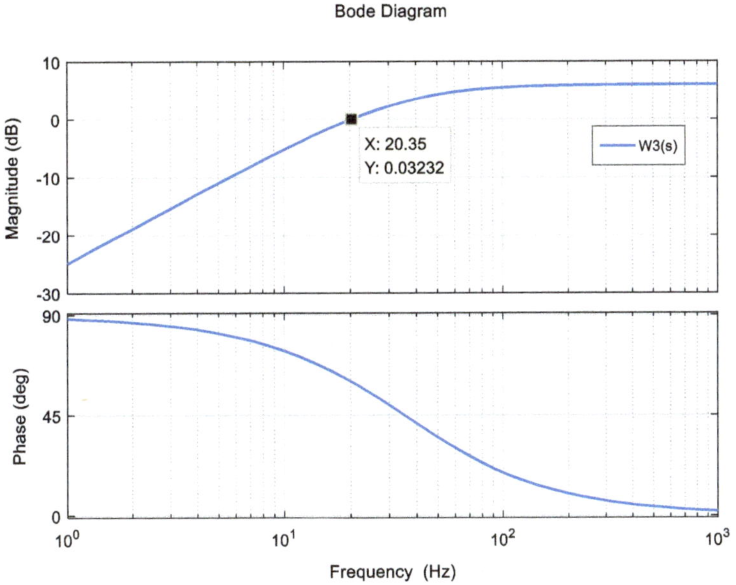

Figure 6. Bode plot of $W_3(s)$.

3.3. Parameter Optimization of Structured Controller

Based on comprehensive considerations of the performance requirements of the system and the implementation cost of the controller, the controller was designed as shown in Formula (8).

$$K(s) = K_P(1 + K_I/s + K_D s) \times R(s) \times S(s) \tag{8}$$

K_P, K_I, K_D is the to-be-optimized parameter.

Given the above analysis, and for the structured H_∞ optimization of the designed controller structure and the weighting function, the control performance requirements of the flexible system was considered comprehensively, and the minimum K_P, K_I, K_D, and the minimum value satisfying Formula (9) can be obtained by optimizing the adjustable parameters.

$$\|H\|_\infty \le \gamma \tag{9}$$

In formula $H = diag(\ W_1T_1 \quad W_2T_2 \quad W_3T_3\ )$, the adjustable parameter yielded is the optimal one for the system controller.

When seeking the optimal parameter of the structured controller H_∞, the linear fractional transformation (LFT) [18–24] was employed with T_1, T_2, and T_3 in Formulas (4)–(6), and the structured controller C with parameters was extracted and expressed in the following linear fraction forms to optimize the parameters:

$$\begin{cases} T_1 = F_l(\; P_1 \quad C \;) \\ T_2 = F_l(\; P_2 \quad C \;) \\ T_3 = F_l(\; P_3 \quad C \;) \end{cases} \tag{10}$$

In this case, the optimal parameters of the structured controller in Figure 1 yielded through repeated iterative calculations are as follows:

$$K_P = 8, K_I = 0.5, K_D = 0.95$$

The final controller is:

$$K(s) = 8(1 + 0.5/s + 0.95s) \times R(s) \times S(s) \tag{11}$$

4. Simulation Performance Analysis

Figure 7 shows the Bode plot of the unified open-loop frequency domain. According to Figure 7, the gain margin of the system is 5.17 dB; the phase margin is 22.5°; and the cut-off frequency is approximately 1.8 Hz, exceeding the required cut-off frequency of 1.5 Hz, which indicates that the stability and bandwidth requirements of the system have been met. For the system flexibility at 13 to 14 Hz, the controller provides about −15 to −55 dB gain suppression, which basically meets the suppression requirements for the inherent flexibility of the system.

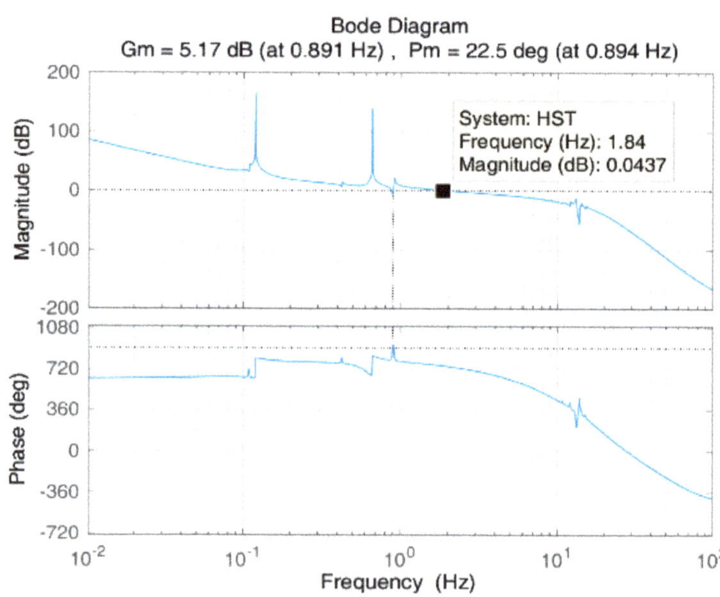

Figure 7. Open-loop Bode plot of the system.

Figure 8 is a closed-loop Bode plot of the system. Figure 9 is the time-domain response of the system. The figure shows that the proposed controller provides gain attenuation far higher than 20 dB for the jitter at 0.12 Hz or 0.66 Hz, which is also proved by the time-domain response of the system shown in Figure 7.

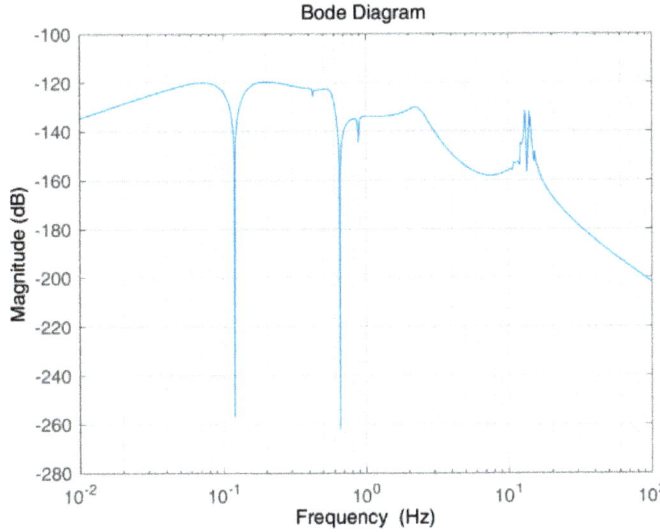

Figure 8. Closed-loop Bode plot of the system.

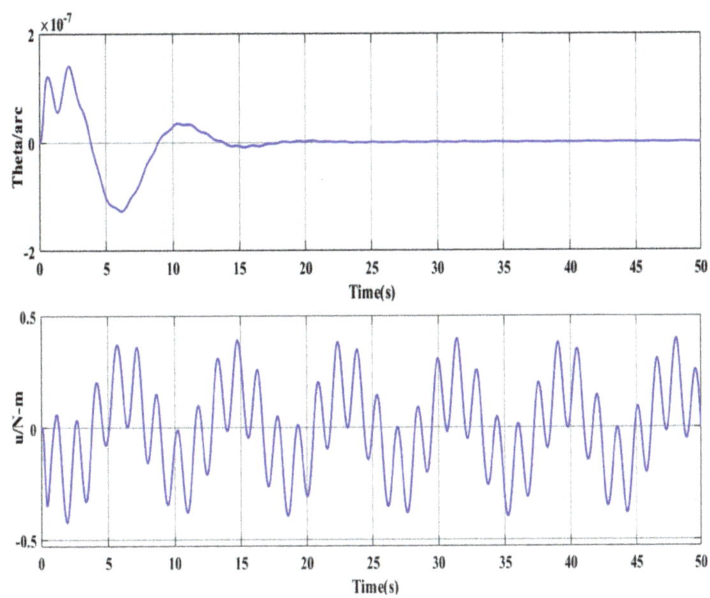

Figure 9. System time-domain response.

With the same weight function, the obtained H-inf controller is as follows:

$$K_h(s) = \frac{3.492 \times 10^7 \times (s + 0.6402)(s^2 + 0.002009s + 1.609 \times 10^{-6})(s^2 + 0.0003605s + 0.002128)}{(s + 0.01)^2(s^2 + 0.00022s + 0.002118)(s^2 + 3357s + 3.763 \times 10^{-6})} \quad (12)$$

The open-loop Bode plot of the H-inf controller is shown in Figure 10.

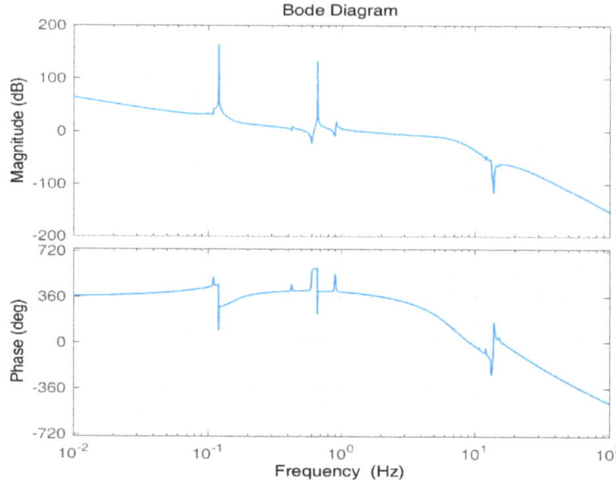

Figure 10. Open-loop Bode plot of H-inf controller.

The frequency domain response shown in Figure 11 shows that the structured integrated controller is excellent in vibration suppression for the flexible structure at 0.12 Hz and 0.66 Hz. The time-domain response shown in Figure 12 shows that under the same weight function, both controllers (structured integrated controller and H-inf controller) can completely suppress solar panel disturbance within a certain period, meet the control requirements, and present stability. For the structured integrated controller, the overshoot of the system is 2.101%; the peak time is 0.711 s; and the adjustment time is 13.562 s. For the traditional H-inf controller, the overshoot is 2.31%; the peak time is 0.876 s; and the adjustment time is 14.227 s. The data clearly show that the structured integrated controller has significantly better dynamic performance than the H-inf controller and the structured integrated controller has a lower order.

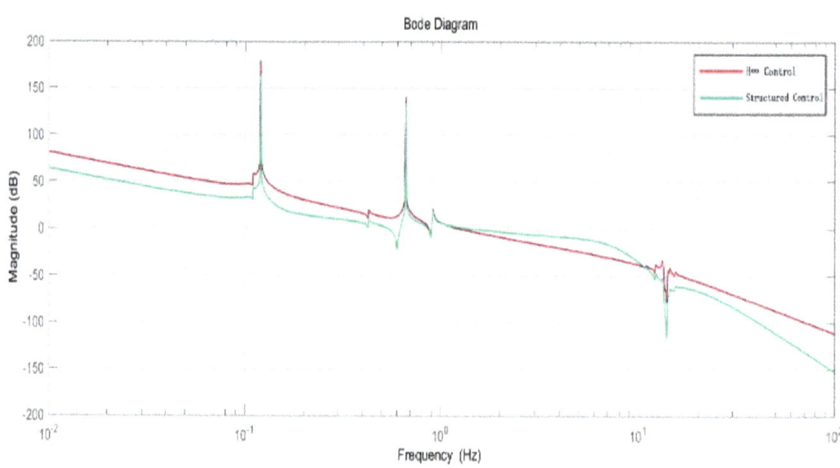

Figure 11. Open-loop simulation comparison between the structured integrated controller and traditional H-inf controller.

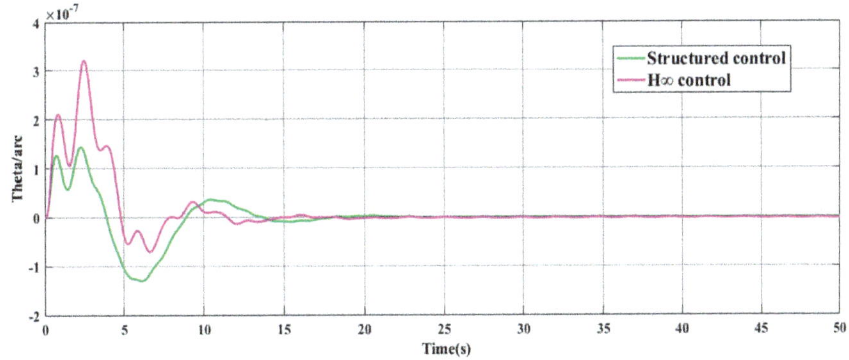

Figure 12. Simulation comparison of time-domain response between the structured integrated controller and traditional H-inf controller.

5. Conclusions

To solve the comprehensive control problems of spacecraft with large flexible appendages such as insufficient bandwidths, low system directivity accuracy, and flexible structure vibration, this paper proposes a structured integrated controller that satisfies control requirements by selecting appropriate weight functions. The simulation results indicate that the proposed controller can effectively suppress the vibration of large flexible appendages at 0.12 Hz and 0.66 Hz. While ensuring high pointing accuracy, the structured integrated controller can meet the requirements of attitude stability and bandwidths. Compared with the traditional H-inf controller, the proposed controller has the advantages of lower complexity and system order as well as lower engineering costs and implementation difficulty, with better control performance.

Author Contributions: Conceptualization, A.P. and Y.Z.; Methodology, A.P. and Y.Z.; Validation, Y.Z. and H.Z.; Data Curation, Y.Z.; Formal analysis Y.Z. and H.F.; Funding acquisition, A.P.; Writing—Original Draft Preparation, Y.Z.; Writing—Review and Editing, Y.Z. and H.F.; Visualization, Y.Z. All authors have read and agreed to the published version of the manuscript.

Funding: This work was funded in part by the Science and Technology Foundation of Guizhou Province (QKH [2020]1Y273, [2018]5781, [2018]5615), Guizhou Provincial Department of Education, Youth Talent Development Project, Qianke [2021] 100, Research on Robust Integrated Control of Flexible System, and the National Natural Science Foundation (NNSF) of China under Grant 51867005.

Conflicts of Interest: The authors declare no conflict of interest. The funders had no role in the design of the study; in the collection, analyses, or interpretation of data; in the writing of the manuscript, or in the decision to publish the results.

References

1. Cao, D.; Bai, K.; Ding, H.; Zhou, X.; Pan, Z.; Chen, L.; Zhan, S. Advances in dynamics and vibration control of large-scale flexible spacecraft. *Chin. J. Theor. Appl. Mech.* **2019**, *51*, 1–13. (In Chinese)
2. Gennaro, S.D. Passive Attitude Control of Flexible Spacecraft from Quaternion Measurements. *J. Optim. Theory Appl.* **2003**, *116*, 41–60. [CrossRef]
3. Song, B.; Yan, G.; Li, B.; Zheng, P. Active Disturbance Rejection Control for Flexible Spacecraft High-Accuracy Attitude Pointing. *Aerosp. Shanghai* **2014**, *31*, 1–7, 36.
4. Wie, B.; Liu, Q.; Bauer, F. Classical and robust H(infinity) control redesign for the Hubble Space Telescope. *J. Guid. Control Dyn.* **1993**, *16*, 1069–1077. [CrossRef]
5. Cao, H.; Zhu, Q.; Sun, J.; Yu, C. Precise Orientation Control of Large Flexible Spacecraft. *Flight Control Detect.* **2019**, *2*, 17–24. (In Chinese)
6. Li, A.; Liu, M.; Shi, Y. Adaptive sliding mode attitude tracking control for flexible spacecraft systems based on the Takagi-Sugeno fuzzy modelling method. *Acta Astronaut.* **2020**, *175*, 570–581. [CrossRef]
7. Tao, J.; Zhang, T.; Liu, Q. Novel finite-time adaptive neural control of flexible spacecraft with actuator constraints and prescribed attitude tracking performance. *Acta Astronaut.* **2021**, *179*, 646–658. [CrossRef]

8. Golestani, M.; Esmaeilzadeh, S.M.; Mobayen, S. Fixed-time control for high-precision attitude stabilization of flexible spacecraft. *Eur. J. Control* **2021**, *57*, 222–231. [CrossRef]
9. Sun, J.; Li, S.; Huang, J.; Zhu, D. Robust coordinated control for large flexible spacecraft based on consensus theory. *J. Frankl. Inst.* **2020**, *357*, 5359–5379. [CrossRef]
10. Tang, L.; Guo, Z.; Guan, X.; Wang, Y.; Zhang, K. Integrated control method for spacecraft considering the flexibility of the spacecraft bus. *Acta Astronaut.* **2020**, *167*, 73–84. [CrossRef]
11. Qian, L.I.; Yang, H.; Zhao, D.; Jiang, B. Fault-tolerant control and vibration suppression of flexible spacecraft: An interconnected system approach. *Chin. J. Aeronaut.* **2020**, *33*, 2014–2023.
12. Hu, Q. Robust adaptive sliding-mode fault-tolerant control with L2-gain performance for flexible spacecraft using redundant reaction wheels. *IET Control Theory Appl.* **2010**, *4*, 1055–1070. [CrossRef]
13. Qiao, J.; Li, X.; Xu, J. A composite disturbance observer and H∞ control scheme for flexible spacecraft with measurement delay and input delay. *Chin. J. Aeronaut.* **2019**, *32*, 1472–1480. [CrossRef]
14. Gahinet, P.; Apkarian, P. Frequency-domain tuning of fixed-structure control systems. In Proceedings of the UKACC International Conference on Control, Cardiff, UK, 3–5 September 2012; pp. 178–183.
15. Apkarian, P.; Noll, D. Structured H-infinity control of infinite dimensional systems. *Int. J. Robust Nonlinear Control* **2017**. [CrossRef]
16. Apkarian, P. Tuning controllers against multiple design requirements. In Proceedings of the 16th International Conference on System Theory, Control and Computing (ICSTCC), Sinaia, Romania, 12–14 October 2012; pp. 3888–3893.
17. Apkarian, P.; Dao, M.N.; Noll, D. Parametric Robust Structured Control Design. *IEEE Trans. Autom. Control* **2015**, *60*, 1857–1869. [CrossRef]
18. Gahinet, P.; Apkarian, P. Structured H∞ Synthesis in MATLAB. *IFAC Proc. Vol.* **2011**, *44*, 1435–1440. [CrossRef]
19. da Silva de Aguiar, R.S.; Apkarian, P.; Noll, D. Structured Robust Control Against Mixed Uncertainty. *IEEE Trans. Control Syst. Technol.* **2017**, *26*, 1771–1781. [CrossRef]
20. Gahinet, P.; Apkarian, P. Decentralized and fixed-structure H∞ control in MATLAB. In Proceedings of the 50th IEEE Conference on Decision and Control and European Control Conference, Orlando, FL, USA, 12–15 December 2011; pp. 8205–8210.
21. Apkarian, P.; Noll, D. Nonsmooth H∞ synthesis. In Proceedings of the WSEAS International Conference on Automatic Control, Modeling and Simulation, Rio de Janeiro, Brazil, 30 May–3 June 2005; pp. 430–435.
22. Apkarian, P.; Noll, D. Non-smooth structured control design with application to PID loop-shaping of a process. *Int. J. Robust Nonlinear Control* **2007**, *17*, 1320–1342. [CrossRef]
23. Apkarian, P.; Noll, D. Nonsmooth Optimization for Multidisk H∞ Synthesis. *Eur. J. Control* **2006**, *12*, 229–244. [CrossRef]
24. Sharkey, J.P. *A SAGA/Ga Controller with Increased Attenuation at 0.6 Hz*; Internal Memo, NASA Marshall Space Flight Center: Huntsville, AL, USA, 1991.

entropy

Article

Improved Adaptive Augmentation Control for a Flexible Launch Vehicle with Elastic Vibration

Aiping Pang [1,2], Hongbo Zhou [1,2], Wenjie Cai [3] and Jing Zhang [1,*]

1 College of Electrical Engineering, Guizhou University, Guiyang 550025, China; appang@gzu.edu.cn (A.P.); hit_zhb@163.com (H.Z.)
2 Guizhou Provincial Key Laboratory of Internet + Intelligent Manufacturing, Guiyang 550025, China
3 China Academy of Launch Vehicle Technology, Beijing 100076, China; caiwenjie217@126.com
* Correspondence: zhangjingguizhou@126.com

Abstract: The continuous development of spacecraft with large flexible structures has resulted in an increase in the mass and aspect ratio of launch vehicles, while the wide application of lightweight materials in the aerospace field has increased the flexible modes of launch vehicles. In order to solve the problem of deviation from the nominal control or even destabilization of the system caused by uncertainties such as unknown or unmodelled dynamics, frequency perturbation of the flexible mode, changes in its own parameters, and external environmental disturbances during the flight of such large-scale flexible launch vehicles with simultaneous structural deformation, rigid-elastic coupling and multimodal vibrations, an improved adaptive augmentation control method based on model reference adaption, and spectral damping is proposed in this paper, including a basic PD controller, a reference model, and an adaptive gain adjustment based on spectral damping. The baseline PD controller was used for flight attitude control in the nominal state. In the non-nominal state, the spectral dampers in the adaptive gain adjustment law extracted and processed the high-frequency signal from the tracking error and control-command error between the reference model and the actual system to generate the adaptive gain. The adjustment gain was multiplied by the baseline controller gain to increase/decrease the overall gain of the system to improve the system's performance and robust stability, so that the system had the ability to return to the nominal state when it was affected by various uncertainties and deviated from the nominal state, or even destabilized.

Keywords: multiplicative adaptation; gain adjustment; spectral damping; robust stability

Citation: Pang, A.; Zhou, H.; Cai, W.; Zhang, J. Improved Adaptive Augmentation Control for a Flexible Launch Vehicle with Elastic Vibration. *Entropy* **2021**, *23*, 1058. https://doi.org/10.3390/e23081058

Academic Editors: Quanmin Zhu, Giuseppe Fusco, Jing Na, Weicun Zhang and Ahmad Taher Azar

Received: 16 June 2021
Accepted: 10 August 2021
Published: 16 August 2021

1. Introduction

As the exploration of the space environment progresses, the missions of spacecraft exploration become more and more diversified, and as the application of polymer materials in the space field progresses, the structure of spacecraft is gradually developing towards large and flexible structures. In order to carry these large and flexible-structure spacecraft, launch vehicles with a large carrying capacity have become an inevitable requirement of space-development strategies [1]. At the same time, the lightweight polymer material used in the body of launch vehicles has increased the flexible mode of the vehicles, leading to the presence of structural deformation, rigid body-elastic vibration coupling, multi-modal vibration, and other characteristics of the body at the same time. These factors make the attitude control of launch vehicles subject to oscillations and difficult to attenuate, or even lead to system instability, which poses a new challenge to the reliability and robustness of launch vehicles [2].

For high-risk aerospace applications, both government and industry rely heavily on classical control theory, and gain-scheduling PID control is still the mainstream control method for current launch vehicles, due to the advantages of its simple structure, good anti-interference, and ease of analysis in the time domain (or frequency domain). Typical applications include the Saturn V and Space Shuttle of the United States, the Ariane of

Europe, and the Long March series of launch vehicles of China [1]. The application of classical control theory in launch vehicles has matured, and is well verified with few failures. Although the classical control methods can usually meet the flight requirements, the traditional gain-scheduling PID control is no longer able to meet the control requirements of launch vehicles due to their increasing mass and aspect ratio, the increase in the flexible structure of their components, and the obvious influence of the flexible mode and elastic vibration of large launch vehicles, and is unable to cope with the control instability problems caused by the excessive interference and modal uncertainty during the flight. Due to the unknown and unmodeled dynamics, external perturbations of the flight environment, and changes in its own parameters, the attitude of a large flexible launch vehicle will inevitably generate errors during flight [3]. Traditional controllers are usually designed with high gain to suppress attitude errors, but excessive gain during full flight will easily cause the control commands to vibrate, and the flexible mode will also generate a series of vibration signals, which will affect the control effect. There are two main solutions to the problem of elastic-vibration suppression in the design of launch vehicle control systems: one is to design controllers for different characteristics using robust control theory, and the second is to suppress the elastic vibration signal by designing a notch filter. Unlike the robust controller design, the notch filter does not require significant changes to the original rocket control system to deal with elastic vibration. The zero point of the notch filter is used to eliminate the high-frequency pole of the elastic rocket system and determine the frequency center according to the system requirements of the filter. The method of using a notch filter to suppress the elastic vibration of a projectile is widely used in the design of the launcher attitude-control system. The determination of the parameters of the notch filter is the key to the design, and the location of the zero point of the notch filter; i.e., the frequency center, can be determined after the transfer function of the elastic launch vehicle is determined with sufficient accuracy in the model [4,5]. In order to solve the low-frequency, dense-frequency elastic vibration modes appearing in the launch vehicle, some scholars adopted the method of attitude control of the flexible launch vehicle by adaptive control of the adaptive notch, and the adaptive controller of the adaptive notch filter successfully stabilized the uncertain and time-varying equations of the launch vehicle model dynamics through thrust vector control [6]. Another scholar designed a bending mode filter for the whole system, which had a better filtering function for low-frequency, dense-frequency modes, and achieved good control results [7].

In response to the limitations of the classical approach, in order to increase the robust stability of the launch vehicle attitude-control system, many scholars began to work on advanced control methods, and since 1990, NASA has developed a variety of launch vehicle control techniques in the Advanced Guidance Control program, including trajectory linearization control methods, neural network adaptive control methods, and higher-order sliding-mode control methods; the development of such advanced controls has the potential to improve system performance and increase robustness [8–10]. Classical adaptive-control concepts were proposed for attitude-control systems applied to rockets [11–13]. However, many adaptive-control concepts are not feasible when applied to high-risk aerospace systems due to the stringent flight environment. In addition, many adaptive techniques appearing in the literature are not applicable to conditionally stable systems with complex flexible dynamics. Therefore, researchers have optimized model-referenced adaptive control for these situations and proposed an Adaptive Augmentation Control (AAC) [14], widely used in launch-vehicle and missile-longitudinal control in recent years. AAC a gain-adjustment method based on a model-referenced adaptive-control design that generates adaptively adjusted gain from the generalized error between the reference model and the actual system as a supplement to the nominal controller. Orr et al. introduced a scheme for adaptive control of multiplicity applicable to rockets [3], and then improved the adaptive-control scheme [15] to improve the performance of the original method with higher sensitivity to external inputs. The method was developed by the NASA Marshall Space Center (MSFC) and became a major part of the U.S. Space Launch System (SLS) to adapt to unpredictable external environmental disturbances and a variety of flight dynamics characteristics (elastic vibration of flexible modes, control structure coupling, servo delay, etc.) and to reduce the probability of flight destabilization [15,16]. NASA included the method in the development of the flight control system for the SLS program in early 2013, and tested the designed method in the F/18-A to verify the resilience of the control system in adverse flight conditions [17,18]. Brinda et al. performed an adaptive gain-adjustment controller design for the longitudinal channel of a two-stage launch vehicle using a Chebyshev high-pass filter to improve the problem of insufficient amplitude of the low-frequency part of the control signal in the original adaptive gain-adjustment structure of the low-order high-pass filter [19,20]; however, there were equal-amplitude ripples in the passband of the Chebyshev filter. Zhang applied a fault-tolerant control method and adaptive vibration frequency recognition method to AAC, and designed a corresponding correction network based on an SMM algorithm to identify each order vibration frequency to improve system control performance and stability [21]. Cui Naigang et al. applied an interference compensation control loop and active load reduction control loop based on the dilated state observer to the adaptive gain adjustment structure, and performed a simulation analysis of the pitch channel control [22]. To enhance the robustness to changes in elastic modal parameters, Domenico Trotta integrated the AAC control architecture with adaptive notch filters and proposed two novel and effective tuning methods for adaptively enhanced control systems, which were optimized by robust design and solved by genetic algorithms to achieve continuous improvement in the performance and robustness of standard launch vehicle single-axis attitude controllers in atmospheric flight [6,23]. Diego Navarro designed two adaptive augmentation control laws using a robust control design (structured H∞ control) as a baseline controller to improve the robust performance of AAC control, while analyzing the effect of the adaptive action on the classical stability margin, and validated this analysis using nonlinear time-domain stability margin evaluation techniques [24]. However, if the expansion state observer is not properly selected, the observer is easily affected by the noise signal, or even diverges when there is additional measurement noise caused by elastic vibration and other additional dynamics. However, advanced nonlinear stabilization techniques to reduce the error by increasing the control gain are not feasible for aerospace systems with high complexity; meanwhile, the above-mentioned adaptive gain-adjustment control scheme features complicated algorithms, costly computation, and challenging implementation, which introduce unknown risks to the actual system.

In this paper, we aim to establish a rigid-bullet coupling model of a large flexible spacecraft with second-order vibration signals and design an improved adaptive augmentation control method based on the reference model adaptive control method to address additional dynamics issues such as increased attitude-tracking errors and flexible-mode elastic vibrations caused by uncertainty (modeling uncertainty, frequency perturbation of the flexible mode) and external environmental interference during ascent of a large launch vehicle with a flexible mode. The scheme first determines the adjustment threshold of forward gain on the basis of the baseline PD controller, and takes the tracking error and control command error signals as the input of the two channels of the adaptive control law. The spectral dampers in the two channels (tracking error and control command elastic vibration) process the error signals (the high-pass filter extracts the high-frequency signal of a specific frequency from the error signal, and the low-frequency filter lowers the frequency to reduce the influence of the high-frequency signal on the actuator) to produce the corresponding suppression gain (error-suppression gain and elastic-suppression gain) to form the overall gain of the AAC, and increase or decrease the forward gain of the system to improve the control performance of the system. The AAC controller will not affect the PD controller when the basic PD controller is able to handle the control tasks better [3,15], whereas the AAC controller will adjust the adaptive gain to achieve the overall gain of the system when the impact of external perturbations and uncertainties is significant, so as to recover the system performance when the system deviates severely from the nominal state and meet the performance requirements of attitude control and robust stability of the large flexible launch vehicle in the flight process.

This paper is organized as follows. Section 2 introduces the rigid-bullet coupling model of a large flexible launch vehicle with a second-order elastic vibration signal. Section 3, on controller design, details the improved adaptive augmented control scheme: (1) based on the rigid-bullet coupling model of a large flexible launch vehicle, the base-line PD controller is given and the maximum critical value of the forward gain tunable is determined after analyzing the flexible mode of the system in the frequency domain; (2) Two error signals are selected between the reference model and the actual system as input signals for the two channels of the adaptive control law, thereby increasing/decreasing the overall forward gain of the system; and (3) we introduce the role of the spectrum damper and parameter selection (mainly the extraction of high-frequency signal and low-frequency output for the error input signal of two channels). Section 4 presents a simulation analysis of the improved adaptive augmentation control designed in this paper, and the simulation results of several launch vehicle runaway scenarios are presented and discussed. By comparing the traditional PD control with the adaptive augmentation controller containing the baseline PD controller, we observed that the adaptive augmentation control improved the performance and robust stability of the large flexible launch vehicle during flight.

2. Mathematical Model of the Launch Vehicle with Second-Order Vibration Modes

The coupling problem between the rigid body motion and elastic vibration of large launch vehicles is more prominent than that of medium-sized and small launch vehicles. The motion process is more complicated due to the large mass of large launch vehicles, the increase in the aspect ratio, and the complex interference and uncertainty during flight, so the elastic vibration mode cannot be neglected. The rigid-bullet coupling mathematical model of the launch vehicle was established based on the forces (gravity, aerodynamic, thrust, control, etc.), moments, and vibration factors during the ascent of the launch vehicle.

In the velocity coordinate system, the translational and rotational motion of the launch vehicle around the centroid (pitch channel) can be expressed as (1) and (2). Considering the plane bending vibration and torsional elastic vibration of the arrow body, the general elastic vibration equation can be obtained by using the modal superposition method and the orthogonality of the vibration pattern, as shown in (3).

$$\begin{aligned} mV\dot{\theta}\cos\sigma = \; & -mg\cos\theta + qS_m C_y^{\alpha}\alpha - qS_m \sum_{i=1}^{n}\left(q_{iy}\int_{x_0}^{x_l} C_{yx}^{\alpha} R_{iz}(x)dx\right) + \tfrac{qS_m}{V} C_y^{\overline{\omega}}(x_d - x_z)\omega_z \\ & - \tfrac{qS_m}{V}\sum_{i=1}^{n}\left(\dot{q}_{iy}\int_{x_0}^{x_l} C_{yx}^{\alpha} U_{iy}(x)dx\right) + P\sin\alpha + \tfrac{P}{2}\delta_\varphi\cos\alpha - P\cos\alpha\sum_{i=1}^{n} R_{iz}(x_R)q_{iy} + 2m_R l_R\ddot{\delta}_\varphi + F_y \end{aligned} \tag{1}$$

$$\begin{aligned} J_z\dot{\omega}_z + (J_y - J_x)\omega_x\omega_y = \; & -2m_R l_R(x_R - x_z)\ddot{\delta}_\varphi - 2m_R l_R \dot{W}_x\delta_\varphi - \tfrac{P}{2}(x_R - x_z)\delta_\varphi - P\sum_{i=1}^{n} U_{iy}(x)q_{iy} \\ & - P(x_R - x_z)\sum_{i=1}^{n} R_{iz}(x)q_{iy} - qS_m C_y^{\alpha}(x_d - x_z)\alpha + qS_m\sum_{i=1}^{n}\left(q_{iy}\int_{x_0}^{x_l} C_{yx}^{\alpha}(x - x_z)R_{iz}(x)dx\right) \\ & - \tfrac{qS_m}{V} m_z^{\overline{\omega}_z} l^2\omega_z + \tfrac{qS_m}{V}\sum_{i=1}^{n}\left(\dot{q}_{iy}\int_{x_0}^{x_l} C_{yx}^{\alpha}(x - x_z)U_{iy}(x)dx\right) + M_z \end{aligned} \tag{2}$$

$$\begin{aligned} \ddot{q}_{iy} + 2\xi_i\omega_i\dot{q}_{iy} + \omega_i^2 q_{iy} = \; & \left(-P\sum_{j=1}^{n} R_{jz}(x)q_{jy} + \tfrac{P}{2}\delta_\varphi\right)U_{iy}(x_R) + 2m_R l_R\ddot{\delta}_\varphi U_{iy}(x_R) \\ & + \left(2m_R l_R\ddot{\delta}_\varphi + 2m_R l_R\dot{W}_x\delta_\varphi\right)R_{iz}(x_R) + qS_m\left(\int_{x_0}^{x_l} C_{yx}^{\alpha} U_{iy}(x)dx\right)\alpha \\ & - qS_m\sum_{j=1}^{n}\left(q_{jy}\int_{x_0}^{x_l} C_{yx}^{\alpha} U_{iy}(x)R_{jz}(x)dx\right) + \tfrac{qS_m}{V}\left(\int_{x_0}^{x_l} C_{yx}^{\alpha}(x - x_z)U_{iy}(x)dx\right)\omega_z \\ & - \tfrac{qS_m}{V}\sum_{j=1}^{n}\left(\dot{q}_{jy}\int_{x_0}^{x_l} C_{yx}^{\alpha} U_{iy}(x)U_{jy}(x)dx\right) \end{aligned} \tag{3}$$

The identification and meaning of the parameters in the above formula are shown in Table 1.

Table 1. Parameters and identification.

Notation	Identification
m	Mass of the launch vehicle
g	Gravitational acceleration
P	Total engine thrust
V	Arrow speed
C_y^α	Lifting coefficient
q	Motive pressure
S_m	Characteristic area of arrows
θ	Trajectory inclination
α	Attack angle
σ	Heading (angle of course)
J_z	Moment of inertia
ω_z	Pitch velocity
$m_z^{\overline{\omega}_z}$	Pitch damping torque factor
δ_φ	Pitch channel motor pendulum
x_R	Distance from thrust point to arrow tip
x_z	Distance from mass centre to tip of arrow body
x_d	Distance from pneumatic core to arrow tip
l	Arrow length
m_R	Quality of each engine
l_R	Distance from engine pendulum to pivot
$\dot{W}_x$	Radial apparent acceleration of rocket
F_y	Interference
M_z	Interference torque
U_{iy}	Micro-displacement of plane bending vibration
R_{iy}	Micro-element corner of distortion elastic vibration
q_{iy}	The ith-order oscillation pattern of the pitch channel
ω_i	The ith-order oscillation angle frequency
ξ_i	The ith-order oscillation damping ratio

The nominal controller of the control system of the large flexible launch vehicle designed in this paper was based on a small perturbation linearization model, so the mathematical model of the rigid-bullet coupling of the large flexible launch vehicle was simplified to a small perturbation linearization model, as in (4) [25,26]:

$$\begin{aligned}
&\Delta\dot{\theta} = c_1\Delta\alpha + c_2\Delta\theta + c_3\delta_\varphi + c_3{}''\ddot{\delta}_\varphi + \sum_{i=1}^{n} c_{1i}\dot{q}_i + \sum_{i=1}^{n} c_{2i}q_i + \overline{F}_y \\
&\Delta\ddot{\varphi} + b_1\Delta\dot{\varphi} + b_2\Delta\alpha + b_3\delta_\varphi + b_3{}''\ddot{\delta}_\varphi + \sum_{i=1}^{n} b_{1i}\dot{q}_i + \sum_{i=1}^{n} b_{2i}q_i = \overline{M}_z \\
&\Delta\varphi = \Delta\theta + \Delta\alpha \\
&\ddot{q}_{iy} + 2\xi_i\omega_i\dot{q}_{iy} + \omega_i^2 q_{iy} = D_{1i}\Delta\dot{\varphi} + D_{2i}\Delta\alpha + D_{3i}\delta_\varphi + D_{3i}{}''\ddot{\delta}_\varphi + \sum_{j=1}^{n} D_{ij}\dot{q}_{jy} + \sum_{j=1}^{n} D_{ij}q_{jy}
\end{aligned} \quad (4)$$

In addition, the period of the arrow body centroid motion is much longer than the period of the pitch attitude angle motion, so the impact of the arrow body centroid motion can be ignored in the study of the arrow body attitude angle motion, and when also ignoring the influence of each oscillation pattern in the elastic vibration equation, then (4) can be further simplified as follows:

$$\begin{aligned}
&\Delta\ddot{\varphi} + b_1\Delta\dot{\varphi} + b_2\Delta\varphi + b_3\delta_\varphi + \sum_{i=1}^{n} b_{1i}\dot{q}_i + \sum_{i=1}^{n} b_{2i}q_i = \overline{M}_z \\
&\Delta\varphi = \Delta\alpha \\
&\ddot{q}_{iy} + 2\xi_i\omega_i\dot{q}_{iy} + \omega_i^2 q_{iy} = D_{1i}\Delta\dot{\varphi} + D_{2i}\Delta\varphi + D_{3i}\delta_\varphi + \overline{Q}_{iy}
\end{aligned} \quad (5)$$

where $\overline{Q}_{iy}$ is the generalized disturbance force on the elastic vibration of the ith order.

When the attitude angle and angular rate signals of the arrow body are obtained through the attitude measuring element, there is an additional elastic vibration signal in the obtained measurement signal influenced by the elastic vibration of the arrow body, and the actual measurement signal is as follows (6):

$$\begin{aligned} \Delta\varphi_T &= \Delta\varphi - \sum_i W'_i(x_T) q_i \\ \Delta\dot{\varphi}_{gT} &= \Delta\dot{\varphi} - \sum_i W'_i(x_{gT}) \dot{q}_i \end{aligned} \tag{6}$$

where $W'_i(x_T)$ is the slope of the ith-order oscillation pattern of the attitude angular measuring element at the mounting position x_s, and $W'_i(x_{gT})$ is the slope of the ith order vibration pattern at the installation of the attitude angular rate measurement element.

In this paper, we considered the design of a nominal controller for a second-order elastic vibration model of a launch vehicle, with the input, state variable, and output defined as:

$$u = \delta_\varphi,\ y = \begin{bmatrix} \Delta\varphi & \Delta\dot{\varphi} \end{bmatrix}^T,\ x = \begin{bmatrix} \Delta\varphi & \Delta\dot{\varphi} & q_1 & q_2 & \dot{q}_1 & \dot{q}_2 \end{bmatrix}^T$$

Here we first ignore the effect of external perturbations and build the standard state space model. External perturbations will be added to the control input signal and explained in Section 4. The state space of the system is described by (7):

$$\begin{aligned} \dot{x} &= Ax + Bu \\ y &= Cx + Du \end{aligned} \tag{7}$$

where the matrix A, B, C, D is given by (1)–(5),

$$A = \begin{bmatrix} 0 & 1 & 0 & 0 & 0 & 0 \\ -b_2 & -b_1 & -b_{21} & -b_{22} & -b_{11} & -b_{12} \\ 0 & 0 & 0 & 0 & 1 & 0 \\ 0 & 0 & 0 & 0 & 0 & 1 \\ D_{21} & D_{11} & -\omega_1^2 & 0 & -2\xi_1\omega_1 & 0 \\ D_{22} & D_{12} & 0 & -\omega_2^2 & 0 & -2\xi_2\omega_2 \end{bmatrix} B = \begin{bmatrix} 0 \\ -b_3 \\ 0 \\ 0 \\ D_{31} \\ D_{32} \end{bmatrix}$$

$$C = \begin{bmatrix} 1 & 0 & -W'_1(X_T) & -W'_2(X_T) & 0 & 0 \\ 0 & 1 & 0 & 0 & -W'_1(X_{gT}) & -W'_2(X_{gT}) \end{bmatrix} D = \begin{bmatrix} 0 \\ 0 \end{bmatrix}$$

The data selected in this paper are shown in Table 2.

Table 2. Parameters and values.

Parameters	Values	Parameters	Values	Parameters	Values
b_1	0.6348	D_{11}	0.6407	ω_1	3.5906
b_2	−0.0286	D_{12}	−27.2466	ω_2	7.6941
b_3	1.1530	D_{21}	−1.6713	ξ_1	0.005
b_{11}	-3.2693×10^{-5}	D_{22}	−3.6153	ξ_2	0.005
b_{12}	0.0015	D_{31}	−3.5752	$W_1'(X_T)$	−0.015
b_{21}	7.2608×10^{-4}	D_{32}	−142.71	$W_2'(X_T)$	2×10^{-4}
b_{22}	0.0029	$W_1'(X_{gT1})$	0.01	$W_2'(X_{gT1})$	0.004

3. Adaptive Augmentation Controller Design

The adaptive augmentation controller combines the adaptive controller with a classically designed linear control system using a multiplicative forward gain that enhances the system by adjusting the total loop gain in real time based on the error between the actual output and the output of the reference model. When the baseline controller performs well, the adaptive augmentation controller produces little enhancement. When the baseline controller is unable to effectively meet the performance requirements, the adaptive augmentation controller adjusts the total gain of the system by increasing/decreasing the adaptive gain to improve the performance of the control system; when the system is in a high degree of uncertainty or deviates from the nominal system, the adaptive controller can compensate the PD baseline controller to a greater extent to avoid system instability.

Figure 1 shows the adaptive augmentation control block diagram, which mainly consists of two parts: the PD-based baseline controller and the adaptive controller composed of the reference model and the adaptive law.

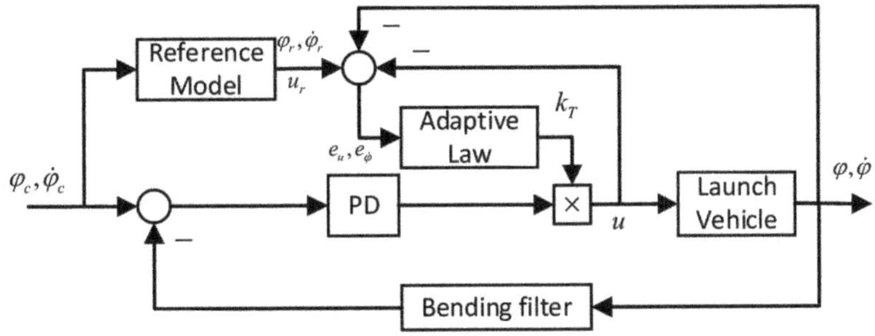

Figure 1. The adaptive augmentation control system.

3.1. Reference Model

The reference model was used to simulate the controlled motion of the rigid body of the launch vehicle in the nominal state, which produced the nominal response to the guidance instruction by adjusting the control parameters, and then the gap with the actual response of the launch vehicle was used as the input of the adaptive control law to adjust the gain of the PD controller. In adaptive gain control, a typical second-order system is used as the reference model [27], and the reference model used in this paper was obtained by neglecting the elastic vibrations in (5); the model's state space is given in (8):

$$\begin{aligned} \dot{x}_r &= A_r x_r + B_r u_r \\ y_r &= C_r x_r + D_r u_r \end{aligned} \tag{8}$$

where $A_r = \begin{bmatrix} 0 & 1 \\ -b_2 & -b_1 \end{bmatrix}$, $B_r = \begin{bmatrix} 0 & -b_3 \end{bmatrix}^T$, $C_r = \begin{bmatrix} 1 & 0 \\ 0 & 1 \end{bmatrix}$, $D_r = \begin{bmatrix} 0 \\ 0 \end{bmatrix}$.

3.2. Baseline PD Controller

In the stability of elastic vibration, there is a difference between amplitude stability and phase stability. The so-called amplitude stability refers to the amplitude $Gain(\omega) < 0$, when the phase-frequency curve crosses $\pm(2\text{n}+1)\pi$; the so-called phase stability refers to the amplitude $Gain(\omega) > 0$, when the phase-frequency curve does not cross $\pm(2\text{n}+1)\pi$ The magnitude of stability of the essence of the engine oscillation control force generated by the excitation is less than the elastic vibration in the inherent damping under the role of attenuation, so the magnitude of stability depends on the inherent damping of elastic vibration and control system on the elastic vibration signal of the sufficient attenuation. The essence of phase stability is to take the elastic vibration signal as part of the control signal through the correction network to obtain the correct phase; for the elastic vibration to produce additional damping to achieve the purpose of stability, the phase stability does not depend on the inherent damping of elastic vibration, but the phase-frequency characteristics of the correction network put forward strict requirements.

Figure 2 shows an open-loop Nichols plot of the pitch channel, in which the open-loop frequency response of the angle (Phi) satisfies the amplitude-stability and phase-stability conditions, while the angle rate (Omega) does not satisfy the amplitude-stability condition (amplitude $Gain(\omega) > 0$ when the phase frequency curve traverses the 180° curve) or the phase-stability condition (the phase frequency curve traverses the 180° line at frequency

7.694 rad/s, while amplitude $Gain(\omega) > 0$). In adaptive augmentation control, the PD controller provides the basic control gain for the launch vehicle and is the basic controller in the AAC control framework. In this paper, we directly selected the PD controller parameters $K_p = -9.2$ and $K_d = -3.8$. For the rigid–flexible coupling model at the 30 s moment, the following notch filter was established as in (9):

$$W(s) = \left(\frac{s^2 + 2 \times 0.005 \times 7.69s + 7.69^2}{s^2 + 2 \times 0.6 \times 9s + 9^2}\right)^2 \left(\frac{s^2 + 2 \times 0.005 \times 7.69s + 7.69^2}{s^2 + 2 * 0.6 \times 6.8s + 6.8^2}\right)^2 \quad (9)$$

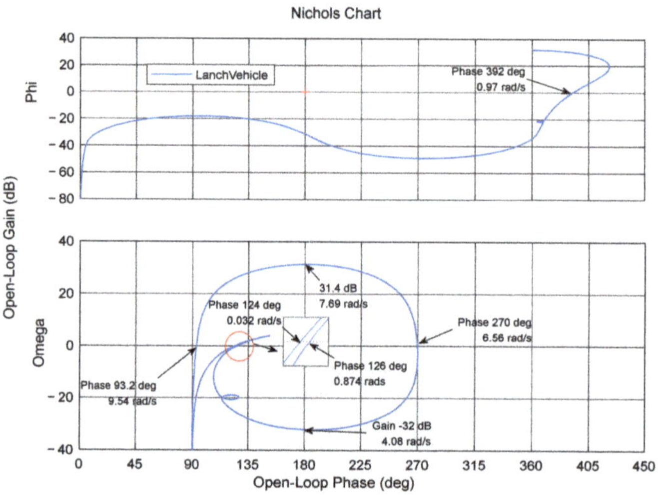

Figure 2. The open-loop frequency characteristics of the system.

In a classical PID feedback control system, a higher forward gain can improve the performance and robustness of the system with a fixed ratio of proportional and differential gain. However, due to the special performance requirements and stability requirements of large flexible launch vehicles (presence of high uncertainty, elastic vibration), forward gain must be limited to a small range to provide better performance and robustness for the baseline controller, and the design requirements usually limit the allowable forward gain to a range not less than 6 dB from the critical stability value to improve the system's ability to cope with uncertainty. The closed-loop spectral characteristics of the system can be reflected to some extent by the open-loop margin of the system, and when the adaptive gain k_T reaches a certain critical value, the closed-loop system exhibits resonance phenomena at certain frequencies (open-loop characteristics cross the $j\omega$ axis in the complex plane), while for any $k_T + \varepsilon$, the closed-loop system exhibits dispersion phenomena. The spectral characteristics of the closed-loop system at different forward gain K_T are shown in Figure 3. From this, the critical value of the octave forward gain in the adaptive augmentation control can be determined.

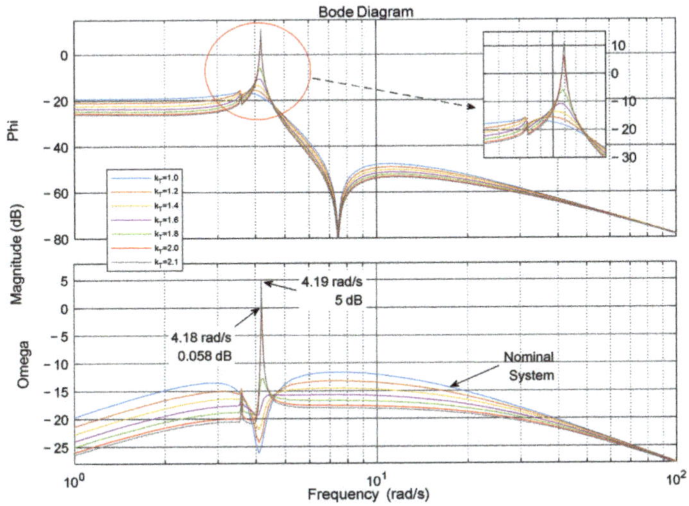

Figure 3. The adjustable range of forward gain.

3.3. *Multiplicative Adaptive Control Law Based on Spectral Damping*

For large launch vehicle elastic vibration, the first-order elastic vibration mode has a low frequency and small phase deviation, and is usually stabilized by the phase-stabilization method, while the second-order and higher-frequency elastic vibration modes have large phase deviations, and are usually stabilized by the amplitude-stabilization method. In general, the first-order vibration mode can be phase-stabilized by selecting the mounting position of the rate gyro, the second-order vibration mode requires amplitude stabilization, and the higher-order vibration mode is amplitude-stabilized by high-frequency filtering of the correction network. Figure 4 shows a block diagram of the improved adaptive broadening control system designed in this paper, where the control command error signal e_u and tracking error $e_{\varphi,\dot{\varphi}}$ between the reference model and the actual model were used as inputs to the adaptive control method, and the adaptive gain k_T was calculated through the two channels of oscillation suppression and error suppression, respectively, to adjust the baseline PD controller gain. The control command error, tracking error, and adjustment gain of adaptive augmentation control are shown below:

$$e_u = u_r - u \tag{10}$$

$$e_{\varphi,\dot{\varphi}} = 0.5e_\varphi + e_{\dot{\varphi}} = 0.5(\varphi_r - \varphi) + (\dot{\varphi}_r - \dot{\varphi}) \tag{11}$$

$$k_T = sat_{k_0}^{k_{\max}}\{k_e y_e - k_s y_s + 1\} \tag{12}$$

where $k_{\max}$ is the upper bounds of the adjustment gain, k_0 is the lower bounds of the adjustment gain, k_e is the adjustment gain of tracking error term, y_e is the output signal of the tracking error signal through the high- and low-pass filters, k_s is the adjustment gain of control command error term, and y_s is the output signal of the control command error signal through the spectrum dampener.

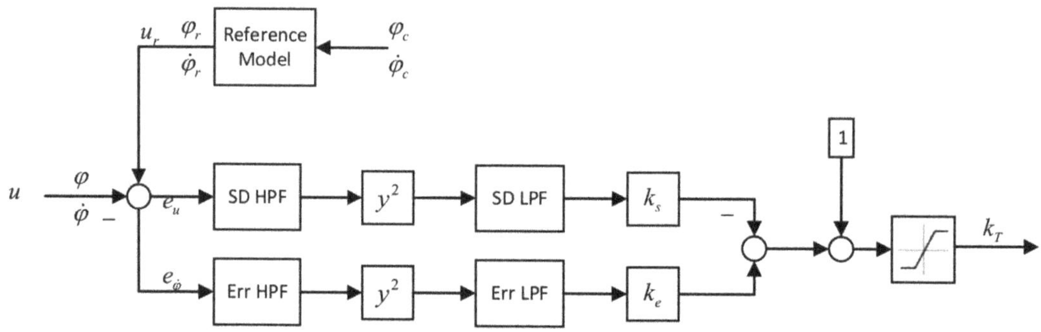

Figure 4. The adaptive control algorithm.

3.4. Spectrum Dampers

The adjustment gains k_e and k_s of the two spectrum dampers adjust the spectrum output signals y_e and y_s of the two channels, which are formed by the tracking error signal and the controller command error signal, as follows:

$$y_e = ErrL_P(s)\left(ErrH_P(s)e_{\varphi,\dot{\varphi}}\right)^2 \tag{13}$$

$$y_s = SDL_P(s)(SDH_P(s)e_u)^2 \tag{14}$$

In general, the DC gain of the designed high-pass filter should be as small as possible (the passband gain is usually set to 1), while the transition band should be as steep as possible (limited to 1.5 rad/s) The forms of the high- and low-pass filters are shown in (15) and (16):

$$H_P(s) = \frac{s^2}{s^2 + 2\xi_{hp}\omega_{hp}s + \omega_{hp}{}^2} \tag{15}$$

$$L_P(s) = \frac{\omega_{lp}{}^2}{s^2 + 2\xi_{lp}\omega_{lp}s + \omega_{lp}{}^2} \tag{16}$$

where ω_{hp}, ω_{lp} are the cutoff frequencies of the high- and low-pass filters; and ξ_{hp}, ξ_{lp} are the damping ratios, with values ranging from 0.5 to 0.8.

In the tracking error channel, the high- and low-pass filters successively process the error signal, and the AAC gain k_T is to be enhanced by increasing the error suppression gain k_e, thus improving the overall forward gain of the system to reduce the system tracking error and improve system performance. In this channel, we set the control frequency near the shear frequency of the rigid-body system (0.87 rad/s), considering the need to compensate the control of the system tracking error and improve the overall gain of the system [5,28]. The shear frequency of the low-pass filter is near this frequency, and the shear frequency of the high-pass filter is one octave above this frequency, so the transfer function of the high and low-pass filters is given accordingly as follows:

$$ErrH_P(s) = \frac{s^2}{s^2 + 2 \times 0.8 \times 8.7s + 8.7^2} \tag{17}$$

$$ErrL_P(s) = \frac{1.2^2}{s^2 + 2 \times 0.6 \times 1.2s + 1.2^2} \tag{18}$$

In the control command error channel, the spectrum damper is mainly used to process the elastic vibration signal in the control command and reduce the AAC gain k_T by setting a specific frequency to adjust the elastic rejection gain k_s, thus reducing the overall gain (excessive gain) of the system to suppress the elastic vibration of the system and reduce its instability. The input of the high- and low-pass filters is the additional instruction error generated by the rigid-body controller instruction and the elastic vibration excitation, where the high-pass filter is used to obtain the elastic vibration signal from the control instruction. The analysis in the previous section showed that the system was prone to modal vibration at 7.69 rad/s, so we designed the control frequency at this frequency point. The cut-off frequency ω_{hp} of the high-pass filter should be taken slightly higher than this frequency. The low-pass filter is used to eliminate the high-frequency components of the signal, which is squared before entering the low-pass filter, so the value of the cut-off frequency ω_{lp} of the low-pass filter should be taken near this frequency. The corresponding parameters of the spectrum dampers in the elastic rejection channel are as follows:

$$SDH_P(s) = \frac{s^2}{s^2 + 2 \times 0.8 \times 24.3s + 24.3^2} \tag{19}$$

$$SDL_P(s) = \frac{7.69^2}{s^2 + 2 \times 0.6 \times 7.69s + 7.69^2} \tag{20}$$

4. Simulation Results and Analysis

In order to illustrate the role of the improved AAC control scheme in the launch vehicle system, the tracking curves and adaptive control gains of the nominal system and two different failure scenarios were presented and analysed in the simulation to verify the effectiveness of the designed algorithm. Assuming at 10 s after the launch vehicle takes off, the angle and rate commands of pitch were given as shown in Figure 5. The gain saturation function of the AAC controller in this example was taken as $K_{\max} = 2$ and $K_{\min} = 0.5$.

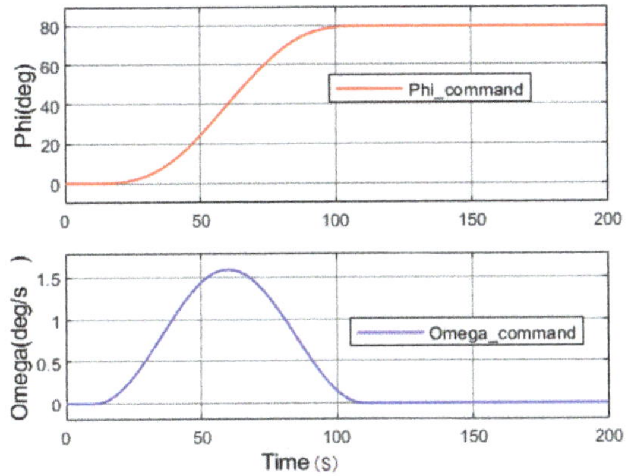

Figure 5. The pitch and angle rate commands.

If the system was in the normal state, the output of the system under PD control and AAC was consistent with the rigid-body nominal system, as shown in Figure 6a. The control commands and control rates of PD and AAC are shown in Figure 6b; there was almost no difference in the visible output curves. At the same time, the adjusted gain k_e and k_S of two channels in the AAC were close to 0, as shown in Figure 6c, and the overall adaptive gain was always kept at a stable value $k_T = 1$, which meant that the AAC did not produce any effect in the normal state. This was in line with our original design requirement that the AAC not be involved in control activities when the baseline PD controller was able to achieve a good performance output.

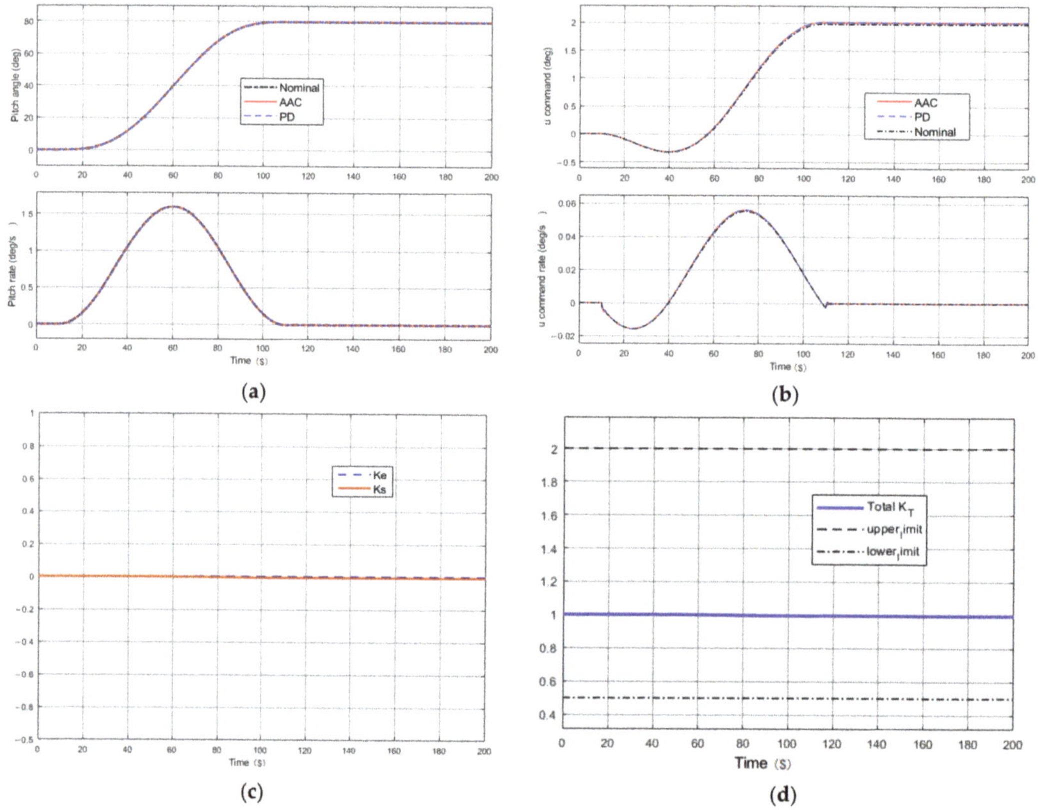

Figure 6. Performance in the nominal state: (**a**) pitching attitude; (**b**) control commands; (**c**) gain adjustment for the tracking error and control command error; (**d**) total gain of adaptive augmentation control.

We assumed that there was a baseline controller gain loading error during operation (the gain of the PD controller was not sufficient to meet the system requirements; in this case, $K_p = -1$ and $K_d = -0.8$); meanwhile, the control commands of the system were perturbed by a square wave with an amplitude of 0.5 and a duration of 5 s. As shown in Figure 7a, the system was able to track the input commands better, but then there was a certain steady-state error (about 2°) in the baseline PD control compared to the nominal system, and in contrast, the steady-state error was reduced to half (less than 1°) under the AAC control adjustment. In addition, in Figure 7b there is a corresponding reduction in the control command error, while the control command error rate only fluctuated significantly when it was just perturbed. Among the two channels, the adjusting gain k_s produced elastic suppression due to the disturbance in 25 s, and then the gain for suppressing elastic vibration fell back to 0; while k_e produced error suppression mainly after 60 s due to the tracking error. The overall gain k_T of the AAC control was less than 1 at 25 s in the elastic suppression channel k_s, and then gradually increased (>1) due to the error suppression gain k_e generated, and the overall AAC gain was always maintained at a saturated value due to the long-term presence of steady-state errors.

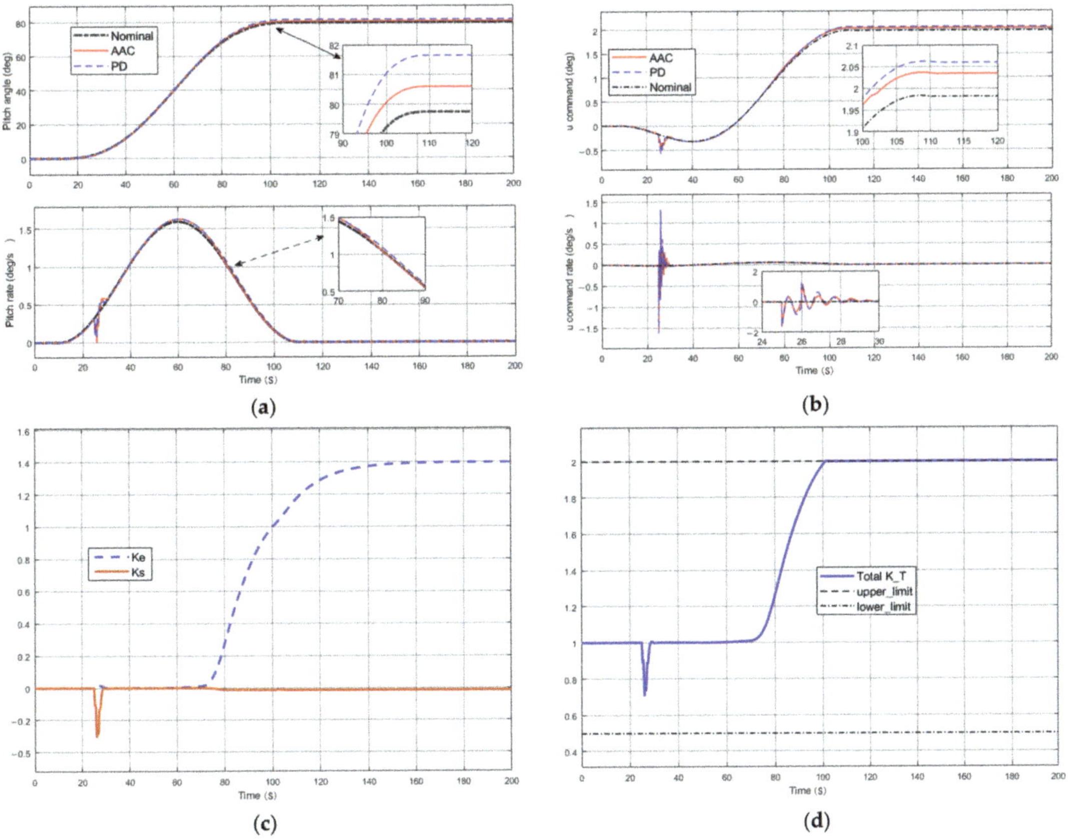

Figure 7. Performance in a state in which the baseline controller gain was misloaded and suffered external perturbations at t = 25 s: (**a**) pitching attitude; (**b**) control commands; (**c**) gain adjustment for the tracking error and control command error; (**d**) total gain of adaptive augmentation control.

Assuming that there was uncertainty in establishing the rigid spring coupling model, the elastic vibration frequency of the model was reduced by 40% while the same perturbation signals described above existed. The pitch angle and pitch angle rate signals at this time are shown Figure 8a. The system was able to follow the control commands to some extent in the early stages when the adaptive channel was closed (i.e., only the baseline PD controller was in action), but with the passage of time and accumulation, the system ended up in a divergent state. When the model parameters were changed substantially, the model was fundamentally changed, and the controlled object deviated from the nominal state. The original PD controller parameters were not suitable for this model, and the excessive forward gain aggravated the elastic vibration of the system. At the same time, the PD controller parameters were not reduced accordingly, which intensified the system oscillation and eventually could not be suppressed, leading to system dispersion. However, with the AAC controller, the system was able to suppress the system oscillation caused by the elastic modal perturbation and could track the reference input. At the same time, as shown in Figure 8b, for the control command and rate output, we can see that the control command of the baseline PD controller began to oscillate and could not be inhibited at the same time by the perturbation signal, while the AAC could well inhibit the oscillation of the control command (the control command error was less than 0.5, and the control command error rate was less than 1), which was beneficial for the actuator in the actual system, and had a good input signal.

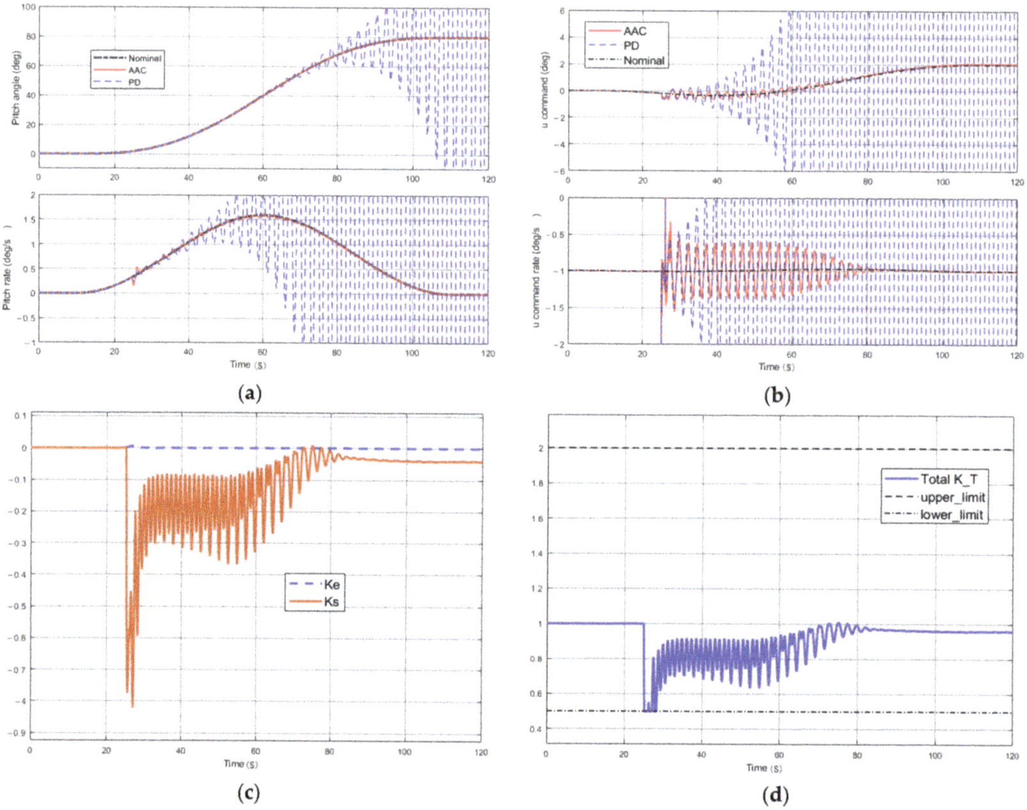

Figure 8. Performance at an elastic vibration frequency of 40% perturbation and external perturbation at t = 25 s: (**a**) pitching attitude; (**b**) control commands; (**c**) gain adjustment for the tracking error and control command error; (**d**) total gain of adaptive augmentation control.

The adjusted gain of the two channels k_e, k_s and the overall gain k_T of the AAC control are shown in Figure 8c,d. We observed that in the case in which the AAC was involved in the control, the baseline PD controller was not able to achieve a good tracking effect due to the ingress of the elastic mode, then the AAC controller generated a corresponding gain value k_s (in this case mainly for the suppression of elastic vibration), and then set the AAC gain to less than 1 to reduce the overall gain of the system and meet the requirements. When the baseline controller can achieve the tracking effect better, then the value of AAC gain KT will fall back to 1. It is obvious from the above analysis that the adaptive control designed in this paper had a good robust stability to the ingress of the elastic mode, and under the adjustment of the AAC control, the launch vehicle could adjust the control gain online and in real time to set the engine swing angle of the servo to keep the rocket stable.

5. Conclusions

Adaptive augmentation control has important research significance and development potential for the control of large flexible launch vehicles, and can increase the robustness of the system, avoid oscillations and even destabilization problems caused by the estimation errors of the flexible mode, and improve the safety and reliability of rocket operation. The improved AAC control scheme designed in this paper had good performance, and the simulation showed that during the nominal state of the system, the AAC control did not affect the baseline PD controller. When the system was subjected to external disturbances or PD controller errors (the controller parameters were loaded at values less than the set value), the AAC control could generate a multiplicative gain greater than 1 to boost the system forward gain and reduce the steady-state error of the system to half of the PD control ($<1^\circ$). With large regression of the flexible modal vibration frequency, the PD controller could cause the system to become unstable (uncontrollable). The AAC control could reduce the system forward gain by generating a multiplicative gain of less than 1 and limiting the control input signal to 0.5° to keep the system in a stable operation.

The simulation results of the above scenarios showed that the enhancement provided by the improved AAC control designed in this paper matched the expected goals and requirements, while the scheme had the same verifiability for the poorer control due to uncertainties caused by large-scale variations in thrust, mass, and atmospheric characteristics. The current simulation results showed suppression of up to 40% of the effect of flexible mode ingress, which greatly improved the robustness of the system. Future work on this program may consider sensor measurement noise and the nonlinear environment of closed-loop guidance [29], which would assist in fine-tuning and improving the extraction and calculation of error signals for use in conjunction with other load-shedding/fault-tolerant controls in the future development of large flexible launch vehicles.

Author Contributions: Conceptualization, A.P. and J.Z.; data curation, H.Z.; funding acquisition, J.Z.; investigation, W.C.; methodology, A.P. and H.Z.; resources, A.P.; validation, H.Z. and W.C.; visualization, H.Z.; writing—original draft, H.Z.; writing—review and editing, A.P. All authors have read and agreed to the published version of the manuscript.

Funding: This work was funded in part by the Science and Technology Foundation of Guizhou Province (QKH [2020] 1Y273, [2018] 5781, [2018] 5615, [2016] 5133), Guizhou Provincial Department of Education, Youth Talent Development Project (Qianke [2021] 100), and the National Natural Science Foundation (NNSF) of China under Grant No. 51867005.

Data Availability Statement: The data presented in this study are available on request from the corresponding author.

Conflicts of Interest: The authors declare no conflict of interest. The funders had no role in the design of the study; in the collection, analyses, or interpretation of data; in the writing of the manuscript; or in the decision to publish the results.

References

1. Ting, C. Development of LM-9 heavy rocket has made great progresses. *China Aerosp.* **2018**, *11*, 29–31.
2. Wu, Y.; He, L. Attitude control technology of new-generation launch vehicles. *J. Beijing Univ. Aeronaut. Astronaut.* **2009**, *35*, 1294–1297.
3. Orr, J.; Van Zwieten, T. Robust, practical adaptive control for launch vehicles. In Proceedings of the AIAA Guidance, Navigation, and Control Conference, Minneapolis, MN, USA, 13–16 August 2012.
4. Gaylor, R.; Schaeperkoetter, R.L.; Cunningham, D.C. An adaptive tracking filter for bending-mode stabilization. *J. Spacecr. Rocket.* **1967**, *4*, 573–577. [CrossRef]
5. Orr, J.S. Optimal recursive digital filters for active bending stabilization. In Proceedings of the 2013 American Astronautical Society (AAS) Guidance, Navigation, and Control Conference, Breckenridge, CO, USA, 1–6 February 2013.
6. Trotta, D.; Zavoli, A.; De Matteis, G.; Neri, A. Adaptive attitude control of launch vehicles in atmospheric flight. In Proceedings of the AAS/AIAA Astrodynamics Specialist Conference, Pittsburgh, PA, USA, 9–12 August 2020.
7. Jang, J.W.; Hall, R.; Bedrossian, N.; Hall, C. Ares-I bending filter design using a constrained optimization approach. In Proceedings of the AIAA Guidance, Navigation and Control Conference and Exhibit, Honolulu, HI, USA, 18–21 August 2008.
8. Khoshnood, A.M.; Moradi, H.M. Robust adaptive vibration control of a flexible structure. *ISA Trans.* **2014**, *53*, 1253–1260. [CrossRef] [PubMed]
9. Yang, F.; Wei, C.; Cui, N.; Xu, J. Adaptive generalized super-twisting algorithm based guidance law design. In Proceedings of the 14th International Workshop on Variable Structure Systems (VSS), Nanjing, China, 1–4 June 2016.
10. Cui, N.G.; Xu, J.; Mu, R.; Han, P. Gain-scheduled reusable launch vehicle attitude controller design. In Proceedings of the International Conference on Mechatronics and Automation, Changchun, China, 9–12 August 2009.

11. Guo, Z.; Zhao, J.; Zhou, M.; Zhou, J. On a new adaptive multivariable twisting sliding mode control approach and its application. In Proceedings of the 3rd International Conference on Control and Robotics Engineering (ICCRE), Nagoya, Japan, 20–23 April 2018.
12. Liu, J.; Yu, X.; Jin, S.; Hou, Z. Model free adaptive attitude control for a launch vehicle. In Proceedings of the Chinese Control Conference (CC(C), Guangzhou, China, 27–30 July 2019.
13. Navarro-Tapia, D.; Marcos, A.; Bennani, S.; Roux, C. Robust-control-based design and comparison of an adaptive controller for the VEGA launcher. In Proceedings of the AIAA Scitech 2019 Forum, San Diego, CA, USA, 7–11 January 2019.
14. Trotta, D.; Zavoli, A.; De Matteis, G.; Neri, A. Opportunities and limitations of adaptive augmented control for launch vehicle attitude control in atmospheric flight. In Proceedings of the AAS/AIAA Astrodynamics Specialist Conference, Portland, MA, USA, 11–15 August 2019.
15. Wall, J.; Orr, J.; VanZwieten, T. Space launch system implementation of adaptive augmenting control. In Proceedings of the 2014 American Astronautical Society (AAS) Guidance, Navigation, and Control Conference, Breckenridge, CO, USA, 31 January–5 February 2014.
16. Orr, J.S.; Wall, J.H.; VanZwieten, T.S.; Hall, C.E. Space launch system ascent flight control design. In Proceedings of the AAS Guidance, Navigation, and Control Conference, Breckenridge, CO, USA, 31 January–5 February 2014.
17. VanZwieten, T.S.; Gilligan, E.T.; Wall, J.H.; Orr, J.S.; Miller, C.J.; Hanson, C.E. Adaptive augmenting control flight characterization experiment on an F/A-18. In Proceedings of the 2014 American Astronautical Society (AAS) Guidance Navigation and Control Conference, Breckenridge, CO, USA, 31 January–5 February 2014.
18. VanZwieten, T.S.; Gilligan, E.T.; Wall, J.H.; Miller, C.J.; Hanson, C.E.; Orr, J.S. In-flight suppression of an unstable F/A-18 structural mode using the space launch system adaptive augmenting control system. In Proceedings of the AIAA Guidance, Navigation, and Control Conference, Kissimmee, FL, USA, 5–9 January 2015.
19. Brinda, V.; Narayanan, A.; Smrithi, U.S.; Kishore, W.A. Classical adaptive augmentation control for a typical second-generation launch vehicle. In Proceedings of the 4th IFAC Conference on Advances in Control and Optimization of Dynamical Systems (ACODS 2016), Tiruchirappalli, India, 1–5 February 2016.
20. Smrithi, U.S.; Brinda, V. Augmentation of Classical and Adaptive Control for Second Generation Launch Vehicles. *Int. J. Eng. Res. Technol.* **2016**, *5*, 432–439. [CrossRef]
21. Zhang, L.; Wei, C.; Jing, L.; Cui, N. Heavy lift launch vehicle technology of adaptive augmented fault tolerant control. In Proceedings of the 2016 IEEE Chinese Guidance, Navigation and Control Conference, Nanjing, China, 12–14 August 2016.
22. Cui, N.; Cheng, C.; Zhe, P.; Wei, C.; He, F. Adaptive augmented disturbance rejection and load-relief control for launch vehicle. *Missile Space Veh.* **2017**, *6*, 1–6.
23. Trotta, D.; Zavoli, A.; De Matteis, G.; Neri, A. Optimal tuning of adaptive augmenting controller for launch vehicles in atmospheric flight. *J. Guid. Control. Dyn.* **2020**, *43*, 2133–2140. [CrossRef]
24. Navarro-Tapia, D.; Marcos, A.; Bennani, S. Envelope extension via adaptive augmented thrust vector control system. *J. Guid. Control. Dyn.* **2021**, *44*, 1044–1052. [CrossRef]
25. Zhao, X. Study on Attitude Control Method 0f Heavy Launch Vehicle Based on Adaptive H∞ Control. Master's Thesis, Deptartment of Aerospace Engineering, Harbin Institute of Technology, Harbin, China, 2012.
26. Tan, S.; Zhou, W.; Wu, Z.; Yang, Y. Adaptive control design of attitude control system for launch vehicle. In Proceedings of the 9th National Conference on Dynamics and Control, Xi'an, China, 18 May 2012.
27. He, F. Research on Model Reference Adaptive Augmenting Control of Heavy lift Launch Vehicle. Master's Thesis, Department of Aerospace Engineering, Harbin Institute of Technology, Harbin, China, 2018.
28. Frost, S.A.; Balas, M.J.; Wright, A.D. Augmented adaptive control of a wind turbine in the presence of structural modes. In Proceedings of the 2010 American Control Conference, Baltimore, MD, USA, 30 June–2 July 2010.
29. Balas, M.J.; VanZwieten, T.; Hannan, M. Nonlinear (Lyapunov) Stability of the Space Launch System Flight Control System with Adaptive Augmenting Control. In Proceedings of the AIAA Scitech 2019 Forum, San Diego, CA, USA, 7 January 2019.

entropy

MDPI

Article

Robust Controller Design for Multi-Input Multi-Output Systems Using Coefficient Diagram Method

Kai Liu, Fanwei Meng *, Shengya Meng and Chonghui Wang

College of Department of Control Science and Engineering, Northeastern University at Qinhuangdao, Qinhuangdao 066004, China; miaomiao@stumail.neu.edu.cn (K.L.); 2001944@stu.neu.edu.cn (S.M.); 2071918@stu.neu.edu.cn (C.W.)

* Correspondence: mengfanwei@neuq.edu.cn; Tel.: +86-157-3213-7808

Abstract: The coupling between variables in the multi-input multi-output (MIMO) systems brings difficulties to the design of the controller. Aiming at this problem, this paper combines the particle swarm optimization (PSO) with the coefficient diagram method (CDM) and proposes a robust controller design strategy for the MIMO systems. The decoupling problem is transformed into a compensator parameter optimization problem, and PSO optimizes the compensator parameters to reduce the coupling effect in the MIMO systems. For the MIMO system with measurement noise, the effectiveness of CDM in processing measurement noise is analyzed. This paper gives the control design steps of the MIMO systems. Finally, simulation experiments of four typical MIMO systems demonstrate the effectiveness of the proposed method.

Keywords: MIMO; coupling; PSO; CDM; measurement noise; robust controller

Citation: Liu, K.; Meng, F.; Meng, S.; Wang, C. Robust Controller Design for Multi-Input Multi-Output Systems Using Coefficient Diagram Method. *Entropy* **2021**, 23, 1180. https://doi.org/10.3390/e23091180

Academic Editors: Quanmin Zhu, Giuseppe Fusco, Jing Na, Weicun Zhang and Ahmad Taher Azar

Received: 3 August 2021
Accepted: 3 September 2021
Published: 8 September 2021

1. Introduction

Multi-input multi-output (MIMO) systems, defined as systems with multiple control inputs and outputs, are widely used in industrial systems. Many common industrial control systems can be modeled as MIMO systems, such as chemical reactors, distillers, generators, and automobile transmission systems [1–5]. A consensus is that the control of the MIMO systems is more complex than the control of the single-input single-output (SISO) systems. In the MIMO systems, the outputs are affected by each input. In other words, there is a coupled interaction between the input and output variables of the MIMO systems. Due to the interaction in the MIMO systems, it is not easy to directly apply the advanced control methods based on the SISO systems.

Currently, the control strategies of the MIMO systems are mainly based on the methods of decoupling. Decoupling strategies can be divided into static decoupling and dynamic decoupling. The former achieves decoupling based on steady-state gain, which can effectively reduce the impact of model uncertainty, but the high-frequency response of MIMO systems is often not ideal [6]. The dynamic decoupling can achieve a trade-off between complexity and decoupling performance. In recent years, various dynamic decoupling strategies have been developed, such as ideal decoupling, simplified decoupling, and reverse decoupling. The ideal decoupler can provide a simple decoupling system, but the ideal decoupler is difficult to realize in practical applications. The opposite is simplified decoupling. Although simplified decoupling can obtain simple decoupling and decoupling, the decoupling system will be very complicated. In [7], Hagglund T proposed a decoupling method that approximates the sum of elements to reduce the system's complexity after decoupling. Reverse decoupling takes into account the advantages of ideal decoupling and simplified decoupling. However, when there is a time-delay element in the MIMO systems, reverse decoupling cannot guarantee the system's stability. In addition, researchers have used intelligent algorithms for MIMO systems and proposed various intelligent decoupling algorithms [8–12]. However, because the design of this kind of method is difficult to understand and the controller is complicated, it is difficult for engineers to adopt.

As an algebraic design method, the CDM proposed by S. Manabe is simple and easy to implement [13]. Compared with other control methods, CDM only requires the designer to define one parameter: the equivalent time constant [14]. At the same time, all algebraic equations in the CDM are expressed in the form of polynomials, which facilitates the elimination of poles and zeros in the design and analysis of the control systems. CDM has been proven to be a method to ensure the robustness of the control system, and its effectiveness has been proven through a series of experiments [15–17]. Therefore, with the continuous improvement of CDM, CDM has been continuously applied to existing control systems. Mohamed T. H. combined CDM with ecological optimization technology (ECO) for load frequency design in multi-regional power systems in [18]. Experimental results show that the proposed method is robust in the presence of disturbance uncertainty. Because CDM is simple, effective and robust, it is also applied in MIMO system control [19]. CDM was be used to design a PI controller with two cone-shaped official position research objects in [20]. The simulation results prove the effectiveness of CDM on disturbance suppression. In [21], CDM was used to solve the controller gain to suppress the vibration in the flexible robot system.

The first problem to be solved in the controller design of the MIMO system is how to achieve decoupling. Compared with other existing results, this article transforms the decoupling problem into the parameter optimization problem and gives an interaction measurement to evaluate the decoupling degree of the MIMO systems. The PSO algorithm is used to optimize the parameters of the compensator to achieve decoupling. After decoupling, the systems tend to have high order. The CDM considers the robustness and interference suppression performance of the system and the simplicity of design. Therefore, motivated by the advantages of CDM, this paper applies the CDM to the field of controller design for MIMO systems. At the same time, considering measurement noise can generate undesired control activity resulting in wear of actuators and reduced performance, this article analyzes the controller's suppression effect on measurement noise based on the CDM. To verify the effectiveness and universality of the proposed method, this paper gives four typical design examples of MIMO systems in the hope of providing engineers and technicians a reference.

The main innovations of this paper are as follows:

(1) Converts the compensator design problem used for decoupling into parameter optimization problems to reduce the difficulty of decoupling.
(2) Gives an interaction measurement to quantify the interaction of coupled systems.
(3) Analyzes the controller's suppression effect on measurement noise based on the CDM.
(4) Research on the application of CDM methods to MIMO systems needs to be promoted. In order to make up for the shortcomings of existing research, this paper presents a design strategy of a robust controller based on CDM, which provides a reference for the design of the MIMO system controller in other articles.

The rest of this article is organized as follows: Section 2 gives an interaction measurement of the coupling interaction and uses the PSO algorithm to design the compensator to achieve decoupling; Section 3 summarizes the design process of the CDM controller and analyzes the controller's suppression effect on measurement noise based on the CDM. Section 4 outlines a set of controller design procedures for MIMO systems; four unique objects are simulated to verify the effectiveness of the proposed method in Section 5. Finally, a conclusion is given.

2. Decoupling Design

At present, there are two solutions to the interaction of MIMO systems. One is to use modern control theory, and the other is to limit the interaction to a certain extent and treat MIMO systems as multiple SISO systems, which is called decoupling control. Generally speaking, decoupling control is simple to operate, so it is often used. This article designs a compensator in the frequency domain to decouple. At the same time, in order to verify whether the designed compensator achieves the expected decoupling effect, this section provides the MIMO system interaction measurement.

2.1. Compensator Design

The schematic diagram of the decoupling design of the $n \times m$ MIMO system is shown in Figure 1.

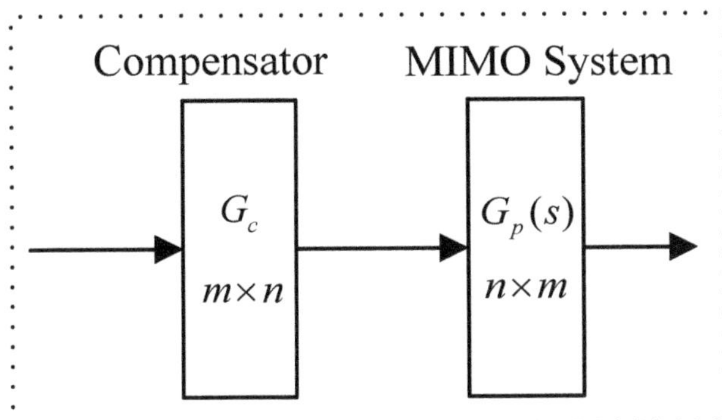

Figure 1. Schematic diagram of the decoupling design of the $n \times m$ MIMO system.

where the model of the MIMO system is represented by the transfer function $G_p(s) \in R^{n\times m}$, which is Equation (1).

$$G_p(s) = \begin{bmatrix} g_{11}(s) & \cdots & g_{1m}(s) \\ \cdots & \ddots & \cdots \\ g_{n1}(s) & \cdots & g_{nm}(s) \end{bmatrix}. \tag{1}$$

where $g_{ij}(s), i = 1, 2, \cdots, n; j = 1, 2, \cdots, m$ is the transfer function element in $G_p(s)$. Design the compensator $G_c(s) \in R^{m\times n}$ as shown in Equation (2).

$$G_c(s) = \begin{bmatrix} h_{11}(s) & \cdots & h_{1n}(s) \\ \cdots & \ddots & \cdots \\ h_{m1}(s) & \cdots & h_{mn}(s) \end{bmatrix}, \tag{2}$$

where $h_{ij}(s)(i = 1, 2, \cdots, m; j = 1, 2, \cdots, n)$ is the transfer function element in $G_c(s)$. In order to reduce the difficulty of the designed compensator $G_c(s)$, set $G_c(s)$ as a constant matrix. When the compensator G_c acts on the MIMO system $G_p(s)$, the decoupling system $Q(s) \in R^{n\times n}$ is obtained as Equation (3).

$$\begin{aligned} Q(s) &= G_p(s)G_c \\ &= \begin{bmatrix} g_{11}(s) & \cdots & g_{1m}(s) \\ \cdots & \ddots & \cdots \\ g_{n1}(s) & \cdots & g_{nm}(s) \end{bmatrix} \begin{bmatrix} h_{11} & \cdots & h_{1n} \\ \cdots & \ddots & \cdots \\ h_{m1} & \cdots & h_{mn} \end{bmatrix} \\ &= \begin{bmatrix} f_{11}(s) & \cdots & f_{1n}(s) \\ \cdots & \ddots & \cdots \\ f_{n1}(s) & \cdots & f_{nn}(s) \end{bmatrix}. \end{aligned} \tag{3}$$

The purpose of designing the compensator is to make the decoupling system $Q(s)$ diagonal in all frequency domains, which means non-diagonal elements $f_{lr}(s) = 0$

$(l \neq r, l = 1, 2, \cdots n, r = 1, 2, \cdots n)$. In this way, the interaction is minimized, and the decoupling effect is the best. However, it is not easy to find such an ideal compensator. Therefore, this paper selects a specific frequency $s = j\omega_0$ to design the compensator. What needs to be explained is that the selection of a specific frequency $s = j\omega_0$ depends on the control object and requires the designer to use design experience to verify it through repeated experiments.

Use $s = j\omega_0$ to denote the element in the $r(r = 1, 2, \cdots n)$ column of $Q(s) = G_p(s)G_c$, we can get

$$f_{lr}(j\omega_0) = g_l(j\omega_0)\hat{h}_r = (\alpha_l + j\beta_l)\hat{h}_r \quad l = 1, 2, \cdots n, \tag{4}$$

where $g_l(j\omega_0)$ is the $G_p(j\omega_0)$ row vector of l, $\hat{h}_r$ is the G_c column column vector of r, $\alpha_l = \mathrm{Re}\{g_l(j\omega_0)\}$, and $\beta_l = \mathrm{Im}\{g_l(j\omega_0)\}$.

In order to achieve $Q(j\omega_0)$ diagonalization, we make the absolute value square of the off-diagonal elements in the $r(r = 1, 2, \cdots n)$ column of $Q(j\omega_0)$ equal to zero, which is

$$|f_{lr}(j\omega_0)|^2 = \hat{h}_r^T(\alpha_l{\alpha_l}^T + \beta_l{\beta_l}^T)\hat{h}_r = 0 \quad l \neq r. \tag{5}$$

Under Equation (5), the optimal solution $\hat{h}_r$ can be obtained, thereby obtaining the compensator G_c and the decoupling system $Q(s)$. However, the decoupling system $Q(s)$ may not meet the decoupling design requirements. One reason is that under the condition of a certain frequency $s = j\omega_0$, Equation (5) can only guarantee that the absolute value square of the off-diagonal elements of $Q(j\omega_0)$ is equal to zero, but the absolute value square of diagonal elements $|f_{lr}(j\omega_0)|^2(l = r)$ is not equal to zero or does not tend to zero. Suppose the designed compensator G_c cannot guarantee that $Q(s)$ at a certain frequency $s = j\omega_0$ achieves diagonalization. In that case, there is no guarantee that $Q(s)$ can be decoupled in the entire frequency domain. The other reason is that the $\hat{h}_r(r = 1,2, \cdots n)$ may be a trivial solution, so the compensator designed is meaningless. To effectively illustrate the above description, we give a concrete example next.

Example 1. *Consider the two-input two-output one-order inertial system, The transfer function is:*

$$G_p = \begin{bmatrix} \dfrac{2}{6s+1} & \dfrac{-3.6}{4s+1} \\ \dfrac{0.4}{9s+1} & \dfrac{4}{42s+1} \end{bmatrix}. \tag{6}$$

we select the frequency $\omega_0 = 1$, and under Equation (5), use PSO to obtain the compensator as equation:

$$G_c = \begin{bmatrix} -1.2499 \times 10^{-16} & 3.2508 \times 10^{-22} \\ 1.7313 \times 10^{-17} & 1.0364 \times 10^{-21} \end{bmatrix}. \tag{7}$$

It can be seen from Equation (7) that without any restrictions, the calculated G_c is meaningless. Therefore, taking into account the above deficiencies, we make the following additions based on the constraint condition of Equation (5): Firstly, we select the square $|f_{lr}(j\omega_0)|^2$ $(l = r)$ of the absolute value of diagonal elements of $Q(j\omega_0)r(r = 1, 2, \cdots n)$ column as the objective function to obtain its maximum value. Secondly, in order to prevent the trivial solution of the obtained $\hat{h}_r$, we add the Equation (8) as the constraint condition:

$$\hat{h}_r^T\hat{h}_r = 1. \tag{8}$$

In summary, the decoupling problem of MIMO systems is transformed into the optimization problem as follows:

$$\max\ f(\hat{h}_r) = \hat{h}_r^T(\alpha_l\alpha_l{}^T + \beta_l\beta_l{}^T)\hat{h}_r \quad l = r,$$
$$\text{s.t. } \hat{h}_r^T(\alpha_l\alpha_l{}^T + \beta_l\beta_l{}^T)\hat{h}_r = 0 \quad l \neq r, \tag{9}$$
$$\hat{h}_r^T\hat{h}_r = 1.$$

2.2. *Interaction Measurement*

The design process of the compensator has been given in Section 2.1. Since the magnitude of the interaction between the variables of the decoupling system $Q(s)$ does not have a specific numerical measurement, it is not clear whether the designed compensator can achieve the desired decoupling effect. Therefore, this section presents an interaction measurement for the MIMO system to evaluate the impact of the decoupling degree of the compensator. The equation is established on the basis that the diagonal elements of the diagonal matrix are equal to the reciprocal of the diagonal elements of its inverse.

Assuming that the controlled variable of the decoupling system $Q(s)$ is $Y = [y_1, y_2, \cdots, y_n]^T$, the manipulated variable is $U = [u_1, u_2, \cdots, u_n]^T$, and u_i $(i = 1 \cdots n)$ controls y_i. For the i-th channel of $Q(s)$, the open-loop gain of the channel is obtained when all other manipulated variables are zero, that is, equality (10). The open-loop gain of the channel is obtained when all other controlled variables are zero, that is, equality (11).

$$\text{other loops are open:} \left(\frac{\partial y_i}{\partial u_i}\right)_{u_n=0, n\neq i} = f_{ii}. \tag{10}$$

$$\text{other loops are closed:} \left(\frac{\partial y_i}{\partial u_i}\right)_{y_n=0, n\neq i} = \tilde{f}_{ii}. \tag{11}$$

Here, measurement for $Q(s)$ interaction in MIMO systems is given:

$$E = \sum_{i=1}^{n} \frac{|f_{ii} - \tilde{f}_{ii}|}{|\tilde{f}_{ii}|} \quad i = 1 \sim n. \tag{12}$$

When the decoupling system $Q(s)$ is diagonalized, $E = 0$. Therefore, when Equation (12) is equal to zero or close to zero, it shows that other channels have no or minimal relationship with the channel, and the decoupling effect is good.

Remark 1. *Equations (10) and (11) are based on the steady-state of the MIMO system, but this situation is usually not maintained at other frequencies. Therefore, it can only be used as the measurement of the interaction size of the MIMO system and cannot be used as the judgment of whether the MIMO systems are decoupled.*

2.3. *Parameter Tuning of Compensator*

In this paper, particle swarm optimization (PSO) is used to optimize the objective function. Firstly, the fitness function is compiled. In order to facilitate programming, $-|f_{lr}(j\omega_0)|^2$ $(l = r)$ is taken as the objective function to obtain its minimum value. The fitness function can be obtained as follows:

$$Fit[f(\hat{h}_r)] = -|f_{lr}(j\omega_0)|^2 = -\hat{h}_r^T(\alpha_l\alpha_l{}^T + \beta_l\beta_l{}^T)\hat{h}_r \quad l = r. \tag{13}$$

Secondly, through the above fitness function and constraints in the frequency domain, $\hat{h}_r$ $(r = 1, 2, \cdots n)$ can be obtained through PSO debugging, and thus the compensator G_c can be obtained. Figure 2 shows the flowchart of PSO.

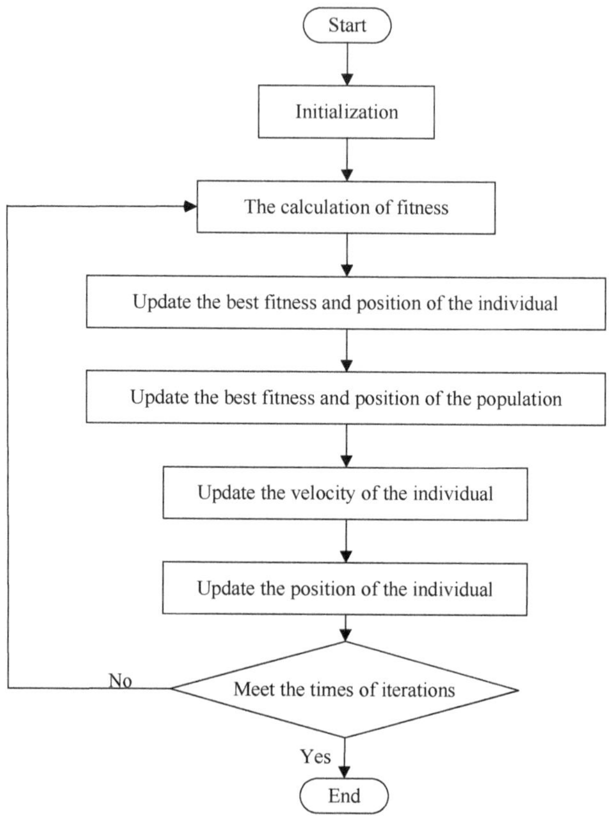

Figure 2. The flowchart of PSO.

PSO is essentially a stochastic algorithm, which has the function of self-organization, evolution, and memory and the strong searching ability and fast optimizing speed. In order to demonstrate the superiority of PSO to other evolutionary algorithms, we execute a number of comparisons between PSO and other evolutionary algorithms, such as Genetic Algorithm (GA), Shuffled Frog Leaping Algorithm (SFLA) and Cuck Search (CS). GA originates from Darwin's idea of natural evolution and follows the natural law of competition and survival of the fittest. GA is characterized by fast search speed, strong randomness, simple process, and robust flexibility. Still, it is easy to fall into the local optimum due to the reduction of population diversity in the evolution process. CS is a new swarm intelligence algorithm based on simulating cuckoo's nesting behavior. The algorithm has been successfully applied to solve various optimization problems due to its fewer parameters and easy realization. A significant feature of the CS is that it uses Levy flight to generate new solutions. The high randomness of Levy flight is that it can make the search process throughout the whole search space so that the global search ability of the algorithm is strong. However, the Levy flight height's randomness causes the CS algorithm's poor ability to perform a refined search in the local area and the slow convergence of the algorithm. SFLA simulates the communication and cooperation behaviors of frog populations in the process of foraging in nature, which has the advantages of fewer control parameters, simple operation, and easy realization. The specific parameter settings of different evolutionary algorithms are proposed in Table 1. The population size of each algorithm is 50, and the times of iterations are 100. The crossover probability and mutation probability of GA are 0.9 and 0.1, respectively. SFLA's moving maximum distance is 0.02,

CS's maximum discovery probability is 0.05. The weight of inertia, the self-learning factor and the population-learning factor of PSO are 0.35, 1.5 and 2.5. respectively.

In order to facilitate a comparison, we randomly select $G_p(s)=[0.2, 0.5; -0.3, 0.6]$, set $G_c=[h_1, h_2; h_3, h_4]$, and only obtain the first column $\hat{h}_1 : h_1$ and h_3 of G_c. Each algorithm is implemented independently 30 times. Table 2 presents the statistical results of each algorithm, including the maximum, minimum, average, standard deviation values of the objective function, and the average computational time. According to Table 2, we can see that PSO has an evident advantage of minimum, average, standard deviation values and average computational time over other algorithms. Figure 3 is the convergence graph of the optimization algorithms. It can be seen that PSO has a fast convergence speed and a good effect in finding the optimal global solution.

Table 1. The parameter settings of different evolutionary algorithms.

Evolutionary Algorithms	Parameter Settings
Genetic Algorithm (GA)	Population size = 50 The times of iterations = 100 Crossover probability = 0.9 Mutation probability = 0.1
Shuffled Frog Leaping Algorithm (SFLA)	Population size = 50 The times of iterations = 100 Moving maximum distance = 0.02
Cuck Search (CS)	Population size = 50 The times of iterations = 100 Maximum discovery probability = 0.05
Particle Swarm Optimization (PSO)	Population size = 50 The times of iterations = 100 The weight of inertia = 0.35 The self-learning factor = 1.5 The population-learning factor=2.5

Table 2. Statistical results of different algorithms.

	Fmax	Fmin	Fave	Fstd	Time (s)
GA	0.8885	−0.2030	0.0859	0.2911	0.2099
SFLA	3.9953	−0.2095	0.5383	1.1147	0.2389
CS	0.1941	−0.2124	−0.1844	0.0845	0.1347
PSO	−0.1908	−0.2128	−0.2117	0.0039	0.0756

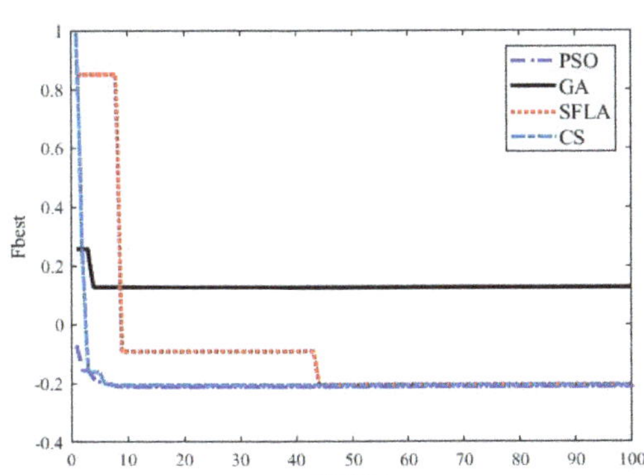

Figure 3. Convergence graphs of the optimization algorithms.

Furthermore, according to the research works with respect to the non-parametric statistical tests for different algorithms [22], some statistical tests have been adopted to compare the performance of GA, SFLA, CS and PSO. Table 3 proposes ranks achieved by Friedman, Friedman aligned and Quade tests for the objective function obtained by different algorithms. It is noticeable from Table 3 that PSO performs best in all statistical tests. Consequently, PSO has the superiority over other evolutionary algorithms in solving unknown parameters of compensator G_c.

Table 3. The ranks achieved by Friedman, Friedman aligned and Quade tests.

	Friedman Ranks	Friedman Aligned Ranks	Quade Ranks
GA	1.9	7.32	1.81
SFLA	2.5	10.7	2.41
CS	1.2	5.6	1.38
PSO	1.1	4.5	1.10

3. CDM Controller Design and Measurement Noise Rejection

In Section 2, the decoupling design can obtain the decoupling system $Q(s)$ with minimized interaction, but its open-loop transfer function is complex, and the order is high. Therefore, when stability, response characteristics, and robustness are considered simultaneously, the designed controller will become more complicated. CDM can effectively solve such problems.

3.1. CDM Controller Design

For the SISO linear systems, the standard block diagram designed by CDM is shown in Figure 4. The CDM control system consists of two parts: the controlled object and the CDM controller.

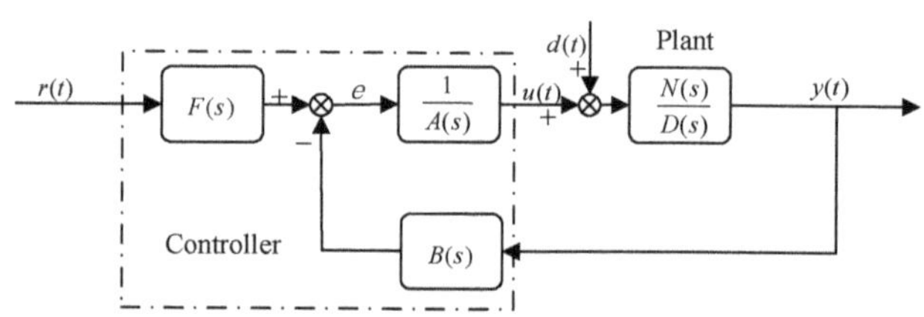

Figure 4. CDM control system standard block diagram.

where $r(t)$, $u(t)$, $y(t)$ and $d(t)$ are reference signal, control quantity, output quantity and disturbance quantity, respectively. The control function of the controller $u(t)$ may be interfered by the interference signal d. $N(s)$ and $D(s)$ are the numerator and denominator polynomials of the controlled object, respectively, defined as follows:

$$\begin{aligned} N(s) &= b_m s^m + b_{m-1} s^{m-1} + \cdots + b_1 s + b_0, \\ D(s) &= d_n s^n + d_{n-1} s^{n-1} + \cdots + d_1 s + d_0, \end{aligned} \tag{14}$$

where $b_m, b_{m-1} \cdots b_0$ and $d_n, d_{n-1} \cdots d_0$ are real coefficients and $m \leq n$. $A(s)$ and $B(s)$ are the denominator and numerator polynomial of the controller, respectively, defined as follows:

$$A(s) = \sum_{i=0}^{p} l_i s^i, \quad B(s) = \sum_{i=0}^{q} k_i s^i, \tag{15}$$

where l_i and k_i are unknown coefficients of the controller and $i \leq n$. There are many criteria for the selection of $A(s)$ and $B(s)$ polynomials. Disturbance is one of the selection criteria. When $l_0 = 0$, the influence of disturbance signal can be well suppressed. $F(s)$ is the reference numerator of the controller, which can ensure that the steady-state error in the performance of the closed-loop system is reduced to zero. The definition form is as follows:

$$F(s) = (\frac{P(s)}{N(s)})_{|s=0}, \tag{16}$$

where $P(s)$ is the characteristic polynomial of the closed-loop system. From Figure 4, we can obtained

$$P(s) = D(s)A(s) + N(s)B(s) = \sum_{i=0}^{n} a_i s^i, a_i > 0, \tag{17}$$

where a_i is the real coefficient. The design parameters of CDM-related characteristic polynomials are equivalent to the time constant τ and stability index γ_i, defined as follows:

$$\begin{cases} \tau = \dfrac{a_1}{a_0}, \\ \gamma_i = \dfrac{{a_i}^2}{a_{i+1}a_{i-1}}, \quad i = 1 \sim n-1, \\ \gamma_0 = \gamma_n = \infty, \\ {\gamma_i}^* = \dfrac{1}{\gamma_{i+1}} + \dfrac{1}{\gamma_{i-1}}, \end{cases} \tag{18}$$

where ${\gamma_i}^*$ denotes the stability limit, which is used to constrain the value of the stability index γ_i, and ${\gamma_i}^*$ is mainly used to ensure that it meets the Lyapunov stability conditions in the actual design process. The equivalent time constant τ is closely related to the setting time and bandwidth, which determines the rapid response of the system. If the setting time is represented by t_s, according to the Manabe standard form [13], its relationship with the equivalent time constant t_s is $\tau = t_s/(2.5 \sim 3)$.

The selection of the stability index γ_i determines the stability and time domain response characteristics of the system. Robustness is different from the system's stability, mainly considering the influence of system parameters on the speed of pole change. Control systems with other structures may have different robustness even if they have the same characteristic equation. The robustness of the system can only be determined when the open-loop system structure is determined. An essential feature of CDM in the application is that the controller structure and the characteristic polynomial can be designed simultaneously, and the robustness of the system can be guaranteed by setting the controller structure.

If Equation (17) of the corresponding system is a third-order system, according to the Routh stability criterion, the stability condition is $a_2 a_1 > a_3 a_0$. According to the expression in formula (18), this is equivalent to requiring the stability index to satisfy $\gamma_1\gamma_2 > 1$. Similarly, the stability conditions of the fourth-order system are $a_2 > (a_1/a_3)a_4 + (a_3/a_1)a_0$ and $\gamma_2 > {\gamma_2}^*$. For the fifth-order and above systems, Lyapunov gives several sufficient conditions for different forms of stability and instability, among which the conditions suitable for the CDM are as follows [23]: if all the fourth-order polynomials of the system are stable and have a margin of 1.12 times, the system is stable. If some third-order polynomials in the system are unstable, the system is unstable. The stability conditions of the system can be described as :

$$\begin{cases} a_i > 1.12(\dfrac{a_{i-1}}{a_{i+1}} a_{i+2} + \dfrac{a_{i+1}}{a_{i-1}} a_{i-2}), \\ \gamma_i > 1.12{\gamma_i}^*, \quad i = 2 \sim (n-1). \end{cases} \tag{19}$$

Manabe has proved that the system can obtain better stability and response characteristics when $\gamma_i > 1.12{\gamma_i}^*$ and ${\gamma_i}'$ values are between 1 and 4. If the stability index is selected according to $\gamma_i > 1.5\gamma_i{}^*$, the system's robustness is improved by sacrificing stability and response characteristics [23]. With the help of some design experience, designers can consider stability, response characteristics and robustness by reasonably selecting the structure and parameters of the controller.

In this article, we use the stability index γ_i values in the Manabe standard form. According to the Manabe standard form, the stability index γ_i is defined as:

$$\gamma_1 = 2.5, \ \ \gamma_0 = \gamma_n = \infty, \ \ \gamma_i = 2; \ \ i = 2 \sim (n-1). \tag{20}$$

Using the equivalent time constant τ and the stability index γ_i, the characteristic polynomial $P(s)$ can be obtained as follows:

$$P(s) = a_0 \left[\left\{ \sum_{i=2}^{n} \left(\prod_{j=1}^{i-1} \frac{1}{\gamma_{i-j}^{j}} \right) (\tau s)^i \right\} + \tau s + 1 \right]. \tag{21}$$

By comparing the coefficients of the characteristic polynomial Equations (17) and (21), the CDM controller parameters can be obtained.

3.2. Measurement Noise Rejection

To meet with the design needs of real-life, we analyze the output effect of the measurement noise in the controlled variable u in Section 3.2. Usually the block diagram presented in Figure 4 is extended by including measuring noise. Measurement noise may have a different character, but it is typically dominated by high frequencies, and low-frequency noise would correspond to drift. High-frequency noise can be suppressed by limiting the bandwidth of the closed-loop system. CDM can restrain the influence of high-frequency noise by selecting the equivalent time constant τ to limit the bandwidth of the closed-loop system. The reason is that the rapid response of the closed-loop system is proportional to the bandwidth, The equivalent time constant τ is closely related to the setting time and bandwidth, which determines the rapid response of the system. Here, we give the analysis of low-frequency noise suppression. Those signals are represented in Figure 5.

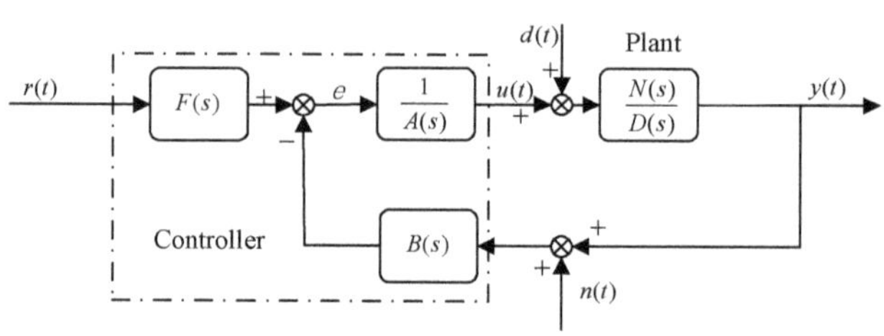

Figure 5. Basic structure of a CDM controller.

The newly added signal $n(t)$ denoting the measurement noise. We assume $n(t)$ is bounded with $|n(t)| \leq \mu \cdot h(t)$, where μ and $h(t)$ are a positive constant and a step-type signal, respectively [24]. In order to analyze the output effect of the measurement noise in the controlled variable $u(t)$, the reference signal $r(t)$ and disturbance signal $d(t)$ is set to zero. This leads to a relationship between $n(t)$ and $u(t)$, $n(t)$ and $y(t)$ given by the following differential equation:

$$\begin{aligned} -n(s) \cdot B(s) \cdot D(s) = (A(s) \cdot D(s) + N(s) \cdot B(s)) \cdot U(s), \\ -n(s) \cdot B(s) \cdot N(s) = (A(s) \cdot D(s) + N(s) \cdot B(s)) \cdot Y(s), \end{aligned} \tag{22}$$

where $n(s)$ is the Laplace transforms of $n(t)$, $U(s)$ is the Laplace transforms of $u(t)$, $Y(s)$ is the Laplace transform of $y(t)$.

Let us impose $b_i \neq 0$ for $i = 0, \cdots, m$,$d_i \neq 0$ for $i = 0, \cdots, n$ in Equation (14). The product $B(s) \cdot N(s)$ is a $q + m$ order polynomial and the product $B(s) \cdot D(s)$ is a $q + n$ order polynomial, respectively. The polynomials will be denoted by $C(s)$ and E(s) defined as:

$$C(s) = \sum_{i=0}^{q+m} g_i' \cdot s^i, \ E(s) = \sum_{i=0}^{q+n} g''_i \cdot s^i. \tag{23}$$

Assuming $r(t) = 0, d(t) = 0$, this steady-state system behaviour will be easily handled in the Laplace domain. By applying the final value theorem, the following equality should hold:

$$\lim_{t \to \infty} y(t) = \lim_{s \to 0} s \cdot Y(s), \tag{24}$$

However, in order to satisfy this equality, all the $Y(s)$ poles must have negative real parts and no more than one pole can be at the origin [25].

Assuming causality and zero initial conditions, the application of Laplace transform to (22) leads to,

$$\begin{aligned} U(s) &= \frac{-E(s)}{A(s) \cdot D(s) + C(s) \cdot B(s)} n(s). \\ Y(s) &= \frac{-C(s)}{A(s) \cdot D(s) + C(s) \cdot B(s)} n(s). \end{aligned} \tag{25}$$

Applying the final value theorem to the above expression then,

$$\lim_{s \to 0} s \cdot -C(s) \cdot n(s) = 0 \tag{26}$$

Due to $|n(t)| \leq \mu \cdot h(t)$ and $h(t)$, the Laplace transform is $\frac{1}{s}$, and we use $\frac{\mu}{s}$ replace $n(s)$, thus expression (26) takes the following format:

$$\lim_{s \to 0} s \cdot -C(s) \cdot \frac{\mu}{s} = -\mu \cdot g_0. \tag{27}$$

Since g_0 is equal to the product of b_0 and k_0 and since $b_0 \neq 0$ then, in order for g_0 to be zero, the controller coefficient k_0 must be equal to zero. Similarly, when the controller coefficient k_0 is equal to 0, the measurement noise has no effect on the control $u(t)$. Therefore, when the controller coefficient k_0 is equal to zero, the measurement noise does not affect performance.

4. Overall Design Ideas

This paper designs a compensator G_c and a centralized CDM controller for the $n \times m$ MIMO system in Figure 6. Systematic design ideas ensure the feasibility of decoupling design methods in large and small systems. At the same time, when the MIMO system interaction is minimized with high accuracy, the controller can achieve good control effects due to the robustness of the CDM. The most considerable advantages of CDM can be listed as follows:

1. A characteristic polynomial and controller are simultaneously designed. The characteristic polynomial specifies stability and response. The structure of the controller guarantees robustness. Thus, a simple controller, which satisfies the stability, response, and robustness requirements, can be designed with ease.
2. Compared with PID control that needed to develop different tuning methods for the process with various properties, it is sufficient to use a single design procedure in the CDM technique. This is an outstanding advantage.

The decoupling control and CDM controller design for the MIMO system can be summarized as the following steps.

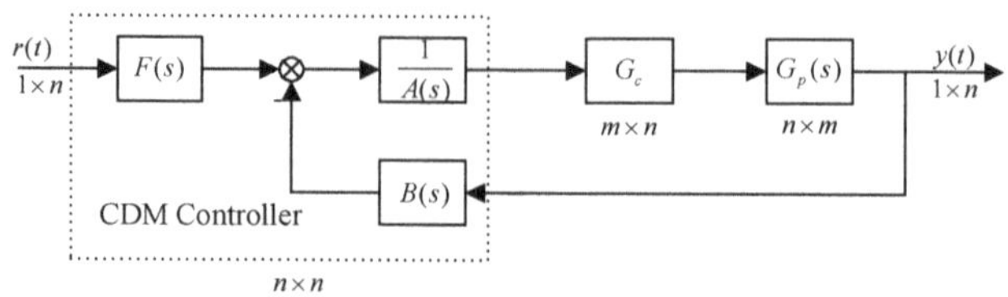

Figure 6. MIMO system control block diagram.

Figure 7 shows the design steps for MIMO systems, where the design process of CDM controllers is shown as the following:

1. Set the SISO controller parameters $A(s)$ and $B(s)$. $k_0 = 0$ is a good choice for measurement noise suppression.
2. Select CDM design parameters. The stability index γ_i in this paper is in the Manabe standard form of Equation (20). As long as the value of the equivalent time constant τ is determined, the controller parameters can be obtained. The τ value mainly determines the response time of the system. Generally, the τ value is determined according to the design requirements of the system setting time and bandwidth.
3. Solve the SISO controller parameters. $A(s)$, $B(s)$ can be obtained by Equations (17) and (21). $F(s)$ can be obtained by Equation (16).

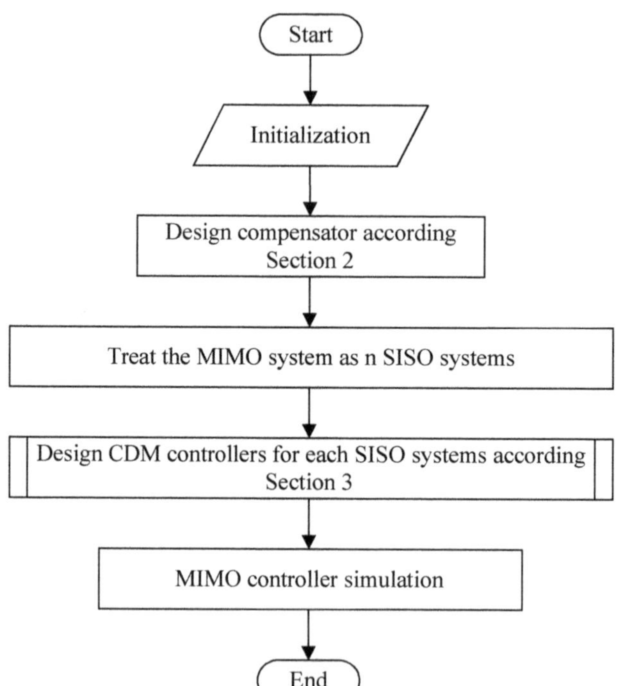

Figure 7. MIMO system control block diagram.

Remark 2. *The necessary condition for designing CDM controllers is that both denominators and molecules of the transfer function of the controlled object need to be expressed by rational polynomials. If there is a delay element in the transfer function of the controlled object, the improved Padé approximation method in reference [26] is used to deal with the delay element. According to the results of [26], the third-order improved Padé approximation is:*

$$e^{-sL} = \frac{60 - 24sL + 3(sL)^2}{60 + 36sL + 9(sL)^2 + (sL)^3}. \tag{28}$$

where L is delay time.

5. Simulation Experiment

This section conducts simulation experiments on four unique control targets to prove the effectiveness of this method. The experiments are evaluated with a step response of 1. The state variable is set to $x_i(i = 1, 2, 3, \ldots, n)$, and the system output is set to $y_i(i = 1, 2, 3, \ldots, m)$. Use the compensator for decoupling. When the interaction of the MIMO system is minimized, treat it as n SISO systems, and set each SISO system as $A_i = (i = 1, 2, 3, \ldots, m)$.

Example 2. *Consider the two-input two-output second-order inertial system (sugar factory model) in [27]. The transfer function is:*

$$G_p = \begin{bmatrix} \frac{0.28}{21s^2 + 10s + 1} & \frac{-0.33}{30s^2 + 11s + 1} \\ \frac{0.4}{270s^2 + 39s + 1} & \frac{0.5}{432s^2 + 42s + 1} \end{bmatrix}. \tag{29}$$

Without the decoupling design, the step response curve is shown in Figure 8.

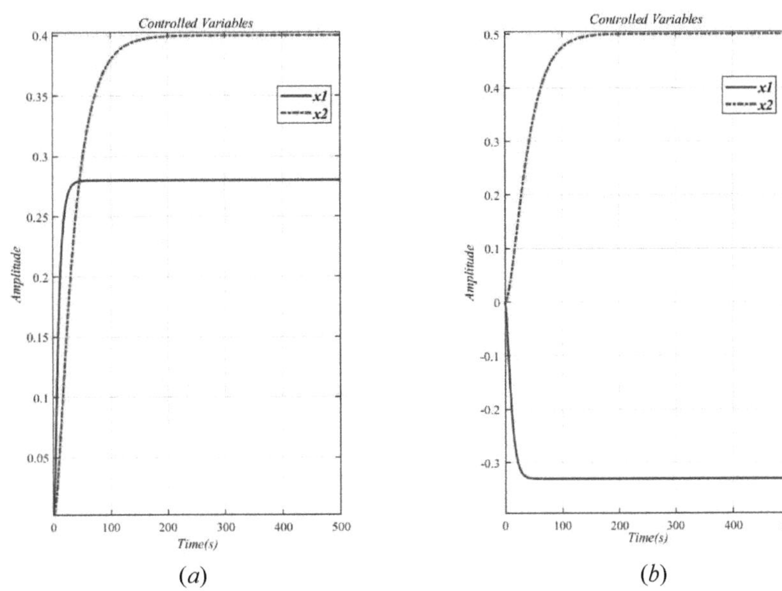

Figure 8. Step response curve of the original system (29) without decoupling design. (**a**) $x_1 \neq 0$; (**b**) $x_2 \neq 0$.

It can be seen from Figure 8a that $x_1 \neq 0$. As shown in Figure 8b, $x_2 \neq 0$. The two loops are obviously related, so the system (29) is a related system to the interaction. Therefore a decoupling design method is used to eliminate the interaction of the original system.

Select the angular frequency $\omega_0 = 0.13$, and obtain the compensator as Equation (30).

$$G_c = \begin{bmatrix} 0.7807 & 0.7606 \\ -0.6256 & 0.6507 \end{bmatrix}. \tag{30}$$

Apply the compensator (30) to the original system (29), then the decoupling system $Q(s)$ is obtained. Figure 9 draws the step response curve of the original system (29) after the decoupling design. As can be seen from Figure 9, the interaction of $Q(s)$ is effectively suppressed, especially in the static response part of the system. However, there is still a weak interaction in the dynamic response part. Overall, the decoupling effect is good.

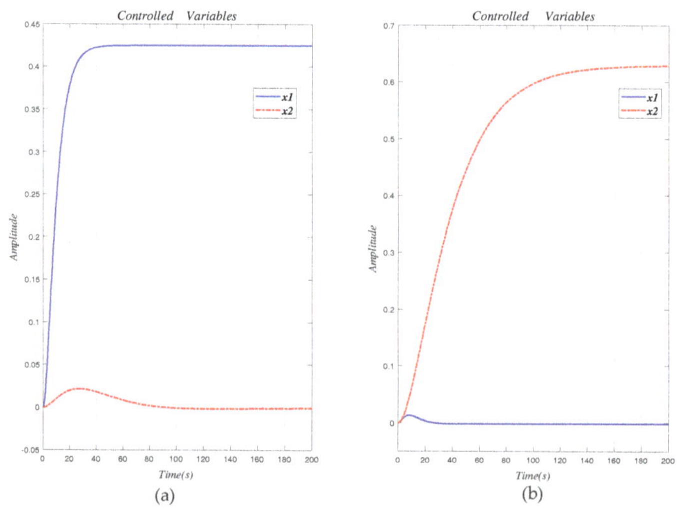

Figure 9. Step response curve of the original system (29) with decoupling design. (**a**) A_1; (**b**) A_2.

The two SISO systems after decoupling are set to A_1 and A_2. For A_1 and A_2, use the stability index γ_i and equivalent time constant τ in Table 4 to calculate CDM control polynomial parameters. Table 4 shows the equivalent time constant τ and CDM control polynomial parameter values.

Table 4. The equivalent time constant τ and CDM control polynomial parameter values.

	System A_1	**System A_2**
τ	11.2	16
$F(s)$	1.6811	0.8358
$A(s)$	$0.0313s^2 + 0.0855s$	$0.0921s^2 + 0.0928s$
$B(s)$	$6.9322s^2 + 6.1273s + 1.681$	$25.3384s^2 + 7.9861s + 0.83576$

Using the CDM control polynomial parameters in Table 4 to control A_1 and A_2, the results are shown in Figure 10.

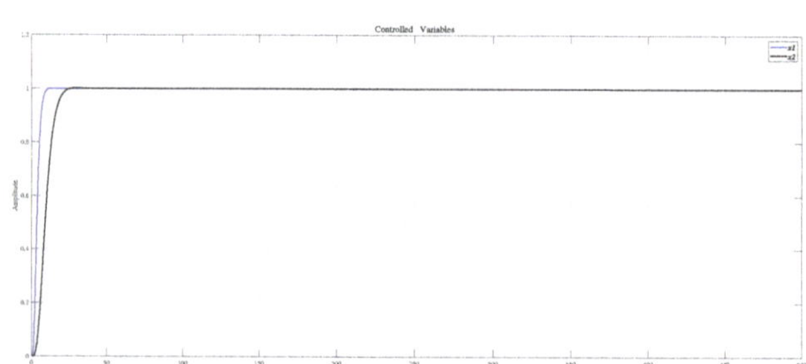

Figure 10. CDM controls A_1 and A_2 step response.

Figure 11 shows the result of inserting the compensator (30) in front of the controlled object (29) and using the CDM parameters in Table 4 for control. Affected by the interaction of the dynamic response part, the system overshoot increases.

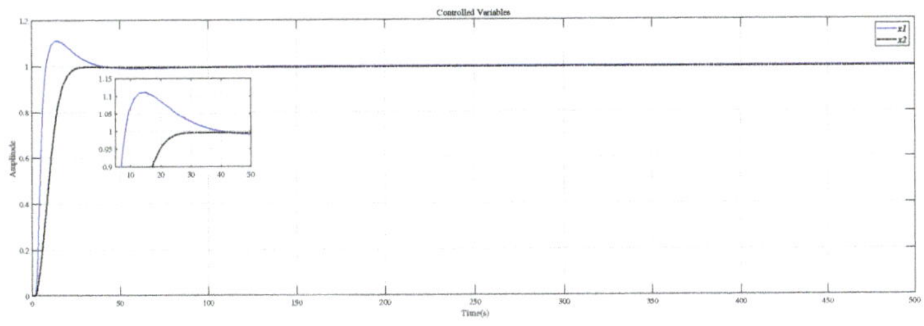

Figure 11. Use this method to control the MIMO system (29) step response.

Here, the compensator (31) designed in [27] is compared with the method in this paper. Table 5 summarizes the results of evaluating the compensators (30) and (31) using formula (12). Table 5 shows that the decoupling effect of the compensator designed by the method in this paper is better.

$$G_c = \begin{bmatrix} 0.174 & 0.479 \\ -0.219 & 0.503 \end{bmatrix}. \quad (31)$$

Table 5. Comparison of evaluation.

	Ref. [27]	Proposed Method
evaluating value	0.048587	0.0000067

In [27], Masaya et al. designed a PID controller according to Shunji's optimization method. We also use the design method proposed in [27] to design the controller for the decoupling system $Q(s)$, which uses the compensator (30), and Figure 12 shows the results of the Masaya-PID controller and the CDM controller in this paper to control the decoupling system $Q(s)$. From this figure and the performance values appearing in Table 6, it is seen that the CDM controller has a more successful time-domain performance.

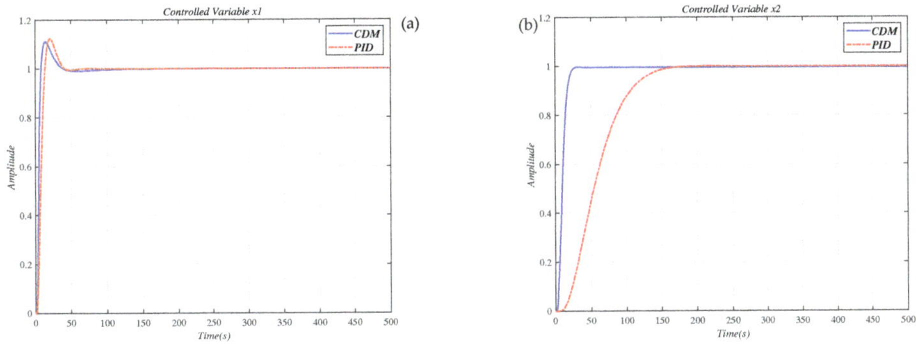

Figure 12. Use the PID controller in [27] and CDM controller to control the decoupling system $Q(s)$ step response. (**a**) controlled variable x_1; (**b**) controlled variable x_2.

Table 6. Performance values of the time response curves shown in Figure 12.

	Settling Time	Max Overshoot %
Masaya-y_1	43	15
CDM-y_1	41	13
Masaya-y_2	172	1.5
CDM-y_2	26	0

In order to verify the robustness of the method in this paper, the system with disturbances and modeling errors is simulated. When there is a step disturbance in the original system, the control result is shown in Figure 13. According to Figure 13, the influence of the disturbance signal subsides in a short time. Suppose that the correct system model is represented by Equation (32), and the system model with errors is represented by Equation (29). The changes in the parameter of Equation (29) are in the interval $\pm 15\%$. The compensator (30) is applied to the correct system model (32), and the CDM controller is used to control. The experimental results are shown in Figure 14. From the results in Figure 14, it can be seen that when the model has measurement errors, the control effect of the method in this paper is good. Figures 13 and 14 show that the method proposed in this paper is robust.

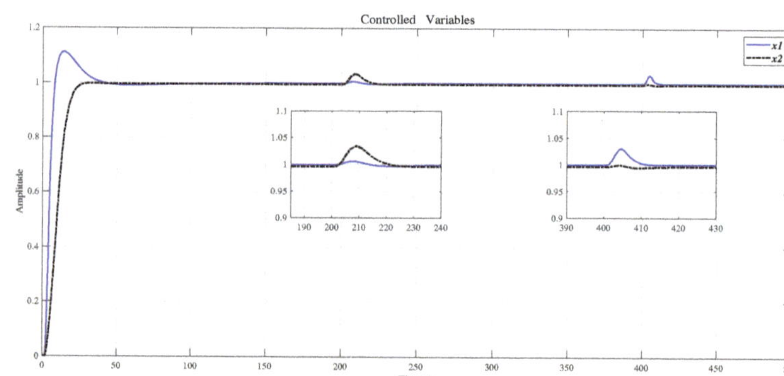

Figure 13. Use the method in this paper to control the step response of the original system (29) with the step disturbance.

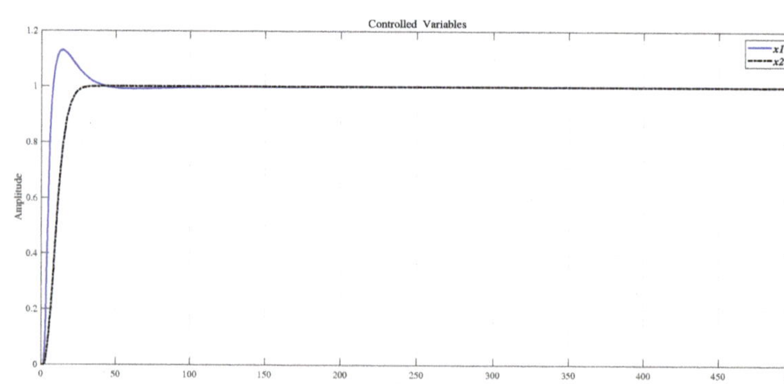

Figure 14. Use the method of this article to control the step response of the system (32).

$$G_P = \begin{bmatrix} \frac{0.238}{21s^2+10s+1} & \frac{-0.3795}{30s^2+11s+1} \\ \frac{0.34}{270s^2+39s+1} & \frac{0.575}{432s^2+42s+1} \end{bmatrix}. \quad (32)$$

Example 3. *The multivariable four-tank system has a tunable transmission of zero [28,29]. With appropriate "tuning", this system will exhibit nonminimum-phase characteristics. Applying the nominal operating parameters given in [28,29] yields the four-tank system model:*

$$G_p = \begin{bmatrix} \frac{0.1987}{65s+1} & \frac{-0.3779}{(65s+1)(34s+1)} \\ \frac{0.4637}{(54s+1)(45.3s+1)} & \frac{0.16194}{54s+1} \end{bmatrix}. \quad (33)$$

For the four-tank system of the controlled object (33), the frequency $\omega_0 = 0.34$ is selected, and the precompensator is obtained.

$$G_c = \begin{bmatrix} -0.3262 & 0.8884 \\ 0.9455 & -0.4597 \end{bmatrix}. \quad (34)$$

Use the evaluation formula (12) to evaluate the decoupling system $Q(s)$ after the compensator formula (34) acts on the controlled object formula (33), and the result is 0.000013. It shows that after the decoupling design, the interaction of the controlled object (33) is weak, and the decoupling effect is well.

The two SISO systems after decoupling are set to A_1 and A_2. In order to suppress measurement noise, we select the controller coefficient $k_0 = 0$. Then, use the stability index γ_i and equivalent time constant τ in Table 7 to calculate CDM control polynomial parameters. Table 7 shows the equivalent time constant τ and CDM control polynomial parameter values.

Table 7. The equivalent time constant τ and CDM control polynomial parameter values.

	System A_1	System A_2
τ	120	100
$F(s)$	3.0833	2.8187
$A(s)$	$69.2884s^2 + 4.1572s + 0.9019$	$75.3466s^2 + 3.8905s + 0.9513$
$B(s)$	$-3231.1s^2 - 109.6204s$	$-3056.7s^2 - 88.0734s$

Use the CDM controller in Table 7 to control A_1 and A_2, the result is shown in Figure 15.

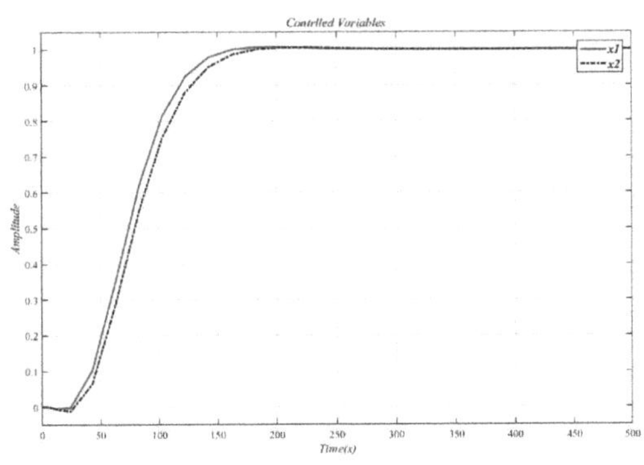

Figure 15. CDM controller controls A_1 and A_2 step response.

Figure 16 shows the result of inserting the compensator (34) in front of the controlled object (33) and using the CDM parameters in Table 7 for control. It can be seen that the system overshoot is slightly increased due to the interaction.

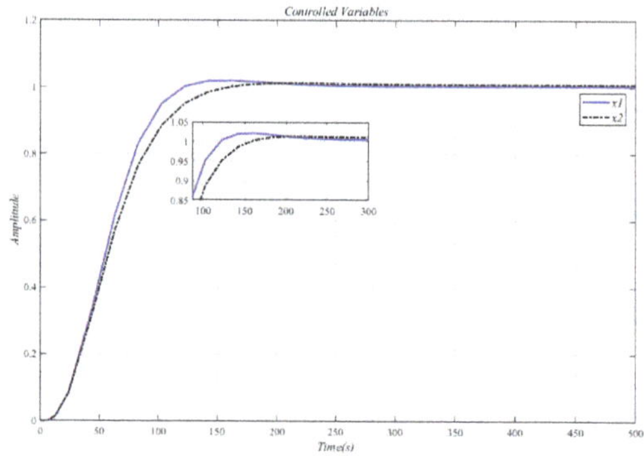

Figure 16. Use this method to control the MIMO system (33) step response.

Figure 17 shows the result of the controlled MIMO system (33) under the measurement noise whose magnitude is limited within [−0.0016, 0.0016]. It can be seen that the system response has not changed, and the measurement noise does not affect performance.

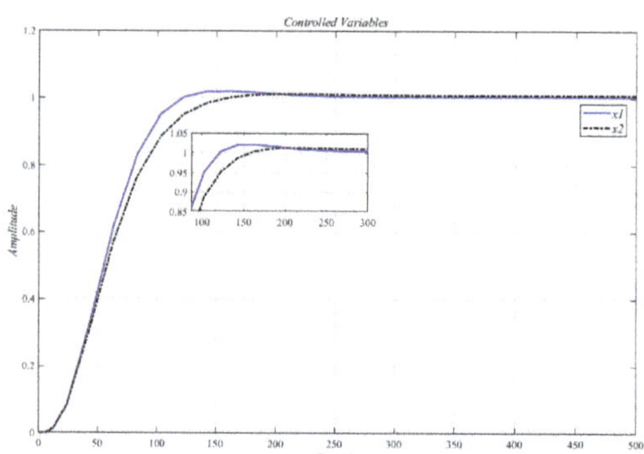

Figure 17. Controlled MIMO system (33) under the measurement noise.

Example 4. *The controlled object (29) increases the delay link to become the accused object (35).*

$$G_P = \begin{bmatrix} \frac{0.28}{21s^2+10s+1}e^{-0.71s} & \frac{-0.33}{30s^2+11s+1}e^{-2.24s} \\ \frac{0.4}{270s^2+39s+1}e^{-0.59s} & \frac{0.5}{432s^2+42s+1}e^{-0.68s} \end{bmatrix}. \tag{35}$$

Using the method in this paper, select the angular frequency $\omega_0 = 0.28$ and obtain the compensator.

$$G_c = \begin{bmatrix} 0.7809 & 0.7625 \\ -0.6247 & 0.6470 \end{bmatrix}. \tag{36}$$

Use Equation (12) to evaluate the effect of the compensator (Equation (36)) on the controlled object (Equation (35)) to obtain the decoupling system $Q(s)$, and the result is 0.000000057. It shows that the system interaction effect of using compensator decoupling is minimal, and the decoupling effect is good.

The two SISO systems after decoupling are set to A_1 and A_2. For A_1 and A_2, the delay link is approximated by the improved Padé approximation method in [26]. Then use the stability index γ_i and Table 8 equivalent time constant τ to calculate the CDM control polynomial parameters. Table 8 shows the equivalent time constant τ and CDM control polynomial parameter values.

Table 8. The equivalent time constant τ and CDM control polynomial parameter values.

	System A_1	**System A_2**
τ	76	64
$F(s)$	0.0017	0.00028
$A(s)$	$0.3665s^2 - 0.0016s$	$0.0015s^2 + 0.000555s$
$B(s)$	$0.2464s^2 + 0.1321s + 0.0017$	$0.0728s^2 + 0.0102s + 0.0002773$

Using the CDM control polynomial parameters in Table 8 to control A_1 and A_2, the results are shown in Figure 18.

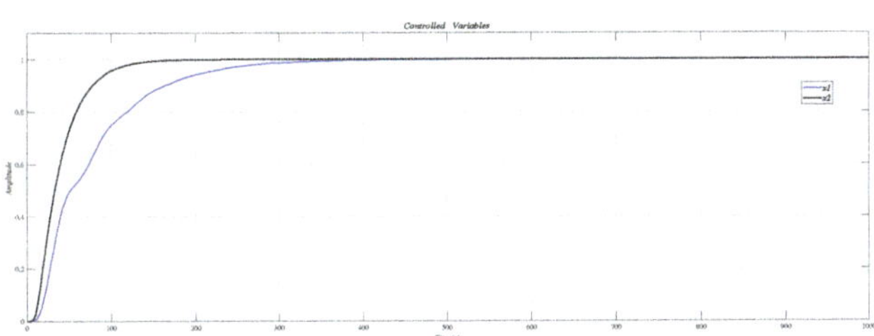

Figure 18. CDM control A_1 and A_2 step response.

Figure 19 is the result of inserting the compensator (36) before the controlled object (35) and using the CDM parameters in Table 8 for control. Figure 18 is the same as Figure 19, which proves that the decoupling effect is well.

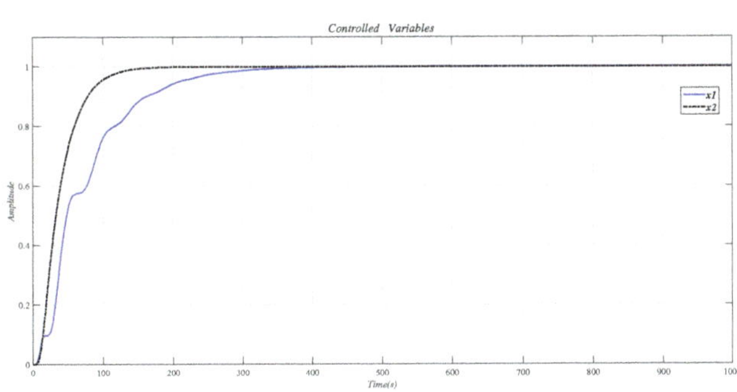

Figure 19. Use this method to control the MIMO system (33) step response.

Example 5. *The controlled object (29) adds a line of input and a column output to become the controlled object (37).*

$$G_p = \begin{bmatrix} \frac{0.28}{21s^2+10s+1} & \frac{-0.33}{30s^2+11s+1} & \frac{0.38}{45s^2+12s+1} \\ \frac{0.4}{270s^2+39s+1} & \frac{0.5}{432s^2+42s+1} & \frac{0.6}{543s^2+68s+1} \\ \frac{0.9}{500s^2+30s+1} & \frac{0.45}{440s^2+45s+1} & \frac{1}{600s^2+89s+1} \end{bmatrix}. \quad (37)$$

Using the method in this paper, select the angular frequency $\omega_0 = 0.13$, and obtain the compensator:

$$G_c = \begin{bmatrix} 0.6251 & -0.7607 & 0.8234 \\ 0.3317 & 0.0941 & 0.0350 \\ -0.7061 & 0.6423 & -0.5667 \end{bmatrix}. \quad (38)$$

Use the evaluation (12) to evaluate the decoupling system $Q(s)$ after the compensator formula (38) acts on the controlled object formula (37), and the result is 0.0005411. The two SISO systems after decoupling are set to A_1, A_2 and A_3. For A_1, A_2 and A_3, use the stability index γ_i and Table 9 equivalent time constant τ to calculate the CDM control polynomial parameters. Table 9 shows the equivalent time constant τ and CDM control polynomial parameter values.

Table 9. The equivalent time constant τ and CDM control polynomial parameter values.

	System A_1	System A_2	System A_3
τ	38.67	100	68
$F(s)$	-5.4186	3.6895	0.5582
$A(s)$	$0.0011s^2 + 0.0169s$	$0.0253s^3 + 0.08397s^2 + 12.9567s$	$0.0027s^2 + 0.0055s$
$B(s)$	$-25.92s^2 - 19.0082s$ -5.4186	$41568.98s^3 + 8205.15s^2$ $+447.8735s + 3.6895$	$9.8979s^2 + 3.4319s$ $+0.5582$

Using the CDM control polynomial parameters in Table 9 to control A_1, A_2 and A_3, the results are shown in Figure 20.

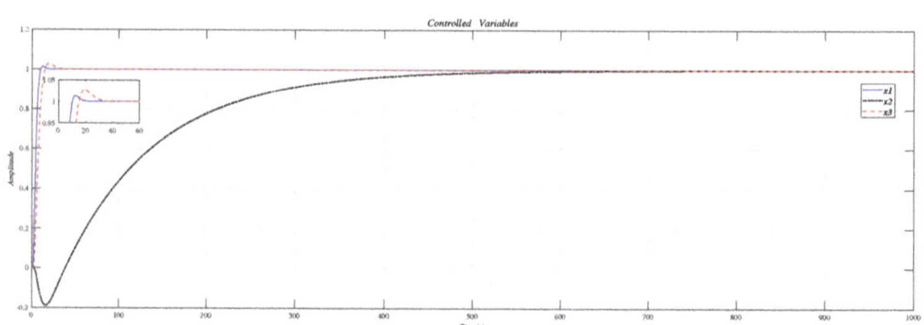

Figure 20. CDM control A_1, A_2 and A_3 step response.

Figure 21 is the result of inserting the compensator (38) before the controlled object (37) and using the CDM parameters in Table 9 for control.

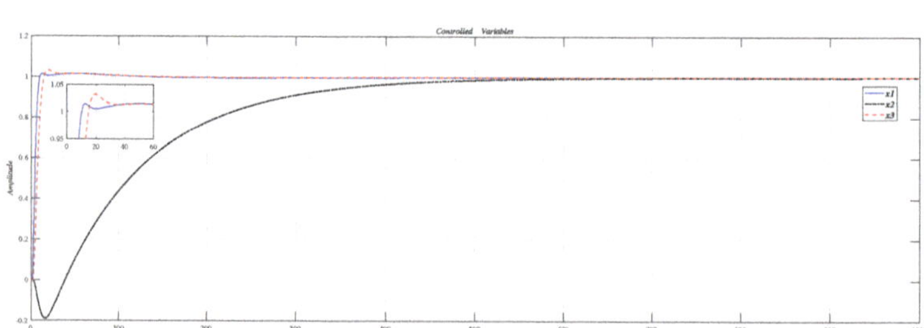

Figure 21. Use this method to control the MIMO system (37) step response.

6. Conclusions

This paper proposes a multivariable system controller design method based on the CDM and analyzes the controller's suppression effect on measurement noise based on the

CDM. The decoupling design is realized by designing the compensator in the frequency domain, and the compensator parameters are optimized through PSO. At the same time, use statistical tests to compare four evolutionary algorithms, including PSO, GA, SFLA, CS, to prove the advantages of PSO. After decoupling, the open-loop transfer function of the system is complex. Therefore, the controller structure design and parameter tuning are based on CDM. Finally, simulation experiments are carried out for four unique control targets. The results show that the decoupling effect of the MIMO system is good, and the designed system can take into account stability, response characteristics, and robustness at the same time, which confirms the effectiveness of the method.

Author Contributions: All authors discussed and agreed on the idea and scientific contribution. Conceptualization, K.L., F.M. and S.M.; methodology, K.L. and F.M.; software, S.M. and C.W.; validation, K.L., F.M. and S.M.; writing—original draft preparation, K.L. and S.M.; writing—review and editing, F.M., S.M. and C.W.; funding and supervision by F.M. All authors have read and agreed to the published version of the manuscript.

Funding: This research was funded by the National Natural Science Foundation of China (12162007), Natural Science Foundation of Hebei Province (F2019501012).

Institutional Review Board Statement: Not applicable.

Informed Consent Statement: Not applicable.

Data Availability Statement: Not applicable.

Conflicts of Interest: The authors declare no conflict of interest.

References

1. Liu, L.; Yin, S.; Zhang, L.; Yin, X.; Yan, H. Improved results on asymptotic stabilization for stochastic nonlinear time-delay systems with application to a chemical reactor system. *IEEE Trans. Syst. Man Cybern. Syst.* **2016**, *47*, 195–204. [CrossRef]
2. Kumar, S.D.; Chandramohan, D.; Purushothaman, K.; Sathish, T. Optimal hydraulic and thermal constrain for plate heat exchanger using multi objective wale optimization. *Mater. Today Proc.* **2020**, *21*, 876–881. [CrossRef]
3. Meng, F.; Song, P.; Liu, K. PID–P compound control of flexible transmission system with sandwich structure. *Control Theory Appl.* **2020**, *37*, 2432–2440.
4. Komareji, M.; Shang, Y.; Bouffanais, R. Consensus in topologically interacting swarms under communication constraints and time-delays. *Nonlinear Dyn.* **2018**, *93*, 1287–1300. [CrossRef]
5. Meng, F.; Wang, D.; Yang, P.; Xie, G. Application of Sum of Squares Method in Nonlinear H infinity Control for Satellite Attitude Maneuvers. *Complexity* **2019**, *2019*, 5124108. [CrossRef]
6. Lee, J.; Hyun Kim, D.; Edgar, T.F. Static decouplers for control of multivariable processes. *AIChE J.* **2005**, *51*, 2712–2720. [CrossRef]
7. Hagglund, T. The one-third rule for PI controller tuning. *Comput. Chem. Eng.* **2019**, *127*, 25–30. [CrossRef]
8. Diaz-Rodriguez, I.D.; Han, S.; Bhattacharyya, S.P. PID Control of Multivariable Systems. *Anal. Des. PID Controll.* **2019**, 217–231. [CrossRef]
9. Coelho, M.S.; da Silva Filho, J.I.; Côrtes, H.M.; de Carvalho, A., Jr.; Blos, M.F.; Mario, M.C.; Rocco, A. Hybrid PI controller constructed with paraconsistent annotated logic. *Control Eng. Pract.* **2019**, *84*, 112–124. [CrossRef]
10. Liao, Q.; Sun, D. Sparse and decoupling control strategies based on takagi-sugeno fuzzy models. *IEEE Trans. Cybern.* **2019**, *51*, 947–960. [CrossRef]
11. Luan, X.L.; Wang, Z.Q.; Liu, F. Centralized PI control for multivariable non-square systems. *Control Decis.* **2016**, *31*, 811–816.
12. Meng, F.; Pang, A.; Dong, X.; Han, C.; Sha, X. H infinity optimal performance design of an unstable plant under bode integral constraint. *Complexity* **2018**, *2018*, 4942906. [CrossRef]
13. Manabe, S. Coefficient diagram method. *IFAC Proc. Vol.* **1998**, *31*, 211–222. [CrossRef]
14. Coelho, J.P.; Pinho, T.M.; Boaventura-Cunha, J. Controller system design using the coefficient diagram method. *Arab. J. Sci. Eng.* **2016**, *41*, 3663–3681. [CrossRef]
15. Hariz, M.B.; Bouani, F.; Ksouri, M. Robust controller for uncertain parameters systems. *ISA Trans.* **2012**, *51*, 632–640. [CrossRef] [PubMed]
16. Abtahi, S.F.; Yazdi, E.A. Robust control synthesis using coefficient diagram method and m-analysis: An aerospace example. *Int. J. Dyn. Control* **2019**, *7*, 595–606. [CrossRef]
17. Kumar, M.; Hote, Y.V. Maximum sensitivity-constrained coefficient diagram method-based PIDA controller design: Application for load frequency control of an isolated microgrid. *Electr. Eng.* **2021**. [CrossRef]
18. Mohamed, T.H.; Shabib, G.; Ali, H. Distributed load frequency control in an interconnected power system using ecological technique and coefficient diagram method. *Int. J. Electr. Power Energy Syst.* **2016**, *82*, 496–507. [CrossRef]

19. Ma, C.; Cao, J.; Qiao, Y. Polynomial-method-based design of low-order controllers for two-mass systems. *IEEE Trans. Ind. Electron.* **2012**, *60*, 969–978. [CrossRef]
20. Banu, U.S.; Aparna, V.; Hussain, M. Coefficient diagram method based control for two interacting conical tank process. In Proceedings of the 2017 Trends in Industrial Measurement and Automation (TIMA), Chennai, India, 6–8 January 2017; pp. 1–5.
21. Mitsantisuk, C.; Nandayapa, M.; Ohishi, K.; Katsura, S. Design for sensorless force control of flexible robot by using resonance ratio control based on coefficient diagram method. *Automatika* **2013**, *54*, 62–73. [CrossRef]
22. Derrac, J.; García, S.; Molina, D.; Herrera, F. A practical tutorial on the use of nonparametric statistical tests as a methodology for comparing evolutionary and swarm intelligence algorithms. *Swarm Evol. Comput.* **2011**, *1*, 3–18. [CrossRef]
23. Manabe, S. Sufficient condition for stability and instability by Lipatov and its application to the coefficient diagram method. In Proceedings of the 9th Workshop on Astrodynamics and Flight Mechanics, Sagamihara, Japan, 22–23 July 1999; Volume 440; p. 449.
24. Ahrens, J.H.; Khalil, H.K. High-gain observers in the presence of measurement noise: A switched-gain approach. *Automatica* **2009**, *45*, 936–943. [CrossRef]
25. Oppenheim, A.V.; Willsky, A.S.; Hamid Nawab, S. *Signals & Systems*; Prentice-Hall: Upper Saddle River, NJ, USA, 1997.
26. Vajta, M. Some remarks on Pade-approximations. In Proceedings of the 3rd TEMPUS-INTCOM Symposium, Veszprém, Hungary, 9–14 September 2000; Volume 242.
27. Sasaki, M.; Mori, K.; Li, X. PID Controller Design of MIMO Systems by using Coefficient Diagram Method. *Essays Soc. Electron.* **2012**, *132*, 1465–1472.
28. Vadigepalli, R.; Gatzke, E.P.; Doyle, F.J., III. Robust control of a multivariable experimental four-tank system. *Ind. Eng. Chem. Res.* **2001**, *40*, 1916–1927. [CrossRef]
29. Viknesh, R.; Sivakumaran, N.; Sarat Chandra, J.; Radhakrishnan, T.K. A critical study of decentralized controllers for a multivariable system. *Chem. Eng. Technol.* **2004**, *27*, 880–889. [CrossRef]

entropy

Article

Observer Based Multi-Level Fault Reconstruction for Interconnected Systems

Mei Zhang [1], Boutaïeb Dahhou [2], Qinmu Wu [1] and Zetao Li [1,*]

[1] Guizhou Provincial Key Laboratory of Internet Plus Collaborative Intelligent Manufacturing Electrical Engineering School, Guizhou University, Guiyang 550025, China; mzhang3@gzu.edu.cn (M.Z.); qmwu@gzu.edu.cn (Q.W.)
[2] UPS, LAAS, University de Toulouse, F-31400 Toulouse, France; boutaib.dahhou@laas.fr
* Correspondence: ztli@gzu.edu.cn

Abstract: The problem of local fault (unknown input) reconstruction for interconnected systems is addressed in this paper. This contribution consists of a geometric method which solves the fault reconstruction (FR) problem via observer based and a differential algebraic concept. The fault diagnosis (FD) problem is tackled using the concept of the differential transcendence degree of a differential field extension and the algebraic observability. The goal is to examine whether the fault occurring in the low-level subsystem can be reconstructed correctly by the output at the high-level subsystem under given initial states. By introducing the fault as an additional state of the low subsystem, an observer based approached is proposed to estimate this new state. Particularly, the output of the lower subsystem is assumed unknown, and is considered as auxiliary outputs. Then, the auxiliary outputs are estimated by a sliding mode observer which is generated by using global outputs and inverse techniques. After this, the estimated auxiliary outputs are employed as virtual sensors of the system to generate a reduced-order observer, which is caplable of estimating the fault variable asymptotically. Thus, the purpose of multi-level fault reconstruction is achieved. Numerical simulations on an intensified heat exchanger are presented to illustrate the effectiveness of the proposed approach.

Keywords: local unknown input; interconnected system; local reconstrucability; global reconstrucability; reduce-order uncertain observer

Citation: Zhang, M.; Dahhou, B.; Wu, Q.; Li, Z. Observer Based Multi-Level Fault Reconstruction for Interconnected Systems. *Entropy* **2021**, 23, 1102. https://doi.org/10.3390/e23091102

Academic Editors: Quanmin Zhu, Giuseppe Fusco, Jing Na, Weicun Zhang and Ahmad Taher Azar

Received: 16 July 2021
Accepted: 18 August 2021
Published: 25 August 2021

1. Introduction

Increasing developments in modern technologies have led to a high complexity of control systems. Thus, either due to physical or analytical purpose, modern control systems are frequently tackled as interconnected systems. Potential faults in interconnected systems have also become inevitable and increasingly complex since faults of the interconnected system can be represented at either the local subsystem level, or at the global system level with the whole system in view, considering faults such as unknown external disturbance, or parameter variations. Faults at either level may not only cause the decline of the performance of both the global system or the local subsystem, but also may trigger a series of fault subsystems. Compared with residual fault diagnosis methodologies, fault reconstruction is capable of identifying the size, location, and dynamics of the fault. In addition, the fault can usually be regarded as an unknown input to the system. The problem of reconstructing the inaccessible inputs from the available measurements is therefore motivated and has attracted remarkable interest in the last decades. Particularly, reconstruction of unknown or inaccessible inputs from noise or indirect measures is very common in many real industrial situations.

In the case of fault diagnosis and unknown input reconstruction for interconnected systems, centralized structure-based fault reconstruction approaches are well investigated, e.g., in Refs. [1–19]. A significant approach of FD and FR for dynamic systems are the observer based methodologies [1–7], with differential geometry-based techniques also representing another attractive method [8–13]. Investigations aimed at solving problems of FD and FR of nonlinear dynamic systems via algebraic and differential techniques can be found in studies such as Refs. [12–17]. These approaches are normally applications of dynamic inversion to achieve the purpose of FD and FR, just as the familiar idea of dynamic inversion is used in the control problem of dynamic systems. Basic notions of this kind of analysis method include the concepts of input reconfigurability [12], left invertibility of dynamic system [9], relative degree of dynamic system and zero dynamics [16].

However, the application of individual system-based methodologies is mainly limited. First, the identification of internal dynamics at local level is incomplete; second, it lacks the dynamics information of the global system. In real applications, it is rather difficult to utilize a centralized scheme to solve the problem of fault reconfiguration in interconnected systems. Luckily, due to advances in computing and communications, it is becoming increasingly popular to directly adopt hierarchical, decentralized, and distributed schemes to deal with fault reconfiguration [20]. In fact, naturally, the architecture of the underlying subsystem is decentralized or distributed, which means that it is necessary to develop distributed FD and FR frameworks. In other words, local fault diagnosis and reconstruction should be performed [21–39]. However, since the interconnected systems are becoming increasingly complex, the problem of system fault reconstruction has also become increasingly difficult, especially problems related to fault propagation, due to the fact that faults occurring in one subsystem influence adjacent subsystems. Therefore, in order to better understand the fault propagation problems, there is research concerning both local and global systems such as in Refs. [20–37]. An important method is to propose a local observer for individual subsystems using its own input and output measurements. All local observers work together to achieve the purpose of estimation and diagnosis of the global system. In this way, the intensive traditional observer design method, based on a single dynamic system, can be employed, such as the high gain observer in Ref. [24], sliding mode observer in Ref. [32], adaptive observer in Ref. [23], etc.

However, the operation of distributed FD and FR approaches greatly relies on reliable information about the full measurement of all subsystems. Such a dependence makes theses methodologies much more challenging, since online measurements available for each subsystem are either difficult to obtain or are inaccurate and or expensive. The matching conditions may be truly too harsh to be satisfied for many physical systems, which makes these methods for unknown input reconstruction not available.

Therefore, it is of great importance to solve the above-mentioned difficulties when analyzing the interconnected systems, which has also motivated us to carry out this research. In this work, system inversion and observer design techniques are combined and extended, aimed at tackling multi-level faults (unknown input) reconstruction problems of the interconnected system. A distributed fault reconstruction scheme is developed and the propagation of the fault effects among interconnected subsystems is investigated. The initial objective is to recognize unknown inputs at the low-level subsystem by using information provided at the global level. A remarkable benefit is that it is capable of reconstructing the system state and local fault signals simultaneously, including incipient faults, for which the fault is considered as an unknown input uncertainty. By introducing the fault as an additional state of the low subsystem, an extended reduced-order observer is developed to produce an estimation of this state. In particular, the output of the lower subsystem is assumed unknown, and is considered as auxiliary output. An inverse-based high order sliding mode observer is developed, aimed at estimating the auxiliary output and its derivatives via measurements of global system. By using this estimation information of auxiliary output, an extended reduced-order observer is generated, aimed at reconstructing the unknown inputs locally. The applicable system categories of this method include systems that depend on polynomial input and its time derivatives. Encouraging numerical simulation results confirm the effectiveness of the proposed multi-level fault reconstruction approach.

The rest of this article is organized as follows: in Section 2, condition of fault reconstructability both locally and globally is given, while in Section 3, a multi-level fault reconstruction scheme for an interconnected system is proposed. First, at the local level, by introducing an auxiliary output to replace its inaccessible output, an extended reduce-order observer is designed to estimate both the states and the fault signals. Second, in order to give an estimation of the auxiliary output and its derivatives, a high order sliding mode observer is introduced. Finally, by gathering all the estimates from both observers, the local fault reconstruction via global information is achieved. In Section 4, the effectiveness of the proposed approach is illustrated by numerical simulations implemented on an intensified heat exchanger. Conclusions and further works are discussed in Section 5.

2. Model Description and Problem Formulation

Analytically, the system can be decomposed into several subsystems, and different control or supervision algorithms can then be developed from both local and global viewpoints, as shown in Figure 1.

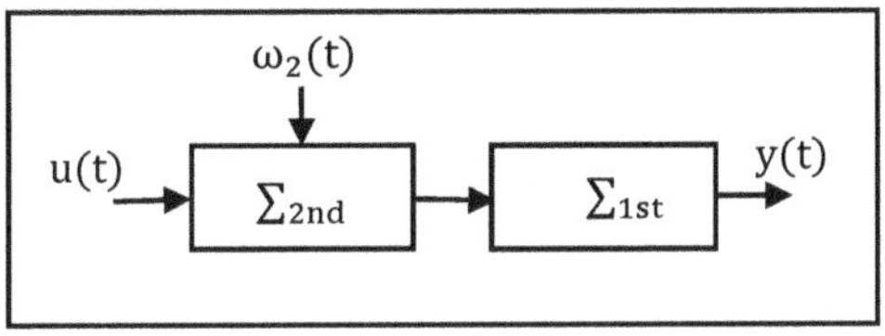

Figure 1. Interconnected system structure.

The important aspect is to develop models of individual subsystems that can describe cause $\omega_2(t)$ and effect $y(t)$ relationships between the 1st and 2nd subsystems. In this case, estimation technologies on states and parameters are capable.

It is supposed that the 1st subsystem can be described with the following state affine form by (1):

$$\sum_{1st}: \begin{cases} \dot{x}_1 = f_1(x_1) + g_1(x_1)u_1 \\ y = h_1(x_1, u_1) \end{cases} \tag{1}$$

where $x_1 \in \Re^n \in M$ is the state of the 1st subsystem, $u_1 \in \Re^m$ is the input of 1st subsystem, which represents elements such as the control input, reference signal, etc., and is also the output of the 2nd subsystem; $y \in \Re^p$ is output of the 1st subsystem, as well as the overall system. f_1, g_1 are smooth vector fields on M. $x_1(t_0) = x_{10}$ is the initial condition. In addition, it is assumed that u_1 is inaccessible and can be recovered through available measures of the global system.

Consider the following nonlinear systems for the 2nd subsystem subject to either actuator or sensor faults ω_2 by (2):

$$\sum_{2nd}: \begin{cases} \dot{x}_2 = f_2(x_2, u, \omega_2) \\ u_1 = h_2(x_2, u, \omega_2) \end{cases} \tag{2}$$

where the state is represented by $x_2 \in \mathcal{R}^n$; $u \in \mathcal{R}^l$ is the input of the 2nd subsystem, as well as the overall system; is the output; $\omega_2 = (\omega_{21}, \omega_{22}, \ldots, \omega_{2k}) \in \mathcal{R}^k$ represents the either actuator or sensor faults of the system. f_2, h_2 are assumed to be analytical vector functions. Specifically, each fault is related to the variables of a specific device and subcomponent. Each of these faults implies an abnormal physical change, such as sticking, leakage or actuator blockage.

In this way, the studied interconnected system is composed of the two local subsystems $\sum_{1st}$ and $\sum_{2nd}$; for the global system, the vector u and y represent its input and output, respectively.

For the interconnected systems described by (1) and (2), the main purpose of the study is to reconstruct fault vector ω_2 at the local level using information at the global level; meanwhile, performance supervision of the global system, as well as individual subsystems, is obliged. A significant objective is to examine whether the unknown inputs ω_2 at local 2nd subsystem can be reconstructed uniquely by output of the 1st subsystem at a global level, given initial states. The initial task is to propose conditions under which both the unknown input and initial state of a known model can be determined from output measurements. For that, the concept of the differential transcendence degree of a differential field extension and the algebraic observability concept of the variable are employed. An interconnected observer based scheme is then developed and analyzed to perform local fault variables reconstruction. A reduced-order uncertainty observer combined with a high order sliding mode observer is developed to achieve this purpose. Finally, the performance of a traditional distributed UIO approach and the proposed multi-level FR approach are compared in detail through numerical simulations, which are presented in an intensified heat exchanger.

3. On Condition of Fault Reconstructability Locally and Globally

In this section, the assumptions and main results on condition of fault reconstructability locally and globally are discussed. An initial task is to prove that the fault vector ω_2 at local level and output vector y at the global are implicitly causal. Moreover, it is also necessary to provide condition to guarantee that local fault impacts on global information are distinguishable. Basic notions are introduced first, and related concepts can be found in Refs. [16,30].

3.1. Fault Reconstructability Condition

To cope with the problem, faults are regarded as local unknown inputs of the interconnected system. Thus, local faults' reconstructability can be treated equivalent with the capability of reconstructing unknown inputs at the local level. In solving the problem of input reconstruction, the primary task is to evaluate the observability of input, so as to distinguish whether the change of input of dynamic systems can be reflected in the change of the output. In order to ensure that the local unknown input can be reconstructed from the global outputs by means of a finite number of ordinary differential equations, there are conditions involving observability and reconstructionability to be met.

From Ref. [30], if any unknown variable x in a dynamic satisfies a differential algebraic equation, the coefficients κ of the equation are greater than in the components of u and y, and the number of its derivatives is finite, then the x is algebraically observable with respect to $\kappa(u_1, \omega_2)$. Any dynamic with output y is said to be algebraically observable if, and only if, any variable has this property. In addition, a fault (unknown input) is defined as a transcendent element over $\kappa(u)$, in which case a faulty system can be viewed as extension of differential transcendence with both fault (unknown input) and its time derivatives. Motivated by this, fault observability of an interconnected system can be defined from multi-level viewpoints:

Definition 1. (Local Algebraic observability). *For subsystem (2), a fault element* $\omega_2 \in \kappa(u, \omega_2)$ *is said to be locally algebraically observable if* ω_2 *satisfies a differential algebraic equation with coefficients over* $\kappa(u, u_1, \omega_2)$.

Definition 2. (Global Algebraic observability). *For interconnected systems depicted by (1) and (2), a fault element* $\omega_2 \in \kappa(u, \omega_2)$ *is said to be globally algebraically observable if* ω_2 *satisfies a differential algebraic equation with coefficients over* $\kappa(u, y, \omega_2)$.

Typically, the problem of observability and left invertibility of dynamic system can be equivalently tackled, while the property of left over invertibility usually means a recontructability of the system input from the output. From Refs. [16,30], if invertibility of the interconnected system, denoted by (1) and (2), can be insured, then it is capable of obtaining the fault element $\omega_{2i}(i = 1, \ldots, k)$ globally from information of overall system output y. Equivalently, if subsystems depicted by (1) and (2) are invertible, respectively, then their inputs and unknown inputs vectors u_1 and ω_2 can be expressed locally by their corresponding local measured outputs y and u_1. To accomplish the aims, the central issue is to provide conditions which can guarantee invertibility of both individual systems and the interconnected system. Luckily, this has been discussed in previous paper in Ref. [30]. It can be seen that a differential output rank is defined to determine invertibility of single dynamic system, while invertibility of all the subsystems are the necessary and sufficient condition for ensuring invertibility of the interconnected system.

Definition 3. (Local reconstructability). *For system (2), it is said to be locally reconstructable if the system is invertible. In this way, it is capable of estimating the unknown input ω_2 from local system information u and u_1.*

For the concept of algebraic observability, it is required that each fault component can be written as the solution of the polynomial equation in ω_{2i} and the finite number of time derivatives of u and u_1 with coefficients in k.

$$\mathcal{H}\left(\omega_{2i}, u, \dot{u}, \ldots, u_1, \dot{u}_1, \ldots\right) = 0 \tag{3}$$

Definition 4. (Global reconstructability). *For the interconnected nonlinear system described by (1) and (2), it is said to be globally reconstructable if the interconnected system is invertible, in this way, it is capable of estimating the unknown input ω_2 from global system information u and y.*

In other words, it is required that the local unknown input vector can be expressed as a solution of a polynomial equation in ω_{2i} and the finite number of time derivatives of u and y with coefficients in k.

$$\mathcal{H}\left(\omega_{2i}, u, \dot{u}, \ldots, y, \dot{y}, \ldots\right) = 0 \tag{4}$$

As mentioned before, requirements of health measurement of all the subsystems increases the difficulty the procedure. In this work, u_1 is supposed to be inaccessible. Therefore, it is also critical for estimating a reliable u_1 and to ensure that reconstructed u_1 has a one-to-one relationship with fault vector ω_{2i}. If it can prove that the reconstructed $\hat{u}_1$ is converged to u_1 with acceptable accuracy, then by substituting u_1 as its estimates $\hat{u}_1$ in (3), the fault vector $(\omega_{2i}, i = 1, \ldots, k)$ is capable of obtaining by a solution of a polynomial equation in ω_{2i} and the finite number of time derivatives of u and $\hat{u}_1$, with coefficients in k.

$$\mathcal{H}\left(\omega_{2i}, u, \dot{u}, \ldots, \hat{u}_1, \dot{\hat{u}}_1, \ldots\right) = 0 \tag{5}$$

In summary, if ω_2 is algebraically observable with respect to u and y, then ω_2 is said to be reconstructable. If, and only if, the interconnected system is invertible both locally and globally, the task of reconstruction of the local unknown input vector ω_2 from global measures, y can be achieved. That is, if the overall interconnected system is invertible, then the impacts of the unknown input ω_2 on the global system, output y is distinguished.

3.2. Minimum Number of Measurements and Reconstructable Unknown Inputs

In this work, accessible measurements are of great importance when implementing the proposed FR method. Therefore, the minimum number of measurements is an essential prerequisite for determining whether a fault in the dynamic system is reconstructable or not. This problem is also related to the problem of invertibility of the dynamic system. According to [16], in order to insure invertibility of the system, the differential output rank of the system should equal to the number of the fault candidates. The differential output rank is also defined as the maximum number of outputs associated with differential polynomial equations with coefficients over $\mathcal{K}$ (independent of x). It means that the available measurable outputs of the system must be greater than or equal to the possible faults.

Remark 1. *For a subsystem described by (1), invertibility cannot be guaranteed if the available outputs are less than the inputs. Conversely, if there are more outputs than inputs, then the redundant outputs are unneeded.*

Remark 2. *For subsystem in failure mode depicted in (2), invertibility of the system cannot be guaranteed if the available outputs is less than the possible faults.*

Proposition 1. *From remarks 1 and 2, the simultaneous reconstructable failure number ($\omega_{2i}, i = 1, \ldots, k$) depends on the number of the measurable outputs.*

Remark 3. *For interconnected system depicted in (1) and (2), the minimum number of available measurements predetermined the reconstructable unknown inputs, thus, equal dimensions of both subsystems and the whole interconnected system is more meaningful.*

4. Observer Design for Unknown Input Reconstruction

As mentioned before, existing observers for fault reconstruction are mainly focused on individual systems. Although there is some research concerning both local and global subsystems, the associated match criteria are usually overly strict to be satisfied in real industrial applications. In order to cope with this difficulty, this work is concerned with the challenges by deriving a fault reconstruction method based on some auxiliary outputs. The architecture of the proposed multi-level fault reconstruction method is shown in Figure 2.

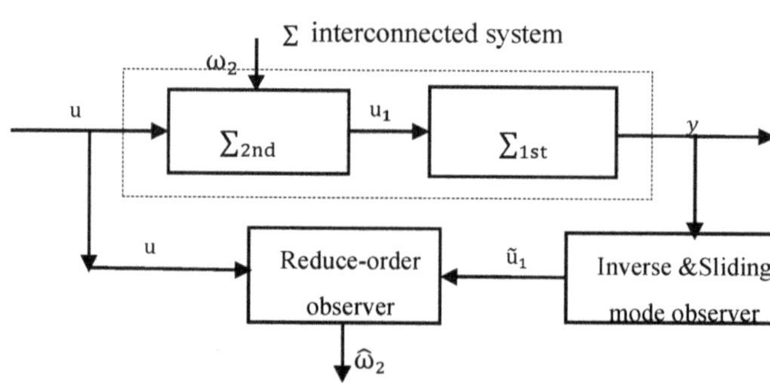

Figure 2. Structure of the proposed local unknown input reconstructor.

The main idea is based on distributed observer design, since distribution resources of dynamic systems are said to be particularly effective for estimation of interconnected systems, due to the fact that they can update internal states using local measurement outputs. However, the significant challenge here is the inaccessible of the interconnection, which is the input of the first subsystem and the output of second subsystem. To cope with this difficulty, first, in order to reconstruct local unknown fault of the first subsystem, the interconnection is extended as an additional state of the first subsystem, an asymptotic reduced-order observer is proposed for the first subsystem, using local input and output measurement information. Then, it is considered the problem that local output is not available directly. An inverse and sliding mode observer based estimator for the second subsystem is then designed to generate an estimation of the local output, and the estimated auxiliary output is applied to the reduced-order observer to replace its measured output. A kind of multi-level fault reconstruction is achieved by gathering estimation of these two observers.

4.1. Asymptotic Reduced-Order Observer Design with Auxiliary Output

Considered system (2), the unknown fault vector $\omega_2(t)$ is assimilated as an extra state of the system with uncertain dynamics. It is expressed according to the states, the unknown inputs (faults) and the known inputs of the system. The dynamics of this new state are unknown. The original system is then converted into an extended system where the dynamics of the extra state are unknown, and it is assumed to be bounded. The original problem is then an observation problem, where the aim is to observe this extra state of the system.

The new extended system is given by:

$$\begin{cases} \dot{x}_2 &= f_2(x_2, u, \omega_2) \\ \dot{\omega}_2 &= \psi(x_2, u, \omega_2) \\ u_1 &= h_2(x_2, u, \omega_2) \end{cases} \tag{6}$$

where $\psi(x_2, u, \omega_2)$ is a bounded uncertain function, $\omega_2(t)$ is algebraically observable over $\kappa(u, u_1)$. It should be noted that a typical structure observer, similarly to a classic Luenberger observer, is not available in the literature because the term $\psi(x_2, u, \omega_2)$ is unknown. Therefore, in order to estimate the unknown input variable ω_2, a proportional reduced-order uncertainty observer using differential algebraic techniques is applied to the fault estimation is constructed to overcome the above problem.

An asymptotic reduced-order observer with a corresponding quadratic-type Lyapunov function can be constructed for system (6):

$$\dot{\hat{\omega}}_{2i} = K_i(\omega_{2i} - \hat{\omega}_{2i}),\ 1 \leq i \leq \varepsilon \tag{7}$$

where $\hat{\omega}_2$ denotes the estimate of the unknown input vector $\omega_2(t)$ and the convergence of the observer is determined by K_i.

Normally, time derivatives of the output are included in the algebraic equation of the unknown input vector, which may enhance computation burden and cause significant computation error even under minor measurement noise, then it is practical and worthwhile to employ an auxiliary variable rather than the computations of the time derivatives.

If the unknown fault vector is algebraically observable and can be written in the following form:

$$\omega_{2i} = \alpha_i \dot{u}_1 + \beta_i(u, u_1) \tag{8}$$

where α_i is a constant vector and $\beta_i(u, u_1)$ is a bounded function.

If a C^1 real-valued function γ exists, such that a proportional asymptotic reduced-order unknown input observer exists, for system (6) it can be written as:

Theorem 1. *Supposed that the auxiliary output vector u_1 is available for measurement, then the system*

$$\begin{cases} \dot{\gamma}_i &= -K_i\gamma_i + K_i\beta_i(u, u_1) - K_i^2\alpha_i u_1 \\ \hat{\omega}_{2i} &= \gamma_i + K_i\alpha_i u_1 \end{cases} \tag{9}$$

is a proportional asymptotic reduced-order unknown input observer for system (6).

The observer (9) can be implemented under assumption that u_1 is measured. However, in our design u_1 is assumed to be unavailable, it is therefore obliged to produce an estimate of the auxiliary output to substitute the measured one.

Remark 4. *By optimizing the observer gain, the optimum tradeoff between the speed of state reconstruction and the robustness to model uncertainty is realized. In this way, the designed observer is not only capable of recovering the system state but also of minimizing the impacts of the measurement noise.*

Remark 5. *It is worth noting that better robustness can be achieved by adding integral action to the proportional asymptotically reduced-order fault observer during the implementation of the observer.*

4.2. Auxiliary Output Estimation

The premise of implementation of Theorem 1 is that the auxiliary output u_l is measurable. It is therefore first required to reconstruct a smooth function of this auxiliary output together with its derivatives from the output data records. To deal with the actual situation, a system inverse-based high order sliding mode observer is considered to accurately estimate the auxiliary output vector u_l and the derivative in the subsystem (1). If this estimation can be well achieved, then the estimated u_l in (9) and its derivatives can be utilized to substitute u_l in (1) to complete the purpose of unknown input reconstruction. In order to achieve this purpose, we first need to construct a dynamic system, which is indeed the realization of the inverse of the original system.

Specifically, for an invertible nonlinear system with the form of (1), a finite relative degree of the output r_i, $i = 1, \ldots, m$ is first defined as the smallest integer as follows:

$$L_{g_1} L_{f_1}^{r_i-1} h_i(x_1) = \left[L_{g_{11}} L_{f_1}^{r_i-1} h_i(x_1) L_{g_{12}} L_{f_1}^{r_i-1} h_i(x_1) \ldots L_{g_{1m}} L_{f_1}^{r_i-1} h_i(x_1) \right] \neq [0, 0, \ldots, 0]$$

Then, by calculating expressions for their derivatives, one gets:

$$\begin{bmatrix} y_1^{(r_1)} \\ \vdots \\ y_m^{(r_m)} \end{bmatrix} = \begin{bmatrix} L_{f_1}^{r_1} h_{11}(x_1) \\ \vdots \\ L_{f_1}^{r_m} h_{1m}(x_1) \end{bmatrix} + \begin{bmatrix} L_{g_{11}} L_{f_1}^{r_1-1} h_{11}(x_1) & \ldots & L_{g_{1m}} L_{f_1}^{r_1-1} h_{11}(x_1) \\ \ldots & \ldots & \ldots \\ L_{g_{11}} L_{f_1}^{r_m-1} h_{1m}(x_1) & \ldots & L_{g_{1m}} L_{f_1}^{r_m-1} h_{1m}(x_1) \end{bmatrix} u_l \tag{10}$$

Although the algebraic polynomial (10) is based on a system inversion, and has already been able to compute u_l, the requirements of calculating of the successive derivative of the output may burden the reconstruction process. In practical applications especially, the measurements often subject to noise, it may result in large overshoot, even failure. An inverse-based high order sliding mode observer is then generated to tackle this problem.

Define the following change of the coordinates:

$$\xi_i = \left[\xi_i^1, \xi_i^2, \ldots, \xi_i^{r_i} \right]^T = \left[\phi_i^1(x_1), \phi_i^2(x_1), \ldots, \phi_i^{r_i}(x_1) \right]^T = \left[h_{1i}(x_1), L_{f1} h_{1i}(x), \ldots, L_{f1}^{r_i-1} h_{1i}(x) \right]^T \quad i = 1, \ldots, m$$

Next, to construct:

$$\begin{aligned} y_i &= \xi_i^1 \dot{\xi}_i^j = \xi_i^{j+1};\ 1 \leq j \leq r_i - 1 \\ \dot{\xi}_i^{r_i} &= L_{f1}^{r_i} h_{1i}(\Phi^{-1}(\xi, \eta) + \textstyle\sum_{j=1}^m L_{g_{1i}} L_{f1}^{r_i-1} h_{1i}\left(\Phi^{-1}(\xi, \eta)\right) u_{lj};\ j = r_i \end{aligned}$$

The expression of input vector u_l is then issued:

$$u_l = A\left(\Phi^{-1}(\xi, \eta)\right)^{-1} \left(\begin{bmatrix} \xi_1^{(r_1)} \\ \vdots \\ \xi_m^{(r_m)} \end{bmatrix} - \begin{bmatrix} L_{f1}^{r_1} h_{11}(\Phi^{-1}(\xi, \eta) \\ \vdots \\ L_{f1}^{r_m} h_{1m}(\Phi^{-1}(\xi, \eta) \end{bmatrix} \right) \tag{11}$$

The inversed based sliding mode observer can then be designed as follows:

$$\begin{aligned} \hat{y}_i &= \hat{\xi}_i^1 \dot{\hat{\xi}}_i^j = \hat{\xi}_i^{j+1} + \lambda_i^j |\hat{y}_i - y_i|^{1/2} \text{sgn}(\hat{y}_i - y_i);\ 1 \leq j \leq r_i - 1 \\ \dot{\hat{\xi}}_i^{r_i} &= \lambda_i^{r_i} |\hat{y}_i - y_i|^{\frac{1}{2}} \text{sgn}(\hat{y}_i - y_I);\ j = r_i \end{aligned} \tag{12}$$

Finally, estimation of ξ_i is achieved finitely:

$$\begin{aligned} \hat{\xi}_i &= \left[\hat{\xi}_i^1, \hat{\xi}_i^2, \ldots, \hat{\xi}_i^{r_i} \right]^T = \left[\hat{\phi}_i^1(x_1), \hat{\phi}_i^2(x_1), \ldots, \hat{\phi}_i^{r_i}(x_1) \right]^T \ i = 1, \ldots, m \\ \hat{\xi} &= [\hat{\xi}_1, \hat{\xi}_2, \ldots, \hat{\xi}_m] = [\hat{\phi}_1(x_1), \hat{\phi}_2(x_1), \ldots, \hat{\phi}_m(x_1)] \end{aligned} \tag{13}$$

4.3. Reconstruction of the Unknown Inputs by Asymptotic Reduced-Order Observer with Auxiliary Output

Now that the estimation of auxiliary output vector u_1 and its derivatives has been achieved, then the unknown input can be reconstructed by this information. The $\hat{u}_1$ is the exact estimate of the auxiliary output vector u_1 in a finite time obtained from the high order sliding mode observer.

Proposition 2. *If it can be insured that reconstructed $\hat{u}_1$ is correctly converged, the conclusion can be obtained that the fault vector ω_2 and $\hat{u}_1$ has one-to-one correspondence.*

Since estimation of the auxiliary output vector is now possible with acceptable accuracy, observer (7) can then be extended in the following form in (14).

Theorem 2. *Supposed that the auxiliary output vector $\hat{u}_1$ is obtained, then an asymptotical reduced-order observer in accordance with the original system (2) can be generated as follows:*

$$\left\{ \begin{array}{rcl} \dot{\gamma}_i &=& -K_i\gamma_i + K_i\beta_i(u, \hat{u}_1) - K_i^2\alpha_i\hat{u}_1 \\ \hat{\omega}_{2i} &=& \gamma_i + K_i\alpha_i\hat{u}_1 \end{array} \right. \tag{14}$$

system (14) is capable of asymptotically reconstructing local unknown input vector finitely.

Proof. Subtracting the first equation of (14) from the first one of original system (2), error dynamic of the observer can be reached. □

While it has been proven that the estimated $\hat{u}_1$ is the accurate estimation of the auxiliary output vector u_1 in a finite time, the convergence of (14) is straightforward because the error dynamic system is not corrupted.

5. Numerical Simulation Implementation on a Pilot Intensified Heat Exchanger

In this section, the effectiveness of our proposed methods is illustrated on a pilot intensified heat exchanger which can be found in Ref. [31] for physical details. Here, the heat exchanger system is regarded as an interconnected system, in which the heat exchanger itself is a subsystem, and the actuator is regarded as the other subsystem cascaded with the heat exchanger. The purpose of the simulation is to prove that the unknown local internal signals of the actuator, like unknown air pressure change, can be recovered by measuring the outlet temperature of the heat exchanger.

5.1. Interconnected System Modelling

Define measured outlet temperatures T_p, T_u of both fluids as two states x_{11}, x_{12} of the heat exchanger subsystem, flow rates F_p, F_u of the two fluids are defined as two inputs u_{11}, u_{12}, which are also the interconnection of the interconnected system, outputs y_1, y_2 are specified as x_{11}, x_{12},

The state space form of heat exchanger subsystem can then be written as:

$$\left\{ \begin{array}{rcl} \dot{x}_{11} &=& \frac{u_{11}}{V_p}(T_{pi} - x_{11}) + \frac{h_pA}{\rho_pC_{pp}V_p}(x_{12} - x_{11}) \\ \dot{x}_{12} &=& \frac{u_{12}}{V_u}(T_{ci} - x_{12}) + \frac{h_uA}{\rho_uC_{pu}V_u}(x_{11} - x_{12}) \end{array} \right. \tag{15}$$

The actuators in this process are two pneumatic control valves, it is to define the stem displacement X_1, X_2 and their derivatives $\frac{dX_1}{dt}, \frac{dX_2}{dt}$ as four states $x_2^T = \begin{bmatrix} x_{21} & x_{22} & x_{23} & x_{24} \end{bmatrix}$ of the actuator subsystem, two local inputs $v^T = \begin{bmatrix} v_1 & v_2 \end{bmatrix}$ are defined as the pneumatic pressure of two valves $u_1^T = \begin{bmatrix} F_1 & F_2 \end{bmatrix}$, two fluid flow rate $F_1 \quad F_2$ are outputs of the subsystem, which correspond to inputs F_p, F_u in the heat exchanger subsystem, and are assumed unmeasured in this subsystem.

The state space form of actuator subsystem can then be written as:

$$\left\{\begin{array}{l} \dot{x}_2 = \begin{bmatrix} 0 & 1 & 0 & 0 \\ -\frac{k_1}{m} & -\frac{\mu_1}{m} & 0 & 0 \\ 0 & 0 & 0 & 1 \\ 0 & 0 & -\frac{k_2}{m} & -\frac{\mu_2}{m} \end{bmatrix} x_2 + \begin{bmatrix} \frac{A_a}{m} & 0 \\ 0 & 0 \\ 0 & \frac{A_a}{m} \\ 0 & 0 \end{bmatrix} v + \begin{bmatrix} 0 \\ \omega_{21} \\ 0 \\ \omega_{22} \end{bmatrix} \\ u_1 = \begin{bmatrix} C_v\sqrt{\frac{\Delta P_1}{sg}} & 0 & C_v\sqrt{\frac{\Delta P_2}{sg}} & 0 \end{bmatrix} x_2 \end{array}\right. \quad (16)$$

ω_{21} ω_{22} are defined as two local fault variables. Each one of these faults represents a variation in the respective control valve gain, which can be originated by an electronic component malfunction, leakage, or an obstruction in the control valve.

5.2. *Observer Design for Unknown Input Reconstruction*

5.2.1. Reduce-Order Observer Design

By calculating output differential rank, it is obvious that both subsystem and the overall system are invertible. Then, it is necessary to verify the condition provided by 3.1 and to construct an algebraic equation for each component of the unknown inputs with coefficients in $\Pi(v, u_1)$.

By obtaining a second time derivative of u_1, it is possible to obtain a differential algebraic polynomial for the unknown inputs whose coefficients are in $\Pi(v, u_1)$.

$$\left\{\begin{array}{l} \omega_{21} = \dot{x}_{22} + \frac{\mu_1}{m}x_{22} + \frac{k_1}{m}u_{11} - \frac{A_a}{m}v_1 \\ \omega_{22} = \dot{x}_{24} + \frac{\mu_2}{m}x_{24} + \frac{k_2}{m}u_{12} - \frac{A_a}{m}v_2 \end{array}\right. \quad (17)$$

Obviously, the time derivates of outputs and the states appear in the algebraic equation of the unknown input, then, according to (13), an auxiliary variable is used to avoid using them.

5.2.2. System Inversion Based Interconnection Reconstruction

The input of the first subsystem can also be represented by means of the output and its derivatives.

Differential all two outputs in (15), and one can obtain:

$$\left\{\begin{array}{l} \dot{y}_1 = \frac{h_pA}{\rho_pC_{pp}V_p}(y_2 - y_1) + \frac{u_{11}}{V_p}(T_{pi} - y_1) \\ \dot{y}_2 = \frac{h_uA}{\rho_uC_{pu}V_u}(y_1 - y_2) + \frac{u_{12}}{V_u}(T_{ui} - y_2) \end{array}\right. \quad (18)$$

Denoted estimates of the two inputs of the heat exchanger subsystem as $\tilde{u}_1 = \begin{bmatrix} \tilde{u}_{11} & \tilde{u}_{12} \end{bmatrix}$, the following expression can be achieved by using above results:

$$\left\{\begin{array}{l} \tilde{u}_{11} = \frac{V_p}{T_{pi}-y_1}\left(\dot{y}_1 - \frac{h_pA}{\rho_pC_{pp}V_p}y_2 + \frac{h_pA}{\rho_pC_{pp}V_p}y_1\right) \\ \tilde{u}_{12} = \frac{V_u}{T_{ui}-y_2}\left(\dot{y}_2 - \frac{h_uA}{\rho_uC_{pu}V_u}y_1 + \frac{h_uA}{\rho_uC_{pu}V_u}y_2\right) \end{array}\right. \quad (19)$$

Obviously, successive derivatives of outputs y_1 and y_2 are required to develop an inversed based second order sliding mode observer to produce exact estimates of them finitely.

Construct new ordinates as:

$$y_1 = T_p = \xi_1^1\ y_2 = T_u = \xi_2^1 \quad (20)$$

The sliding observer of Formula (10) is obtained. Then, the estimated $\tilde{u}_{11}$ and$\tilde{u}_{12}$ can be used to obtain observer of (21).

By construction:

$$\begin{cases} \gamma_1 = \hat{\omega}_{21} + K_1 x_{22} \\ \gamma_2 = \hat{\omega}_{24} + K_2 x_{24} \end{cases} \quad (21)$$

The following reduce-order observer are obtained:

$$\begin{cases} \dot{\gamma}_1 = -K_1\gamma_1 + K_1\left(\frac{\mu_1}{m}x_{22} + \frac{k_1}{m}\tilde{u}_{11} - \frac{A_a}{m}v_1\right) - K_1^2 x_{22} \\ \dot{\gamma}_2 = -K_2\gamma_2 + K_2\left(\frac{\mu_2}{m}x_{24} + \frac{k_2}{m}\tilde{u}_{12} - \frac{A_a}{m}v_2\right) - K_2^2 x_{24} \end{cases} \quad (22)$$

Then, an asymptotic observer is constituted.

5.3. Simulation Results and Discussion

Aimed at illustrating the effectiveness of the proposed multi-level fault reconstruction method, two numerical simulations are carried out in this section. Two kinds of faults are considered, containing sudden changes and incipient variations. In addition, a simulation comparison between the well-known UIO proposed in [30] and the proposed FR is also provided. Detailed values of the variables used for the simulation can be found in [30].

Case 1. *Abrupt fault situation.*

In this simulation, the fault variables are considered to be abrupt ones. The simulation is implemented with initial conditions $\gamma_1 = \gamma_2 = 0$, and the observer gains are given by $K_1 = K_2 = 5$. Two unknown inputs ω_{21}, ω_{22} are considered. Dynamics of ω_{21} remains zero from the beginning, and at t = 50 s, it changes to 10 and never comes back. The value of ω_{22} jump to 60 at 120 s and drops back down at 160 s. Simulation results are reported in Figures 3–8.

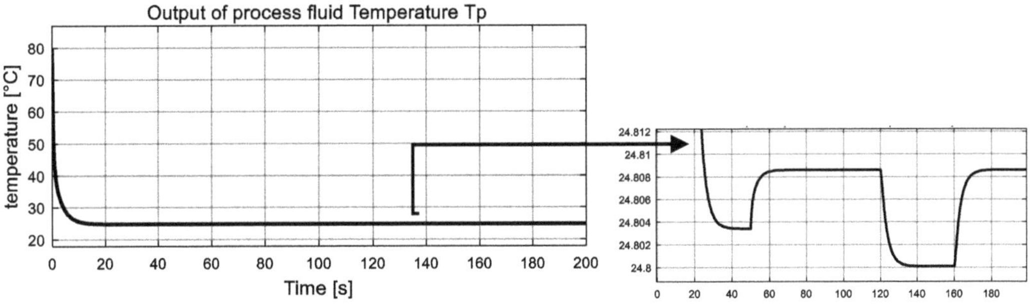

Figure 3. Measured global output: process fluid temperature T_p.

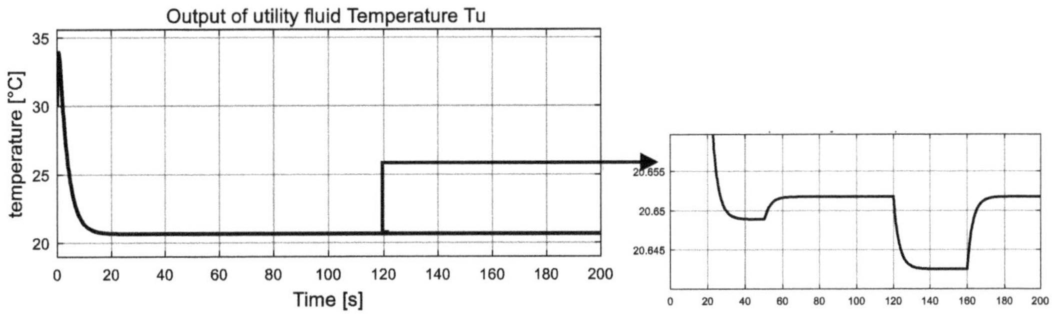

Figure 4. Measured global output: utility fluid temperature T_u.

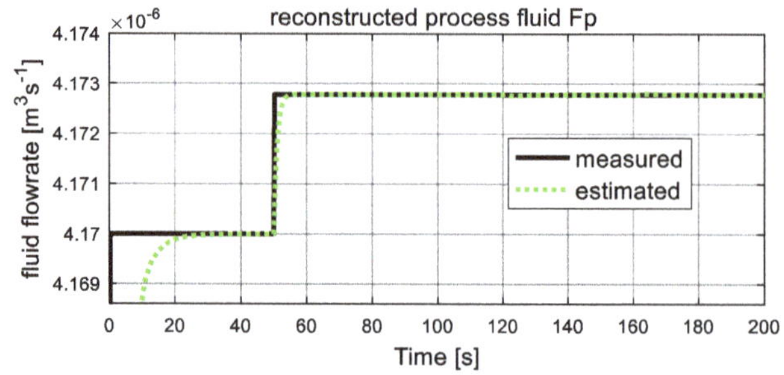

Figure 5. Auxiliary output: process fluid flowrate F_p.

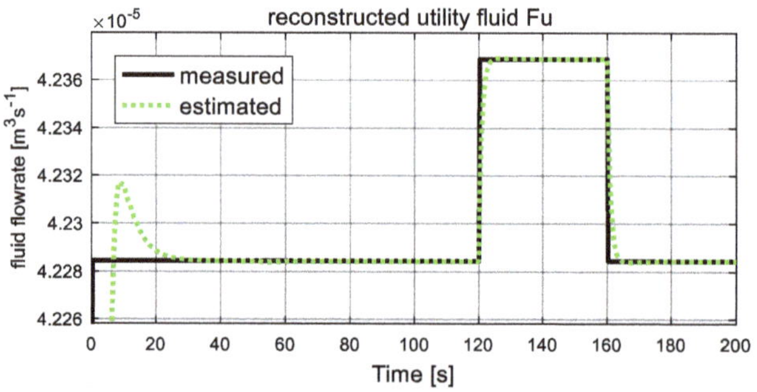

Figure 6. Auxiliary output: utility fluid flowrate F_u.

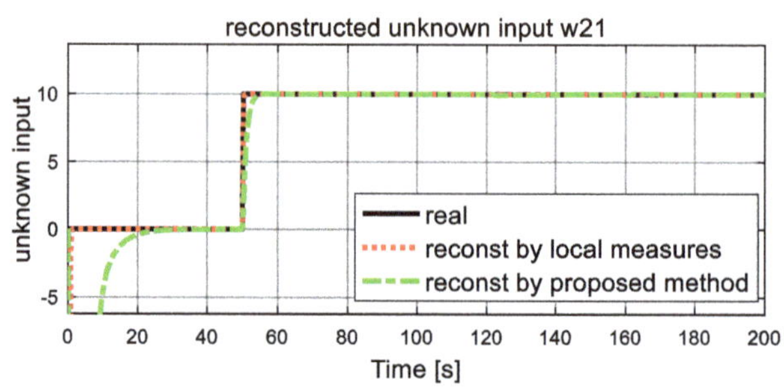

Figure 7. Simulated and reconstructed unknown input ω_{21}.

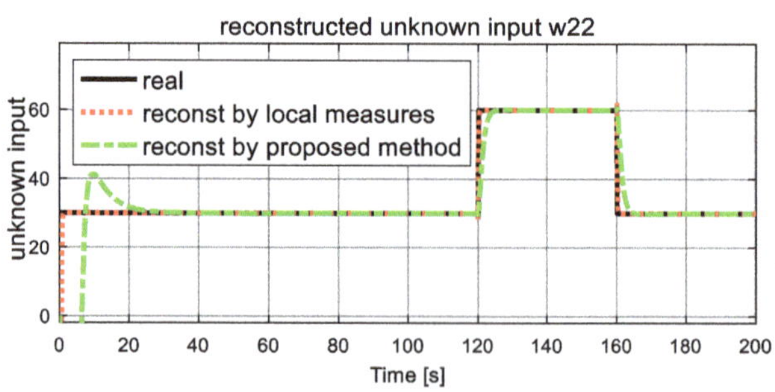

Figure 8. Simulated and reconstructed unknown input ω_{22}.

The measured global outputs, temperature of both fluid T_p and T_u, are shown in Figures 3 and 4. It can be seen that both temperature curves change abruptly at 50 s, 120 s and 160 s. Interestingly, these changes coincide with the changes of two unknown inputs. The measured information is fed to the inverse-based sliding mode observer to correctly estimate the interconnection of the two subsystems, which are also the auxiliary outputs of the low subsystem.

As shown in Figures 5 and 6, the computed fluid flowrates are denoted by the black solid lines, and the dash green lines represent the estimated values. The two figures verify the tracking capacities of designed sliding mode observer. It can be seen that after a short transient time, the estimated curves converge to the computed lines with ready accuracy. From Figure 5, at t = 50 s, the process fluid flowrate F_p increases suddenly and stables at a new level after a short transient time again, which is in accordance with the assumption. Figure 6 shows the computed and estimated result of utility fluid flow rate F_u. It is obvious that the value of computed F_u and its estimated value $\hat{F}_u$ converged adequately after a relatively short transient period. Then, at 120 s, it jumps abruptly and drops to the original value at 160 s, and the estimated dash line tracks the computed solid line again after about 2 s. These variations are influenced by variation of unknown input ω_{22}. Since both estimated fluid flowrates give accurate estimation values to the computed values, they can be used as auxiliary outputs to reduce-order observer to recover the local fault variables.

Dynamics of the fault (unknown inputs) are shown in Figures 7 and 8. The real simulated values are denoted by the black solid lines, and the dash lines and dash-dot lines represent the reconstructed values by a traditional unknown UIO and the proposed FR, respectively, where local measures are available for UIO. From Figures 7 and 8, it is clear that both reconstructed unknown inputs follow closely their corresponding true values. After a short transient time, the reconstructed unknown inputs ω_{21} and ω_{22} in both dash lines and dash-dot lines give accurate estimation values to the simulated real values in solid line. From Figure 7, at 50 s, the estimated ω_{21} unexpectedly increase, and finally it stabilizes at a new level, and an increase of 10 is observed. These changes satisfy the assumption of the unknown inputs ω_{21} correctly. It is also obvious that the traditional UIO method converges quickly than the proposed FR. The similar result is obtained in the estimated ω_{22} of unknown input in Figure 8. At time 120 s, as expected, both simulated and reconstructed curves of the unknown inputs ω_{22} jump with corresponding to the assumption, an increase of 60 is observed, then another drop happens at t = 160 s and it returns to zero with a −60 reduction. It also proves that the reconstructed value in dash line and dash-dot line track well the real simulated value in the solid line. Again, they demonstrate that traditional UIO has better rapidity for fault reconstruction than the proposed FR, and they have the same accuracy as fault reconstruction. However, the proposed FR is more suitable for real engineering world since it does not need local output measures.

The simulation curves indicate that the proposed observer is proper for reconstructing the dynamics of the local unknown inputs with acceptable accuracy, using global measurements.

Case 2. *Incipient fault situation.*

The safe and reliable operation of dynamic systems through the early detection of a small fault before it becomes a serious failure is a crucial component of the overall system's performance and sustainability. In this case, an incipient variation is considered on individual unknown inputs. The simulation is implemented with initial conditions $\gamma_1 = \gamma_2 = 0$, and the observer gains are given by $K_1 = 10$, $K_2 = 5$. Two unknown inputs ω_{21}, ω_{22} are considered. The dynamics of ω_{21} is generated by $10\left[1+\sin\left(0.2te^{-0.05t}\right)\right]$. Dynamics of ω_{22} is generated by $3\left[1+\sin\left(0.5te^{-0.1t}\right)\right]$. Simulation results are reported in Figures 9–14.

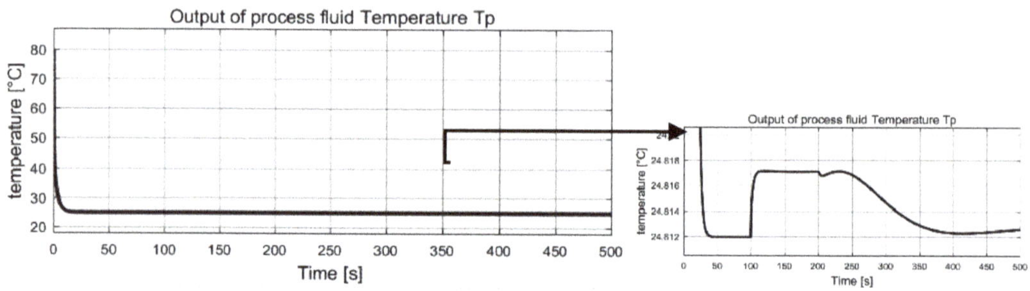

Figure 9. Measured global output: process fluid temperature T_p.

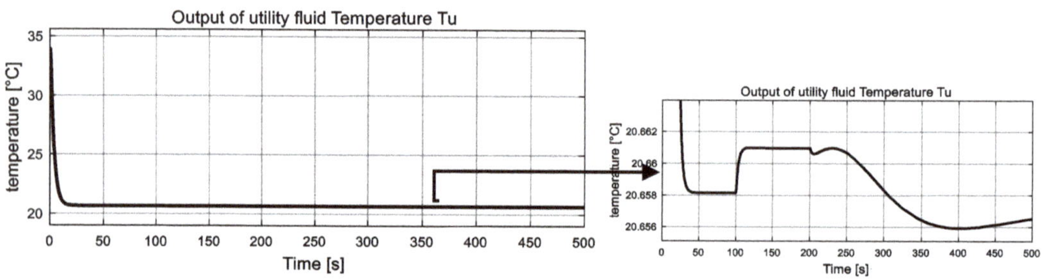

Figure 10. Measured global output: utility fluid temperature T_u.

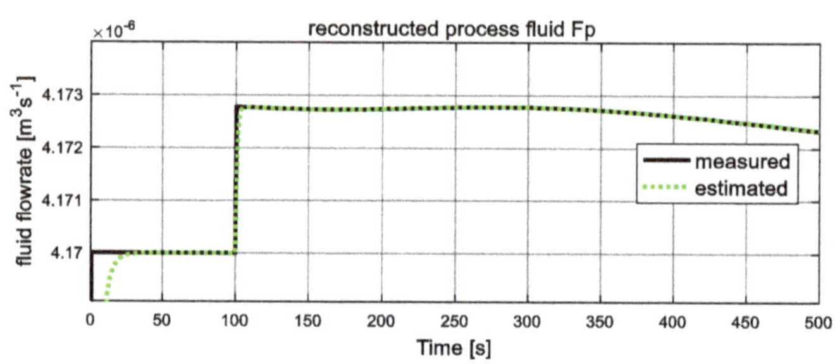

Figure 11. Auxiliary output: process fluid flowrate F_p.

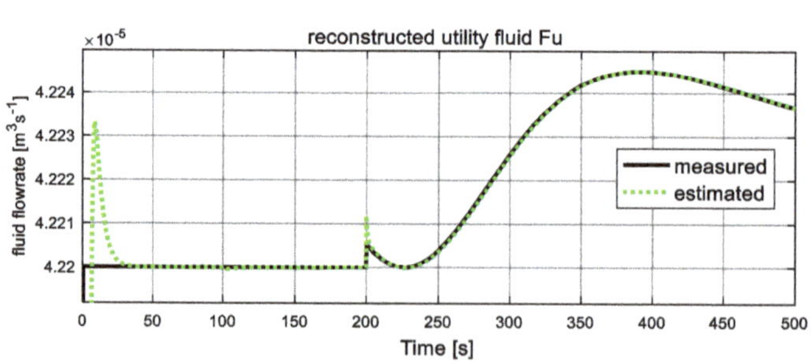

Figure 12. Auxiliary output: utility fluid flowrate F_u.

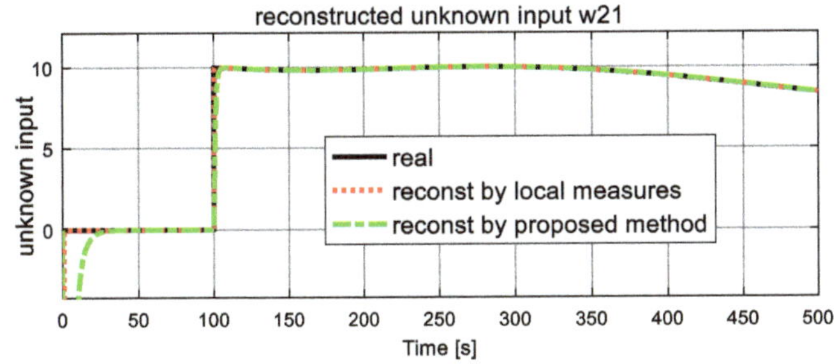

Figure 13. Simulated and reconstructed unknown input ω_{21}.

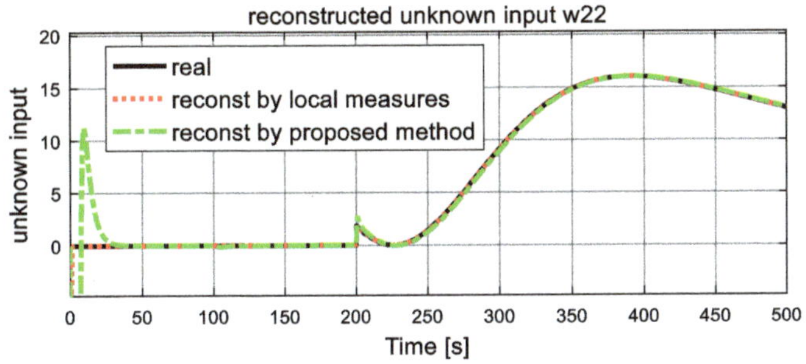

Figure 14. Simulated and reconstructed unknown input ω_{22}.

The measured global outputs, temperature of both fluid T_p and T_u, are shown in Figures 9 and 10. It can be seen that both temperature curves change irregularly and incipiently, with these changes coinciding with the changes of two unknown inputs. This measured information are fed to the sliding mode observer to estimate the interconnection of the two subsystems, which are also the auxiliary outputs of the low subsystem.

As shown in Figures 11 and 12, the computed fluid flowrates are denoted by the black solid lines, and the dash lines represent the estimated values. The two figures verify the tracking capacities of designed sliding mode observer. It can be seen that after a short transient time, the estimated curves converge to the computed lines with ready accuracy. Both estimated fluid flowrates give accurate estimation values to the computed values, they can be used as auxiliary outputs to reduce-order observer to recover the unknown inputs.

Dynamics of the unknown inputs are shown in Figures 13 and 14. The real simulated values are denoted by the black solid lines, and the dash lines represent the reconstructed values. From Figures 13 and 14, it is clear that the reconstructed unknown input follows closely their corresponding true values. After a short transient period, the reconstructed unknown inputs ω_{21} and ω_{22} in dash lines produce an accurate estimation value to the simulated real values indicated by the solid line. It can also illustrate that the reconstructed value in dash line tracks well the real simulated value as shown by the solid line.

The obtained results clearly put forward the following features. The results demonstrate that traditional UIO has a faster speed of fault reconstruction than the proposed FR, and both methods can obtain high accuracy in incipient fault reconstruction procedure. Therefore, the proposed multi-level local fault (unknown input) reconstruction approach is effective for an interconnected system with unmeasured information.

6. Conclusions and Discussion

This paper addresses the multi-level local fault (unknown input reconstruction) problem of interconnected nonlinear systems. By introducing the local fault as an additional state and auxiliary outputs of the low subsystem, then the extended states, the auxiliary outputs and their derivatives are then accurately estimated by combining functions of an asymptotical reduce-order observer and an inverse-based second order sliding mode observer. Effectiveness of the proposed schemes is verified by using simulations on an intensified heat exchanger system, and the satisfactory performances are validated by good simulation results. However, large bias and computation errors are observed when significant measured output noise is involved. The applicable system categories of this method include systems that depend on polynomial input and its time derivatives. In addition, the results of this work can easily explore the application scenarios, such as fault detection and fault reconstruction.

In this paper, model uncertainty and external disturbances are not taken into consideration during the FR designing process. Therefore, enhancing the robustness to model uncertainty and external disturbance is a meaningful direction for further research, and relevant investigation has already been started. Moreover, the reconstructed information by the proposed FR could be used in active fault tolerant control of dynamic system for better achieving its effectiveness, and could be another focus of further research.

Author Contributions: M.Z. and Z.L. conceived and designed the study. M.Z. and Q.W. carried out simulations, and wrote the original draft. B.D. reviewed and edited the manuscript. All authors have read and agreed to the published version of the manuscript.

Funding: This research was funded by the National Natural Science Foundation of China, Grant No. 62003106 and No. 51867006, Provincial Natural Science Foundation of Guizhou, Grant No. ZK (2021) 321. Talent Project of GZU (2018) 02. Key Lab construction project of Guizhou Province (2016) 5103.

Conflicts of Interest: The authors declare no conflict of interest.

References

1. Ding, S.X. *Model-Based Fault Diagnosis Techniques-Design Schemes, Algorithms, and Tools*; Springer: Berlin/Heidelberg, Germany, 2008.
2. Gao, Z.; Cecati, C.; Ding, S.X. A Survey of Fault Diagnosis and Fault-Tolerant Techniques Part II: Fault Diagnosis with Knowledge-Based and Hybrid/Active Approaches. *IEEE Trans. Ind. Electron.* **2015**, *62*, 1. [CrossRef]
3. Mironova, A.; Mercorelli, P.; Zedler, A. A multi input sliding mode control for Peltier Cells using a cold–hot sliding surface. *J. Frankl. Inst.* **2018**, *355*, 9351–9373. [CrossRef]
4. Traoré, D.; Leon, J.D.; Glumineau, A. Adaptive Interconnected Observer-Based Backstepping Control Design For Sen-sorless PMSM. *Automatica* **2012**, *48*, 682–687. [CrossRef]
5. Jia, Q.; Wu, L.; Li, H. Robust Actuator Fault Reconstruction for Takagi-Sugeno Fuzzy Systems with Time-varying Delays via a Synthesized Learning and Luenberger Observer. *Int. J. Control Autom. Syst.* **2021**, *19*, 799–809. [CrossRef]
6. Yan, X.-G.; Edwards, C. Nonlinear robust fault reconstruction and estimation using a sliding mode observer. *Automatica* **2007**, *43*, 1605–1614. [CrossRef]
7. Jia, Q.; Chen, W.; Zhang, Y.; Li, H. Fault reconstruction for continuous-time systems via learning observers. *Asian J. Control* **2014**, *18*, 549–561. [CrossRef]
8. Vijay, P.; Tadé, M.O.; Ahmed, K.; Utikar, R.; Pareek, V. Simultaneous estimation of states and inputs in a planar solid oxide fuel cell using nonlinear adaptive observer design. *J. Power Sources* **2014**, *248*, 1218–1233. [CrossRef]
9. Chan, J.C.L.; Lee, T.H.; Tan, C.P.; Trinh, H.; Park, J.H. A Nonlinear Observer for Robust Fault Reconstruction in One-Sided Lipschitz and Quadratically Inner-Bounded Nonlinear Descriptor Systems. *IEEE Access* **2021**, *9*, 22455–22469. [CrossRef]
10. De Persis, C.; Isidori, A. A geometric approach to nonlinear fault detection and isolation. *IEEE Trans. Automat. Control* **2001**, *46*, 853–865. [CrossRef]
11. Edelmayer, A.; Bokor, J.; Szabo, Z.; Szigeti, F. Input reconstruction by means of system inversion: A geometric approach to fault detection and isolation in nonlinear systems. *Int. J. Appl. Math. Comput. Sci.* **2004**, *14*, 189–199.
12. Fang, H.; De Callafon, R.A.; Cortes, J. Simultaneous input and state estimation for nonlinear systems with applications to flow field estimation. *Automatica* **2013**, *49*, 2805–2812. [CrossRef]
13. Pan, S.; Xiao, D.; Xing, S.; Law, S.; Du, P.; Li, Y. A general extended Kalman filter for simultaneous estimation of system and unknown inputs. *Eng. Struct.* **2016**, *109*, 85–98. [CrossRef]
14. Hou, M.; Patton, R. Input Observability and Input Reconstruction. *Automatica* **1998**, *34*, 789–794. [CrossRef]
15. Chakrabarty, A.; Corlessb, M. Estimating unbounded unknown inputs in nonlinear systems. *Automatica* **2019**, *104*, 57–66. [CrossRef]
16. Martínez-Guerra, R.; Mata-Machuca, J.; Rincón-Pasaye, J. Fault diagnosis viewed as a left invertibility problem. *ISA Trans.* **2013**, *52*, 652–661. [CrossRef]

17. Li, J.; Raïssi, T.; Wang, Z.H.; Wang, X.S.; Shen, Y. Interval estimation of state and unknown input for linear discrete-time systems. *J. Frankl. Inst.* **2020**, *357*, 9045–9062. [CrossRef]
18. Illonetto, G.; Saccomani, M.P. Input estimation in nonlinear dynamical systems using differential algebra techniques. *Automatica* **2006**, *42*, 2117–2129. [CrossRef]
19. Fridman, L.; Shtessel, Y.; Edwards, C.; Yan, X.G. Higher-order sliding-mode observer for state estimation and input reconstruction in nonlinear systems. *Int. J. Robust Nonlinear Control* **2008**, *18*, 399–412. [CrossRef]
20. Chakrabarty, A.; Sundaram, S.; Corless, M.J.; Buzzard, G.T.; Rundell, A.E.; Zak, S.H. Distributed Unknown Input Observers For Interconnected Nonlinear Systems. In Proceedings of the American Control Conference, Boston, MA, USA, 6–8 July 2016; pp. 101–106.
21. Dashkovskiy, S.; Naujok, L. Quasi-ISS/ISDS observers for interconnected systems and applications. *Syst. Control Lett.* **2015**, *77*, 11–21. [CrossRef]
22. Yang, J.; Zhu, F.; Yu, K.; Bu, X. Observer-based state estimation and unknown input reconstruction for nonlinear complex dynamical systems. *Commun. Nonlinear Sci. Numer. Simul.* **2015**, *20*, 927–939. [CrossRef]
23. Zhang, M.; Wu, Q.-M.; Chen, X.-P.; Dahhou, B.; Li, Z.-T. Observer Design for Nonlinear Invertible System from the View of Both Local and Global Levels. *Appl. Sci.* **2020**, *10*, 7966. [CrossRef]
24. Qi, Z.; Zhang, X. A distributed detection scheme for process faults and sensor faults in a class of interconnected nonlinear uncertain systems. In Proceedings of the Decision and Control (CDC), 2012 IEEE 51st Annual Conference, Maui, HI, USA, 10–13 December 2012.
25. Hamida, M.A.; Leon, J.D.; Glumineau, A.; Boisliveau, R. An adaptive interconnected observer for sensorless control of PM synchronous motors with online parameter identification. *IEEE Trans. Ind. Electron.* **2013**, *60*, 739–748. [CrossRef]
26. Keliris, C.; Polycarpou, M.M.; Parisini, T. Distributed fault diagnosis for process and sensor faults in a class of interconnected input–output nonlinear discrete-time systems. *Int. J. Control* **2015**, *88*, 1472–1489. [CrossRef]
27. Ali, H.S.; Alma, M.; Darouach, M. Controllers design for two interconnected systems via unbiased observers. *IFAC Proc. Vol.* **2014**, *47*, 10808–10813. [CrossRef]
28. Xi, X.; Zhao, J.; Liu, T.; Yan, L. Distributed-observer-based fault diagnosis and fault-tolerant control for time-varying discrete interconnected systems. *J. Ambient. Intell. Humaniz. Comput.* **2020**, *11*, 459–482. [CrossRef]
29. Methnani, S.; Gauthier, J.P.; Lafont, F. Sensor fault reconstruction and observability for unknown inputs, with an application to wastewater treatment plants. *Int. J. Control* **2011**, *84*, 822–833. [CrossRef]
30. Zhang, M.; Dahhou, B.; Li, Z.T. On Invertibility of an Interconnected System Composed of Two Dynamic Subsystems. *Appl. Sci.* **2021**, *11*, 596. [CrossRef]
31. Indra, S.; Travé-Massuyès, L.; Chanthery, E. A decentralized fault detection and isolation scheme for spacecraft: Bridging the gap between model-based fault detection and isolation research and practice. *EDP Sci.* **2013**, *6*, 281–298.
32. Boem, F.; Ferrari, R.M.; Parisini, T. Distributed Fault Detection and Isolation of Continuous-Time Non-Linear Systems. *Eur. J. Control* **2011**, *17*, 603–620. [CrossRef]
33. Zhang, M.; Li, Z.T.; Cabassud, M.; Dahhou, B.E. Root cause analysis of actuator fault based on invertibility of interconnected system. *Int. J. Model. Identif. Control* **2017**, *27*, 256–270. [CrossRef]
34. Ng, M.K.Y.; Tan, C.P.; Oetomo, D. Disturbance decoupled fault reconstruction using cascaded sliding mode observers. *Automatica* **2012**, *48*, 794–799. [CrossRef]
35. Lendek, Z.; Babuška, R.; De Schutter, B. Distributed Kalman filtering for cascaded systems. *Eng. Appl. Artif. Intell.* **2008**, *21*, 457–469. [CrossRef]
36. Su, Y.; Zheng, C.; Mercorelli, P. Global finite-time stabilization of planar linear systems with actuator saturation. *IEEE Trans. Circuits Syst. II Express Briefs* **2016**, *64*, 947–951. [CrossRef]
37. Benmansour, K.; Tlemani, A.; Djemai, M.; De Leon, J. A new interconnected observer design in power converter: Theory and experimentation. *Nonlinear Dyn. Syst. Theory* **2010**, *10*, 211–224.
38. Bartyś, M.; Patton, R.; Syfert, M.; Heras, S.D.L.; Quevedo, J. Introduction to the DAMADICS actuator FDI benchmark study. *Control Eng. Pract.* **2006**, *14*, 577–596. [CrossRef]
39. Djeziri, M.A.; Toubakh, H.; Ouladsine, M. Fault prognosis based on fault reconstruction: Application to a mechatronic system. In Proceedings of the 3rd International Conference on Systems and Control, Algiers, Algeria, 29–31 October 2013; pp. 383–388.

entropy

MDPI

Article

Cluster-Delay Mean Square Consensus of Stochastic Multi-Agent Systems with Impulse Time Windows

Huan Luo [1,3], Yinhe Wang [1], Ruidian Zhan [1,2,4,*], Xuexi Zhang [1], Haoxiang Wen [3] and Senquan Yang [3]

1 School of Automation, Guangdong University of Technology, Guangzhou 510006, China; luohuan@sgu.edu.cn (H.L.); yinhewang@gdut.edu.cn (Y.W.); zxxnet@gdut.edu.cn (X.Z.)
2 School of Advanced Manufacturing, Guangdong University of Technology, Guangzhou 510006, China
3 School of Intelligent Engineering, Shaoguan University, Shaoguan 512026, China; xiangxiang_0@163.com (H.W.); sqyangg@sgu.edu.cn (S.Y.)
4 Chipeye Microelectronics Foshan Ltd., Foshan 528225, China
* Correspondence: zhanruidian@chipeye.cn; Tel.: +86-020-3932-2190

Abstract: This paper investigates the cluster-delay mean square consensus problem of a class of first-order nonlinear stochastic multi-agent systems with impulse time windows. Specifically, on the one hand, we have applied a discrete control mechanism (i.e., impulsive control) into the system instead of a continuous one, which has the advantages of low control cost, high convergence speed; on the other hand, we considered the existence of impulse time windows when modeling the system, that is, a single impulse appears randomly within a time window rather than an ideal fixed position. In addition, this paper also considers the influence of stochastic disturbances caused by fluctuations in the external environment. Then, based on algebraic graph theory and Lyapunov stability theory, some sufficiency conditions that the system must meet to reach the consensus state are given. Finally, we designed a simulation example to verify the feasibility of the obtained results.

Keywords: cluster-delay mean square consensus; multi-agent systems; stochastic disturbances; impulse time windows; impulsive control

Citation: Luo, H.; Wang, Y.; Zhan, R.; Zhang, X.; Wen, H.; Yang, S. Cluster-Delay Mean Square Consensus of Stochastic Multi-Agent Systems with Impulse Time Windows. *Entropy* **2021**, 23, 1033. https://doi.org/10.3390/e23081033

Academic Editors: Quanmin Zhu, Giuseppe Fusco, Jing Na, Weicun Zhang and Ahmad Taher Azar

Received: 29 June 2021
Accepted: 8 August 2021
Published: 11 August 2021

1. Introduction

In today's era, automation and intelligence are the mainstream directions of technological development. As a typical representative among them, multi-agent systems (MASs) [1] are widely used in epidemiology [2,3], sociology [4,5], engineering circles [6–8], and other fields with their powerful distributed integration capabilities. In [9,10], a concept called Holonic MAS was proposed, and subsequent researchers have achieved a series of meaningful results on this basis. As a key subject in the field of distributed collaborative control, the research on the consensus of MASs has also received increasingly more attention from the academic community, including group or cluster consensus [11–13], leader-following consensus [14–16], H_∞ consensus [17–19], finite-time or fixed-time consensus [20–22], etc. In practical applications, MASs are required to simultaneously tend to multiple consensus states according to different task requirements. Specifically, MASs is divided into multiple clusters (i.e., subgroups) based on the degree of association between agents, and the states of all individuals included in each cluster eventually tend to be the same.

In particular, if a virtual state is selected as the consensus state of a certain cluster, and the remaining clusters' consensus states are different delay states corresponding to the virtual state, such a case is called cluster-delay consensus, and it is also a special case of the group consensus. In [23], for a class first-order nonlinear MASs, the authors proposed the cluster-delay consensus problem for the first time and studied it through a continuity control strategy. Furthermore, in [24], a new type of pinning consensus protocol with intermittent effect was designed to ensure that the system can achieve the cluster-delay consensus. Moreover, by using the pinning leader-following approach, the cluster-delay consensus of first-order nonlinear MASs with aperiodic intermittent communication was studied in [25]. On the basis of these research work, the cluster-delay consensus problem with intermittent effects and layered intermittent communication was studied in [26] through tracking approach. In [27], the authors extend the research work on the first-order integrator system to more complex second-order system, and investigated the cluster-delay consensus problem of a class of second-order nonlinear MASs.

However, the above-mentioned works are all based on the continuity control protocol, which requires the agent to maintain continuous communication with its neighbors. First, it has higher requirements for communication guarantee capability. Second, it also increases the control cost. In applications, the agent may not be able to obtain the neighbor's information continuously, and the above research results will no longer be applicable. At this time, it is conservative. Different from the traditional continuous control method, impulsive control has the advantages of low control cost, high control efficiency, and strong adaptability. Consequently, it is widely used in the research on leader-following consensus or group consensus of MASs [28–30]. Therefore, it is necessary to study the cluster-delay consensus of MASs via impulsive control [31]. In addition, there are some other interesting control mechanisms, such as the fuzzy control-based on sampled data [32,33], which is widely used in the consensus or synchronization problems research of MASs. Actually, the impulsive controller may not accurately act on the system at an ideal fixed impulse instant, it may be earlier or later. Therefore, the impulse appears randomly within a time window that is defined as an impulse time window in [34], and the window must be known. In order to obtain more general results, it is undoubtedly necessary to introduce the concept of impulse time window into the study of cluster-delay consensus. In general, MASs is also affected by stochastic disturbances caused by fluctuations in the external environment. Therefore, it is also necessary to study the cluster-delay consensus of nonlinear stochastic MASs (SMASs) [35].

Inspired by the above discussion, based on impulsive control strategy, we study the cluster-delay consensus of a class of SMASs with impulse time windows. The main contributions are as follows.

- In this paper, the cluster-delay consensus problem of MASs is studied based on the concept of the impulse time window for the first time. From this perspective, our contribution is mainly reflected in solving the problem of how to reasonably preset the impulsive time sequence under the new application background. In other words, setting the corresponding impulse time window layout according to our research results can ensure that MASs achieve cluster-delay consensus under the action of non-fixed position impulsive control signals.
- This paper studies the cluster-delay mean square consensus problem of MASs based on the uncertainty model for the first time, and gives a sufficient mean square consensus criterion through the Itô formula, which deepens and expands the current research jobs to a certain extent.

The organization of the rest of this paper is shown below. Section 2 introduces the commonly used symbols and the content of algebraic graph theory. In Section 3, the research problem is described and the corresponding system model is constructed. In Section 4, the corresponding consensus criterion is derived through the analysis method. Then, numerical simulation is given in Section 5 to verify the validity of the obtained results. Section 6 summarizes the work of the full text.

2. Notation and Preliminaries

The symbols $\mathbb{R}$, $\mathbb{R}^{m\times n}$, and $\mathbb{N}$ denote the sets of real numbers, $m \times n$ matrices, and natural numbers, respectively. $\mathbb{R}^n$ denotes n-dimensional Euclidean space. $\mathbb{N}^+$ denotes the set of positive integers. Symbols $|x|$ and $||x||$ represent the absolute value and the Euclidean norm for $x \in \mathbb{R}$ and $x \in \mathbb{R}^n$, respectively. The Kronecker product and the Kolmogorov operator are denoted by $\otimes$ and $\mathcal{L}$, respectively. For $\varrho \in \mathbb{R}^{m\times n}$, $(\varrho)^T$ and $\lambda_{max}(\varrho)$ denote the transpose and the maximal eigenvalue of the matrix ϱ, respectively. $E(\cdot)$ denotes the mathematic expectation of corresponding variable. Let $w(t)$ be the Wiener process with m-dimensional, which defined on the complete probability space $(\Omega, \mathcal{F}, \{\mathcal{F}_t\}_{t\geq 0}, P)$ with filtration $\{\mathcal{F}_t\}_{t\geq 0}$. $\text{diag}(\cdot)$ represents a diagonal matrix.

Consider a class of MASs of N agents, and the system's communication topology can be denoted by digraph $\mathcal{G} = (\mathcal{D}, \mathcal{E}, A)$ without self-circulation, where $\mathcal{D} = \{\mathcal{D}_1, \dots, \mathcal{D}_N\}$ is the set of nodes, $\mathcal{E} = \{(\mathcal{D}_j, \mathcal{D}_i) : i, j = 1, \dots, N\} \subset \mathcal{D} \times \mathcal{D}$ is the set of edges, $A = [a_{ij}]$ is the weighted adjacency matrix with order $N \times N$. If $\mathcal{D}_i$ receives the state information of $\mathcal{D}_j$, the weight of edge $(\mathcal{D}_j, \mathcal{D}_i)$ is greater than 0, for convenience, let $a_{ij} = 1$. Otherwise, $a_{ij} = 0$. The degree matrix is denoted by $D = \text{diag}(d_i, i = 1, \dots, N)$, where $d_i = \sum_{j=1, j\neq i}^{N} a_{ij}$. Then, $L = D - A = [l_{ij}]$ denotes the Laplacian matrix, where $l_{ij} = \begin{cases} -a_{ij}, i \neq j \\ -\sum_{j=1, j\neq i}^{N} l_{ij}, i = j \end{cases}$. If MASs contains a leader $\mathcal{D}_0$, then the connection matrix is denoted by $\mathcal{C} = \text{diag}(c_1, \dots, c_N)$. When agent i receives the leader's information, for convenience, let the weight of edge $(\mathcal{D}_0, \mathcal{D}_i)$ be $c_i = 1$. Otherwise, $c_i = 0$. If all agents can receive the leader's information, the leader is called a globally reachable node (i.e., $\mathcal{C}$ is an N-dimensional identity matrix).

Similar to the work in [23], we give the explanation and description of the following concepts in advance to facilitate the understanding of the cluster-delay consensus. If MASs is divided into multiple clusters labeled by $\hat{\mathcal{D}}_1, \dots, \hat{\mathcal{D}}_Q$, respectively, and let the index sets of Q clusters be $\hat{\mathcal{D}}_1 = \{1, 2, \cdots, m_1\}, \dots, \hat{\mathcal{D}}_{\hat{i}} = \{m_1 + m_2 + \cdots + m_{\hat{i}-1} + 1, \dots, m_1 + \cdots + m_{\hat{i}-1} + m_{\hat{i}}\}, \dots, \hat{\mathcal{D}}_Q = \{m_1 + m_2 + \cdots + m_{Q-1} + 1, \dots, N\}$, where $N = m_1 + \cdots + m_Q$, $\hat{i} \in \{1, 2, \cdots, Q\}$, $Q \in \mathbb{N}^+$, $m_{\hat{i}} \in \mathbb{N}^+$. If the i-th agent belongs to a certain cluster, let the subscript of the index set of the cluster be $\hat{i}$, that is, $i \in \hat{\mathcal{D}}_{\hat{i}}$ and $\hat{i} = 1, \dots, Q$. As for why these concepts are introduced, we will describe them in detail in the following part.

3. Problem Description and Model Construction

We consider a first-order nonlinear SMASs composed of N agents, the i-th agent's dynamic is defined by

$$\begin{aligned} dx_i(t) =& [f(t, x_i(t)) + \mathcal{A}x_i(t) - \rho_{\hat{i}}(S_{\hat{i}}(t) - S_1(t - \tau_{\hat{i}})) \\ &+ u_i(t)]dt + \xi(t, x_i(t))dw(t), \end{aligned} \tag{1}$$

where $x_i(t) \in \mathbb{R}^n$ is the state vector (or displacement state vector in some physical systems), $\mathcal{A}$ is a known constant matrix, $f : \mathbb{R} \times \mathbb{R}^n \to \mathbb{R}^n$ is a continuous nonlinear function, $u_i(t) \in \mathbb{R}^n$ is the control input, $S_{\hat{i}}(t) \in \mathbb{R}^n$ is the state vector of the virtual leader of the cluster where the i-th agent belongs, $S_1(t - \tau_{\hat{i}})$ is the delay state of the virtual leader of the first cluster, $\tau_{\hat{i}}$ is the time delay, and $\tau_1 = 0$, $\rho_{\hat{i}}$ is the coupling strength, $\xi : \mathbb{R} \times \mathbb{R}^n \to \mathbb{R}^{n\times m}$ stands for the noisy intensity function.Besides, $w(t)$ is an m-dimensional Wiener process defined on the complete probability space $(\Omega, \mathcal{F}, \{\mathcal{F}_t\}_{t\geq 0}, P)$ with filtration $\{\mathcal{F}_t\}_{t\geq 0}$ which satisfies the usual conditions (i.e., $\mathcal{F}_0$ contains all P-null sets and $\mathcal{F}_t$ is right continuous), and $w_i(t)$ and $w_j(t)$ are independent of each other when $i \neq j$.

Assumption 1. *Each agent has a communication connection with the virtual leader of the cluster to which it belongs, and the first cluster's virtual leader has a communication connection with the virtual leaders of all other clusters.*

Different from continuous control strategy, the following impulsive controller is designed.

$$\begin{aligned} u_i(t) = & \sum_{k=1}^{+\infty} \delta(t - t_k)(K(\alpha \sum_{j=1}^{N} a_{ij}(x_j(t) - S_{\hat{j}}(t) \\ & - (x_i(t) - S_{\hat{i}}(t))) - \beta(x_i(t) - S_{\hat{i}}(t)))), \end{aligned} \tag{2}$$

where $\delta(t)$ is the Dirac function, K is an impulsive gain matrix, $\alpha \in (0, 1)$ and $\beta \in (0, 1)$ are the coupling strengths, $\{t_k\}$ satisfies $0 \leq t_0 < \cdots < t_k$ and $\lim_{k\to+\infty} t_k = +\infty$, $x_i(t)$ is right continuous at each t_k, i.e., $\lim_{h\to 0^+} x_i(t_k + h) = x_i(t_k)$, $k \in \mathbb{N}^+$.

Remark 1. *In [23], the continuity control protocol was designed as* $u_i(t) = -k_{\hat{i}}(s_i(t) - s_1(t - \tau_{\hat{i}})) - \sigma_i(x_i(t) - s_{\hat{i}}(t)) + \sum_{j \notin V_{\hat{i}}} l_{ij}(x_j(t) - x_i(t))$. *It is easy to see that the i-th agent needs to continuously obtain the state information of its neighbors j to update the control signal so that the control cost, as well as the communication burden, are higher. In other words, once the communication between agents cannot be maintained continuously, the above-mentioned controller will lose its effectiveness. However, the impulsive controller is shown in (2) only acts on the system at a series of discrete-time points, which reduces the control cost and the communication volume effectively in the control process. Therefore, the impulsive control mechanism is suitable for some actual environments with a limited communication load, and its adaptability is stronger. In addition, when the state error between the agent and its leader is large, the agent's state will have a large instantaneous jump via the impulsive control, so the response speed is faster than that in other methods.*

For the impulsive control mechanism, we need to preset an impulsive time sequence and assume that the impulse acts on the system at these given ideal moments. However, due to the limitations of physical equipment and objective environments, in practical applications, the real instant of impulse appearance is earlier or later than the ideal moment. In [34], the authors proposed a concept called impulse time window to describe this common phenomenon, as shown in Figure 1, where Z_k^l and Z_k^r are the left and right end points of the k-th window, respectively, t_k is the real impulsive control moment, $t_0 \leq Z_1^l < t_1 < Z_1^r < Z_2^l < \cdots$. It can be seen from Figure 1 that impulse appears randomly in the window, and each window corresponds to only one impulse.

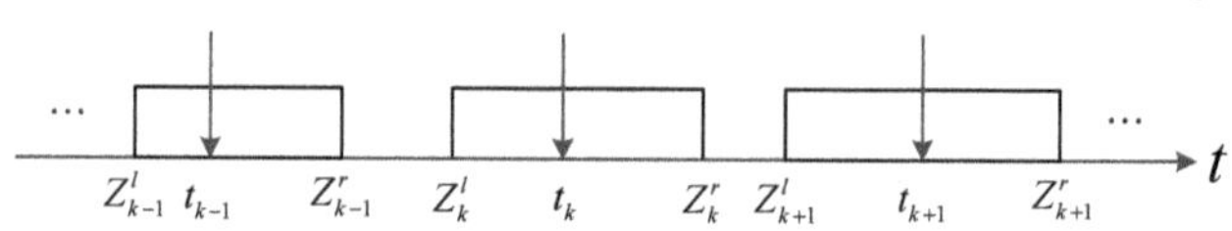

Figure 1. Impulse time windows on the time axis.

We introduce the corresponding virtual leaders into each cluster of SMASs, and their dynamic equations are described by

$$dS_y(t) = [f(t, S_y(t)) + \mathcal{A}S_y(t) - \rho_y(S_y(t) - S_1(t - \tau_y))]dt + \xi(t, S_y(t))dw(t), \tag{3}$$

where $y = 1, \ldots, Q$.

Remark 2. *Because our research object is a SMASs without real leaders, in order to facilitate group control, we assign corresponding virtual leaders to each cluster in the system. Note that to make it easier to construct an error system, the virtual leader and the follower agent have the same dynamics. As the follower agents in each cluster need to reach their respective consensus states, the number of virtual leaders is the same as the number of clusters in the system. At the same time, suppose there is a coupling relationship of state information between some virtual leaders, as shown in Figure 2.*

Let $e_y(t) = S_y(t) - S_1(t - \tau_y)$, $e(t) = (e_1^T(t), \cdots, e_Q^T(t))^T$, $F(t, e_y(t)) = f(t, S_y(t)) - f(t - \tau_y, S_1(t - \tau_y))$, $\bar{F}(t, e(t)) = (F^T(t, e_1(t)), \cdots, F^T(t, e_Q(t)))^T$, $\tilde{\xi}(t, e_y(t)) = \xi(t, S_y(t)) - \xi(t - \tau_y, S_1(t - \tau_y))$, $\bar{\xi}(t, e(t)) = (\tilde{\xi}^T(t, e_1(t)), \cdots, \tilde{\xi}^T(t, e_Q(t)))^T$. Then, according to (3), we can get the following error system.

$$de(t) = [(I_Q \otimes I_n)\bar{F}(t, e(t)) + (I_Q \otimes \mathcal{A})e(t) - (\Lambda \otimes I_n)e(t)]dt + \bar{\xi}(t, e(t))dw(t), \tag{4}$$

where $\Lambda = \text{diag}(\rho_1, \rho_2, \cdots, \rho_Q)$.

Next, based on (1) and (2), we have the system model with impulse time windows as follows.

$$\begin{cases} dx_i(t) = [f(t,x_i(t)) + \mathcal{A}x_i(t) - \rho_{\hat{i}}(S_{\hat{i}}(t) - S_1(t-\tau_{\hat{i}}))]dt + \xi(t,x_i(t))dw(t), t \in [t_0, Z_1^l] \\ \qquad \cup [Z_k^l, t_k) \cup (t_k, Z_k^r], \\ \Delta x_i(t) = x_i(t) - x_i(t^-) \\ \qquad = K(\alpha \sum_{j=1}^{N} a_{ij}(x_j(t^-) - S_{\hat{j}}(t^-) - (x_i(t^-) - S_{\hat{i}}(t^-))) - \beta(x_i(t^-) - S_{\hat{i}}(t^-))), t = t_k. \end{cases} \tag{5}$$

Let $\hat{x}_i(t) = x_i(t) - S_{\hat{i}}(t)$, $\check{f}(t,\hat{x}_i(t)) = f(t,x_i(t)) - f(t,S_{\hat{i}}(t))$, $\check{\xi}(t,\hat{x}_i(t)) = \xi(t,x_i(t)) - \xi(t,S_{\hat{i}}(t))$. Then, error system (6) can be obtained as

$$\begin{cases} d\hat{x}_i(t) = [\check{f}(t,\hat{x}_i(t)) + \mathcal{A}\hat{x}_i(t)]dt + \check{\xi}(t,\hat{x}_i(t))dw(t), t \in [t_0, Z_1^l] \cup [Z_k^l, t_k) \cup (t_k, Z_k^r], \\ \Delta\hat{x}_i(t) = \hat{x}_i(t) - \hat{x}_i(t^-) \\ \qquad = K(\alpha \sum_{j=1}^{N} a_{ij}(\hat{x}_j(t^-) - \hat{x}_i(t^-)) - \beta\hat{x}_i(t^-)), t = t_k. \end{cases} \tag{6}$$

Let $\hat{x}(t) = (\hat{x}_1^T(t), \dots, \hat{x}_N^T(t))^T$, $\check{F}(t,\hat{x}(t)) = (\check{f}^T(t,\hat{x}_1(t)), \cdots, \check{f}^T(t,\hat{x}_N(t)))^T$, $\hat{\xi}(t,\hat{x}(t)) = (\check{\xi}^T(t,\hat{x}_1(t)), \cdots, \check{\xi}^T(t,\hat{x}_N(t)))^T$. Therefore, system (6) can be rewritten as

$$\begin{cases} d\hat{x}(t) = [(I_N \otimes \mathcal{A})\hat{x}(t) + (I_N \otimes I_n)\check{F}(t,\hat{x}(t))]dt + \hat{\xi}(t,\hat{x}(t))dw(t), t \in [t_0, Z_1^l] \cup [Z_k^l, t_k) \\ \qquad \cup (t_k, Z_k^r], \\ \hat{x}(t) = \Omega\hat{x}(t^-), t = t_k, \end{cases} \tag{7}$$

where $\Omega = I_{Nn} - (\beta I_N + \alpha L) \otimes K$, I_N and I_{Nn} are the identity matrices with N-order and Nn-order, respectively.

Remark 3. *In [23–26], the authors have adopted continuity control strategies to study the cluster-delay consensus problem of deterministic MASs. Obviously, this control method will greatly increase control costs and risks [36]. In contrast, this paper is characterized in that the influence of stochastic disturbances is considered, and what is more, it adopts a more advantageous impulsive control strategy. Therefore, the results obtained in this paper are suitable for actual scenarios in the presence of stochastic disturbances and limited communication load. Compared with the work in [31], the system model researched in this paper is more complicated, that is, the concepts of stochastic disturbances and impulse time window are introduced in the construction of the model and the controller, respectively. When the impulse signal appears jitter or drift, the obtained results effectively solve the problem of how to preset the impulse time sequence. In addition, compared with the research work related to the impulse time window, this paper studies the cluster-delay consensus problem of a class of nonlinear SMASs for the first time, and our work is mainly to explore the feasibility of combining these two different research fields. Although the authors considered the influence of random noises in [35,37], the continuity control strategy they applied may bring a great communication burden to the actual control. In this regard, by applying impulsive control mechanism, our paper avoids this problem well.*

For the subsequent consensus analysis, we give the following necessary lemma, assumption, and definitions.

Lemma 1 ([38]). *For vectors $x, \hat{y} \in \mathbb{R}^n$ and constant $\sigma > 0$, we can get $x^T\hat{y} + \hat{y}^T x \le \sigma x^T x + \sigma^{-1}\hat{y}^T\hat{y}$.*

Assumption 2. *$\forall x_i, x_j \in \mathbb{R}^n$, there exist Lipschtiz constants ϕ and $\hat{\phi}$ such that $\|f(t,x_i) - f(t,x_j)\| \le \phi\|x_i - x_j\|$ and $\|\xi(t,x_i) - \xi(t,x_j)\| \le \hat{\phi}\|x_i - x_j\|$.*

Definition 1 ([23]). *The SMASs with (3) and (5) are said to reach cluster mean square consensus, if there exist the solutions of (3) and (5) such that* $\lim_{t\to+\infty} \mathrm{E}(\|\hat{x}_i(t)\|^2) = 0$, *where* $\hat{x}_i(t) = x_i(t) - S_{\hat{i}}(t)$.

Definition 2 ([23]). *The SMASs with (3) are said to reach delay mean square consensus, if there exist the solutions of (3) such that* $\lim_{t\to+\infty} \mathrm{E}(\|e_y(t)\|^2) = 0$, *where* $e_y(t) = S_y(t) - S_1(t-\tau_y)$.

Definition 3. *The SMASs with (3) and (5) are said to reach cluster-delay mean square consensus, if there exist the solutions of (3) and (5) such that* $\lim_{t\to+\infty} \mathrm{E}(\|\hat{x}_i(t)\|^2) = 0$, *and* $\lim_{t\to+\infty} \mathrm{E}(\|e_y(t)\|^2) = 0$, *where* $\hat{x}_i(t) = x_i(t) - S_{\hat{i}}(t)$ *and* $e_y(t) = S_y(t) - S_1(t-\tau_y)$.

Remark 4. *As mentioned above, for ease of understanding, we have provided three different definitions of consensus. Obviously, only when the given conditions in Definitions 1 and 2 are met at the same time, Definition 3 related to cluster-delay consensus needed in this paper can be established. In other words, Definition 3 includes Definitions 1 and 2, and Definitions 1 and 2 are independent of each other. This also facilitates the step-by-step proof of the following consensus analysis part.*

4. Consensus Analysis

In this section, based on the Lyapunov stability theory and combined with the Itô formula, we conduct a theoretical analysis of the cluster-delay consensus problem of the uncertain MASs and give the corresponding consensus criterion. The core idea of the proof is to transform the consensus problem of the original system into the stability analysis problem of the error system. According to Definition 3, the work of this part needs to be divided into two parts, namely, the proof of cluster mean square consensus and delay mean square consensus.

Theorem 1. *Under Assumptions 1–2, for the involved scalars* $\sigma > 0$, $\phi > 0$, *and* $\hat{\phi} > 0$ *satisfying the following conditions (1)–(2), if there exist the solutions of (3) and (5) such that* $\lim_{t\to+\infty} \mathrm{E}(\|\hat{x}_i(t)\|^2) = 0$, *and* $\lim_{t\to+\infty} \mathrm{E}(\|e_y(t)\|^2) = 0$, *then the SMASs with (3) and (5) will achieve cluster-delay mean square consensus.*

(1) There exists a constant $\vartheta > 1$ *such that* $\ln(\vartheta\lambda^*) + \hat{\rho}\left(Z_{k+1}^l - Z_k^l\right) \le 0$, *where* $\lambda^* = \lambda_{\max}(\Omega^T\Omega)$, $\Omega = I_{Nn} - (\beta I_N + \alpha L) \otimes K$, $\hat{\rho} = \gamma + \sigma + \sigma^{-1}\phi^2 + \hat{\phi}^2$, *and* $\gamma = \lambda_{\max}(I_N \otimes (\mathcal{A} + \mathcal{A}^T))$.

(2) There exists a negative definite matrix $I_Q \otimes \mathcal{A} - \Lambda \otimes I_n$ *such that* $\tilde{\rho} = \sigma + \sigma^{-1}\phi^2 + \hat{\phi}^2 + 2\lambda_{\max}(I_Q \otimes \mathcal{A} - \Lambda \otimes I_n) < 0$, *where* $\Lambda = \mathrm{diag}(\rho_1, \rho_2, \cdots, \rho_Q)$.

Proof. *(a): cluster mean square consensus*

Construct the following Lyapunov function:

$$V(t, \hat{x}(t)) = \hat{x}^T(t)\hat{x}(t). \tag{8}$$

The stochastic derivative of (8) is derived by the *Itô* formula along the trajectory of system (7) as follows.

$$dV(t, \hat{x}(t)) = \mathcal{L}V(t, \hat{x}(t)) + 2\hat{x}^T(t)\hat{\xi}(t, \hat{x}(t))dw(t), \tag{9}$$

$$\mathcal{L}V(t, \hat{x}(t)) = 2\hat{x}^T(t)[(I_N \otimes \mathcal{A})\hat{x}(t) + (I_N \otimes I_n)\check{F}(t, \hat{x}(t))] + \mathrm{trace}[\hat{\xi}^T(t, \hat{x}(t))\hat{\xi}(t, \hat{x}(t))]. \tag{10}$$

According to Assumption 2 and Lemma 1, from (10), we have

$$2\hat{x}^T(t)(I_N \otimes \mathcal{A})\hat{x}(t) \le \gamma V(t, \hat{x}(t)), \tag{11}$$

$$\begin{aligned} &2\hat{x}^T(t)(I_N \otimes I_n)\check{F}(t, \hat{x}(t)) \\ &\le \sigma\hat{x}^T(t)\hat{x}(t) + \sigma^{-1}\check{F}^T(t, \hat{x}(t))\check{F}(t, \hat{x}(t)) \\ &\le (\sigma + \sigma^{-1}\phi^2)V(t, \hat{x}(t)) \end{aligned} \tag{12}$$

and

$$\text{trace}[\hat{\xi}^T(t,\hat{x}(t))\hat{\xi}(t,\hat{x}(t))] \leq \hat{\phi}^2 V(t,\hat{x}(t)). \tag{13}$$

For $t \in [t_{k-1}, t_k)$, assume that Δt is a small enough positive constant such that $t + \Delta t \in (t_{k-1}, t_k)$, then one has

$$EV(t+\Delta t, \hat{x}(t+\Delta t)) - EV(t,\hat{x}(t)) = \int_t^{t+\Delta t} E\mathcal{L}V(s,\hat{x}(s))ds. \tag{14}$$

By (11)–(14), we can obtain

$$D^+EV(t,\hat{x}(t)) = E\mathcal{L}V(t,\hat{x}(t)) \leq \hat{\rho}EV(t,\hat{x}(t)). \tag{15}$$

When $t \in \left[t_0, Z_1^l\right)$ and $t \in \left[Z_k^r, Z_{k+1}^l\right)$, from (15), we have

$$EV\left(Z_1^l, \hat{x}\left(Z_1^l\right)\right) \leq EV(t_0,\hat{x}(t_0))\exp\left(\hat{\rho}\left(Z_1^l - t_0\right)\right), \tag{16}$$

$$EV(Z_{k+1}^l, \hat{x}(Z_{k+1}^l)) \leq EV(Z_k^r, \hat{x}(Z_k^r))\exp(\hat{\rho}(Z_{k+1}^l - Z_k^r)). \tag{17}$$

Let $k = 1$. For $t \in \left[Z_1^l, t_1\right)$, it holds that

$$EV(t_1^-, \hat{x}(t_1^-)) \leq EV(t_0,\hat{x}(t_0))\exp(\hat{\rho}(t_1 - t_0)). \tag{18}$$

When $t = t_k$, one has

$$EV(t_k) = E\left(\hat{x}^T(t_k^-)\Omega^T\Omega\hat{x}(t_k^-)\right) \leq \lambda^* EV(t_k^-, \hat{x}(t_k^-)). \tag{19}$$

Thus, from (19), we have

$$EV(t_1,\hat{x}(t_1)) \leq \lambda^* EV(t_1^-, \hat{x}(t_1^-)). \tag{20}$$

For $t \in (t_1, Z_1^r)$, we can get

$$\begin{aligned} &EV(Z_1^r, \hat{x}(Z_1^r)) \\ &\leq EV(t_1,\hat{x}(t_1))\exp(\hat{\rho}(Z_1^r - t_1)) \\ &\leq \lambda^* EV(t_1^-, \hat{x}(t_1^-))\exp(\hat{\rho}(Z_1^r - t_1)) \\ &\leq \lambda^* EV(t_0,\hat{x}(t_0))\exp(\hat{\rho}(Z_1^r - t_0)). \end{aligned} \tag{21}$$

When $t \in \left[Z_1^r, Z_2^l\right)$, by (17) and (21), it follows that

$$\begin{aligned} EV\left(Z_2^l, \hat{x}\left(Z_2^l\right)\right) &\leq EV(Z_1^r, \hat{x}(Z_1^r))\exp\left(\hat{\rho}\left(Z_2^l - Z_1^r\right)\right) \\ &\leq \lambda^* EV(t_0,\hat{x}(t_0))\exp\left(\hat{\rho}\left(Z_2^l - t_0\right)\right). \end{aligned} \tag{22}$$

Let $k = 2$. When $t \in \left[Z_2^r, Z_3^l\right)$, it yields

$$EV(Z_3^l, \hat{x}(Z_3^l)) \leq (\lambda^*)^2 EV(t_0,\hat{x}(t_0))\exp(\hat{\rho}(Z_3^l - t_0)).$$

By analogy, for $t \in \left[Z_k^l, Z_{k+1}^l\right)$, if there exists a constant $\vartheta > 1$ such that $\ln(\vartheta\lambda^*) + \hat{\rho}\left(Z_{k+1}^l - Z_k^l\right) \leq 0$, then we have

$$\begin{aligned} EV(t,\hat{x}(t)) &\leq (\lambda^*)^k EV(t_0,\hat{x}(t_0))\exp(\hat{\rho}(t-t_0)) \\ &\leq EV(t_0,\hat{x}(t_0))\exp\left(\hat{\rho}\left(t-Z_k^l\right)\right)\lambda^*\exp(\hat{\rho}(Z_k^l-Z_{k-1}^l))\cdots\lambda^*\exp(\hat{\rho}(Z_1^l-t_0)) \\ &\leq \frac{1}{\vartheta^k}EV(t_0,\hat{x}(t_0))\exp\left(\hat{\rho}\left(t-Z_k^l\right)\right). \end{aligned} \tag{23}$$

From (23), it can be seen that $EV(t,\hat{x}(t)) \to 0$ when $t \to \infty$. That is, $\lim_{t\to\infty} E(\|x_i(t) - S_{\tilde{i}}(t)\|^2) = 0$. Therefore, the SMASs with (3) and (5) can achieve the cluster mean square consensus.

(b): delay mean square consensus

Construct the following Lyapunov function:

$$V(t,e(t)) = e^T(t)e(t). \tag{24}$$

The stochastic derivative of (24) is derived by the *Itô* formula along the trajectory of system (4) as follows.

$$dV(t,e(t)) = \mathcal{L}V(t,e(t)) + 2e^T(t)\bar{\xi}(t,e(t))dw(t), \tag{25}$$

$$\begin{aligned} \mathcal{L}V(t,e(t)) =& 2e^T(t)[(I_Q\otimes\mathcal{A}-\Lambda\otimes I_n)e(t)+(I_Q\otimes I_n)\bar{F}(t,e(t))] \\ &+\mathrm{trace}[\bar{\xi}^T(t,e(t))\bar{\xi}(t,e(t))]. \end{aligned} \tag{26}$$

Similar to (12) and (13), we have

$$\begin{aligned} &2e^T(t)(I_Q\otimes I_n)\bar{F}(t,e(t)) \\ &\leq \sigma e^T(t)e(t)+\sigma^{-1}\bar{F}^T(t,e(t))\bar{F}(t,e(t)) \\ &\leq (\sigma+\sigma^{-1}\phi^2)V(t,e(t)). \end{aligned} \tag{27}$$

and

$$\mathrm{trace}\left[\bar{\xi}^T(t,e(t))\bar{\xi}(t,e(t))\right] \leq \hat{\phi}^2 V(t,e(t)). \tag{28}$$

Furthermore, one has

$$2e^T(t)(I_Q\otimes\mathcal{A}-\Lambda\otimes I_n)e(t) \leq 2\lambda_{\max}(I_Q\otimes\mathcal{A}-\Lambda\otimes I_n)V(t,e(t)). \tag{29}$$

In the same way, we can get the following inequality similar to (14).

$$EV(t+\Delta t,e(t+\Delta t)) - EV(t,e(t)) = \int_t^{t+\Delta t} E\mathcal{L}V(s,e(s))ds. \tag{30}$$

According to (27)–(30), we can obtain

$$D^+EV(t,e(t)) = E\mathcal{L}V(t,e(t)) \leq \tilde{\rho}EV(t,e(t)). \tag{31}$$

From (31), one has

$$EV(t,e(t)) \leq EV(t_0,e(t_0))\exp(\tilde{\rho}(t-t_0)). \tag{32}$$

At this time, if matrix $I_Q\otimes\mathcal{A}-\Lambda\otimes I_n$ is negative definite and its maximum eigenvalue satisfying $\tilde{\rho} < 0$, then it can be known from (32) that $EV(t,e(t)) \to 0$ when $t \to \infty$. That is, $\lim_{t\to\infty} E(\|S_y(t) - S_1(t-\tau_y)\|^2) = 0$. Consequently, the SMASs with (3) can reach the delay mean square consensus.

According to parts (a) and (b), we can say that the SMASs with (3) and (5) can reach the cluster-delay mean square consensus. This completes the proof. □

Remark 5. *By condition (1), we have $Z_{k+1}^l - Z_k^l \leq \frac{-\ln(\vartheta\lambda^*)}{\hat{\rho}}$, where parameters $\lambda^* < 1$ and $\hat{\rho}$ can be obtained by simple calculations. Without loss of generality, ϑ can be equivalently regarded as an adjustable variable that satisfies $\vartheta\lambda^* < 1$. Obviously, the artificial preset of the impulse time* *windows and the selection of the value of ϑ influence each other. When the interval between adjacent impulse time windows is designed to be larger, this means that ϑ needs to be larger to ensure that $\vartheta\lambda^* < 1$ holds. At this time, it can be seen from (23) that the convergence speed of the error system will decrease. Reflected in the actual control, the impulsive interval may become larger due to the above-mentioned design changes, and the number of impulses within a certain period of time will be reduced, resulting in a slower system convergence speed, and vice versa.*

Remark 6. *We know that $\mathcal{A}$ is a known real matrix, and its value depends on the inherent dynamic behavior of SMASs. In other words, for a particular system, the value of $\mathcal{A}$ cannot be adjusted. Therefore, to satisfy condition (2), we can only adjust the diagonal matrix Λ composed of virtual coupling strengths $\rho_1, \cdots$, and ρ_Q. According to condition (2), we can see that the stronger the coupling strengths are, the easier the inequality $\tilde{\rho} < 0$ is satisfied. At the same time, the delay mean-square consensus of SMASs may be realized faster.*

Remark 7. *Different from the general literature, the proof method in this paper combines the characteristics of multiple current methods and has been successfully applied to the study of the SMASs' cluster-delay mean square consensus problem. In a sense, this is an extension of current research methods. Moreover, how to construct the dynamic equation of the virtual leaders, how to design an impulsive controller and adjust its parameters, how to design a reasonable impulse time sequence, and how to design a reasonable simulation program to verify the effectiveness of the research method are challenging jobs. In addition, through the above research, we can reasonably preset the impulse time sequence to avoid the possible adverse effects of the digital signal's jitter or drift on the system when the MASs are facing stochastic disturbances. In practical applications, the target MASs studied in this paper can be cluster drones flying in formation, numerous unmanned vehicles on the road, or a network of multiple power stations.*

5. Numerical Simulation

Next, we design a simulation example to verify the validity of the obtained results.

Example 1. *Consider a first-order nonlinear SMASs composed of 9 agents, and its topology graph is shown in Figure 2. In order to easily identify the state trajectory of each agent in the simulation diagram, we choose a class of one-dimensional variable as the agent's state, namely, $n = 1$.*

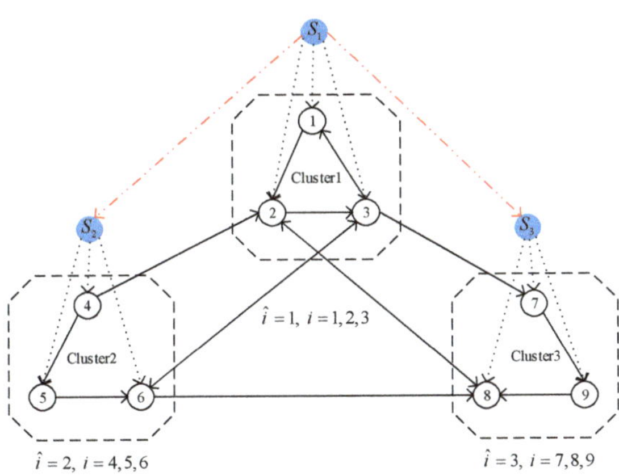

Figure 2. Multi-agent systems with virtual leaders.

Let the initial states $x_1(t_0) = 1$, $x_2(t_0) = 10$, $x_3(t_0) = -12$, $x_4(t_0) = -3$, $x_5(t_0) = -16$, $x_6(t_0) = 5$, $x_7(t_0) = 15$, $x_8(t_0) = -7$, $x_9(t_0) = 9$, $S_1(t_0) = -1$, $S_2(t_0) = -6$, and $S_3(t_0) = 3$. Let functions $f(t, x_i(t)) = x_i(t)\sin(\tan t)$, $f(S_y(t), t) = S_y(t)\sin(\tan t)$, and $\xi(t, x_i(t)) = 0.16|\cos(t)|x_i(t)$. Obviously, we can choose Lipschtiz constants $\phi = 1$ and $\hat{\phi} = 0.16$. Furthermore, let $\mathcal{A} = 1$, $\alpha = 0.2$, $\beta = 0.8$, $K = \text{diag}(0.3, \cdots, 0.3)$, $\rho_2 = \rho_3 = 20$,

$\tau_2 = 0.2$, and $\tau_3 = 0.3$. Based on the above parameters, we have $\lambda^* = 0.6111$, $\gamma = 2$. We can choose parameters $\vartheta = 1.2$ and $\sigma = 1$. Then, it can be calculated according to condition 1) in Theorem 1 that $Z^l_{k+1} - Z^l_k \leq 0.103$. In addition, it is clear that the matrix $I_Q \otimes \mathcal{A} - \Lambda \otimes I_n$ is negative definite and satisfies the condition $\tilde{\rho} < 0$.

Finally, for convenience, we designed a class of layout of the impulse time windows as shown in Figure 3. Specifically, we stipulate that the width of each window in the figure is 0.05 and that the impulse appears in the center point of each window. In other words, $\forall k \in \mathbb{N}^+$, we have $Z^l_{k+1} - Z^l_k = 0.05$ and $t_{k+1} - t_k = 0.05$.

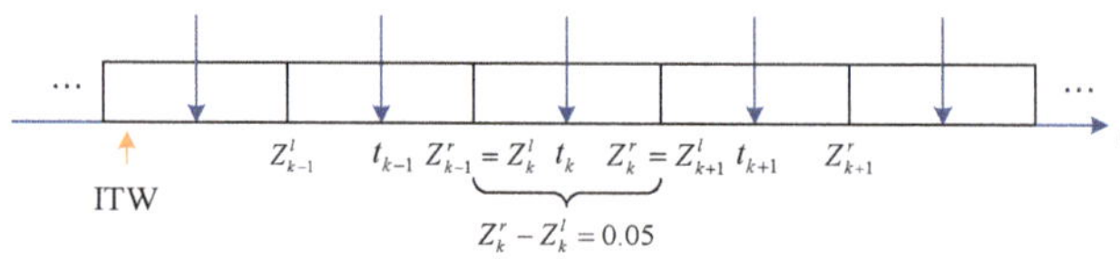

Figure 3. Design layout of impulse time window (ITW).

Based on the above work, Figures 4–6 are obtained by Matlab platform as follows.

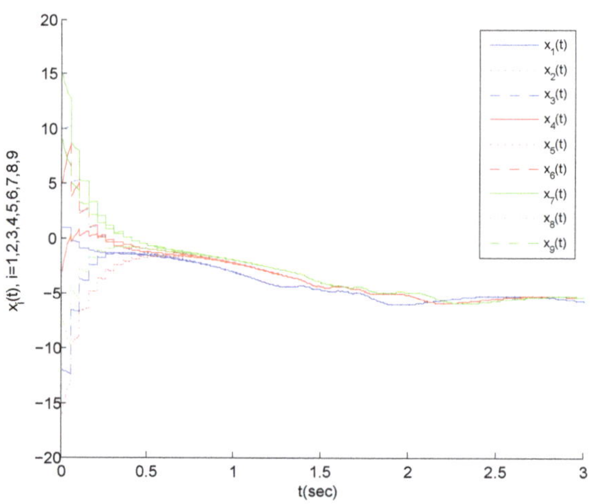

Figure 4. State trajectory of each agent under impulsive control.

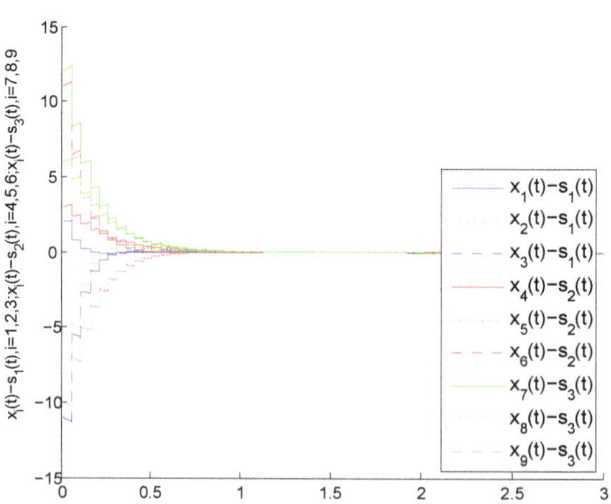

Figure 5. The state error trajectory for each agent in clusters.

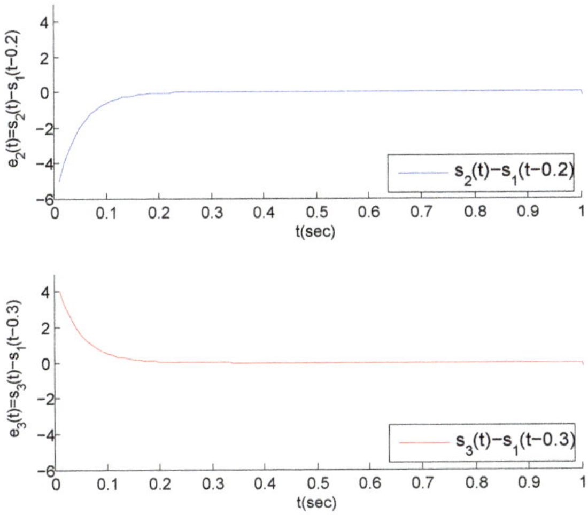

Figure 6. The error trajectories between $S_2(t)$, $S_3(t)$ and their virtual leaders' states when $\rho_2 = \rho_3 = 20$.

According to Figure 4, we can find that the MASs are divided into three clusters, and the states of the three agents in each cluster gradually tend to a common state (i.e., the virtual leader's state). Correspondingly, the state error between each agent and its virtual leader also gradually tends to 0, as shown in Figure 5. Thus, the cluster mean square consensus of system (5) can be achieved. Obviously, it can be seen from Figure 6 that system (3) has achieved the delay mean square consensus. In summary, based on the impulse time windows, the cluster-delay mean square consensus of the SMASs with (3) and (5) can be realized.

We know that by adjusting the size of the impulsive interval during the simulation process and observing the impact of this operation on the speed of the multi-agent systems to achieve cluster mean square consensus, it can verify the dynamic relationship between the selection of parameter ϑ and the preset layout of the impulse time windows. We assume that there exists a parameter $1 < \vartheta < 1.2$ such that $Z_{k+1}^l - Z_k^l = 0.12$ and $t_{k+1} - t_k = 0.12$, and Figure 7 is obtained. By Figure 7, as described in Remark 5, it takes longer for SMASs to achieve cluster mean square consensus. Thus, the discussions in Remark 5 are reasonable.

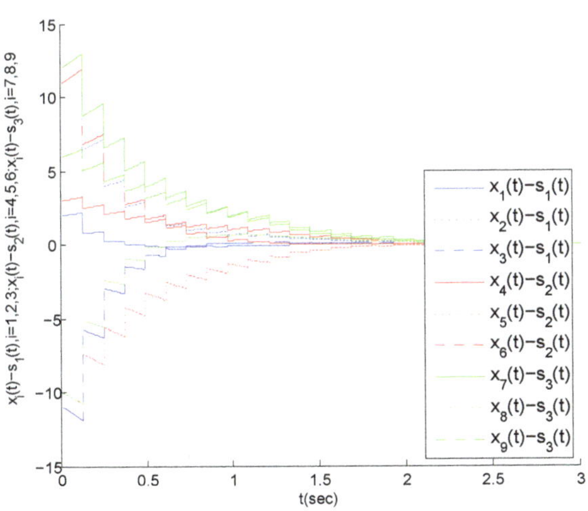

Figure 7. The state error trajectory for each agent in clusters.

To verify the correctness of the theoretical analysis in Remark 6, we increase the value of the coupling strengths. That is, let $\rho_2 = \rho_3 = 30$, and Figure 8 is obtained. According to Figure 8, it can be found that the two error trajectories in the figure can approximately converge to 0 at about 0.15. This convergence speed is obviously faster than that in Figure 6. Therefore, the obtained results in Remark 6 are correct.

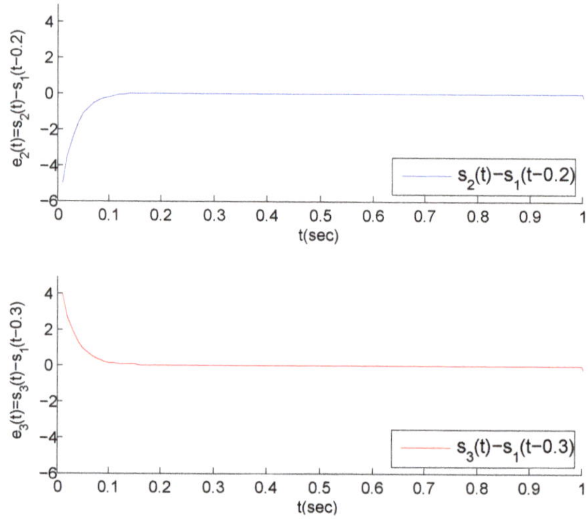

Figure 8. The error trajectories between $S_2(t)$, $S_3(t)$ and their virtual leaders' states when $\rho_2 = \rho_3 = 30$.

6. Conclusions

Based on the discrete impulsive control strategy, this paper studies the cluster-delay mean square consensus problem of a class of SMASs with impulse time windows. According to the algebraic graph theory and Lyapunov stability theory, sufficient consensus criteria are given, and the obtained results are more general than the existing work. Moreover, according to the obtained conditions, the upper bound of the interval between the left endpoints of the two adjacent windows can be derived, which is conducive to the reasonable setting of the windows, so as to ensure that the cluster-delay consensus of SMASs in the mean square sense can be realized. Finally, a simulation example is designed to analyze and verify the feasibility of the relevant results. However, the research work in this paper still has some shortcomings. For instance, the dynamic model of each agent is homogeneous, and there are fewer objective factors considered in the system. Due to the wide application of heterogeneous MASs in practical applications, it is necessary to extend existing research work to heterogeneous MASs. In addition, considering the influence of factors such as time delay and switching topology in this paper is also a meaningful direction for work in the future.

Author Contributions: Conceptualization, H.L. and Y.W.; methodology, H.L. and Y.W.; software, H.W. and S.Y.; validation, Y.W., R.Z. and X.Z.; formal analysis, H.L.; investigation, H.L.; resources, R.Z.; data curation, X.Z.; writing—original draft preparation, H.L.; writing—review and editing, Y.W. and H.L.; visualization, H.W.; supervision, Y.W. and X.Z.; project administration, Y.W.; funding acquisition, R.Z. All authors have read and agreed to the published version of the manuscript.

Funding: This research work was funded by the Key-Area Research and Development Program of Guangdong Province (2019B010142001 and 2019B010140002), and the National Natural Science Foundation of China (61803104 and 61673120).

Institutional Review Board Statement: Not applicable.

Informed Consent Statement: Not applicable.

Data Availability Statement: Not applicable.

Conflicts of Interest: The authors declare no conflict of interest.

Abbreviations

The following abbreviations are used in this manuscript:

MASs	Multi-agent systems
SMASs	Stochastic multi-agent systems
ITM	Impulse time window

References

1. Wooldridge, M. *An Introduction to Multiagent Systems;* John Wiley & Sons: Hoboken, NJ, USA, 2009.
2. Roche, B.; Guégan, J.F.; Bousquet, F. Multi-agent systems in epidemiology: A first step for computational biology in the study of vector-borne disease transmission. *BMC Bioinform.* **2008**, *9*, 1–9. [CrossRef] [PubMed]
3. Cisse, P.A.; Dembele, J.M.; Lo, M.; Cambier, C. Multi-agent Systems for Epidemiology: Example of an agent-based simulation platform for schistosomiasis. In *Agents and Multi-Agent Systems for Health Care;* Springer: Berlin/Heidelberg, Germany, 2017; pp. 131–153.
4. Petruzzi, P.E.; Pitt, J.; Busquets, D. Electronic social capital for self-organising multi-agent systems. *ACM Trans. Auton. Adapt. Syst. TAAS* **2017**, *12*, 1–25. [CrossRef]
5. Vinogradova, M.; Maloletko, A.; Shatsky, A. Social entrepreneurship within the framework of multi-agent system. In *Economic and Social Development: Book of Proceedings;* Varazdin Development and Entrepreneurship Agency (VADEA): Varazdin, Croatia, 2017; pp. 632–643.
6. Olszewski, R.; Pałka, P.; Turek, A.; Kietlińska, B.; Płatkowski, T.; Borkowski, M. Spatiotemporal modeling of the smart city residents' activity with multi-agent systems. *Appl. Sci.* **2019**, *9*, 2059. [CrossRef]
7. Herrera, M.; Pérez-Hernández, M.; Kumar Parlikad, A.; Izquierdo, J. Multi-agent systems and complex networks: Review and applications in systems engineering. *Processes* **2020**, *8*, 312. [CrossRef]
8. Li, D.; Ge, S.S.; He, W.; Ma, G.; Xie, L. Multilayer formation control of multi-agent systems. *Automatica* **2019**, *109*, 108558. [CrossRef]
9. Fisher, K. Agent-based design of holonic manufacturing systems. *Robot. Auton. Syst.* **1999**, *27*, 3–13. [CrossRef]
10. Fischer, K.; Schillo, M.; Siekmann, J. Holonic multiagent systems: A foundation for the organisation of multiagent systems. In Proceedings of the International Conference on Industrial Applications of Holonic and Multi-Agent Systems, Prague, Czech Republic, 1–3 September 2003; pp. 71–80.
11. Li, X.; Yu, Z.; Li, Z.; Wu, N. Group consensus via pinning control for a class of heterogeneous multi-agent systems with input constraints. *Inf. Sci.* **2021**, *542*, 247–262. [CrossRef]
12. Pu, X.; Ren, L.; Liu, Y.; Ji, L. Group consensus of multi-agent systems with cooperative-competitive interaction and communication delay in switching topologies networks based on the delta operator method. *Neurocomputing* **2020**, *390*, 57–68. [CrossRef]
13. Yaghoubi, Z.; Talebi, H.A. Cluster consensus for nonlinear multi-agent systems. *J. Intell. Robot. Syst.* **2020**, *100*, 1069–1084. [CrossRef]
14. Zhu, W.; Wang, D.; Zhou, Q. Leader-following consensus of multi-agent systems via adaptive event-based control. *J. Syst. Sci. Complex.* **2019**, *32*, 846–856. [CrossRef]
15. Zhang, J.; Zhang, H.; Lu, Y.; Sun, S. Event\self-triggered leader-following consensus of multi-agent systems with general dynamics. *IET Control Theory Appl.* **2020**, *14*, 1209–1219. [CrossRef]
16. Zhao, G.; Hua, C.C. Leader-following consensus of multi-agent systems via asynchronous sampled-data control: A hybrid systems approach. *IEEE Trans. Autom. Control* **2021**. [CrossRef]
17. Wang, Q.; Wang, J.L.; Wu, H.N.; Huang, T. Consensus and H-infinity consensus of nonlinear second-order multi-agent systems. *IEEE Trans. Netw. Sci. Eng.* **2020**, *7*, 1251–1264. [CrossRef]
18. He, M.; Mu, J.; Mu, X. H-infinity leader-following consensus of nonlinear multi-agent systems under semi-Markovian switching topologies with partially unknown transition rates. *Inf. Sci.* **2020**, *513*, 168–179. [CrossRef]
19. Xue, X.; Yuan, C.; Wu, F. Convexified H-infinity output-feedback consensus synthesis for linear multi-agent systems. *IET Control Theory Appl.* **2019**, *13*, 1619–1628. [CrossRef]
20. Du, H.; Wen, G.; Wu, D.; Cheng, Y.; Lu, J. Distributed fixed-time consensus for nonlinear heterogeneous multi-agent systems. *Automatica* **2020**, *113*, 108797. [CrossRef]
21. Wu, Y.; Wang, Z. Fuzzy adaptive practical fixed-time consensus for second-order nonlinear multiagent systems under actuator faults. *IEEE Trans. Cybern.* **2020**, *51*, 1150–1162. [CrossRef]
22. Cai, Y.; Zhang, H.; Zhang, J.; Wang, W. Fixed-time leader-following/containment consensus for a class of nonlinear multi-agent systems. *Inf. Sci.* **2021**, *555*, 58–84. [CrossRef]
23. Ma, Z.; Wang, Y.; Li, X. Cluster-delay consensus in first-order multi-agent systems with nonlinear dynamics. *Nonlinear Dyn.* **2016**, *83*, 1303–1310. [CrossRef]

24. Huang, D.; Jiang, H.; Yu, Z.; Chen, X. Cluster-delay consensus in multi-agent systems via pinning leader-following approach with intermittent effect. *Int. J. Control* **2018**, *91*, 2261–2272. [CrossRef]
25. Huang, D.; Jiang, H.; Yu, Z.; Kang, C.; Hu, C. Leader-following cluster consensus in multi-agent systems with intermittence. *Int. J. Control Autom. Syst.* **2018**, *16*, 437–451. [CrossRef]
26. Huang, D.; Jiang, H.; Yu, Z.; Hu, C.; Fan, X. Cluster-delay consensus in MASs with layered intermittent communication: A multi-tracking approach. *Nonlinear Dyn.* **2019**, *95*, 1713–1730. [CrossRef]
27. Huang, J.; Wen, G.; Peng, Z.; Zhang, Y. Cluster-delay consensus for second-order nonlinear multi-agent systems. *J. Syst. Sci. Complex.* **2020**, *33*, 333–344. [CrossRef]
28. Qin, J.; Ma, Q.; Shi, Y.; Wang, L. Recent advances in consensus of multi-agent systems: A brief survey. *IEEE Trans. Ind. Electron.* **2016**, *64*, 4972–4983. [CrossRef]
29. Ma, L.; Wang, Z.; Han, Q.L.; Liu, Y. Consensus control of stochastic multi-agent systems: A survey. *Sci. China Inf. Sci.* **2017**, *60*, 1–15. [CrossRef]
30. Li, Y.; Tan, C. A survey of the consensus for multi-agent systems. *Syst. Sci. Control Eng.* **2019**, *7*, 468–482. [CrossRef]
31. Zhang, Z.; Peng, S.; Chen, T. The cluster-delay consensus of nonlinear multi-agent systems via impulsive control. *J. Phys. Conf. Ser.* **2020**, *1449*, 012099. [CrossRef]
32. Gunasekaran, N.; Saravanakumar, R.; Joo, Y.H.; Kim, H.S. Finite-time synchronization of sampled-data T–S fuzzy complex dynamical networks subject to average dwell-time approach. *Fuzzy Sets Syst.* **2019**, *374*, 40–59. [CrossRef]
33. Gunasekaran, N.; Zhai, G.; Yu, Q. Sampled-data synchronization of delayed multi-agent networks and its application to coupled circuit. *Neurocomputing* **2020**, *413*, 499–511. [CrossRef]
34. Wang, X.; Li, C.; Huang, T.; Pan, X. Impulsive control and synchronization of nonlinear system with impulse time window. *Nonlinear Dyn.* **2014**, *78*, 2837–2845. [CrossRef]
35. Huang, J.; Huang, M. Cluster-delay consensus for the second-order nonlinear multi-agent systems with random noises. In Proceedings of the 2020 39th Chinese Control Conference (CCC), Shenyang, China, 27–29 July 2020; pp. 683–688.
36. Ali, M.S.; Palanisamy, L.; Gunasekaran, N.; Alsaedi, A.; Ahmad, B. Finite-time exponential synchronization of reaction-diffusion delayed complex-dynamical networks. *Discret. Contin. Dyn. Syst. S* **2021**, *14*, 1465.
37. Syed Ali, M.; Usha, M.; Kwon, O.; Gunasekaran, N.; Kumar Thakur, G. ∞/passive non-fragile synchronisation of Markovian jump stochastic complex dynamical networks with time-varying delays. *Int. J. Syst. Sci.* **2021**, *52*, 1270–1283. [CrossRef]
38. Wang, Z.; Lauria, S.; Liu, X. Exponential stability of uncertain stochastic neural networks with mixed time-delays. *Chaos Solitons Fractals* **2007**, *32*, 62–72. [CrossRef]

entropy

Article

Adaptive Fixed-Time Neural Network Tracking Control of Nonlinear Interconnected Systems

Yang Li [1], Jianhua Zhang [1,*], Xinli Xu [1] and Cheng Siong Chin [2]

[1] School of Information and Control Engineering, Qingdao University of Technology, Qingdao 266525, China; yang_li@qut.edu.cn (Y.L.); xuxinli@qut.edu.cn (X.X.)
[2] Faculty of Science, Agriculture, and Engineering, Newcastle University Singapore, Singapore 599493, Singapore; cheng.chin@ncl.ac.uk
* Correspondence: jianhuazhang@qut.edu.cn

Abstract: In this article, a novel adaptive fixed-time neural network tracking control scheme for nonlinear interconnected systems is proposed. An adaptive backstepping technique is used to address unknown system uncertainties in the fixed-time settings. Neural networks are used to identify the unknown uncertainties. The study shows that, under the proposed control scheme, each state in the system can converge into small regions near zero with fixed-time convergence time via Lyapunov stability analysis. Finally, the simulation example is presented to demonstrate the effectiveness of the proposed approach. A step-by-step procedure for engineers in industry process applications is proposed.

Keywords: adaptive fixed-time; neural network; nonlinear interconnected systems

Citation: Li, Y.; Zhang, J.; Xu, X.; Chin, C.S. Adaptive Fixed-Time Neural Network Tracking Control of Nonlinear Interconnected Systems. *Entropy* **2021**, *23*, 1152. https://doi.org/10.3390/e23091152

Academic Editor: José A. Tenreiro Machado

Received: 31 July 2021
Accepted: 29 August 2021
Published: 1 September 2021

1. Introduction

In actual industrial processes, after decades of development, several control strategies based on classical control theory and modern control theory have been developed. However, most of these control methods are based on single-input single-output linear systems. There are many nonlinear, uncertain, unmodeled dynamic problems in actual industrial processes that pose great challenges to the design of control systems. With the development of engineering automation requirements, the research of control strategies based on multi-input multi-output nonlinear systems has attracted growing attention. In recent decades, many control schemes have been proposed for stability analysis and the control for nonlinear systems, such as the adaptive technique [1–3], backstepping technique [4–6], U model control [7–9], sliding mode control [10–12], super twisting algorithm [13,14], neural network technique [6,15,16], etc. In particular, neural network technology has attracted many researchers' attention because of the following aspects: (1) a neural network has the strong ability to learn any function and can approximate any nonlinear system, and (2) because of the self-learning ability of neural networks, the controller does not need much system model and parameter information, so neural network control can be widely used to solve the control problems caused by uncertain models [17]. In [18], the control problem of time-varying output constraints was investigated using the neural network technique. The use of adaptive neural network control for an uncertain nonlinear system with external disturbance was presented in [19]. In [20], neural network controller designs were presented for several classes of nonlinear systems, including single-input single-output nonlinear systems, strict feedback nonlinear systems, nonaffine nonlinear systems, and multi-input multi-output triangular nonlinear systems.

Most of the above research was proven based on Lyapunov stability theory. However, actual systems often have various disturbances that cannot strictly meet the definition of Lyapunov stability. To solve this problem, scholars have mainly introduced new concepts based on two aspects. On the one hand, the concept of input-to-state stability has been introduced; on the other hand, the concept of practical stability has been introduced with the aim of making systems stable in finite time. If a closed-loop system reaches a stable state in a limited time, it is called finite-time stability. Further, if a system meets the convergence time and does not depend on the initial parameters, it is called fixed-time stability. Compared with the traditional finite-time control method, the convergence time of the fixed-time control method is independent of the initial conditions. Exponential stability, finite-time stability, and fixed-time stability are all concepts related to the convergence rate of a system. They are very important for many control applications, such as the explosion of missiles. In [21], taking the stability analysis of a sliding mode control system as an example, input–output stability, finite-time stability, and fixed-time stability are introduced in detail. In [14], the convergence time of the super-twisting algorithm for a nonaffine nonlinear system was calculated. In [22], the finite time input–output stability of nonlinear systems is studied. In [23], a finite-time adaptive controller is designed for interconnected systems with time-varying output constraints to make the system stable in finite time, but the fixed-time convergence problem is not considered in this paper. The use of adaptive fixed-time tracking control for a strict feedback nonlinear system was studied in [24]. However, there are still many problems to be solved in these existing control strategies, such as state constraints, the adaptive backstepping "explosion" problem, and so on.

In the actual production process, many physical models, such as power systems, process control systems, and manipulator models, can be modeled as nonlinear interconnected systems. Because the interconnected terms between nonlinear interconnected subsystems are unknown, it is physically difficult to obtain this information through sensors, and it requires significant computer resources. Therefore, designing a reliable control scheme for nonlinear interconnected systems is a challenging task. The adaptive decentralized control scheme for interconnected nonlinear systems was discussed in [25]. In [26], the robust adaptive tracking control scheme was proposed for uncertain interconnected nonlinear systems. In [27], the use of adaptive neural control for high-order interconnected systems was examined.

Based on the above analysis, the main goal of this article is to design an adaptive fixed-time neural network tracking controller for nonlinear interconnected systems. The main contributions of this paper are as follows:

(1) The combination of fixed-time control and neural network adaptive control for nonlinear interconnected systems.
(2) A fixed-time low pass filter is designed to solve the "explosion of complexity" based on backstepping control technology.
(3) A fixed-time controller is designed, which contains the convergence time of the error system, weights of neural networks, and a low pass filter system.

The article is organized into the following sections. A nonlinear interconnected mathematical description of the problem is presented in Section 2, the adaptive fixed-time neural network control scheme for a class of nonlinear interconnected systems is proposed in Section 3, two simulation examples are provided to show the reliability of the presented control scheme in Section 4, and finally, some conclusions are given in Section 5.

2. Problem Formation and Preliminaries

Consider the interconnected nonlinear system:

$$\begin{aligned} \dot{x}_{i,m} &= x_{i,m+1} + f_{i,m}(\overline{x}_{i,m}) \\ \dot{x}_{i,n} &= u_i + f_{i,n}(\overline{x}_{i,n}) + h_{i,n}(\overline{x}_{1,n}, \overline{x}_{2,n}, \cdots, \overline{x}_{N,n}) \\ y_i &= x_{i,1} \end{aligned} \tag{1}$$

where $x_{i,m} \in R$ is the state of the interconnected nonlinear system; $\overline{x}_{i,m} = [x_{i,1}, \ldots, x_{i,m}]^T \in R^m$ is the state vector of the system; $f_{i,m}(\overline{x}_{i,m}) : R^m \to R$ is the known smooth function; $h_{i,n}(\overline{x}_{1,n}, \overline{x}_{2,n}, \cdots, \overline{x}_{N,n}) : R^{n \times N} \to R$ is the unknown smooth function; $y_i \in R$ is the output of the system; $u_i \in R$ is the corresponding control input of the system; the desired trajectory $y_{i,d}$ and its derivative are continuous and bounded.

Remark 1: *In the next section, we introduce a neural network adaptive control method based on the fixed-time stability theory. The objective of the method is for the nonlinear interconnected system output to be able to track the desired signal and maintain fixed-time stability based on the adaptive fixed-time neural network controller. The designed setting time does not rely on the initial parameters and can be realized only by adjusting the controller parameters.*

3. Adaptive Fixed-Time Tracking Control System Design

3.1. Control System Design

In this section, the design of a fixed-time adaptive law for the error systems of neural networks will be presented. The tracking control system's objective is to drive the error system to fixed-time stability. To solve the tracking control problem, the neural network adaptive controller, based on fixed-time Lyapunov stability theory for nonlinear interconnected systems, is presented. The adaptive fixed-time laws were designed to update the weights of the neural networks for the error systems. Neural networks were used to approximate unknown functions. The parameters of the neural networks were iteratively based on the Lyapunov fixed-time stability theorem. The convergence time can be designed by choosing controller parameters without the initial condition. Based on the controller, the error-closed loop system achieves Lyapunov fixed-time bounded stability, which means the output trajectory can track to the desired trajectory in fixed time.

Remark 2: *The control structure's design for the closed loop system is shown in Figure 1. The states of the error system can be determined from the minus between the setting reference function and the actual output function of the nonlinear interconnected system.*

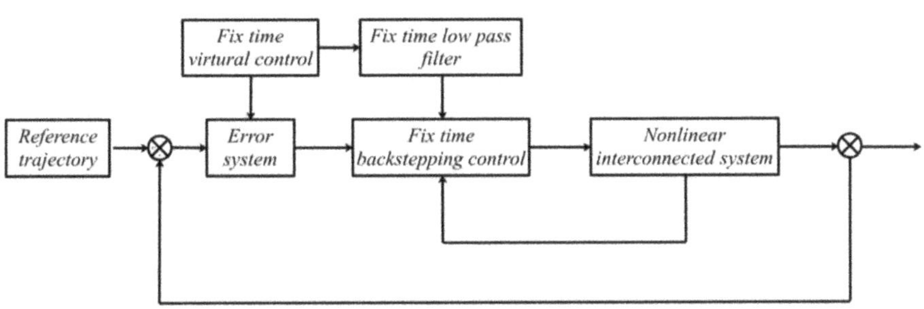

Figure 1. Control structure of the closed system.

3.2. Control System Analysis

Step 1: First, for the system i, the following variables are selected:

$$z_{i,1} = x_{i,1} - y_{i,d} \tag{2}$$

The dynamics of $z_{i,1}$ can be obtained as

$$\dot{z}_{i,1} = x_{i,2} + f_{i,1}(x_{i,1}) - \dot{y}_{i,d} \tag{3}$$

Design the ideal virtual control as

$$\overline{\alpha}_{i,1} = -p_{i,1} z_{i,1}^{p} - q_{i,1} z_{i,1}^{q} - f_{i,1}(x_{i,1}) + \dot{y}_{i,d} \tag{4}$$

where $p_{i,1} > 0, q_{i,1} > 0$ Select the virtual control $\alpha_{i,1}$ and design the adaptive law as

$$\dot{\alpha}_{i,1} = -l_{i,p,1} y_{i,1}^{p} - l_{i,q,1} y_{i,1}^{q} - k_{i,\alpha,1} y_{i,1} - z_{i,1}, \alpha_{i,1}(0) = 0 \tag{5}$$

where, $l_{i,p,1} > 0, l_{i,q,1} > 0$ and the error virtual control is

$$y_{i,1} = \alpha_{i,1} - \overline{\alpha}_{i,1} \tag{6}$$

Therefore, based on system (3) and virtual control

$$\begin{aligned} \dot{z}_{i,1} &= x_{i,2} + f_{i,1}(x_{i,1}) - \dot{y}_{i,d} - p_{i,1}z_{i,1}^{p} - q_{i,1}z_{i,1}^{q} - f_{i,1}(x_{i,1}) + \dot{y}_{i,d} - \overline{\alpha}_{i,1} \\ &= x_{i,2} - p_{i,1}z_{i,1}^{p} - q_{i,1}z_{i,1}^{q} - \alpha_{i,1} + \alpha_{i,1} - \overline{\alpha}_{i,1} \\ &= -p_{i,1}z_{i,1}^{p} - q_{i,1}z_{i,1}^{q} + z_{i,2} + y_{i,1} \end{aligned} \tag{7}$$

where

$$z_{i,2} = x_{i,2} - \alpha_{i,1} \tag{8}$$

and the dynamic error virtual control is

$$\begin{aligned} \dot{y}_{i,1} &= \dot{\alpha}_{i,1} - \dot{\overline{\alpha}}_{i,1} \\ &= -l_{i,p,1}y_{i,1}^{p} - l_{i,q,1}y_{i,1}^{q} - k_{i,\alpha,1}y_{i,1} - z_{i,1} - \dot{\overline{\alpha}}_{i,1} \end{aligned} \tag{9}$$

The Lyapunov candidate functional is chosen as

$$V_1 = \frac{1}{2}z_{i,1}^2 + \frac{1}{2}y_{i,1}^2 \tag{10}$$

Differentiating V_1 with respect to time t yields

$$\begin{aligned} \dot{V}_1 &= z_{i,1}\dot{z}_{i,1} + y_{i,1}\dot{y}_{i,1} \\ &= -p_{i,1}z_{i,1}^{p+1} - q_{i,1}z_{i,1}^{q+1} + z_{i,1}z_{i,2} \\ &\quad -l_{i,p,1}y_{i,1}^{p+1} - l_{i,q,1}y_{i,1}^{q+1} - k_{i,\alpha,1}y_{i,1}^2 - y_{i,1}\dot{\overline{\alpha}}_{i,1} \end{aligned} \tag{11}$$

Step m: the form of the tracking error is

$$z_{i,m} = x_{i,m} - \alpha_{i,m-1} \tag{12}$$

The dynamics of $z_{i,m}$ can be obtained as

$$\dot{z}_{i,m} = x_{i,m+1} + f_{i,m}(x_{i,m}) - \dot{\alpha}_{i,m-1} \tag{13}$$

Design the ideal virtual control as

$$\overline{\alpha}_{i,m} = -p_{i,m}z_{i,m}^{p} - q_{i,m}z_{i,m}^{q} - f_{i,m}(x_{i,m}) + \dot{\alpha}_{i,m-1} \tag{14}$$

where $p_{i,m} > 0, q_{i,m} > 0$ Select the virtual control $\alpha_{i,m}$, and the adaptive law can be obtained as

$$\dot{\alpha}_{i,m} = -l_{i,p,m}y_{i,m}^{p} - l_{i,q,m}y_{i,m}^{q} - k_{i,\alpha,m}y_{i,m} - z_{i,m}, \alpha_{i,m}(0) = 0 \tag{15}$$

where, $l_{i,p,m} > 0, l_{i,q,m} > 0$ and the error virtual control is

$$y_{i,m} = \alpha_{i,m} - \overline{\alpha}_{i,m} \tag{16}$$

Therefore, based on system (13) and virtual control

$$\begin{aligned} \dot{z}_{i,m} &= x_{i,m+1} + f_{i,m}(x_{i,m}) - \dot{\alpha}_{i,m-1} - p_{i,m}z_{i,m}^{p} - q_{i,m}z_{i,m}^{q} - f_{i,m}(x_{i,m}) + \dot{y}_{i,m-1} - \dot{\alpha}_{i,m-1} \\ &= x_{i,m+1} - p_{i,m}z_{i,m}^{p} - q_{i,m}z_{i,m}^{q} - \alpha_{i,m} + \alpha_{i,m} - \overline{\alpha}_{i,m} \\ &= -p_{i,m}z_{i,m}^{p} - q_{i,m}z_{i,m}^{q} + z_{i,m+1} + y_{i,m} \end{aligned} \tag{17}$$

where

$$z_{i,m+1} = x_{i,m+1} - \alpha_{i,m} \tag{18}$$

and the dynamic error virtual control is

$$\begin{aligned}\dot{y}_{i,m} &= \dot{\alpha}_{i,m} - \dot{\overline{\alpha}}_{i,m} \\ &= -l_{i,p,m}y_{i,m}^{p} - l_{i,q,m}y_{i,m}^{q} - k_{i,\alpha,m}y_{i,m} - z_{i,m} - \dot{\overline{\alpha}}_{i,m}\end{aligned} \tag{19}$$

The Lyapunov candidate functional is chosen as

$$V_m = \frac{1}{2}z_{i,m}^2 + \frac{1}{2}y_{i,m}^2 \tag{20}$$

Differentiating V_m with respect to time t yields

$$\begin{aligned}\dot{V}_m &= z_{i,m}\dot{z}_{i,m} + y_{i,m}\dot{y}_{i,m} \\ &= -z_{i,m-1}z_{i,m} - p_{i,m}z_{i,m}^{p+1} - q_{i,m}z_{i,m}^{q+1} + z_{i,m}z_{i,m+1} \\ &-l_{i,p,m}y_{i,m}^{p+1} - l_{i,q,m}y_{i,m}^{q+1} - k_{i,\alpha,m}y_{i,m}^2 - y_{i,m}\dot{\overline{\alpha}}_{i,m}\end{aligned} \tag{21}$$

Step n: the time derivative of $z_{i,n}$ can be described as

$$z_{i,n} = x_{i,n} - \alpha_{i,n-1} \tag{22}$$

Based on dynamics and tracking errors, the dynamics of $z_{i,n}$ can be obtained as

$$\dot{z}_{i,n} = u_i + f_{i,n}(\overline{x}_{i,n}) + h_{i,n}(\overline{x}_{1,n}, \overline{x}_{2,n}, \cdots, \overline{x}_{N,n},) - \dot{\alpha}_{i,n-1} \tag{23}$$

Design the NNs' approximate nonlinear systems as

$$h_{i,n}(\overline{x}_{1,n}, \overline{x}_{2,n}, \cdots, \overline{x}_{N,n}) = W_i^T\Psi(Z) + \varepsilon_i(Z) \tag{24}$$

where $Z = [\overline{x}_{1,n}, \overline{x}_{2,n}, \cdots, \overline{x}_{N,n}]^T$, $\theta_i = \|W_i\|$, and is estimated by $\hat{\theta}_i$. The controller is designed as

$$u_i = -z_{i,n-1} - p_{i,n}z_{i,n}^{p} - q_{i,n}z_{i,n}^{q} - k_{i,n}z_{i,n} - f_{i,n}(\overline{x}_{i,n}) - sign(z_{i,n})\hat{\theta}_i\|\Psi\| + \dot{\alpha}_{i,n-1} \tag{25}$$

where $k_{i,n} > \frac{1}{2}$. Design the adaptive law as

$$\dot{\hat{\theta}}_i = \mu_i\left(|z_{i,n}||\Psi_i| - \varsigma_i\hat{\theta}_i^p - \chi_i\hat{\theta}_i^q\right) \tag{26}$$

where $\mu_i > 0, \varsigma_i > 0, \chi_i > 0$ Therefore, based on system (23) and controller

$$\dot{z}_{i,n} = W_i^T\Psi(Z) + \varepsilon_i(Z) - z_{i,n-1} - p_{i,n}z_{i,n}^{p} - q_{i,n}z_{i,n}^{q} - k_{i,n}z_{i,n} - sign(z_{i,n})\hat{\theta}_i\|\Psi\| \tag{27}$$

the Lyapunov candidate functional is chosen as

$$V_n = \frac{1}{2}z_{i,n}^2 + \frac{1}{2\mu_i}\widetilde{\theta}_i^2 \tag{28}$$

Differentiating V_n with respect to time t yields

$$\begin{aligned}\dot{V}_n &= z_{i,n}\dot{z}_{i,n} + \tfrac{1}{\mu_i}\widetilde{\theta}_i\dot{\hat{\theta}}_i \\ &= z_{i,n}W_i^T\Psi(Z) + z_{i,n}\varepsilon_i(Z) - z_{i,n-1}z_{i,n} - p_{i,n}z_{i,n}^{p+1} - q_{i,n}z_{i,n}^{q+1} \\ &-k_{i,n}z_{i,n}^2 - \hat{\theta}_i|z_{i,n}|\| \Psi \| + \widetilde{\theta}_i\left(|z_{i,n}||\Psi_i| - \varsigma_i\hat{\theta}_i^p - \chi_i\hat{\theta}_i^q\right) \\ &\leq \theta_i|z_{i,n}|\| \Psi \| - \left(k_{i,n} - \tfrac{1}{2}\right)z_{i,n}^2 + \tfrac{1}{2}\overline{\varepsilon}_i - z_{i,n-1}z_{i,n} - p_{i,n}z_{i,n}^{p+1} - q_{i,n}z_{i,n}^{q+1} \\ &-\hat{\theta}_i|z_{i,n}|\| \Psi \| + \widetilde{\theta}_i|z_{i,n}||\Psi_i| - \varsigma_i\widetilde{\theta}_i\hat{\theta}_i^p - \chi_i\widetilde{\theta}_i\hat{\theta}_i^q\end{aligned} \tag{29}$$

where $\varepsilon_i^2(Z) \le \bar{\varepsilon}_i$. Based on Lemma 4, we have

$$\begin{aligned} -\varsigma_i\widetilde{\theta}_i\hat{\theta}_i^p &\le -\gamma_i\widetilde{\theta}_i^{p+1} + \lambda_i\theta_i^{p+1} \\ -\chi_i\widetilde{\theta}_i\hat{\theta}_i^q &\le -\nu_i\widetilde{\theta}_i^{q+1} + \upsilon_i\theta_i^{q+1} \end{aligned} \tag{30}$$

where $\varsigma_i, \gamma_i, \lambda_i, \chi_i, \nu_i, \upsilon_i$ are real numbers, γ_i, λ_i is determined by ς_i, p, and ν_i, υ_i is determined by χ_i, q. Therefore, we have

$$\begin{aligned} \dot{V}_n \le\; & -\left(k_{i,n} - \tfrac{1}{2}\right)z_{i,n}^2 + \tfrac{1}{2}\bar{\varepsilon}_i - z_{i,n-1}z_{i,n} - p_{i,n}z_{i,n}^{p+1} - q_{i,n}z_{i,n}^{q+1} \\ & -\gamma_i\widetilde{\theta}_i^{p+1} + \lambda_i\theta_i^{p+1} - \nu_i\widetilde{\theta}_i^{q+1} + \upsilon_i\theta_i^{q+1} \end{aligned} \tag{31}$$

Theorem 1. *For the interconnected nonlinear system (1), based on the feasible virtual control signal (5), (15) actual controller (25), and adaptive law (26), the error state between the system output and the desired function is fixed-time Lyapunov stability, and the setting time does not rely on the initial parameters.*

Proof. Based on the Lyapunov candidate functionals (10), (20), and (28), differentiating the Lyapunov functional with respect to time t yields (11), (21), and (31). Choosing the Lyapunov candidate functional as

$$V = \frac{1}{2}\sum_{j=1}^{n} z_{i,j}^2 + \frac{1}{2}\sum_{j=1}^{n-1} y_{i,j}^2 + \frac{1}{2\mu_i}\widetilde{\theta}_i^2 \tag{32}$$

and differentiating the Lyapunov functional with respect to time t yields

$$\begin{aligned} \dot{V} \le\; & -\sum_{j=1}^{n} p_{i,j}z_{i,j}^{p+1} - \sum_{j=1}^{n} q_{i,j}z_{i,j}^{q+1} - \sum_{j=1}^{n} l_{i,p,j}y_{i,j}^{p+1} - \sum_{j=1}^{n} l_{i,q,j}y_{i,j}^{q+1} \\ & -\sum_{j=1}^{n-1} k_{i,\alpha,j}y_{i,j}^2 - \sum_{j=1}^{n-1} y_{i,j}\dot{\bar{\alpha}}_{i,j} - \left(k_{i,n} - \tfrac{1}{2}\right)z_{i,n}^2 + \tfrac{1}{2}\bar{\varepsilon}_i \\ & -\gamma_i\widetilde{\theta}_i^{p+1} + \lambda_i\theta_i^{p+1} - \nu_i\widetilde{\theta}_i^{q+1} + \upsilon_i\theta_i^{q+1} \end{aligned} \tag{33}$$

where $\left|\dot{\bar{\alpha}}_{i,1}\right|^2 \le p_i$, and

$$\begin{aligned} \dot{V} \le\; & -\sum_{j=1}^{n} p_{i,j}z_{i,j}^{p+1} - \sum_{j=1}^{n} q_{i,j}z_{i,j}^{q+1} - \sum_{j=1}^{n} l_{i,p,j}y_{i,j}^{p+1} - \sum_{j=1}^{n} l_{i,q,j}y_{i,j}^{q+1} \\ & -\gamma_i\widetilde{\theta}_i^{p+1} - \nu_i\widetilde{\theta}_i^{q+1} - \sum_{j=1}^{n-1}\left(k_{i,\alpha,j} - \tfrac{1}{2}\right)y_{i,j}^2 - \left(k_{i,n} - \tfrac{1}{2}\right)z_{i,n}^2 \\ & +\sum_{j=1}^{n-1} m_j + \tfrac{1}{2}\bar{\varepsilon}_i + \lambda_i\theta_i^{p+1} + \upsilon_i\theta_i^{q+1} \end{aligned} \tag{34}$$

By choosing the control parameters, $k_{i,\alpha,j} > \frac{1}{2}, k_{i,n} > \frac{1}{2}$ we have

$$\begin{aligned} \dot{V} \le\; & -\sum_{j=1}^{n} p_{i,j}z_{i,j}^{p+1} - \sum_{j=1}^{n} q_{i,j}z_{i,j}^{q+1} - \sum_{j=1}^{n} l_{i,p,j}y_{i,j}^{p+1} \\ & -\sum_{j=1}^{n} l_{i,q,j}y_{i,j}^{q+1} - \gamma_i\widetilde{\theta}_i^{p+1} - \nu_i\widetilde{\theta}_i^{q+1} \\ & +\sum_{j=1}^{n-1} m_j + \tfrac{1}{2}\bar{\varepsilon}_i + \lambda_i\theta_i^{p+1} + \upsilon_i\theta_i^{q+1} \\ \le\; & -aV^{\frac{p+1}{2}} - bV^{\frac{q+1}{2}} + c \end{aligned} \tag{35}$$

where

$$\begin{aligned} a &= \frac{\min\left\{p_{i,j\in N}, l_{i,p,j\in N}, \gamma_i\right\}}{\left(\max\left\{\frac{1}{2}, \frac{1}{2\mu_i}\right\}\right)^{\frac{p+1}{2}}}, b = \frac{(2n)^{\frac{1-q}{2}} \min\left\{q_{i,j\in N}, l_{i,q,j\in N}, \nu_i\right\}}{\left(\max\left\{\frac{1}{2}, \frac{1}{2\mu_i}\right\}\right)^{\frac{q+1}{2}}} \\ c &= \sum_{j=1}^{n-1} m_j + \frac{1}{2}\bar{\varepsilon}_i + \lambda_i \theta_i^{p+1} + v_i \theta_i^{q+1} \end{aligned} \tag{36}$$

Therefore, the error system is fixed-time, practical, and stable. □

Remark 3: *From the definition of Lyapunov stability, Lyapunov stability, asymptotic stability, and finite-time stability are the three most basic concepts, and their definitions are progressive.*

At present, the concepts of stability related to the convergence rate mainly include exponential stability, finite-time stability, and fixed-time stability. Exponential stability is the realization of asymptotic stability, while fixed-time stability is the generalization of finite-time stability. From the point of view of standard definitions, they are strictly distinguished, and from the point of view of controller design, they also have different forms.

From the perspective of system convergence time, Lyapunov stability can be divided into infinite-time stability and finite-time stability. In the theoretical analysis of stability, the most common exponential stability is infinite-time stability. Infinite-time stability means that when time tends to infinity, the system state can converge exponentially by designing the parameters of the controller. Finite-time stability means that the system can be stable at a certain time and continue to maintain a stable state by designing the parameters of the controller. For practical engineering control, finite-time stability is obviously more practical than infinite-time stability, but there are some limitations to finite-time stability; for instance, the convergence time of the designed system depends on the initial state. Therefore, in this section, another scheme was introduced: the tracking control method for nonlinear interconnected systems based on fixed-time stability theory. The setting time does not depend on the initial parameters and can be realized only by adjusting the controller parameters. In other words, under the condition that the controlled system is stable in fixed time, even if the initial parameters are changed, the controlled system can still be stable within the originally designed fixed time without redesigning the controller. This greatly improves the potential of the control method to be used in practical applications.

Remark 4: *The design details are summarized in Figure 2. The above-listed step procedure is detailed below.*

Step 1: Design of the ideal virtual control laws (4) and (14), based on backstepping control technology.

Step 2: Design of the virtual control laws (5) and (15), based on the fixed-time low pass filter.

Step 3: The actual controller (25) is obtained recursively through the virtual control signal and the adaptive parameter (26).

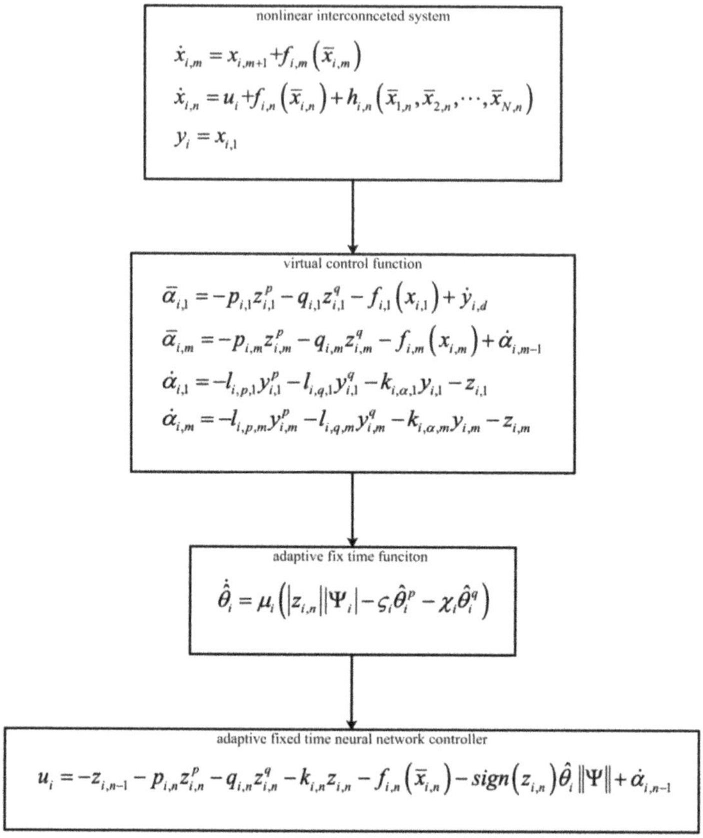

Figure 2. Design procedure.

4. Numerical Examples

The main purposes of the simulation studies include: (1) validating the effectiveness of the adaptive fixed-time neural network tracking controller for nonlinear interconnected systems; (2) showing a step-by-step procedure for building the adaptive fixed-time tracking control system.

A. Numerical Example

Consider the following nonlinear interconnected system:

$$\begin{cases} \dot{x}_{1,1} = x_{1,2} \\ \dot{x}_{1,2} = \sin(x_{1,1}) + u_1 + h_{1,2} \\ y_1 = x_{1,1} \\ \dot{x}_{2,1} = x_{2,2} \\ \dot{x}_{2,2} = \sin(x_{2,1}) + u_2 + h_{2,2} \\ y_2 = x_{2,1} \end{cases} \tag{37}$$

where the function, $h_{1,2} = \sin(x_{2,1}) - \sin(x_{1,1})$. $h_{2,2} = \sin(x_{1,1}) - \sin(x_{2,1})$.

Step 1: Design of the fixed-time ideal virtual control law and fixed-time adaptive law of neural networks:

$$\bar{\alpha}_{i,1} = -2z_{i,1}^{\frac{1}{3}} - 2z_{i,1}^{3} \tag{38}$$

$$\dot{\hat{\theta}}_i = 0.01\left(|z_i||\Psi_i| - 0.1\hat{\theta}_i^{\frac{5}{3}} - 0.1\hat{\theta}_i^{\frac{1}{3}}\right) \tag{39}$$

Step 2: Design of the fixed-time low pass filter:

$$\dot{\alpha}_{i,1} = -2y_{i,1}^{\frac{1}{3}} - 2y_{i,1}^{3} - 2y_{i,1} - z_{i,1}, \alpha_{i,1}(0) = 0 \tag{40}$$

Step 3: The actual controller is obtained recursively through the virtual control signal and the adaptive parameter, and the control input is designed as

$$u_i = -z_{i,1} - 2z_{i,2}^{\frac{1}{3}} - 2z_{i,2}^{3} - 2z_{i,2} - sign(z_{i,2})\hat{\theta}_i\|\Psi_i\| + \dot{\alpha}_{i,1} \tag{41}$$

The initial condition is selected as $x_{11}(0) = -2, x_{21}(0) = 2, \hat{\theta}_1(0) = 0, \hat{\theta}_2(0) = 0$ The neural network consists of seven nodes, the centers $c = [-3, -2, -1, 0, 1, 2, 3]$, and the widths $b = 1$, respectively.

Figure 1 shows the control structure of the closed-loop system. Figure 2 shows the step-by-step design procedure. Figures 3–5 are the simulation results. Figure 3 shows the output of the interconnected system and the output of the interconnected system's convergence to the origin point in finite time, which indicates the control performance of the fixed-time neural network adaptive control. Figure 4 shows the controller of the interconnected system, which is bound and realizable. Figure 5 shows the trajectories of error of the low pass filter, and the error between the virtual control and ideal virtual control indicates that the virtual control is close to the ideal virtual control and that the differential coefficient exists. The simulation results show that the controlled system can become stable in 14 s, and the convergence time can be designed. Even if the states' initial parameters change, the controlled system can still become stable in 14 s. The system includes unknown nonlinear functions, which verify that the neural network control scheme has strong adaptive ability and approximation ability.

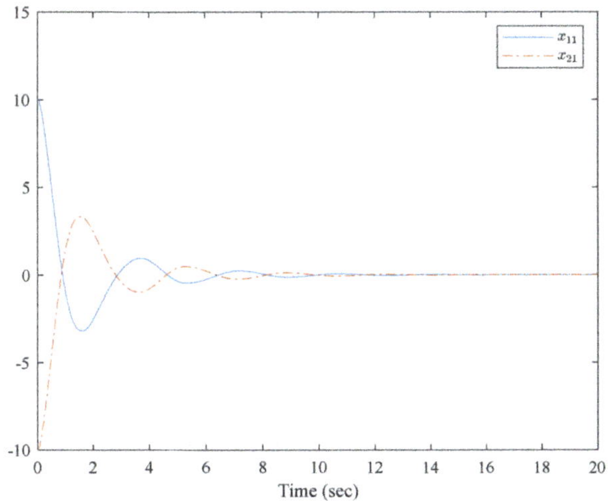

Figure 3. Trajectories of $x_{1,1}$ and $x_{2,1}$ of the interconnected system.

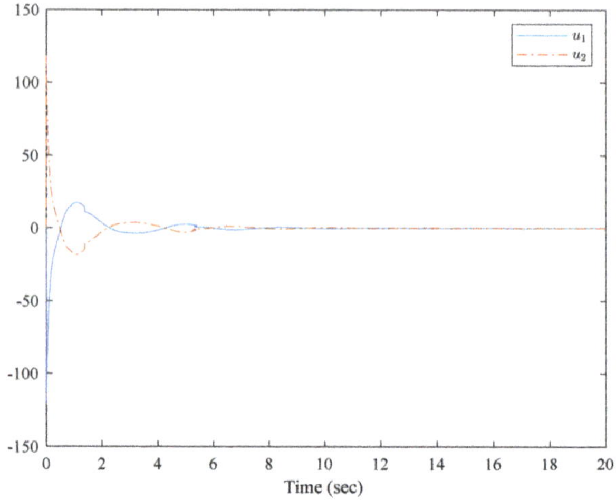

Figure 4. Trajectories of the controller of the interconnected system.

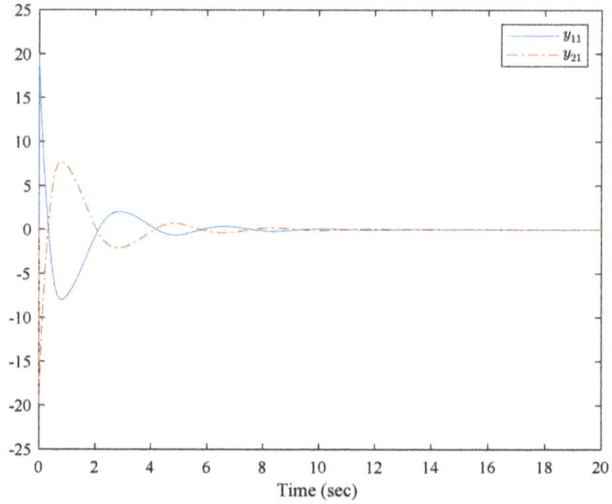

Figure 5. Trajectories of error of the low pass filter.

B. Application Example

Consider a pendulum system [25], with two degrees, as an interconnected system (Figure 6), which has been used as an example of decentralized neural control.

$$\begin{cases} \dot{x}_{1,1} = x_{1,2} \\ \dot{x}_{1,2} = \frac{1}{J_1}u_1 + \frac{kr}{2J_1}(l-b) + \left(\frac{m_1 gr}{J_1} - \frac{kr^2}{4J_1}\right)\sin(x_{1,1}) + \frac{kr^2}{4J_1}\sin(x_{2,1}) \\ y_1 = x_{1,1} \\ \dot{x}_{2,1} = x_{2,2} \\ \dot{x}_{2,2} = \frac{1}{J_2}u_2 + \frac{kr}{2J_2}(l-b) + \left(\frac{m_2 gr}{J_2} - \frac{kr^2}{4J_2}\right)\sin(x_{2,1}) + \frac{kr^2}{4J_2}\sin(x_{2,1}) \\ y_2 = x_{2,1} \end{cases} \tag{42}$$

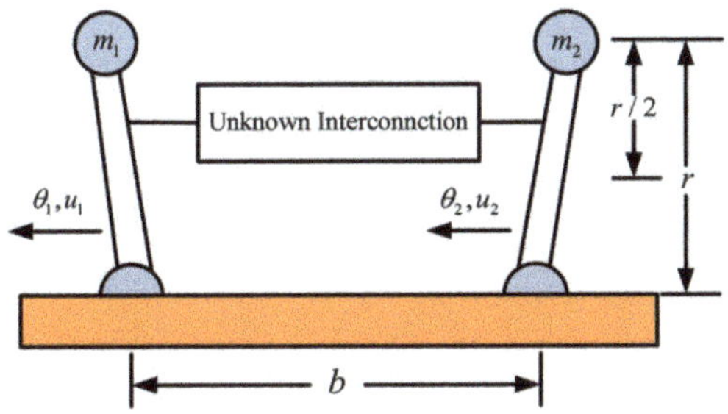

Figure 6. Two inverted pendulums connected by an unknown device.

Step 1: Design of the fixed-time ideal virtual control law and fixed-time adaptive law of neural networks:

$$\overline{\alpha}_{i,1} = -2z_{i,1}^{\frac{1}{3}} - 2z_{i,1}^{3} \tag{43}$$

$$\dot{\hat{\theta}}_i = 0.01\left(|z_i||\Psi_i| - 0.1\hat{\theta}_i^{\frac{5}{3}} - 0.1\hat{\theta}_i^{\frac{1}{3}} \right) \tag{44}$$

Step 2: Design of the fixed-time low pass filter:

$$\dot{\alpha}_{i,1} = -2y_{i,1}^{\frac{1}{3}} - 2y_{i,1}^{3} - 2y_{i,1} - z_{i,1}, \alpha_{i,1}(0) = 0 \tag{45}$$

Step 3: The actual controller is obtained recursively through the virtual control signal and the adaptive parameter, and the control input is designed as

$$u_i = -z_{i,1} - 2z_{i,2}^{\frac{1}{3}} - 2z_{i,2}^{3} - 2z_{i,2} - sign(z_{i,2})\hat{\theta}_i||\Psi_i|| + \dot{\alpha}_{i,1} \tag{46}$$

The initial condition is selected, and $m_1 = 2$ kg , $m_2 = 2.5$ kg, $J_1 = 0.5$ kg, $J_2 = 0.625$ kg, $k = 100$ N/M, $r = 0.5$ m, $l = 0.5$ m, $g = 9.81$ m/s^2, $b = 0.4$ m and the function, $y_{1,d} = 0.5\sin(t)$. $y_{2,d} = 0.5\cos(t)$ The neural network consists of seven nodes, the centers, $c = [-3, -2, -1, 0, 1, 2, 3]$ and the widths, respectively.

The simulation results are shown in Figures 7–10. From the simulation results, it can be seen that the system output can track the desired signal in 2.5 and 5 s, respectively, and the setting time can be designed. Even if the states' initial parameter change, the controlled system can still become stable in the same time. The pendulum system includes unknown nonlinear functions, which verify that the neural network control scheme has strong adaptive ability and approximation ability. The pendulum system also includes interconnected items, which verify that the control scheme can be applied to interconnected nonlinear control systems.

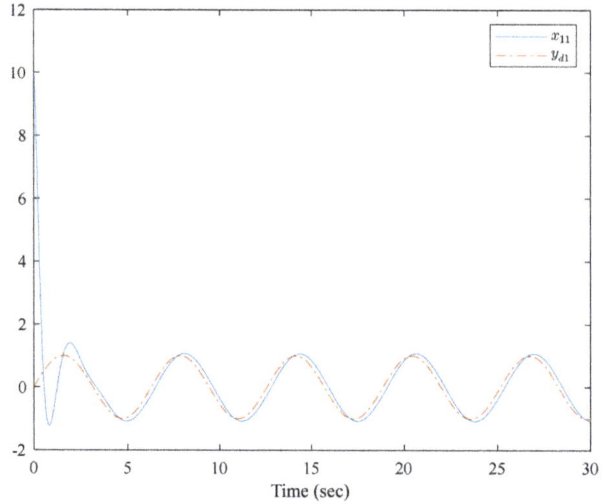

Figure 7. Trajectories of $x_{1,1}$ and $y_{1,d}$ of the interconnected system.

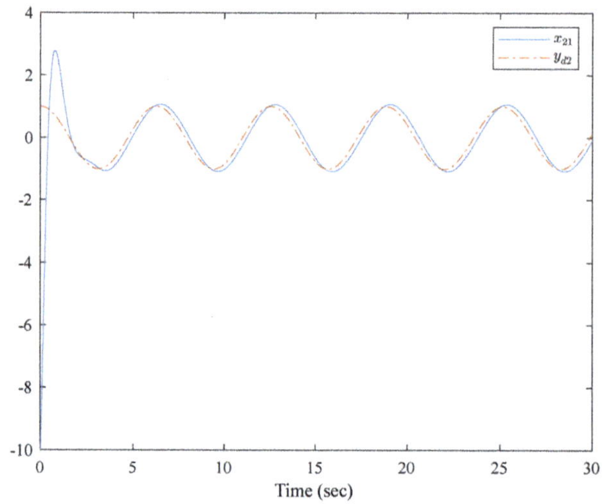

Figure 8. Trajectories of $x_{2,1}$ and $y_{2,d}$ of the interconnected system.

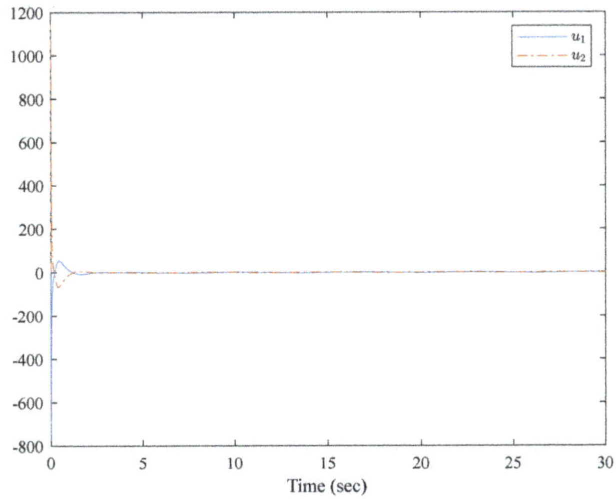

Figure 9. Trajectories of the controller of the interconnected system.

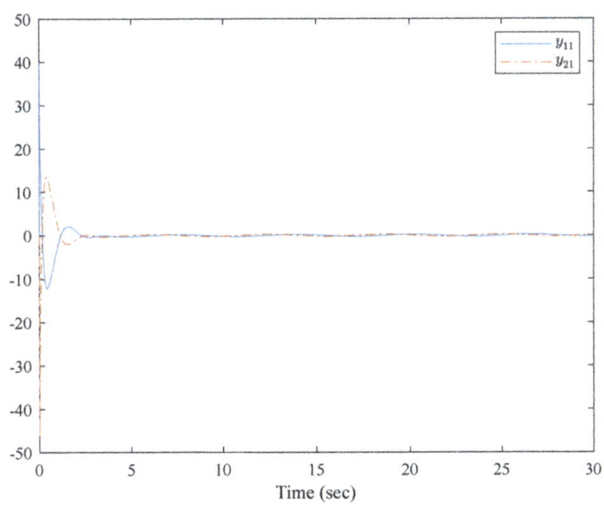

Figure 10. Trajectories of error of the low pass filter.

5. Conclusions

In this paper, through the design of a timing adaptive law for a neural network error system based on the Lyapunov fixed-time stability theorem, the unknown parameters of neural networks are iterated in fixed time. The convergence time can be designed only by modifying the parameters of the system controller and the adaptive rate, and it does not depend on the initial conditions. The fixed-time Lyapunov stability theorem is proposed, and the strict mathematical proof is completed, which will have more practical significance than the ideal stability analysis based on Lyapunov stability theory. Furthermore, the algorithm flow chart is given, which can be used by engineers to realize the proposed tracking control method by a computer for practical engineering. However, the control scheme based on neural networks also has some limitations. Firstly, because it cannot guarantee the asymptotic stability of the system, it is suitable for the controlled system, whose control objective is bounded stability. Secondly, the pure feedback structure is more general than strict feedback, and pure feedback has more application value; therefore, pure feedback interconnected system fixed-time neural network control results should be presented in future works.

Author Contributions: Software, X.X. and C.S.C.; Writing—original draft, Y.L.; Writing—review & editing, J.Z. All authors have read and agreed to the published version of the manuscript.

Funding: This research received no external funding.

Institutional Review Board Statement: Not applicable.

Informed Consent Statement: Not applicable.

Data Availability Statement: Not applicable.

Conflicts of Interest: The authors declare no conflict of interest.

References

1. Gao, W.; Jiang, Y.; Jiang, Z.P.; Chai, T. *Output-Feedback Adaptive Optimal Control of Interconnected Systems Based on Robust Adaptive Dynamic Programming*; Automatica Oxford: Oxford, UK, 2016.
2. Gao, W.; Jiang, Z.P. Adaptive Dynamic Programming and Adaptive Optimal Output Regulation of Linear Systems. *IEEE Trans. Autom. Control.* **2016**, *61*, 4164–4169. [CrossRef]
3. Yan, X.G.; Edwards, C.; Spurgeon, S.K. Decentralised robust sliding mode control for a class of nonlinear interconnected systems by static output feedback. *Automatica* **2004**, *40*, 613–620. [CrossRef]
4. Zhang, J.H.; Li, Y.; Fei, W.B. Neural Network-Based Nonlinear Fixed-Time Adaptive Practical Tracking Control. for Quadrotor Unmanned Aerial Vehicles. *Complexity* **2020**, *2020*, 13. [CrossRef]
5. Tong, S.C.; Min, X.; Li, Y.X. Observer-Based Adaptive Fuzzy Tracking Control. for Strict-Feedback Nonlinear Systems with Unknown Control. Gain Functions. *IEEE Trans. Cybern.* **2020**, *50*, 3903–3913. [CrossRef]
6. Zhang, J.; Zhu, Q.; Li, Y.; Wu, X. Homeomorphism Mapping Based Neural Networks for Finite Time Constraint Control of a Class of Nonaffine Pure-Feedback Nonlinear Systems. *Complexity* **2019**, *2019*, 1–11. [CrossRef]
7. Zhang, J.; Li, Y.; Fei, W.; Wu, X. U-Model Based Adaptive Neural Networks Fixed-Time Backstepping Control for Uncertain Nonlinear System. *Math. Probl. Eng.* **2020**, *2020*, 1–7. [CrossRef]
8. Zhu, Q.M.; Zhao, D.; Zhang, J. A general U-block model-based design procedure for nonlinear polynomial control systems. *Int. J. Syst. Sci.* **2016**, *47*, 3465–3475. [CrossRef]
9. Li, R.; Zhu, Q.; Narayan, P.; Yue, A.; Yao, Y.; Deng, M. U-Model-Based Two-Degree-of-Freedom Internal Model Control of Nonlinear Dynamic Systems. *Entropy* **2021**, *23*, 169. [CrossRef]
10. Yu, X.; Man, Z. Fast terminal sliding-mode control design for nonlinear dynamical systems. *IEEE Trans. Circuits Syst. I Fundam. Theory Appl.* **2009**, *49*, 261–264.
11. Muñoz, D.; Sbarbaro, D. An adaptive sliding-mode controller for discrete nonlinear systems. *IEEE Trans. Ind. Electron.* **2000**, *47*, 574–581. [CrossRef]
12. Da, F. Decentralized sliding mode adaptive controller design based on fuzzy neural networks for interconnected uncertain nonlinear systems. *IEEE Trans. Neural Netw.* **2000**, *11*, 1471–1480.
13. Moreno, J.A.; Osorio, M. Strict Lyapunov Functions for the Super-Twisting Algorithm. *IEEE Trans. Autom. Control* **2012**, *57*, 1035–1040. [CrossRef]
14. Zhang, J.; Zhu, Q.; Li, Y. Convergence Time Calculation for Supertwisting Algorithm and Application for Nonaffine Nonlinear Systems. *Complexity* **2019**, *2019*, 1–15. [CrossRef]
15. Ge, S.S.; Cong, W. Adaptive neural control of uncertain MIMO nonlinear systems. *IEEE Trans. Neural Netw.* **2004**, *15*, 674–692. [CrossRef]
16. Zhang, J.; Zhu, Q.; Wu, X.; Li, Y. A generalized indirect adaptive neural networks backstepping control procedure for a class of non-affine nonlinear systems with pure-feedback prototype. *Neurocomputing* **2013**, *121*, 131–139. [CrossRef]
17. Pan, J.; Qu, L.; Peng, K. Sensor and Actuator Fault Diagnosis for Robot Joint Based on Deep CNN. *Entropy* **2021**, *23*, 751. [CrossRef] [PubMed]
18. He, W.; Huang, H.; Ge, S.S. Adaptive Neural Network Control of a Robotic Manipulator With Time-Varying Output Constraints. *IEEE Trans. Cybern.* **2017**, *47*, 3136–3147. [CrossRef] [PubMed]
19. Wu, Y.; Huang, R.; Li, X.; Liu, S. Adaptive neural network control of uncertain robotic manipulators with external disturbance and time-varying output constraints. *Neurocomputing* **2019**, *323*, 108–116. [CrossRef]
20. Ge, S.S.; Hang, C.C.; Lee, T.H.; Zhang, T. *Stable Adaptive Neural Network Control*; Springer: New York, NY, USA, 2002.
21. Polyakov, A.; Fridman, L. Stability notions and Lyapunov functions for sliding mode control systems. *J. Frankl. Inst.* **2014**, *351*, 1831–1865. [CrossRef]
22. Li, G.; Ji, H. A three-dimensional robust nonlinear terminal guidance law with ISS finite-time convergence. *Int. J. Control* **2016**, *89*, 938–949. [CrossRef]

23. Du, P.; Liang, H.; Zhao, S.; Ahn, C.K. Neural-Based Decentralized Adaptive Finite-Time Control for Nonlinear Large-Scale Systems With Time-Varying Output Constraints. *IEEE Trans. Syst. Man Cybern. Syst.* **2021**, *51*, 3136–3147. [CrossRef]
24. Li, Y.; Zhang, J.; Ye, X.; Chin, C. Adaptive Fixed-Time Control of Strict-Feedback High-Order Nonlinear Systems. *Entropy* **2021**, *23*, 963. [CrossRef] [PubMed]
25. Wang, H.; Liu, W.; Qiu, J.; Liu, P.X. Adaptive Fuzzy Decentralized Control for a Class of Strong Interconnected Nonlinear Systems with Unmodeled Dynamics. *IEEE Trans. Fuzzy Syst.* **2018**, *26*, 836–846. [CrossRef]
26. Li, X.; Yang, G. Neural-Network-Based Adaptive Decentralized Fault-Tolerant Control for a Class of Interconnected Nonlinear Systems. *IEEE Trans. Neural Netw. Learn. Syst.* **2018**, *29*, 144–155. [CrossRef] [PubMed]
27. Si, W.; Dong, X.; Yang, F. Decentralized adaptive neural control for high-order interconnected stochastic nonlinear time-delay systems with unknown system dynamics. *Neural Netw.* **2018**, *99*, 123–133. [CrossRef]

entropy

Article

Constrained Active Fault Tolerant Control Based on Active Fault Diagnosis and Interpolation Optimization

Kezhen Han *, Changzhi Chen, Mengdi Chen and Zipeng Wang

School of Electrical Engineering, University of Jinan, Jinan 250022, China; 202021100375@mail.ujn.edu.cn (C.C.); 202021200742@mail.ujn.edu.cn (M.C.); cse_wangzp@ujn.edu.cn (Z.W.)
* Correspondence: cse_hankz@ujn.edu.cn

Abstract: A new active fault tolerant control scheme based on active fault diagnosis is proposed to address the component/actuator faults for systems with state and input constraints. Firstly, the active fault diagnosis is composed of diagnostic observers, constant auxiliary signals, and separation hyperplanes, all of which are designed offline. In online applications, only a single diagnostic observer is activated to achieve fault detection and isolation. Compared with the traditional multi-observer parallel diagnosis methods, such a design is beneficial to improve the diagnostic efficiency. Secondly, the active fault tolerant control is composed of outer fault tolerant control, inner fault tolerant control and a linear-programming-based interpolation control algorithm. The inner fault tolerant control is determined offline and satisfies the prescribed optimal control performance requirement. The outer fault tolerant control is used to enlarge the feasible region, and it needs to be determined online together with the interpolation optimization. In online applications, the updated state estimates trigger the adjustment of the interpolation algorithm, which in turn enables control reconfiguration by implicitly optimizing the dynamic convex combination of outer fault tolerant control and inner fault tolerant control. This control scheme contributes to further reducing the computational effort of traditional constrained predictive fault tolerant control methods. In addition, each pair of inner fault tolerant control and diagnostic observer is designed integratedly to suppress the robust interaction influences between estimation error and control error. The soft constraint method is further integrated to handle some cases that lead to constraint violations. The effectiveness of these designs is finally validated by a case study of a wastewater treatment plant model.

Keywords: active diagnosis; active reconfiguration; constrained systems; fault tolerance; interpolation control; linear programming

Citation: Han, K.; Chen, C.; Chen, M.; Wang, Z. Constrained Active Fault Tolerant Control Based on Active Fault Diagnosis and Interpolation Optimization. *Entropy* **2021**, 23, 924. https://doi.org/10.3390/e23080924

Academic Editors: Quanmin Zhu, Giuseppe Fusco, Jing Na, Weicun Zhang and Ahmad Taher Azar

Received: 24 June 2021
Accepted: 15 July 2021
Published: 21 July 2021

1. Introduction

Fault tolerance is already a common design property to be considered for most control systems. In terms of the system structure, faults can be classified as sensor faults, actuator faults, and component/parameter faults [1,2]. In general, the first two do not directly affect the intrinsic stability of the system, while the component faults tend to directly change the dynamic characteristics of the system. In the literature, the methods to handle these types of faults can be divided into active fault tolerant control (AFTC) and passive fault tolerant control (PFTC) [3]. PFTC draws on robust control theory to suppress the effects of faults, while AFTC uses fault information to adjust or reconfigure control actions to match the dynamics of the faulty system. Due to such matching adjustments, AFTC typically provides better reliability than PFTC. Many representative results can be found in the survey papers [4–7].

Recently, the design of optimal AFTC for systems with state/input constraints has been received a lot of attention. Unlike the design of unconstrained FTC, the design of constrained FTC has to take into account more requirements, including robust stability, feasibility, optimization efficiency, etc. Particularly, the faults occurring in constrained systems often cause constraint violations, and the unconstrained FTC designed without considering feasibility may result in an empty set of feasible solutions for a given control objective. This often further leads to the eventual loss of closed-loop stability. In the literature, some typical design methods for constrained FTC have been reported, such as Barrier Lyapunov function method [8], command governor [9,10], saturation control [11], model predictive control (MPC) [12], etc. Among these methods, the MPC-based FTC method is widely considered, since MPC has the inherent and flexible capacity to address constrained optimization problems. The representative studies include FTC based on min-max MPC [13], FTC based on explicit MPC [14], multi-actuator/sensor FTC based on set theoretic MPC [15–18], FTC based on dual model MPC [19,20], etc.

Most of the above mentioned MPC-based FTC designs are developed for actuator and sensor faults, whereas relatively few results are reported for component/parameter faults. Since the component faults often change the structural parameters of the system, determining the real-time operating mode of the system is a prerequisite for achieving fault tolerance. A common approach to this problem is to use multiple observers to first discriminate the fault modes, and then activate the corresponding control law of the isolated mode to achieve switching control reconfiguration [21]. Such an approach can be viewed as a passive fault diagnosis (PFD)-based AFTC scheme. Actually, due to the potential lack of diagnostically relevant information in the input–output data, the PFD method may fail to isolate a fault or may isolate a fault incorrectly. Moreover, for high-dimension systems, the multiple observers for parallel applications usually occupy a large amount of memory, and the involved modal discriminant optimization problem is generally computationally demanding. One promising way is to integrate the active fault diagnosis (AFD) methods into FTC, i.e., the AFD-based AFTC. The central idea in AFD is to design a small harmful test/auxiliary input signal that can ensure maximal or full separation among the model predictions corresponding to the different modes of operation [22]. According to different design methods of AFD, some representative results have been presented, such as AFTC based on Youla–Kucera parametrization [23], AFTC based on set detection and isolation [24], AFTC based on performance transformation [25], AFTC based on distributed fault isolation [26], etc.

The above studies have provided different ideas for the construction of AFD and AFTC. Inspired by these results, we find two more problems whose handling can be further improved:

(i) In terms of the implementation of AFD in AFTC, the test inputs used for modal separation are usually optimized online. For the small-scale systems, such computational requirements can be satisfied. However, as the number of system dimensions increases, the computational burden tends to become heavier, which often results in much longer delays of correct fault mode isolation. Recently, an effective solution was proposed in [27], where an implicit expression of the residual limit set is adopted and a constant auxiliary signal and the associated separation hyperplane used to separate the potential system modes are constructed offline. After a fault is detected, only the constant test signal is injected into the system and the current diagnostic observer. Then, the true system mode can be isolated by discriminating the position of the generated residuals in relation to the previously computed separation hyperplane. Given its advantages, such as simple implementation and fast isolation, this approach can provide an effective perspective for the design of control reconfigurations. Therefore, this paper will first attempt to adapt this active fault isolation approach to be integrated into the framework of AFTC to provide critical modal update information for timely regulation of constrained systems.

(ii) In terms of the design of constrained active reconfiguration FTC, most MPC-based methods need to solve computationally intensive optimization problems online. Generally, this often places stringent requirements on the system scale, sample interval, and hardware controller performance. As an alternative solution to constrained optimization, the interpolation control (IC) methods exhibit excellent features [28–30].

The main idea is to optimize an interpolation coefficient in real time based on the updated system states and use this coefficient to make a smooth convex combination of a outer controller and a inner controller. The outer controller is used to enlarge the controllable feasible domain, while the inner controller is used to satisfy the given control performance requirements. In general, the inner controller is optimally designed offline, while the outer controller is determined online simultaneously when the interpolation coefficient is optimized. This method of offline designing some parameters of the controller in advance helps to reduce the online calculation burden. Moreover, the optimized interpolation coefficient enables a smooth transition between the inner–outer controllers and ensures a fast convergence of the states to the set point under the constraints. In particular, the associated optimization problem belongs to standard linear programming (LP), which can be readily solved in the practical implementation. Given these characteristics, the IC-based optimization can provide a good compromise among computational load, feasible region size, performance, etc. Therefore, the development of the IC strategy to solve the constrained AFTC problem would be very promising. To the authors' knowledge, no relevant results have been reported.

Motivated by the above observation, we seek to further push the development of the field of constrained FTC for component/actuator faults by proposing a new AFD-based interpolating FTC synthesis scheme. The central ideas of the technical route are: (1) the passive fault detection (FD) is firstly designed by using a diagnostic observer in the current mode; (2) after a fault is detected, the active fault isolation (FI) and mode identification are then achieved by using a constant test signal and a separated hyperplane; and (3) after the actual mode is isolated, the constrained AFTC is finally determined by virtue of optimizing the interpolation coefficient to combine the inner FTC and outer FTC. How to comprehensively solve the problems involved in this technical route is the main research content of this paper.

Compared with the recent results on constrained AFTC studies (e.g., [13,16,20]), our main contributions can be reflected in the follows aspects: (i) A new and efficient AFD-based AFTC approach for component/actuator faults is proposed. In this work, only one observer is applied in real time to achieve FD and FI, while most of the existing studies use multiple observers for online parallel diagnosis; the fault mode separation is achieved by using auxiliary signals and separating hyperplanes designed offline, rather than by solving receding horizon optimizations and set membership discriminations online; the real-time control reconfiguration-based AFTC is achieved by solving simple LP problems instead of solving quadratic or semi-definite positive programming problems. (ii) When designing diagnostic observers and FTCs, the interaction influences between estimation error and control error is further handled based on integrated design and constraint tightening so as to improve the robust feasibility of AFTC optimization algorithm. (iii) The soft constraints IC-based AFTC strategy is also designed to address some infeasible scenarios, such as, the deviation of states from the maximum controllable invariant set after fault isolation, or the constraints violation caused by some unanticipated factors.

The remainder of this paper is structured as follows. Section 2 provides the problem formulation. In Section 3, the proposed AFD-based interpolating AFTC scheme is explained in detail and an integrated algorithm is also given to summarize the involved offline design and online application steps. In Section 4, the algorithm verification is given. Some conclusion and future work are discussed in Section 5.

Notation 1. *$diag\{X_1, X_2, X_3\}$ is a diagonal matrix with diagonal elements X_1, X_2, and X_3. $A^TP(*) = A^TPA$. $\mathbf{1}_m$ is a m-dimensional column vector with all elements of 1, while I_m is a m-dimensional unitary matrix. Let $p \in P$ and $q \in Q$ be two sets of $\mathbb{R}^n$. Then, $P \oplus Q = \{p+q | p \in P,\ q \in Q\}$ is the Minkowsi sum of two sets. For two sets satisfying $Q \subset P$, $x \in P \sim Q$ represents $x \in P$, but $x \notin Q$. A polyhedron is the intersection of a finite number of open and/or closed half-spaces, and a polytope is a closed and bounded polyhedron.*

2. System Description and Problem Formulation

Consider the following uncertain discrete-time systems affected by unknown component faults, actuator faults and disturbances:

$$x_{k+1} = A^l x_k + B^l u_k + d_k,\ y_k = Cx_k + v_k \tag{1}$$

where $x_k \in \mathcal{X} \subset \mathbb{R}^n$ is the state vector; $u_k \in \mathcal{U} \subset \mathbb{R}^{n_u}$ is the actuator input vector; $d_k \in \mathcal{D} \subset \mathbb{R}^{n_d}$ is the unknown process disturbance vector; $v_k \in \mathcal{V} \subset \mathbb{R}^{n_v}$ is the unknown measurement disturbance vector; $y_k \in \mathbb{R}^{n_y}$ is the measurement output vector. The matrices A^l, B^l and C are constant and have appropriate dimensions. The index l is associated with the configuration in which the system is actually operating, i.e., $(A^l, B^l) \in \{(A^0, B^0),\ (A^1, B^1),\ \cdots, (A^{n_f}, B^{n_f})\}, l \in [0, n_f]$. Without loss of generality, we assume that $l = 0$ corresponds to the healthy condition (A^0, B^0) while any other $l \geq 1$ corresponds to a faulty condition. In addition, $\mathcal{X}, \mathcal{U}, \mathcal{D}, \mathcal{V}$ are defined as the bounded polyhedral constraint sets [30,31]: $\mathcal{X} = \{x \in \mathbb{R}^n : H_x x \leq b_x\}, \mathcal{U} = \{u \in \mathbb{R}^{n_u} : H_u u \leq b_u\}, \mathcal{D} = \{d \in \mathbb{R}^{n_d} : H_d d \leq b_d\}, \mathcal{V} = \{v \in \mathbb{R}^{b_v} : H_v v \leq b_v\}$, where $H_x, H_u, H_d, H_v, b_x, b_u, b_d, b_v$ are predetermined.

Remark 1. *The model* (1) *can represent some uncertainties. Firstly, the changes in the configuration of the system (i.e., $l \in [0, n_f]$) due to the appearance or disappearance of faults are essentially a description of the uncertainty of the system [1]. Secondly, the disturbance terms (d_k, v_k) included in the model can directly reflect the multiple uncertainties in the system. For instance, let $A^l = A^0$, $d_k = \Delta_A x_k + \delta_k$ with unknown but bounded term Δ_A,* (1) *can represent a class of additive parametric uncertainty models; let $A^l = A^0, d_k = A^0(I - \Delta_A)x_k + \delta_k$,* (1) *can represent a class of multiplicative parametric uncertainty models; let d_k be a time-varying/time-invariant uncertainty term only,* (1) *can represent the uncertainty case for a class of mechanistic models with bounded offsets of modeling error, etc. All of these scenarios can be used to reflect a mismatch between the model and the reality.*

Remark 2. *The model* (1) *can represent both component and actuator faults [1]. For example, $A^l = A^0 + \sum_{i=1}^{n} A^i \theta_k^i$ with unknown faulty factor θ_k^i can represent some component/parameter faults; $B^l = B^0 diag\{\theta_k^1, \theta_k^2, \cdots, \theta_k^{n_u}\}$ with $\theta_k^i \in [0, 1]$ can describe some actuator effectiveness loss faults.*

For the sake of simplicity, the dynamics of the l-th system configuration can be rewritten as

$$\begin{aligned} &x_{k+1}^l = A^l x_k^l + B^l u_k^l + d_k,\ y_k^l = Cx_k^l + v_k \\ &s.t.\ \ x_k \in \mathcal{X},\ u_k \in \mathcal{U},\ d_k \in \mathcal{D},\ v_k \in \mathcal{V} \end{aligned} \tag{2}$$

The following assumptions are given for systems (1) and (2).

Assumption 1. *The typical system configurations of concern can be modeled in advance, and these system configurations are controllable.*

Remark 3. *We recognize that not all systems and faults can be tolerant by only one FTC method. Therefore, we make the above assumptions to explain the situations in which the proposed method can be applied.*

Definition 1. *Let $\mathcal{S}$ be a neighborhood of the origin. The closed-loop trajectory of* (1) *is said to be Uniformly Ultimately Bounded (UUB) in $\mathcal{S}$, if $\forall\, x_0$, $\exists T(x_0) > 0$ such that $x_k \in \mathcal{S}$ for $k \geq T(x_0)$.*

The control objective is to construct an AFD-based robust and feasible AFTC strategy such that the states of the controlled system (1) can be steered inside a neighborhood of origin (i.e., UUB) in a way of minimizing the following optimization problem

$$\begin{aligned} \min_{u_k} \mathcal{J}_{x_k,u_k} &= \sum_{t=0}^{\infty} U(x_{k+t},\ u_{k+t}) \\ s.t.\ x_{k+1}^l &= A^l x_k^l + B^l u_k^l + d_k,\ y_k^l = C x_k^l + v_k \\ & x_k \in \mathcal{X},\ u_k \in \mathcal{U},\ d_k \in \mathcal{D},\ v_k \in \mathcal{V} \\ & l \in [0,\ n_f] \end{aligned} \tag{3}$$

where $U(x_k,\ u_k) = x_k^T \Xi x_k + u_k^T \Theta u_k$, $\Xi > 0, \Theta > 0$ is a utility function.

3. Main Results

3.1. The Overall Scheme of the Proposed AFD-Based Interpolation AFTC Method

The overall scheme of the proposed AFD-based interpolation AFTC method is shown in Figure 1. In the subsequent analysis, we let that the index $l \in [0, n_f]$ denotes the unknown actual system operating condition and the index $i \in [0, n_f]$ denotes the recently identified system operating condition. Then, according to the flowchart in Figure 1, the AFTC method works as explained below. First, the I/O data of the practical system (i.e., the lth model) is collected by the ith estimator to give the state estimates $\hat{x}_k^i$ and generate the residuals r_k^i. Second, the fault detection unit performs change detection based on the estimator outputs. When there is no change (i.e., $l = i$), the interpolation control algorithm currently in use continues to regulate the system. When a change/fault is detected (i.e., $l \neq i$), the fault isolation unit is activated, and in this case the pre-designed auxiliary test signal u_{FI}^i is injected into the system and the estimator to perform modal discrimination. Next, after the practical system condition is isolated (i.e., $i = l$), the decision results of the fault isolation unit will update the operating condition index of the estimator and the reconfiguration controller. Next, the suitable interpolation optimization should be selected according to the location of states in relation to the feasible set of controller (i.e., robust control invariant set). Namely, if the states belong to the feasible set of the isolated controller, the general interpolation control is applied; otherwise, the relaxed interpolation control should be activated. Finally, these control actions will adjust the system states to the desired operating region. The design of each unit in this flowchart is given in detail below.

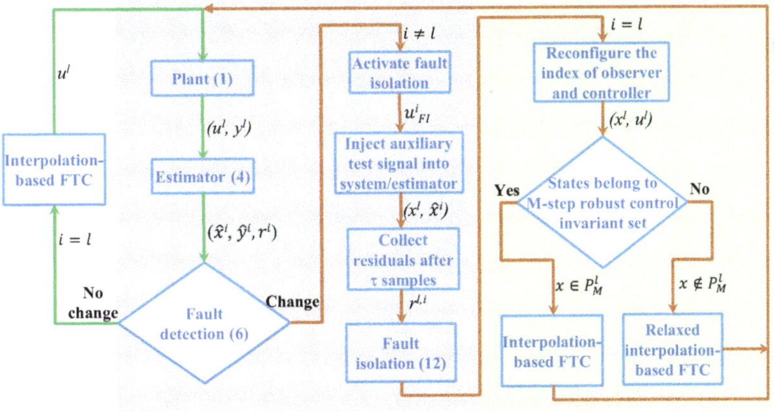

Figure 1. Scheme of AFD-based interpolation AFTC. The fault detection and active fault isolation constitute AFD, which will be designed in Section 3.2; AFTC consists of outer FTC, inner FTC and interpolation optimization, where outer FTC and interpolation optimization are designed in Section 3.4 and inner FTC with observer is designed in Section 3.3.

3.2. AFD: Fault/Mode Change Detection and Isolation

Without loss of generality, the following i-th observer is adopted to estimate states and generate residuals

$$\begin{aligned} \hat{x}^i_{k+1} &= (A^i + L^i C)\hat{x}^i_k + B^i u^i_k - L^i y^i_k \\ \hat{y}^i_k &= C\hat{x}^i_k, r^i_k = y^i_k - \hat{y}^i_k \end{aligned} \tag{4}$$

where $\hat{x}^i_k \in \mathbb{R}^n$ denotes the estimated state vector; $\hat{y}^i_k \in \mathbb{R}^{n_y}$ is the estimated output vector; $r^i_k \in \mathbb{R}^{n_y}$ is the generated residual signal that is used to provide key information of abnormal condition for achieving AFD. L^i is the observer gain.

Assumption 2. *For the sake of discussion, we assume that the observer* (4) *for each* $i \in [0,\ n_f]$ *has been designed in advance, and* $(A^i + L^i C)$ *is Schur stable. The detailed design conditions of* L^i *are given in Theorem 1.*

Remark 4. *In a cycle of AFD, the FI is always triggered by the FD [22]. Moreover, when a fault is detected at time* k_d*, the closed-loop FTC controller that is currently being used should preferably be put on standby to avoid that the feedback function hides the effect of the fault. In this setting, only the auxiliary input is used to stimulate the faulty system. In principle, the design of such an auxiliary input should (1) minimize the harmful influence to the currently matched system operation and (2) accurately identify and isolate the real system operating condition* l.

In Figure 1, there are two cases about the generated residual signal r^i_k. One is that the ith observer currently in use is matched to the real system mode l, and the other is the opposite. In the sequel, we will discuss the characteristics of the corresponding residuals for each of these two cases.

(1) ***Case I*** *(design of FD logic for* $i = l$*)*: First, based on Remark 4 and (1)–(4), the following estimation error system can be established:

$$e_{x^i,k+1} = (A^i + L^i C)e_{x^i,k} + L^i v_k + d_k,\ r^i_k = Ce_{x^i,k} + v_k \tag{5}$$

where $e_{x^i,k} = x^i_k - \hat{x}^i_k$. Given $i \in [0, n_f]$, the relevant disturbance term $L^i v_k + d_k$ is bounded by a deterministic set $\Delta^{i,i}_e = (L^i \mathcal{V}) \oplus \mathcal{D}$. Then, based on a series of finite set iterations along (5) using $\Delta^{i,i}_e$, an approximate maximal RPI set $\Omega^{i,i}_e$ (see Definition 2) can be computed and the limit set of residual r^i can be directly obtained as $\mathcal{R}^{i,i}_{FD} = C\Omega^{i,i}_e \oplus \mathcal{V}$. According to Figure 1 and Remark 4, the detection of mode changes and the triggered action can be formulated as

$$\begin{cases} r^i_k \notin \mathcal{R}^{i,i}_{FD} \Rightarrow \text{Mode change} \Rightarrow \text{Activate FI} \\ r^i_k \in \mathcal{R}^{i,i}_{FD} \Rightarrow \text{No change} \quad \Rightarrow \text{Continue detection} \end{cases} \tag{6}$$

Remark 5. *Considering the possibility of fault occurrence, transformation, or recovery, we uniformly use mode change in* (6) *to indicate any phenomenon that causes a change in the system behavior.*

(2) ***Case II*** *(design of FI logic for* $i \neq l$*)*: The case $i \neq l$ implies that the real status of system has changed and it generally leads to $r^i_k \notin \mathcal{R}^{i,i}_{FD}$. In this case, the fault/mode isolation should be activated. According to the analysis method in [27] and Remark 4, an auxiliary input u^i_{FI} will be used to replace the AFTC input $u^i_{C,k}$. A relevant augmentation representation is firstly constructed as

$$\chi^{l,i}_{k+1} = A^{l,i}_\chi \chi^{l,i}_k + B^{l,i}_\chi u^i_{FI} + E^i_\chi \sigma_k,\ r^{l,i}_k = C_\chi \chi^{l,i}_k + D_\chi \sigma_k \tag{7}$$

where $\chi_k^{l,i} = \left[(x_k^l)^T \quad (\hat{x}_k^i)^T\right]^T$, $A_\chi^{l,i} = \begin{bmatrix} A^l & 0 \\ -L^iC & A^i + L^iC \end{bmatrix}$, $\sigma_k \in \mathcal{E} = \{\left[d_k^T \quad v_k^T\right]^T : d \in \mathcal{D}, v \in \mathcal{V}\}$, $B_\chi^{l,i} = \begin{bmatrix} B^l \\ B^i \end{bmatrix}$, $E_\chi^i = \begin{bmatrix} I & 0 \\ 0 & -L^i \end{bmatrix}$, $C_\chi = \begin{bmatrix} C & -C \end{bmatrix}$, $D_\chi = \begin{bmatrix} 0 & I \end{bmatrix}$. Clearly, the term $E_\chi^i\sigma_k$ lies in the set $\Delta_\chi^i = E_\chi^i\mathcal{E}$. Then, given u_{FI}^i and based on Assumption 1, an approximate maximal RPI set $\Omega_\chi^{l,i}$ for each pair $(l,\ i)$, $i \neq l$, can be determined by finite set iterations along (7). Accordingly, the limit set that is used to achieve modal isolation can be obtained as $\mathcal{R}_{FI}^{l,i} = C_\chi\Omega_\chi^{l,i} \oplus D_\chi\mathcal{E}$. The approximated calculation method of $\mathcal{R}_{FI}^{l,i}$ is given in Appendix A.

A crucial condition for the existence of u_{FI}^i that discriminates between configurations ζ and η in finite time is $\mathcal{R}_{FI}^{\zeta,i} \cap \mathcal{R}_{FI}^{\eta,i} = \emptyset$, $\zeta \neq \eta$. According to [27], such discrimination can be achieved by checking whether the distance between the two sets is positive. Without loss of generality, the following distance metric is defined as

$$dis_{\zeta,\eta}^i = \inf_{(q_\zeta \in \mathcal{R}_{FI}^{\zeta,i}, p_\eta \in \mathcal{R}_{FI}^{\zeta,i})} \| q_\zeta - p_\eta \|_2 \tag{8}$$

Clearly, for each pair $(\zeta,\ \eta) \in \{[0, n_f] \sim i\}, \zeta \neq \eta$, we need to solve (8) to determine a suitable auxiliary input u_{FI}^i such that the distance metric $dis_{\zeta,\eta}^i$ is positive. The distance metric (8) has the following properties.

Lemma 1. *[27] The distance metric function $dis_{\zeta,\eta}^i$ is convex and hence its maximum is reached on certain vertices of the input constraint set.*

Based on Remark 4 and Lemma 1, the optimization design problem of auxiliary input signal u_{FI}^i, $\forall i \in [0, n_f]$ can then be formulated as

$$\begin{cases} \min \ \gamma \\ s.t. \ dis_{\zeta,\eta}^i > 0;\ u_{FI}^i \in vert(\gamma\mathcal{U});\ \sigma_k \in \mathcal{E}; \\ \zeta,\ \eta \in \{[0, n_f] \sim i\};\ \zeta \neq \eta. \end{cases} \tag{9}$$

Once the problem in (9) is solved for each i, the corresponding separation hyperplane (denoted as $\Pi_{\zeta,\eta}^i$) that is used to isolate the new mode can be further calculated through

$$\begin{aligned} \Pi_{\zeta,\eta}^i &= \{r : (r - \check{r}^\eta)^T(r - \check{r}^\eta) = (r - \check{r}^\zeta)^T(r - \check{r}^\zeta)\} \\ &= \{r : (\check{r}^\zeta - \check{r}^\eta)^T r = \frac{(\check{r}^\zeta - \check{r}^\eta)^T(\check{r}^\zeta + \check{r}^\eta)}{2}\} \end{aligned} \tag{10}$$

where $\check{r}^\zeta \in \mathcal{R}_{FI}^{\zeta,i}$ and $\check{r}^\eta \in \mathcal{R}_{FI}^{\eta,i}$ are two points at minimum distance from $\Pi_{\zeta,\eta}^i$, and they can be determined when solving (9). Then, these off-line designed separation hyperplanes will be used for real-time isolation. For simplicity, the isolation function is constructed as

$$Iso_{\zeta,\eta}^i = sign[(\check{r}^\zeta - \check{r}^\eta)^T r_k - \frac{(\check{r}^\zeta - \check{r}^\eta)^T(\check{r}^\zeta + \check{r}^\eta)}{2}] \tag{11}$$

Then, for the residual signals generated in real time, the online FI logic can be designed as

$$\begin{cases} Iso_{\zeta,\eta}^i > 0 \Rightarrow \text{Mode } \zeta \text{ is effective} \\ Iso_{\zeta,\eta}^i < 0 \Rightarrow \text{Mode } \eta \text{ is effective} \end{cases} \tag{12}$$

The current system mode can thus be discerned by making no more than n_f comparisons using (11) and (12).

3.3. Integrated Design of Observer and Unconstrained Controller

When the practical system mode index $l \in [0, n_f]$ is isolated, the control reconfiguration should be activated immediately, i.e., the control action $u^l_{C,k}$ is reconfigured with the new isolated mode index l. Now we will design the control policy $u^l_{C,k}$. Here, we consider for now the case where the constraints ($x \in \mathcal{X}$, $u \in \mathcal{U}$) are not triggered and $u^l_{C,k}$ can then be designed only as an estimator-based robust feedback control policy $u^l_{C,k} = K^l \hat{x}^l_k$, $\forall l \in [0, n_f]$. Under such settings, the closed-loop system dynamics can be obtained as

$$\begin{aligned} x^l_{k+1} &= A^l x^l_k + B^l K^l \hat{x}^l_k + d_k \\ &= (A^l + B^l K^l) x^l_k - B^l K^l e_{x^l,k} + d_k \\ &= \bar{A}^l x^l_k + \bar{B}^l e_{x^l,k} + d_k \end{aligned} \tag{13}$$

where $\bar{A}^l = A^l + B^l K^l$, and $\bar{B}^l = -B^l K^l$.

On the other hand, by defining a virtual output variable vector $z^l_k = \begin{bmatrix} \Xi^{1/2} \\ 0 \end{bmatrix} x^l_k + \begin{bmatrix} 0 \\ \Theta^{1/2} \end{bmatrix} u^l_k$, the unity function of cost function (3) can be represented by $U(x^l_k, u^l_k) = (z^l_k)^T z^l_k$. Then, the closed-loop virtual output by $u^l_{C,k}$ can be deduced as

$$z^l_k = \bar{C}^l x^l_k + \bar{D}^l e_{x^l,k} \tag{14}$$

where $\bar{C}^l = \begin{bmatrix} \Xi^{1/2} \\ 0 \end{bmatrix} + \begin{bmatrix} 0 \\ \Theta^{1/2} \end{bmatrix} K^l$ and $\bar{D}^l = -\begin{bmatrix} 0 \\ \Theta^{1/2} \end{bmatrix} K^l$.

According to [20,32], there may exist robustness interaction influences between estimation accuracy and unconstrained control performance, since the estimation error $e_{x^l,k}$ disturbs the closed-loop system (13) and (14) whilst the unmodeled dynamics d_k usually containing states can affect the estimation system (5). Hence, an integrated design of composite closed-loop system (5), (13) and (14) must be adopted to obtain the satisfactory observer gain L^l and control gain K^l, $\forall l \in [0, n_f]$. The following composite closed-loop system is firstly established:

$$\begin{aligned} \psi^l_{k+1} &= \tilde{A}^l \psi^l_k + \tilde{B}^l \varrho_k \\ z^l_k &= \tilde{C}^l \psi^l_k \end{aligned} \tag{15}$$

where $\psi^l_k = \begin{bmatrix} e^T_{x^l,k} & (x^l_k)^T \end{bmatrix}^T$, $\varrho_k = \begin{bmatrix} v^T_k & d^T_k \end{bmatrix}^T$, $\tilde{A}^l = \begin{bmatrix} \tilde{A}^l_{11} & 0 \\ \tilde{A}^l_{21} & \tilde{A}^l_{22} \end{bmatrix}$, $\tilde{B}^l = \begin{bmatrix} \tilde{B}^l_1 \\ \tilde{B}^l_2 \end{bmatrix}$, $\tilde{A}^l_{11} = A^l + L^l C$, $\tilde{A}^l_{21} = -B^l K^l$, $\tilde{A}^l_{22} = A^l + B^l K^l$, $\tilde{B}^l_1 = \begin{bmatrix} L^l & I \end{bmatrix}$, $\tilde{B}_2 = \begin{bmatrix} 0 & I \end{bmatrix}$, and $\tilde{C}^l = \begin{bmatrix} \bar{D}^l & \bar{C}^l \end{bmatrix}$.

The following theorem presents the integrated design conditions of observer gain and unconstrained feedback gain.

Theorem 1. *For each $l \in [0, n_f]$, a robust observer* (5) *and associated robust feedback control policy $u_{C,k} = K^l \hat{x}^l_k$ can be integratedly determined, if some decision variables $\alpha > 0$, $\beta > 0$,*

$P_1^l = (P_1^l)^T > 0$, $P_2^l = (P_2^l)^T > 0$, Y_1^l, Y_2^l, $\bar{K}^l$, $\bar{L}^l$ *exist as the solutions to the following optimization problem:*

$$\min_{P_1^l, P_2^l, Y_1^l, Y_2^l, \bar{K}^l, \bar{L}^l} \varsigma\alpha + (1-\varsigma)\beta$$

$$s.t. \begin{bmatrix} I - P_1^l & \diamond & \diamond \\ 0 & -\alpha^2 I & \diamond \\ \Gamma_{31l} & \Gamma_{32l} & P_1^l - Y_1^l - (Y_1^l)^T \end{bmatrix} < 0 \tag{16a}$$

$$\begin{bmatrix} P_2^l - Sym(Y_2^l) & \diamond & \diamond & \diamond & \diamond \\ 0 & P_2 - Sym(Y_2^l) & \diamond & \diamond & \diamond \\ 0 & 0 & -\beta^2 I & \diamond & \diamond \\ Y_{41l} & Y_{42l} & \tilde{B}_2 & -P_2^l & \diamond \\ Y_{51l} & Y_{52l} & 0 & 0 & -I \end{bmatrix} < 0 \tag{16b}$$

where $Sym(Y_2^l) = Y_2^l + (Y_2^l)^T$, $\Gamma_{31l} = Y_1^l A^l + \bar{L}^l C$, $\Gamma_{32l} = \begin{bmatrix} \bar{L}^l & Y_1^l \end{bmatrix}$, $Y_{41l} = -B^l \bar{K}^l$, $Y_{42l} = A^l (Y_2^l)^T + B^l \bar{K}^l$, $\tilde{B}_2^l = \begin{bmatrix} 0 & I \end{bmatrix}$, $Y_{51l} = -\begin{bmatrix} 0 \\ \Theta^{1/2} \end{bmatrix} \bar{K}^l$, $Y_{52l} = \begin{bmatrix} \Xi^{1/2} \\ 0 \end{bmatrix} (Y_2^l)^T + \begin{bmatrix} 0 \\ \Theta^{1/2} \end{bmatrix} \bar{K}^l$, $\varsigma \in (0,1)$. *Once the above optimization is solved, the parameters of observer and feedback gain can be calculated by* $K^l = \bar{K}^l ((Y_2^l)^T)^{-1}$ *and* $L^l = (Y_1^l)^{-1} \bar{L}^l$, *respectively.*

Proof. The proof of Theorem 1 is given in Appendix B. □

3.4. Constrained AFTC: Reconfigured Interpolating Control

Based on the set-theoretic concepts in [28,31], several invariant sets are defined.

Definition 2. *Given the controller* $u_{C,k}^l = K^l \hat{x}_k^l$, *the set* $\Omega_{RPI}^l \subseteq \mathcal{X}$ *is a robust positive invariant set (RPI-set) for closed-loop system* (13) *subject to constraint* $x_k^l \in \mathcal{X}$ *if for any* $x_0^l \in \Omega_{RPI}^l$ *we have* $x_k^l \in \Omega_{RPI}^l$ *for all* $\bar{B}^l e_{x^l,k} + d_k$, $k > 0$. *Moreover,* Ω_{MRPI}^l *is the maximal RPI-set if* Ω_{MRPI}^l *contains all the RPI-sets of constrained closed-loop system* (13) *in* $\mathcal{X}$. *For simplicity,* Ω_{MRPI}^l *is represented in the polyhedral form of* $\Omega_{MRPI}^l = \{x^l : F_I^l x^l \leq g_I^l\}$.

The following enlarged invariant set is further defined for some constrained allowable control inputs.

Definition 3. *Given the lth model of* (2) *and the constraints* $(\mathcal{X}, \mathcal{U})$, *the set* $\Omega_{RCI}^l \subseteq \mathcal{X}$ *is a robust control invariant set (RCI-set), if for any* $x_0^l \in \Omega_{RCI}$ *there exists an admissible control input* $u_k^l \in \mathcal{U}$ *such that all the state updates satisfy* $x_k^l \in \Omega_{RCI}$ *for all* d_k *and* $e_{x^l,k}$, $k > 0$. *Similarly, the maximal RCI-set* Ω_{MRCI} *contains all robust RCI-sets.*

Generally, the determination of Ω_{MRCI} is computationally demanding, in particular for high-dimension systems. As an alternative, the M-step robust control invariant set can be used.

Definition 4. *The set* $P_M^l \subseteq \mathcal{X}$ *is defined as a M-step robust control invariant set for the lth model of* (2) *with respect to the constraints* $(\mathcal{X}, \mathcal{U})$, *if there exists an admissible control sequence such that all states* $x_k^l \in P_M^l$ *can be steered into* Ω_{MRPI}^l *in no more than M steps. For simplicity,* P_M^l *is described as* $P_M^l = \{x^l : F_M^l x^l \leq g_M^l\}$.

In general, two cases exist for the location of the states of system after the active FI is completed, namely $x_k^l \in P_M^l$ and $x_k^l \notin P_M^l$. In the sequel, we will construct an interpolating FTC strategy for each of these two cases.

(1) ***Case I*** *($x_k^l \in P_M^l$ after FI)*: Firstly, in order to get Ω_{MRPI}^l of (13), the bounded set of $\bar{B}^l e_{x^l,k} + d_k$ should be determined. By a series of finite set iterations along (5), the

disturbance invariant set of e_{x^l} subject to $v_k \in \mathcal{V}$ and $d_k \in \mathcal{D}$ has been computed as $\Omega_{e'}^{l,l}$. Then, we have $\bar{B}^l e_{x^l,k} + d_k \in (\bar{B}^l \Omega_{e'}^{l,l} \oplus \mathcal{D})$. Further, the *Procedure 2.1* in [28] can be referred to calculate Ω^l_{MRPI} of (13).

In order to describe the control actions that can regulate the state x^l_k from P^l_M back to Ω^l_{MRPI} in no more than M steps, an augmented control sequence $U^l_{M,k} = \left[(u^l_{IC,k})^T \quad (u^l_{IC,k+1})^T \quad \cdots \quad (u^l_{IC,k+M-1})^T \right]^T$ is defined. In fact, these actions are expected to regulate the dynamic behavior of the system in the following manner:

$$
\begin{aligned}
x^l_{k+1} &= A^l x^l_k + B^l u^l_{IC,k} + d_k \in P^l_M \\
&\vdots \\
x^l_{k+M-1} &= A^l x^l_{k+M-2} + B^l u^l_{IC,k+M-2} + d_{k+M-2} \in P^l_M \\
x^l_{k+M} &= A^l x^l_{k+M-1} + B^l u^l_{IC,k+M-1} + d_{k+M-1} \in \Omega^l_{MRPI}
\end{aligned}
\tag{17}
$$

Obviously, in (17) we can observe that the migration process of states can be approximately deduced by the current initial state x^l_k and a sequence of inputs $U^l_{M,k}$. Considering the constraints with Definition 4, we can further describe the maximal admissible control domain of the system (1) with respect to the corresponding control inputs in terms of the following half-space representation for the augmented state space $Q^l_M = \{x^l, U^l_M\}$:

$$
Q^l_M = \{x^l, U^l_M : \bar{F}^l_M \begin{bmatrix} x^l \\ U^l_M \end{bmatrix} \leq \bar{g}^l_M\}
\tag{18}
$$

Remark 6. *Given the previously obtained Ω^l_{MRPI} and certain M, the augmented set Q^l_M can be calculated by following the algorithm in [28]. In addition, by comparing the definition in* (18) *and Definition 4, it can be seen that P^l_M is a projection of Q^l_M onto the state space.*

Without loss of generality, any state vector $x^l_k \in P^l_M$ can be decomposed as a convex combination form

$$
x^l_k = s^l_k x^l_{O,k} + (1 - s^l_k) x^l_{I,k}
\tag{19}
$$

where $x^l_{I,k} \in \Omega^l_{MRPI}$ denotes an inner state vector while $x^l_{O,k} \in P^l_M \sim \Omega^l_{MRPI}$ denotes an outer state vector. $s^l_k \in [0,1]$ is the so-called interpolation coefficient. Since $x_{I,k}$ has already inside Ω^l_{MRPI}, the previously designed unconstrained optimal control law by $K^l x^l_{I,k}$ can be directly adopted to achieve UUB regulation of $x^l_{I,k}$ robustly. Thus, for $x^l_k \notin \Omega^l_{MRPI}$, (19) means that the problem of finding $U^l_{M,k}$ to regulate state x^l_k back to Ω^l_{MRPI} can be transformed into the problem of solving $U^l_{M,k}$ to regulate state $x^l_{O,k}$ back into Ω^l_{MRPI}.

In line with the above state decomposition (19), the following interpolated FTC strategy for the lth model is constructed

$$
u^l_{C,k} = s^l_k u^l_{IC,k} + (1 - s^l_k) u^l_{I,k}
\tag{20}
$$

where $u^l_{I,k} = K^l x^l_{I,k}$ is the inner FTC law while $u^l_{IC,k}$ is the outer FTC law to be determined. It should be noted that $u^l_{I,k}$ is the optimal unconstrained terminal control law, and it generally presents high control performance. However, for $x^l_{O,k} \in P^l_M \sim \Omega^l_{MPRI}$, the constraints will be activated and the performance might be poor. Thus, in order to make the high-performance inner controller as dominant as possible and minimize the constraint activation influence simultaneously, it is desirable to set s_k^l as small as possible. This can be achieved by solving the following optimization problem:

$$\begin{cases} \tilde{s}_k^l = \min\limits_{s_k^l, x_{I,k}^l, x_{O,k}^l, U_{M,k}^l} s_k^l \\ s.t.\ F_I^l x_{I,k}^l \le g_I^l;\ \tilde{F}_M^l \begin{bmatrix} x_{O,k}^l \\ U_{M,k}^l \end{bmatrix} \le \tilde{g}_M^l; \\ x_k^l = s_k^l x_{O,k}^l + (1 - s_k^l) x_{I,k}^l;\ 0 \le s_k^l \le 1. \end{cases} \tag{21}$$

The first constraint in (21) is used to ensure $x_{I,k}^l \in \Omega_{MRPI}^l$; the second inequality is used to ensure that there exists $U_{M,k}^l$ such that $x_{O,k}^l \in \tilde{P}_M^l \subseteq P_M^l$ and $x_{O,k+M}^l \in \Omega_{MRPI}^l$; the third equation guarantees a smooth convex interpolation between $x_{I,k}^l$ and $x_{O,k}^l$ and also achieves a smooth interpolation between the associated two control laws.

Remark 7. *In view of the influence of estimation error on the feasibility of optimization, we have contracted the constraint condition in* (18)*, and obtained the second constraint condition in* (21)*. Specifically, by setting* $u_{IC,k}^l = K^l x_{O,k}^l + c_k^l$*, there is* $u_{C,k}^l = s_k^l u_{IC,k}^l + (1 - s_k^l) u_{I,k}^l = s_k^l (K^l x_{O,k}^l + c_k^l) + (1 - s_k^l) u_{I,k}^l = s_k^l K^l x_{O,k}^l + (1 - s_k^l) K^l x_{I,k}^l + s_k^l c_k^l = K^l (s_k^l x_{O,k}^l + (1 - s_k^l) x_{I,k}^l) + s_k^l c_k^l = K^l x_k^l + \bar{c}_k^l$*, where* $\bar{c}_k^l = s_k^l c_k^l$*. Then, following the augmentation analysis technique in dual-mode predictive control [20], we can calculate a disturbance invariant set of* $[x_k; U_{M,k}^l]$ *that is driven by* $e_{x^l,k}$ *and* d_k*. Further, based on the constraint tightening, a conservative constraint set* $\tilde{Q}_M^l = \{x^l, U_M^l : \tilde{F}_M^l \begin{bmatrix} x^l \\ U_M^l \end{bmatrix} \le \tilde{g}_M^l\}$ *can be determined, where* $\tilde{P}_M^l$ *is a projection of* $\tilde{Q}_M^l$ *onto the state space.*

Since s_k^l, $x_{I,k}^l$, $x_{O,k}^l$ are unknown, the optimization (21) is nonlinear. Let $b_{O,k}^l = s_k^l x_{O,k}^l$, $b_{I,k}^l = (1 - s_k^l) x_{I,k}^l$, and $T_{M,k}^l = s_k^l U_{M,k}^l$, the above optimization problem (21) can be then simplified as a linear programming problem:

$$\begin{cases} \tilde{s}_k^l = \min\limits_{s_k^l, b_{O,k}^l, T_{M,k}^l} s_k^l \\ s.t.\ F_I^l (x_k^l - b_{O,k}^l) \le (1 - s_k^l) g_I^l; \\ \tilde{F}_M^l \begin{bmatrix} b_{O,k}^l \\ T_{M,k}^l \end{bmatrix} \le s_k^l \tilde{g}_M^l;\ 0 \le s_k^l \le 1. \end{cases} \tag{22}$$

When the optimal solution of (22) is obtained, the reconfigured interpolation FTC can then be constructed as $u_{C,k}^l = T_{M1,k}^l + K^l (x_k^l - b_{O,k}^l)$, where $T_{M1,k}^l$ is the first control input in $T_{M,k}^l$.

(2) ***Case II*** *($x_k^l \notin P_M^l$ after FI)*: The soft constraint methods are employed to ensure that states outside P_M^l can also be steered into Ω_{MRPI}^l after the fault is isolated. Depending on the requirements of the actual system for state constraints and input constraints, there exist two general ways to design soft constraints [33,34]. The first is that the input constraints must not be violated while the boundaries of the state constraints can be relaxed appropriately. The other is that the boundaries of both constraints can be adjusted. In either case, the relaxation variable introduced by the soft constraints is non-zero only when the original constraints are violated. Once the original constraints are restored, the relaxation variable must be zero. For the sake of simplicity, the second strategy is adopted and we design the following soft constrained interpolating control algorithm. First of all, we suppose that the maximal admissible control domain (18) can be relaxed to contain states $x_k^l \notin P_M^l$ as follows:

$$\tilde{Q}_{\varsigma_k, M}^l = \{x^l, U_M^l : \tilde{F}_M^l \begin{bmatrix} x^l \\ U_M^l \end{bmatrix} \le \tilde{g}_M^l + \varsigma_k^l \Lambda\} \tag{23}$$

where $\varsigma_k^l \geq 0$ is the relaxation variable and Λ can be a column vector of ones or an arithmetic progression vector with the first term 1 and common difference $-\kappa \in [-1, 0]$. Note that the soft constraints by (23) implicitly define an enlarged $M-$step robust control invariant set $\tilde{P}^l_{\varsigma_k,M}$ for systems (1) with relaxed constraints of states and inputs.

Then, in a similar way to formulate (19) and (20), we can also update the interpolations of states and inputs for $x^l_{I,k} \in \Omega^l_{MRPI}$ and $x^l_{O,k} \in \tilde{P}^l_{\varsigma_k,M} \sim \Omega^l_{MRPI}$. Slightly different from the optimization objective of (21), here the slack variable ς_k also needs to be minimized, i.e., the degree of constraint violation of $\tilde{P}^l_M$ should be minimized. To this point, we can further establish the following optimization problem through the same design of variables as (22):

$$\begin{cases} \tilde{\mu}^l_k = \min\limits_{s^l_k, b^l_{O,k}, T^l_{M,k}, \bar{\varsigma}^l_k} \quad \varepsilon_1 \bar{\varsigma}^l_k + \varepsilon_2 s^l_k \\ s.t.\ F^l_I(x^l_k - b^l_{O,k}) \leq (1 - s^l_k) g^l_I;\ 0 \leq s^l_k \leq 1; \\ \tilde{F}^l_M \begin{bmatrix} b^l_{O,k} \\ T^l_{M,k} \end{bmatrix} \leq s^l_k \tilde{g}^l_M + \bar{\varsigma}^l_k \Lambda;\ \bar{\varsigma}^l_k = s^l_k \varsigma^l_k. \end{cases} \tag{24}$$

where $\varepsilon_1 + \varepsilon_2 = 1$. In order to highlight the function of soft constraint FTC, ε_1 is generally set to be larger than ε_2.

3.5. The AFD-Based Reconfigured Interpolation FTC Algorithm

A binary parameter ε_3 is introduced to unify the optimization problems of (22) and (24):

$$\begin{cases} \tilde{\mu}^l_k = \min\limits_{s^l_k, b^l_{O,k}, T^l_{M,k}, \bar{\varsigma}^l_k} \quad \varepsilon_3(\varepsilon_1 \bar{\varsigma}^l_k + \varepsilon_2 s^l_k) + (1 - \varepsilon_3) s^l_k \\ s.t.\ F^l_I(x^l_k - b^l_{O,k}) \leq (1 - s^l_k) g^l_I;\ 0 \leq s^l_k \leq 1; \\ \tilde{F}^l_M \begin{bmatrix} b^l_{O,k} \\ T^l_{M,k} \end{bmatrix} \leq s^l_k \tilde{g}^l_M + \varepsilon_3 \bar{\varsigma}^l_k \Lambda;\ 0 \leq \bar{\varsigma}^l_k. \end{cases} \tag{25}$$

By setting $\varepsilon_3 = 1$, (25) reduces to (24), which is used to achieve soft constrained interpolating control for the case $x^l_k \notin \tilde{P}^l_M$. By setting $\varepsilon_3 = 0$, (25) reduces to (22) and the standard interpolating control based FTC can then be achieved. All the above developments allow us to write down Algorithm 1.

Algorithm 1 AFD-based interpolation AFTC.

Off-line Design: Given system (1), objective (3) and sets $(\mathcal{X},\ \mathcal{U},\ \mathcal{D},\ \mathcal{V})$. Complete the following designs: Solve Theorem 1 to obtain $K^l,\ L^l,\ \forall l = 0,1,\cdots,n_f$; Based on (5), calculate the limit set of residual $\mathcal{R}^{i,i}_{FD}$, $\forall i = 0,1,\cdots,n_f$; Solve the optimization problem (9) to obtain u^i_{FI} and associated points $\bar{r}^\zeta$, $\bar{r}^\eta$, $\forall i,\zeta,\eta = 0,1,2,\cdots,n_f, i \neq \zeta$, $i \neq \eta$, $\zeta \neq \eta$, and construct separation hyperplane (10); Formulate the FD logic (6) using $\mathcal{R}^{i,i}_{FD}$ and construct the FI logic (12) using (11); Calculate Ω^l_{MRPI} according to Definition 2 and calculate $\tilde{Q}^l_M$ based on (18) and Remarks 6 and 7.

On-line Implementation: Set ς, ε_1, ε_2, $\mu \in [0,1]$, state $x_0 = \hat{x}_0 \in P^0_M$ and $u^0_{FI} = 0$. Solve (25) with $\varepsilon_3 = 0$ to obtain $u^0_{C,0} \leftarrow T^0_{M1,0} + K^0(\hat{x}_0 - b^0_{O,0})$. Let $i \leftarrow 0$, $k \leftarrow 0$ $u_0 \leftarrow u^0_{C,0}$ and perform the following actions:

1: $(\hat{x}_k, r_k) \leftarrow (4)$; ◁ Using (4) to estimate $(\hat{x}_k, r_k)$
2: **if** $r_k \in \mathcal{R}^{i,i}_{FD}$ **then**
3: Solve (25) with $\varepsilon_3 = 0$ to compute $u^i_{C,k} \leftarrow T^i_{M1,k} + K^i(\hat{x}_k - b^i_{O,k})$; Set $u^i_{FI} \leftarrow 0$, $u_k \leftarrow u^i_{C,k}$, $k \leftarrow k+1$, and go to step 1;
4: **else** ◁ A fault is detected
5: Set $u^i_{C,k} \leftarrow 0$, $u_k \leftarrow u^i_{FI}$;
6: **end if**
7: **do**
8: $(\hat{x}_{k+\tau}, r_{k+\tau}) \leftarrow (4)$; ◁ Generate τ residuals
9: Find ζ such that $Iso^i_{\zeta,\eta} > 0, \forall \eta \in [0, n_f] \sim i, \zeta \neq \eta$;
10: **end do**
11: $i \leftarrow \zeta$, $k \leftarrow k+\tau$;
12: **if** $\hat{x}_k \in \tilde{P}^i_M$ then
13: Go to step 3;
14: **else** ◁ $\hat{x}_k \notin \tilde{P}^i_M$
15: Solve (25) with $\varepsilon_3 = 1$ to compute $u^i_{C,k} \leftarrow T^i_{M1,k} + K^i(\hat{x}_k - b^i_{O,k})$; Set $u^i_{FI} \leftarrow 0$, $u_k \leftarrow u^i_{C,k}$, $k \leftarrow k+1$, $(\hat{x}_k, r_k) \leftarrow (4)$, and go to step 12;
16: **end if**

4. Algorithm Verification by a Wastewater Treatment Plant Model

4.1. System Model and Parameters

The purpose of a wastewater treatment plant is to purify the sewage and return clean water to the river. Activated sludge process (ASP) is a very important part of the cleaning procedure [35]. Generally, ASP systems usually consist of a bioreactor and a settler. Bioreactors mainly rely on suspended microorganisms for biodegradation of dissolved substrate. After that, the suspended micro-organisms are completely separated in the settler. Some of the degraded biomass will be recycled to the bioreactor for further purification, while the remaining biomass will be discharged to maintain the balance of limited organisms in the ASP system. The energy needed for the reaction is provided by the dissolved oxygen, and the resulting carbon dioxide is in turn released. In [36], a simplified state-space error model describing the mass balances in ASP systems is built around the equilibrium point $(X_P, U_P) = ([122.7342\ 49.4714\ 196.3750\ 6.8300]^T, [0.06\ 1.35])$. Here, to achieve the fault tolerant mass balance of ASP systems, some uncertain parameters along the model in [36] are additionally considered as follows:

$$
\begin{aligned}
A &= \begin{bmatrix} 0.7685-\Delta_A & 0.1551 & 0.0576 & 0.1273 \\ -0.1438 & 0.4137+\Gamma_A & -0.0859 & -0.013 \\ -0.0109 & -0.0175 & 0.0026+\Delta_A & -0.0018 \\ 0.3396 & 0.0377 & 0.0253 & 0.8335-\Gamma_A \end{bmatrix} \\
B &= \begin{bmatrix} -250.0774 & 0.9268 \\ 398.0189 & -1.4129 \\ -13.8515 & 2.0454 \\ 102.6287 & 0.18 \end{bmatrix} \begin{bmatrix} \Delta_B & 0 \\ 0 & \Gamma_B \end{bmatrix},\ C = \begin{bmatrix} 0 & 0 \\ 1 & 0 \\ 0 & 1 \\ 0 & 0 \end{bmatrix}^T
\end{aligned} \tag{26}
$$

We assume that two types of faults can appear: $(\Delta_A=0.2,\ \Delta_B=0.7,\ \Gamma_A=\Gamma_B=0)$ and $(\Delta_A=\Delta_B=0,\ \Gamma_A=0.3,\ \Gamma_B=0.6)$. The former is identified as fault $l=1$ (faulty mode 1) and the latter is identified as fault $l=2$ (faulty mode 2). Clearly, the health condition $l=0$ (healthy mode 0) is indicated when $(\Delta_A=\Gamma_A=\Delta_B=\Gamma_B=0)$. The other parameters are: $\Xi=0.1I_4, \Theta=0.02I_2,\ n_f=2, d_k\in[-0.5\mathbf{1}_4\ 0.5\mathbf{1}_4], v_k\in[-0.5I_2\ 0.5I_2]$, and

$$\begin{bmatrix} -90.0342 \\ -37.4714 \\ -143.8750 \\ -5.9300 \end{bmatrix} \le x_k \le \begin{bmatrix} 137.2658 \\ 15.5286 \\ 13.6250 \\ 73.1700 \end{bmatrix},\ \begin{bmatrix} -0.0600 \\ -1.3500 \end{bmatrix} \le u_k \le \begin{bmatrix} 2.3400 \\ 13.6500 \end{bmatrix} \tag{27}$$

4.2. Offline Design of AFD and AFTC According to Algorithm 1 and Relevant Validation

According to Algorithm 1, the following parameters of AFTC policy are designed.

Firstly, by solving Theorem 1, the integrated parameters of observer L^l and inner FTC gain matrix K^l are obtained as

$$\begin{aligned} L^0 &= \begin{bmatrix} 4.7216 & 0.2884 \\ -1.2215 & -0.0229 \\ -0.0518 & -0.0051 \\ 9.4863 & 0.6130 \end{bmatrix},\ K^0 = \begin{bmatrix} -0.0016 & -0.0010 & 0.0002 & -0.0021 \\ -0.8410 & -0.2855 & -0.0381 & -0.7295 \end{bmatrix} \\ L^1 &= \begin{bmatrix} 0.1650 & -0.0290 \\ -0.4078 & 0.0669 \\ 0.0095 & -0.1675 \\ 0.7874 & 0.0267 \end{bmatrix},\ K^1 = \begin{bmatrix} -0.0031 & -0.0017 & -0.0002 & -0.0048 \\ -0.6755 & -0.2408 & -0.0910 & -0.8303 \end{bmatrix} \\ L^2 &= \begin{bmatrix} 0.6487 & -0.0128 \\ -0.7432 & 0.0605 \\ 0.0036 & -0.0024 \\ 0.6966 & 0.0089 \end{bmatrix},\ K^2 = \begin{bmatrix} -0.0014 & -0.0021 & 0.0001 & -0.0008 \\ -0.5591 & -0.2930 & 0.0015 & -0.2738 \end{bmatrix} \end{aligned} \tag{28}$$

Secondly, by using the disturbance set $\Delta_e^{i,i}$ for 3-step set iteration along (5), the limit sets of residual for each $i=0,1,2$ are approximately calculated, where the H-representations of $\mathcal{R}_{FD}^{0,0}$, $\mathcal{R}_{FD}^{1,1}$, and $\mathcal{R}_{FD}^{2,2}$ have 23, 38, and 47 inequalities, respectively. Due to the page limit, they are not listed here.

Thirdly, by solving optimization problem (9), some suitable choices of test input signals are determined as $u_{FI}^0=1.5\times\mathcal{U}.V(1)$, $u_{FI}^1=1.3\times\mathcal{U}.V(1)$, $u_{FI}^2=1.1\times\mathcal{U}.V(1)$, respectively. Here, $\mathcal{U}.V(1)$ is used to denote the first vertex of the V-representation of set $\mathcal{U}$. In order to clearly describe the relationship between the FD limit set and the FI separation line, we simulated the residual responses by injecting the above test input signal excitation in different modes of the system. As shown in Figure 2, the AFD can be successfully achieved as long as the residual value exceeds the relevant separation line. Here, the isolation can be accomplished in a maximum of six steps.

Next, the robust invariant sets Ω_{MRPI}^l and P_M^l are calculated for $l=0,1,2$, respectively. In order to describe the relationship among the interpolating AFTC, the controlled states and the corresponding invariant set for each mode, the evolution of an arbitrary initial state $x_0=[-20\ 10\ -10\ -1.83]^T$ is simulated. The results of the first three states are shown in Figure 3. It can be seen from Figure 3a,b that x_0 belongs to $P_M^l\sim\Omega_{MRPI}^l, l=0,1$. Therefore, as shown in sub-Figure 3d, the corresponding interpolation coefficients are not zero and x_0 is adjusted back to Ω_{MRPI}^l in 2-3 steps. Figure 3c illustrates that x_0 belongs to Ω_{MRPI}^2. Hence, the associated interpolation coefficient in Figure 3d is zero.

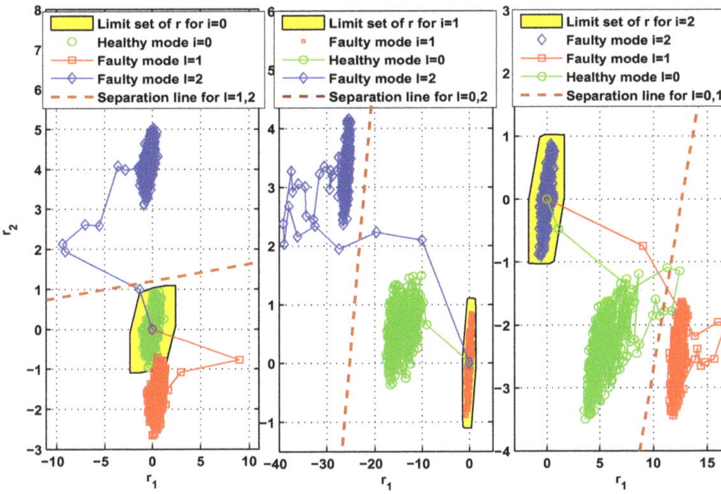

Figure 2. Test of the isolation effect of the constructed active fault isolation method in three scenarios

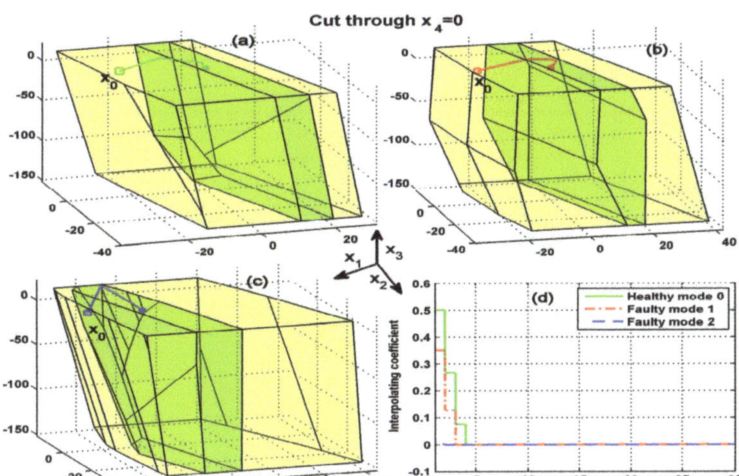

Figure 3. Test of the control effect of the developed interpolating AFTC. The yellow area represents the set Ω^l_{MRPI} and the green area represents the set P^l_M. (**a**) Invariant sets for health mode $l = 0$; (**b**) invariant sets for fault mode $l = 1$; (**c**) invariant sets for fault mode $l = 2$; (**d**) interpolation coefficient $\tilde{s}_k$.

4.3. Simulation Results and Analysis of the above Designed AFD-Based AFTC Method

Based on the parameters obtained above, we next perform performance tests on the proposed AFD-based AFTC method. First, the following fault scenarios are considered:

Fault scenarios: The system initially works in a healthy condition; when $k \in [160\ 550)$, the first kind of fault occurs in the system. For $k \geq 550$, the previous fault disappears and the second type of fault appears.

Then, the online AFTC strategy described in Algorithm 1 is implemented to deal with the above fault situations. The simulation results are collected and depicted in Figures 4–6, where the occurrence and duration of different faults have been marked using different color areas, i.e., green area for healthy condition ($l = 0$), yellow area for type I faults ($l = 1$)

and gray area for type II faults ($l = 2$). As depicted in Figure 4, it takes some time after a fault occurs to achieve the state regulation to track the equilibrium point X_P. The reason is that the fault detection, isolation, and control reconfiguration need to be completed during this time. Taking the fault-tolerant process for the first type of fault as an example, Figure 4 firstly depicts that the estimated values of the states can quickly deviate from their actual values in the moments after the fault occurs. Their estimation errors caused by the presence of the fault further generate large residual values, thus facilitating the timely triggering of FI. In fact, the interpolation coefficient in Figure 6 appears to increase rapidly at $k > 160$, which also indicates the occurrence of abnormal system conditions. The inputs of the corresponding constant value auxiliary test signals are further shown in Figure 5. It should be noted that both variables in Figure 6 are zero at this time. After a few steps, it can be seen in Figure 4 that the first three states have been accurately estimated, which indicates that the FI is completed. However, the estimation of the fourth state still deviates from the actual value. The reason is that the auxiliary signal injected during FI drives it to a large deviation (as shown in Figure 2). Hence, additional time is required to achieve its unbiased tracking.

After FI, the corresponding control reconfiguration is further activated. As shown in Figure 6, the soft constraint FTC (24) is triggered first, which also leads to a sharp increase of the control input in Figure 5. When the states are adjusted into P_M^1 by the soft constraint FTC, the interpolation FTC (22) is activated timely. At the same time, as illustrated in Figure 5, the control inputs subsequently become smaller. The decreasing interpolation coefficient in Figure 6 also indicates that the system states are gradually tuned into Ω_{MRPI}^1. After that, the states are gradually regulated to track the equilibrium point.

Remark 8. *The above process constitutes a complete cycle of AFD and AFTC. Clearly, the decreasing interpolation coefficients and relaxation variables in Figure 6 fully illustrate the convergence of the proposed Algorithm 1. Correspondingly, the state and control variables in Figures 4 and 5 are also adjusted to the equilibrium point (X_P, U_P), which further illustrates that the control system under the influence of the fault is stabilized and the tracking target is achieved.*

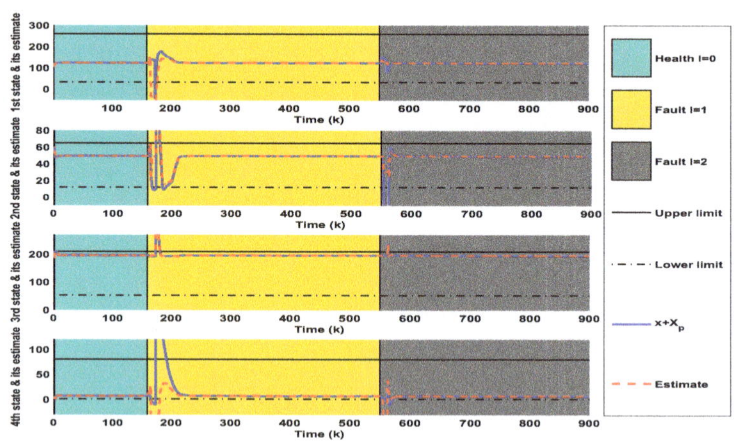

Figure 4. Simulation results of state evolution and estimation under the control of Algorithm 1.

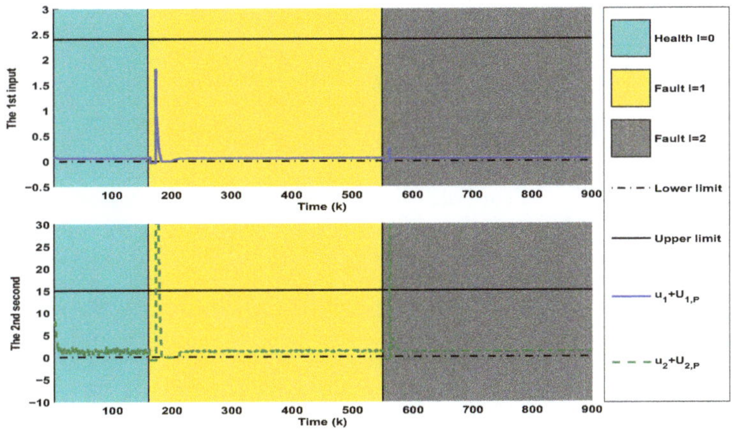

Figure 5. Simulation results of interpolation-based AFTC input obtained from Algorithm 1.

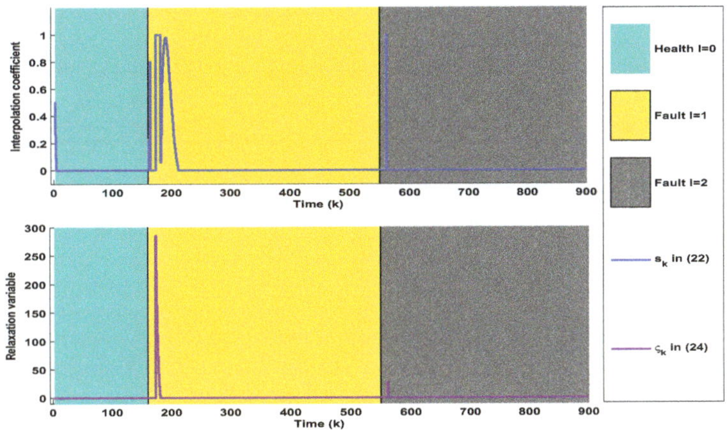

Figure 6. Simulation results of interpolation coefficient and relaxation variable obtained from Algorithm 1.

4.4. Multi-Performance Comparison and Discussion of Active Fault-Tolerant Control Methods

Some qualitative comparisons with the recently reported AFTC methods are given in Table 1.

Table 1. Performance comparison of different FTC methods.

Method / Performance	Algorithm 1	[20]	[13]	[16]
Types of faults that can be handled	Component/actuator faults	Actuator fault	Actuator fault	Actuator fault
Can active fault diagnosis be realized	Yes	-	-	-
Number of observers used in real time	1	-	-	3
Design principle of fault tolerant control	P_M-based IC	Dual-mode MPC	LMI-based MPC	P_M-based MPC
Expression of fault tolerant feasible domain	Polyhedral set	Polyhedral set	Ellipsoidal set	Ellipsoidal set
Optimization problems to be solved	LP	QP	SDP	QP
Can active constraint relaxation be achieved	Yes	-	-	-
Extensibility of FTC method	General	General	High	General

Note: interpolating control (IC), model predictive control (MPC), linear matrix inequality (LMI), linear programming (LP), quadratic programming (QP), semi-positive definite programming (SDP), M-step robust control invariant set (P_M).

The involved comparisons in Table 1 are explained from the following aspects. Firstly, as shown in the second row of Table 1, both component faults and actuator faults are considered in this paper, while only actuator faults are considered in [13,16,20]. In general, the component faults can significantly affect the system dynamics. In this paper, an AFD method is embedded to identify the system operating mode in real time in order to achieve fault tolerance for component faults. Secondly, unlike the multiple-observers-based real-time diagnosis approach in [16], here only one observer needs to be employed at each moment to achieve fault mode identification. Theoretically, this facilitates the diagnosis efficiency and it is also another implicit advantage of using AFD.

In terms of the design and implementation of fault-tolerant methods (i.e., rows 5–7 in Table 1), the MPC optimization problems in [13] are constructed by relying on ellipsoidal constraint sets and LMI, which belongs to SDP and whose solution tends to be more time-consuming. In addition, approximating the feasible domain with ellipsoidal sets is generally more conservative than polyhedral sets. In [20], the dual-mode prediction mechanism is adopted to construct a predictive FTC, whose optimization problem belongs to QP and can be solved relatively efficiently. However, this FTC method is only used to handle actuator additive offset faults and is not suitable for addressing fault tolerance problems of multiplicative faults and component faults. Relatively, the receding horizon set theoretic FTC method in [16] is appealing. This method provides a way to perform the state figure using switching M-step controllable ellipsoidal sets under different fault conditions. However, it may be computationally demanding and takes up a large storage space because of the need to solve real-time QP when the states do not belong to the corresponding maximum allowable invariant set. In this paper, the interpolation methods are employed to combine M-step controllable polyhedral sets and inner feedback control laws to achieve the state figure, and the corresponding fault-tolerant optimization is formed as LP. Compared to the sets that need to be stored by the FTC method in [16], Algorithm 1 only needs to store the maximum M-step controllable polyhedral set for each operating condition, which helps to reduce the storage burden.

The penultimate row of Table 1 illustrates that the soft-constrained FTC method is further integrated into Algorithm 1 and used to deal with some unanticipated situations, such as uncertain fault amplitudes, system parameter drifts, disturbance overruns, etc. The last row of Table 1 implies that the design of the FTC method in [13] is more intuitive and better scalable than the FTC methods in Algorithm 1, [16,20]. It should be noted that the above comparisons are discussed mainly for the characteristics of the involved fault-tolerant methods and not for the contents of the overall studies in [13,16,20]. Clearly, they have different system models and control objectives, and therefore different innovations.

Remark 9. *According to Remark 7, the FTC law based on dual-mode predictive control constructed in [20] can be considered as a special form of the interpolation AFTC developed in this paper. Hence, the interpolation-based AFTC theoretically has a higher degree of design freedom as well as a more efficient optimization capability. To verify this, a further numerical comparison was made. Let the system operate sequentially in two scenarios: scenario I (health $l = 0$) for $1 \leq k < 160$ and scenario II (fault $l = 2$) for $160 \leq k \leq 500$. To be fair, the same active fault diagnosis and integration design were used. Table 2 gives the comparisons of these two methods in terms of interval cost function* (3) *and running time. It can be seen that the interpolation-based AFTC method runs faster and provides better tracking accuracy for scenario I. In scenario II, the developed interpolation-based AFTC remains feasible and continues to optimize the cost function, however the FTC method of [20] will no longer be feasible after $k = 170$. Based on the above numerical comparisons, the effectiveness of the method constructed in this paper can be further verified.*

Table 2. Comparisons of interval cost function (3) and running time.

Scenario	Performance	[20]	Interpolation-Based AFTC
Scenario I	Running time Evaluation of (3)	33.2890 (s) 128.8576	24.6050 (s) 81.7340
Scenario II	Running time Evaluation of (3)	Infeasible for $k \geq 170$ -	126.7660 (s) 359.1910

5. Conclusions

In this paper, a novel activate fault tolerant control scheme is proposed to address the component/actuator faults for the uncertain systems with state/input constraints. Its significant merits are that (1) it relies on only one diagnostic observer for online fault detection and isolation, which helps to reduce the internal memory consumption of the hardware controller; (2) the auxiliary inputs and separation hyperplanes for fault isolation are designed offline in advance, which helps to reduce the online computational burden and increase the freedom of fault isolation decisions; (3) the overall fault tolerant control is reconfigured by optimizing the interpolation coefficient to dynamically regulate the convex combination of inner and outer fault tolerant control laws, which can further reduce the online optimization effort; (4) the inner fault tolerant control and the diagnostic observer are designed offline in advance, and by such design the robust interaction influence on the feasibility of the reconfigured fault tolerant control algorithm can be reduced; (5) the soft constraint method is embedded to achieve a relaxed fault tolerance, which can handle some cases that lead to infeasible constrained optimization in an emergency. The simulation with detailed discussions is given to demonstrate the above benefits of the proposed method.

Some issues need to be further addressed in the future work. For instance, the application of semi-active fault diagnosis to enhance the design flexibility of auxiliary signals; the combination of soft constraint theory and period invariant sets to construct an outer fault tolerant control with flexible and adjustable feasible domains; the construction of parametrization method of interpolated coefficient to avoid solving linear programming problems, etc.

Author Contributions: Conceptualization, K.H.; methodology, K.H.; software, C.C.; validation, C.C. and M.C.; formal analysis, K.H.; investigation, K.H.; resources, K.H.; data curation, K.H.; writing—original draft preparation, K.H.; writing—review and editing, K.H.; visualization, C.C.; supervision, Z.W.; project administration, K.H.; funding acquisition, K.H. All authors have read and agreed to the published version of the manuscript.

Funding: This work was supported by the National Natural Science Foundation of China under Grant number 61803178, and Shandong Provincial Natural Science Foundation under Project number ZR2019BF036.

Institutional Review Board Statement: Not applicable.

Informed Consent Statement: Not applicable.

Data Availability Statement: Not applicable.

Acknowledgments: The authors are thankful to the reviewers for their comments and suggestions to improve the quality of the manuscript.

Conflicts of Interest: The authors declare no conflict of interest.

Appendix A

The approximated calculation method of $\mathcal{R}_{FI}^{l,i}$ is given below.

By [31], $\mathcal{R}_{FI}^{l,i}$ can be explicitly represented as $\mathcal{R}_{FI}^{l,i} = \{C_\chi(I - A_\chi^{l,i})^{-1}B_\chi^{l,i}u_{FI}^i\} \oplus C_\chi O_{\chi,\infty}^{l,i} \oplus D_\chi\mathcal{E}$, where $O_{\chi,\infty}^{l,i} = \{\chi : \sum_{j=0}^{\infty}(A_\chi^{l,i})^j E_\chi^i \sigma_k,\ \sigma_k \in \mathcal{E}\}$. Generally, $O_{\chi,\infty}^{l,i}$ is difficult to determine, especially for high-dimensional systems. In [27], an external approximation method is proposed to enable $O^{l,i}_{\chi,\infty} \subseteq (1+\mu_T)O^{l,i}_{\chi,T}$, where $O^{l,i}_{\chi,T} = \{\chi : \sum_{j=0}^{T}(A^{l,i}_{\chi})^j E^i_{\chi}\sigma_k,\ \sigma_k \in \mathcal{E}\}$ can be calculated in a finite time. Then, for given u^i_{FI}, the internal point of residual limit set $\mathcal{R}^{l,i}_{FI}$ can be parameterized as $C_{\chi}(I - A^{l,i}_{\chi})^{-1}B^{l,i}_{\chi}u^i_{FI} + (1+\mu_T)\sum_{j=0}^{T} C_{\chi}(A^{l,i}_{\chi})^j E^i_{\chi}\sigma_{1,k} + D_{\chi}\sigma_{2,k}$, $\forall \sigma_{1,k}, \sigma_{2,k} \in \mathcal{E}$.

Appendix B

The proof of **Theorem 1** is given below.

Proof. Let $V^l_{1,k} = e^T_{x^l,k}P^l_1 e_{x^l,k}$ and $V^l_{2,k} = (x^l_k)^T(P^l_2)^{-1}x^l_k$ be the Lyapunov functions of (5), (13) and (14), respectively. Equivalently, $V^l_{1,k} + V^l_{2,k}$ is a Lyapunov function of (15). Define $K^l(Y^l_2)^T = \bar{K}^l$ and $Y^l_1 L^l = \bar{L}^l$. Then, by using $-Y^l_1(P^l_1)^{-1}(Y^l_1)^T \le P^l_1 - Y^l_1 - (Y^l_1)^T$ and congruence transformation $diag\{I, I, (Y^l_1)^{-1}\}$ to (16a), the inequality $-diag\{P^l_1 - I, \alpha^2 I\} + \begin{bmatrix}\tilde{A}^l_{11} & \tilde{B}^l_1\end{bmatrix}^T P^l_1(*) < 0$ can be deduced. It further implies that the relation $V^l_{1,k+1} - V^l_{1,k} + e^T_{x^l,k}e_{x^l,k} - \alpha^2\varrho^T_k\varrho_k < 0$ holds.

Similarly, by using the inequality $-Y^l_2(P^l_2)^{-1}(Y^l_2)^T \le P^l_2 - Y^l_2 - (Y^l_2)^T$ and congruence transformation factor $diag\{(Y^l_2)^{-1}, (Y^l_2)^{-1}, I, I, I\}$ to (16b), we get the inequality $-diag\{(P^l_2)^{-1}, (P^l_2)^{-1}, \beta^2 I\} + \begin{bmatrix}\tilde{A}^l_{21} & \tilde{A}^l_{22} & \tilde{B}^l_2\end{bmatrix}^T(P^l_2)^{-1}(*) + \begin{bmatrix}\bar{D}^l & \bar{C}^l & 0\end{bmatrix}^T(*) < 0$. Based on the Lyapunov function $V^l_{2,k}$ and the system model (13) and (14), we can further derive $V^l_{2,k+1} - V^l_{2,k} + (z^l_k)^T z^l_k - e^T_{x^l,k}(P^l_2)^{-1}e_{x^l,k} - \beta^2\varrho^T_k\varrho_k < 0$.

Under zero initial conditions, the summation of $V^l_{2,k+1} - V^l_{2,k} + (z^l_k)^T z^l_k - e^T_{x^l,k}(P^l_2)^{-1} e_{x^l,k} - \beta^2\varrho^T_k\varrho_k < 0$ over $k = 0$ to $k = \infty$ can be bounded in the form of $\| z^l_k \|^2_2 \le \epsilon^2 \| e_{x^l,k} \|^2_2 + \beta^2 \| \varrho_k \|^2_2$, where $e^T_{x^l,k}(P^l_2)^{-1}e_{x^l,k}$ is relaxed by $e^T_{x^l,k}(P^l_2)^{-1}e_{x^l,k} \le eig_{max}((P^l_2)^{-1}) e^T_{x^l,k}e_{x^l,k} = \epsilon^2 e^T_{x^l,k}e_{x^l,k}$. Similarly, $\sum_{k=0}^{\infty}\{V^l_{1,k+1} - V^l_{1,k} + e^T_{x^l,k}e_{x^l,k} - \alpha^2\varrho^T_k\varrho_k\} < 0$ leads to $\| e_{x^l,k} \|^2_2 \le \alpha^2 \| \varrho_k \|^2_2$ under zero initial conditions. Then, we further have $\| z^l_k \|^2_2 \le (\epsilon^2\alpha^2 + \beta^2) \| \varrho_k \|^2_2$. Clearly, the integrated optimization of α and β contributes to improving the synthesized H_∞ performances of observer (5) and unconstrained robust control policy $u^l_{C,k} = K^l\hat{x}^l_k$. The proof is completed. □

References

1. Ding, S. *Model-Based Fault Diagnosis Techniques: Design Schemes, Algorithms, and Tools*, 2nd ed.; Springer: London, UK, 2013.
2. Zhu, C.; Li, C.; Chen, X.; Zhang, K.; Xin, X.; Wei, H. Event-triggered adaptive fault tolerant control for a class of uncertain nonlinear systems. *Entropy* **2020**, *22*, 598. [CrossRef] [PubMed]
3. Blanke, M.; Kinnaert, M.; Lunze, J.; Staroswiecki, M. *Diagnosis and Fault-Tolerant Control*; Springer: Berlin, Germany, 2006.
4. Gao, Z.; Cecati, C.; Ding, S. A survey of fault diagnosis and fault tolerant techniques-Part I: Fault diagnosis with model-based and signal-based approaches. *IEEE Trans. Ind. Electron.* **2015**, *62*, 3757–3767. [CrossRef]
5. Yin, S.; Xiao, B.; Ding, S.; Zhou, D. A review on recent development of spacecraft attitude fault tolerant control system. *IEEE Trans. Ind. Electron.* **2016**, *63*, 3311–3320. [CrossRef]
6. Jiang, J.; Yu, X. Fault-tolerant control systems: A comparative study between active and passive approaches. *Annu. Rev. Control* **2012**, *36*, 60–72. [CrossRef]
7. Arslan, A.; Khalid, M. A review of fault tolerant control systems: Advancements and applications. *Measurement* **2019**, *143*, 58–68.
8. Habibi, H.; Rahimi, N.; Howard, I. Backstepping nussbaum gain dynamic surface control for a class of input and state constrained systems with actuator faults. *Inf. Sci.* **2019**, *482*, 27–46. [CrossRef]
9. Zare, I.; Aghaei, S.; Puig, V. A supervisory active fault tolerant control framework for constrained linear systems. In Proceedings of the 2020 European Control Conference (ECC), St. Petersburg, Russia, 12–15 May 2020; pp. 2027–2032
10. Ferranti, L.; Wan, Y.; Keviczky, T. Fault-tolerant reference generation for model predictive control with active diagnosis of elevator jamming faults. *Int. J. Robust Nonlinear Control* **2019**, *29*, 5412–5428. [CrossRef]
11. Shen, Q.; Yue, C.; Goh, C.; Wang, D. Active fault-tolerant control system design for spacecraft attitude maneuvers with actuator saturation and faults. *IEEE Trans. Ind. Electron.* **2019**, *66*, 3763–3772. [CrossRef]
12. Yang, X.; Maciejowski, J.M. Fault-tolerant model predictive control of a wind turbine benchmark. *IFAC Proc. Vol.* **2012**, *45*, 337–342. [CrossRef]
13. Khan, O.; Mustafa, G.; Khan, A.; Abid, M.; Ali, M. Fault-tolerant robust model predictive control of uncertain time-delay systems subject to disturbances. *IEEE Trans. Ind. Electron.* **2020**. [CrossRef]

14. Sheikhbahaei, R.; Alasty, A.; Vossoughi, G. Robust fault tolerant explicit model predictive control. *Automatica* **2018**, *97*, 248–253. [CrossRef]
15. Yetendje, A.; Seron, M.; De Dona, J. Robust multiactuator fault-tolerant MPC design for constrained systems. *Int. J. Robust Nonlinear Control* **2013**, *23*, 1828–1845. [CrossRef]
16. Franze, G.; Francesco, T.; Domenico, F. Actuator fault tolerant control: A receding horizon set-theoretic approach. *IEEE Trans. Autom. Control* **2014**, *60*, 2225–2230. [CrossRef]
17. Lucia, W.; Domenico, F.; Franze, G. A set-theoretic reconfiguration feedback control scheme against simultaneous stuck actuators. *IEEE Trans. Autom. Control* **2017**, *63*, 2558–2565. [CrossRef]
18. Famularo, D.; Franze, G.; Lucia, W.; Manna, C. A reconfiguration control framework for constrained systems with sensor stuck faults. *Int. J. Robust Nonlinear Control* **2019**, *29*, 1150–1164. [CrossRef]
19. Witczak, M.; Buciakowski, M.; Aubrun, C. Predictive actuator fault tolerant control under ellipsoidal bounding. *Int. J. Adapt. Control Signal Process.* **2016**, *30*, 375–392. [CrossRef]
20. Han, K.; Feng, J.; Zhao, Q. Robust estimator-based dual-mode predictive fault tolerant control for constrained linear parameter varying systems. *IEEE Trans. Ind. Inf.* **2021**, *17*, 4469–4479. [CrossRef]
21. Lunze, J.; Richter, J. Reconfigurable fault-tolerant control: A tutorial introduction. *Eur. J. Control* **2008**, *14*, 359–386. [CrossRef]
22. Heirung, T.; Mesbah, A. Input design for active fault diagnosis. *Ann. Rev. Control* **2019**, *47*, 35–50. [CrossRef]
23. Wang, J.; Zhang, J.; Qu, B.; Wu, H.; Zhou, J. Unified architecture of active fault detection and partial active fault-tolerant control for incipient faults. *IEEE Trans. Syst. Man Cybern. Syst.* **2017**, *47*, 1688–1700. [CrossRef]
24. Raimondo, D.; Marseglia, G.; Braatz, R.; Scott, J. Fault-tolerant model predictive control with active fault isolation. In Proceedings of the 2013 Conference on Control and Fault-Tolerant Systems (SysTol), Nice, France, 9–11 October 2013; pp. 444–449.
25. Puncochar, I.; Siroky, J.; Simandl, M. Constrained active fault detection and control. *IEEE Trans. Autom. Control* **2015**, *60*, 253–258. [CrossRef]
26. Boem, F.; Gallo, A.; Raimondo, D.; Parisini, T. Distributed fault-tolerant control of large-scale systems: An active fault diagnosis approach. *IEEE Trans. Control Netw. Syst.* **2019**, *7*, 288–301. [CrossRef]
27. Blanchini, F.; Casagrande, D.; Giordano, G.; Miani, S.; Olaru, S.; Reppa, V. Active fault isolation: A duality-based approach via convex programming. *SIAM J. Control Optim.* **2017**, *55*, 1619–1640. [CrossRef]
28. Nguyen, H. *Constrained Control of Uncertain, Time-Varying, Discrete-Time Systems*; Springer: Cham, Switzerland, 2014.
29. Nguyen, H.; Gutman, P.; Olaru, S.; Hovd, D. Implicit improved vertex control for uncertain, time-varying linear discrete-time systems with state and control constraints. *Automatica* **2013**, *49*, 2754–2759. [CrossRef]
30. Rubin, D.; Mercader, P.; Gutman, P.; Nguyen, H.; Bemporad, A. Interpolation based predictive control by ellipsoidal invariant sets. *IFAC J. Syst. Control* **2020**, *12*, 100084. [CrossRef]
31. Blanchini, F.; Miani, S. *Set-Theoretic Methods in Control*; Birkhauser: Boston, MA, USA, 2008.
32. Lan, J.; Patton, R. A new strategy for integration of fault estimation within fault-tolerant control. *Automatica* **2016**, *69*, 48–59. [CrossRef]
33. Zeilinger, M.; Morari, M.; Jones, C. Soft constrained model predictive control with robust stability guarantees. *IEEE Trans. Autom. Control* **2014**, *59*, 1190–1202. [CrossRef]
34. Knudsen, B.; Brusevold, J.; Foss, B. An exact penalty-function approach to proactive fault-tolerant economic MPC. In Proceedings of the 2016 European Control Conference (ECC), Aalborg, Denmark, 29 June–1 July 2016; pp. 758–763.
35. Francisco, M.; Skogestad, S.; Vega, P. Model predictive control for the self-optimized operation in wastewater treatment plants: Analysis of dynamic issues. *Comput. Chem. Eng.* **2015**, *82*, 259–272. [CrossRef]
36. El, H.; Vega, P.; Tadeo, F.; Francisco, M. A constrained closed loop MPC based on positive invariance concept for a wastewater treatment plant. *Int. J. Syst. Sci.* **2018**, *49*, 1–15.

entropy

Article

A Comprehensive Diagnosis Method of Rolling Bearing Fault Based on CEEMDAN-DFA-Improved Wavelet Threshold Function and QPSO-MPE-SVM

Yi Wang, Chuannuo Xu, Yu Wang and Xuezhen Cheng *

College of Electrical Engineering and Automation, Shandong University of Science and Technology, Qingdao 266590, China; skd992631@sdust.edu.cn (Y.W.); xuchuannuo@sdust.edu.cn (C.X.); wangyu007@sdust.edu.cn (Y.W.)
* Correspondence: zhenxc6411@163.com or chengxuezhen@sdust.edu.cn; Tel.: +86-135-0532-4619

Abstract: A comprehensive fault diagnosis method of rolling bearing about noise interference, fault feature extraction, and identification was proposed. Based on complete ensemble empirical mode decomposition with adaptive noise (CEEMDAN), detrended fluctuation analysis (DFA), and improved wavelet thresholding, a denoising method of CEEMDAN-DFA-improved wavelet threshold function was presented to reduce the distortion of the noised signal. Based on quantum-behaved particle swarm optimization (QPSO), multiscale permutation entropy (MPE), and support vector machine (SVM), the QPSO-MPE-SVM method was presented to construct the fault-features sets and realize fault identification. Simulation and experimental platform verification showed that the proposed comprehensive diagnosis method not only can better remove the noise interference and maintain the original characteristics of the signal by CEEMDAN-DFA-improved wavelet threshold function, but also overcome overlapping MPE values by the QPSO-optimizing MPE parameters to separate the features of different fault types. The experimental results showed that the fault identification accuracy of the fault diagnosis can reach 95%, which is a great improvement compared with the existing methods.

Keywords: rolling bearing fault; CEEMDAN; DFA; improved wavelet threshold; QPSO; MPE; SVM

Citation: Wang, Y.; Xu, C.; Wang, Y.; Cheng, X. A Comprehensive Diagnosis Method of Rolling Bearing Fault Based on CEEMDAN-DFA-Improved Wavelet Threshold Function and QPSO-MPE-SVM. *Entropy* **2021**, *23*, 1142. https://doi.org/10.3390/e23091142

Academic Editors: Giuseppe Fusco, Quanmin Zhu, Jing Na, Weicun Zhang and Ahmad Taher Azar

Received: 30 July 2021
Accepted: 27 August 2021
Published: 31 August 2021

1. Introduction

Rolling bearings are the most important parts of rotating machinery and are widely used in modern mechanized equipment [1,2]. The fault of rolling bearing components can cause serious damage to the running status of the machine and the production process. Therefore, it is important to explore a new fault diagnosis technique. However, in practical engineering, the collected bearing vibration signals contain various interference signals (e.g., white noise, harmonic interference), and have nonlinear and nonstationary properties, which makes it difficult to distinguish the bearing fault types and to identify them [3]. Therefore, to improve the fault diagnosis accuracy of rolling bearings, this paper researched three aspects of signal denoising, fault feature extraction, and fault type identification of rolling bearings.

Common vibration signal denoising methods include empirical mode decomposition (EMD) [4], discrete wavelet transform (DWT) [5], singular-value decomposition (SVD) [6–8], and so on. Abdelkader et al. [9], based on EMD algorithm, proposed a method to determine the travel point based on the energy values of the IMF components of each order, and an improved wavelet threshold was used to denoise the selected IMF component-bearing vibration signals. Zhao et al. [10] fused the EMD algorithm and selected and tested (ST) the algorithm to denoise the signal, solving the low accuracy in the low-speed operation of the bearing of the EMD denoising method based on the number of interrelationships and the kurtosis criterion. Gao et al. [11] added the CEEMDAN algorithm to adaptive noise, further reducing the modal effect, yielding better convergence, and realizing the fault diagnosis of rolling bearings.

The hard and soft thresholding functions in wavelet thresholding denoising are widely used, but both functions have their shortcomings and can cause a certain degree of distortion when reconstructing the signal. Tajeddini et al. [12] proposed a generic threshold function based on an unbiased risk estimation method to carry out wavelet-packet transform preprocessing of the vibration signal, followed by generic threshold denoising. Kumar et al. [13] proposed an improved modified kurtosis-mixing threshold rule to denoise the vibration signal, which improved the signal-to-noise ratio. Yang et al. [14] used a particle-swarm algorithm to perform adaptive optimal processing of the initial parameters of the wavelet threshold function, which solved the problem of the initial parameters being influenced by empirical values and overcame the defects such as signal noise removal not being complete enough or removing the useful information. Chegini et al. Ref. [15] proposed a new method of bearing vibration signal denoising based on empirical wavelet transform (EWT) as well as the threshold function, which better overcame the defect of constant deviation of the traditional soft threshold function. However, the denoising performance of the wavelet threshold denoising method is closely related to the wavelet basis function and the threshold function, and is not adaptive. Although the CEEMDAN algorithm is adaptive, the definition of noise and useful signal is relatively vague, resulting in high distortion of the denoised signal. Shi et al. [3] proposed a de-trending fluctuation analysis technique for the critical values of the noise component and the useful signal component. This method ensures the accuracy of signal correlation discrimination to a certain extent. Based on the CEEMDAN algorithm decomposition of the vibration signal, Chaabi et al. [16] used wavelet analysis, principal component analysis, and order tracking analysis to perform multi-method denoising. In order to improve identification accuracy of rolling bearings with nonlinear and nonstationary vibration signals, Chen et al. [17] proposed a novel fault diagnosis method based on wavelet thresholding denoising and CMEEMDAN with adaptive noise. To verify the denoising effectiveness of combined wavelet thresholding and CEEMDAN, Bie et al. [18] added random Gaussian white noise to the bearing fault signal to simulate the actual noise disturbance of rolling bearings, and adopted the CEEMDAN-wavelet threshold function to the denoising method. The experimental results showed that the combined denoising method could effectively remove the interference of noise. Therefore, based on the application characteristics of each denoising method, this paper proposed a denoising method based on a CEEMDAN-DFA-improved wavelet threshold function. The method uses the CEEMDAN algorithm to decompose the vibration signal, performs detrended fluctuation analysis (DFA) on the obtained eigenmode function (IMF), calculates the scalar function value of each IMF component, selects the noise-dominated IMF component, and applies an improved wavelet threshold function to denoise it.

The current methods of fault feature extraction for rolling bearings are mainly time-domain analysis, frequency domain analysis, and time-frequency domain analysis; among them, the time-frequency domain analysis method is most commonly used. Zhen et al. [19] decomposed the vibration signal using wavelet packets and selected a suitable bandwidth according to the Kurtosis spectrum correlation theory, and finally applied the envelope spectrum analysis to extract the eigenfrequencies. Han et al. [20] used the Teager energy operator to enhance the signal after wavelet denoising and then extracted the bearing fault features by the CEEMD algorithm. Li et al. [21] used exchange entropy (PE) for bearing fault feature extraction, but could not measure multi-scale signals. Multi-scale variational entropy (MPE) [22,23] was introduced into fault diagnosis by Yin et al. [24]. Du et al. [25] used MPE to extract fault features and combined them with a self-organizing fuzzy classifier based on the harmonic mean difference (HMDSOF) to classify the fault features. MPE can respond to the changes of vibration signals very well, but its parameters have a great influence on the calculation of entropy values. If the parameters are not selected properly, it will cause the arrangement entropy values of multiple signals of bearings to be mixed, and thus the type of bearing failure cannot be identified. In order to achieve complete separation of arrangement entropy values under different operating conditions of the bearings, this paper adopted quantum particle swarm optimization (QPSO) to optimize the initial parameters of MPE, and then selected the appropriate MPE entropy values to construct the rolling bearing fault feature vector.

Common fault classification and identification methods include artificial neural networks (ANN) [26], extreme learning machines (ELM) [27,28], and support vector machines (SVM) [29,30]. The ANN has made possible many achievements in the field of pattern identification, but its identification performance is strongly influenced by parameters and it is easy to fall into local minima during the optimization process. Although ELM runs fast, its generalization performance is poor. SVM has fewer adjustable parameters and runs stably. It can obtain higher diagnostic accuracy under the condition of fewer training samples [24]. Therefore, this paper used a SVM for fault identification of rolling bearings. This paper also proposed a comprehensive rolling bearing fault feature extraction and identification method based on the combination of QPSO-MPE-SVM.

The main work and contributions of this paper are summarized as follows:

(1) A CEEMDAN-DFA-improved wavelet thresholding denoising method was proposed. The method uses the CEEMDAN algorithm to decompose the vibration signal, performs DFA on the obtained IMF, calculates the scalar function value of each IMF component, selects the noise-dominated IMF component, and applies an improved wavelet threshold function to denoise it.
(2) Combining QPSO-MPE-SVM into an effective fault diagnosis method can accurately extract fault features and improve the identification accuracy of bearing faults.
(3) Experimental cases were used to illustrate the effectiveness of the proposed method in bearing vibration signal denoising, fault feature extraction, and fault identification.

The rest of this paper is organized as follows. The CEEMDAN-DFA-improved wavelet thresholding denoising method is introduced, and its validity is verified in Section 2. The QPSO-MPE-SVM fault feature extraction and diagnosis method is presented, and its validity is verified in Section 3. The effectiveness of the proposed method is proven by experimental example in Section 4. In addition, contrastive analysis among the different methods is conducted. Finally, some conclusions are summarized in Section 5.

2. Denoising Algorithm of the CEEMDAN-DFA-Improved Wavelet Threshold Function

2.1. Basic Algorithm Related to CEEMDAN-DFA-Improved Wavelet Threshold Function

2.1.1. CEEMDAN Algorithm

To overcome the high computing time and the residue of added noise present in the IMFs, and to better deal with the modal mixing problem of empirical modal decomposition (EMD), the ensemble empirical mode decomposition (EEMD) was proposed by Liu et al. in 2009 [31]. The EEMD algorithm is effective in suppressing modal aliasing, but a certain degree of distortion is produced when the signal is reconstructed. The improved CEEMDAN algorithm was proposed by Colominas et al. in 2014 [32]. The improved CEEMDAN algorithm is summarized as follows.

(1) The jth IMF component generated by the signal decomposition by the EMD is defined as $E_j(\cdot)$. The jth IMF by the CEEMDAN is defined as $IMF_j{}'$. $n^i(t)$ is for the Gaussian white noise. The CEEMDAN performs I EMD decomposition on the noisy signal $x(t) + \varepsilon_0 \cdot n^i(t)$ formed by the combination of the original signal and the white noise. Then the first IMF component decomposed by CEEMDAN can be expressed as:

$$IMF_1'(t) = \frac{1}{I}\sum_{i=1}^{I} IMF_1^i(t) \tag{1}$$

(2) First residual sequence of the first stage (j = 1) is expressed as:

$$r_1(t) = x(t) - IMF_1'(t) \tag{2}$$

(3) The $r_1(t) + \varepsilon_1 E_1(n^i(t))(i = 1, 2, \cdots)$ is processed several times using the EMD algorithm until the first IMF component is generated. The second IMF component is expressed as:

$$IMF'_2(t) = \frac{1}{I}\sum_{i=1}^{I} E_1\left(r_1(t) + \varepsilon_1 E_1\left(n^i(t)\right)\right) \tag{3}$$

(4) Perform step (3) above for the other remaining stages ($j = 2, 3, \cdots J$), then the $j + 1$ IMF component is expressed as:

$$r_j(t) = r_{j-1} - IMF_j(t) \tag{4}$$

$$IMF'_{j+1}(t) = \frac{1}{I}\sum_{i=1}^{I} E_1\left(r_j(t) + \varepsilon_j E_j\left(n^i(t)\right)\right) \tag{5}$$

(5) Add 1 for j and repeat step (4) until the residual sequence cannot be processed. The number of IMF components is J. The final calculated residual sequence is expressed as:

$$r(t) = x(t) - \sum_{j=1}^{J} IMF_j{}'(t) \tag{6}$$

(6) The original signal $x(t)$ represented by the IMF component and the residual component is expressed as:

$$x(t) = \sum_{j=1}^{J} IMF_j'(t) + r(t) \tag{7}$$

The CEEMDAN algorithm introduces a segment of positive and negative white noise for each stage processing, and overcomes the reconstruction error defect of the EEMD algorithm by processing only one residual component to find each IMF component. The CEEMDAN algorithm is able to reconstruct the decomposed signal with close to zero deviation. However, it also has shortcomings, for example, the definition of noise and useful signal in IMF components containing more noise is relatively ambiguous, and direct removal of these components can cause signal distortion.

2.1.2. DFA Algorithm

To solve the problem of CEEMDAN for noise and useful signal demarcation ambiguity, the DFA algorithm is more accurate [33].

(1) $\overline{x}(i)$ is defined as the average of the time series $x(i)$ in the time intervals [1,N], and denoted as:

$$\overline{x} = \frac{1}{N}\sum_{i=1}^{N} x(i) \tag{8}$$

where N is the number of segments of length n.

(2) The time series $y(k)$ is segmented into segments of length n. It is denoted as:

$$y(k) = \sum_{i=1}^{k}[x(i) - \overline{x}], k = 1, 2, 3, \cdots \tag{9}$$

(3) The trend $y_s(i)$ of each series segment is calculated as:

$$y_s(i) = \sum_{n=0}^{k} a_n i^n \tag{10}$$

(4) After removing the uncertain trend in each series segment, the second-order fluctuation coefficient of the segment series is expressed as:

$$F^2(n,s) = \frac{1}{n}\sum_{i=1}^{n}(y[(s-1)n+i] - y_s(i))^2 \tag{11}$$

$$F_q(n) = \left[\frac{1}{N_n}\sum_{s=1}^{N_s} F^2(n,s)\right]^{1/2} \tag{12}$$

(5) Change the segment length n in step (1), and repeat steps (2) and (3) to obtain the change in the fluctuation function of the time series. The correlation of the time series represented by the Hurst function is expressed as:

$$H = \frac{\log_2 F_q(n)}{\log_2(n)} = \frac{\log_2\left[\frac{1}{N_n}\sum_{s=1}^{N_n} F^2(n,s)\right]}{\log_2(n)} \tag{13}$$

(6) The relationship between the scalar function α and the time series fluctuation function $F(n)$ is expressed as:

$$F(n) \propto n^{\alpha} \tag{14}$$

It can be seen that α is related to $F(n)$, and $F(n)$ is related to H. Therefore, α is related to H. The value of α is proportional to the smoothness of the signal. When $0 < \alpha < 0.5$, it indicates that the proportion of noise in the signal is very large; $\alpha = 0.5$ indicates the signal is not correlated; the signal has a large correlation when $\alpha > 0.5$ [34]. By comparing the values of α in the IMF components, the noise-dominated IMF components and the useful information-dominated IMF components can be distinguished.

DFA is an algorithm for correlation discrimination of non-smooth signals. It is used to confirm the critical value of the noise component and the useful signal component, which can guarantee the accuracy of signal correlation discrimination to a certain extent.

2.1.3. Improved Wavelet Threshold Function

To solve the distortion problem of the signal caused by the CEEMDAN, a wavelet threshold function is introduced to denoise the dominant component of the noise. Traditional threshold functions include hard threshold functions and soft threshold functions. The hard threshold function can maintain the signal characteristics well, but there is a discontinuity at the set threshold λ, which will cause serious oscillations when the signal is reconstructed. The soft threshold function does not have the discontinuity at the set threshold, but the wavelet coefficients after threshold quantization will have a constant deviation, which leads to a large deviation for the signal characteristics. Therefore, Based on the advantages of the hard thresholding function and soft thresholding function, an improved wavelet threshold function is presented to reduce reconstruction error.

The improved threshold function must meet the following conditions.

(1) A good continuity is maintained at the set threshold;
(2) The threshold function has monotonicity and continuity when the wavelet coefficient is greater than the set threshold;
(3) The threshold function should have an asymptote, and the curve $y(x) = x$ can overlap with the asymptote.

The improved wavelet threshold function is expressed as:

$$\hat{\Psi}_{j,k} = \begin{cases} sign\left(\Psi_{j,k}\right)\left[\left|\Psi_{j,k}\right|^2 - \left(\lambda e^{-(|\Psi_{j,k}|-\lambda)/k}\right)^2\right]^{1/2} & \left|\Psi_{j,k}\right| \geq \lambda \\ a\Psi_{j,k} & \left|\Psi_{j,k}\right| < \lambda \end{cases} \tag{15}$$

where k and α in the expression are adjustable, $k > 0$, $a \in (0.05, 0.5)$.

The characteristics of the improved wavelet threshold function are as follows:

(1) When $\psi_{j,k} \to \lambda$, $e^{-(|\psi_{j,k}|-\lambda)/k} \to 1$ and $\widehat{\psi}_{j,k} \to 0$, the improved wavelet threshold function is continuous at the threshold λ. When $\psi_{j,k} \to \infty$, $e^{-(|\psi_{j,k}|-\lambda)/k} \to 0$ and $\widehat{\psi}_{j,k} \to \psi_{j,k}$, the improved wavelet threshold function is an asymptote, which makes the reconstructed signal closer to the actual value.

(2) When $k \to 0$, the properties of the improved wavelet threshold function are close to the hard threshold function. When $k \to \infty$, the properties of the improved wavelet threshold function are close to the soft threshold function.

(3) When $\left|\psi_{j,k}\right| \leq \lambda$, a very large part of the wavelet coefficient is noise, and the a value is adjusted to be as small as possible to remove the noise interference.

The selection of the threshold value has a large impact on the signal denoising effect. The formula for the threshold selection rule used in this paper is expressed as:

$$\lambda_i = \frac{\delta\sqrt{2\ln(N)}}{\ln(i+1)} \tag{16}$$

where δ is for noise intensity. i is the number of decomposition layers. N is the signal length. The threshold λ_i varies with the number of decomposition layers. The wavelet coefficients of a noisy signal are inversely related to the number of decomposition layers. To ensure that the reconstruction error of the signal after denoising is small, the threshold λ_i should be dynamically smaller as the number of decomposition layers increases.

2.2. *The Validation of CEEMDAN-DFA-Improved Wavelet Threshold Function Denoising Algorithm*

The operational flow of the CEEMDAN-DFA-improved wavelet threshold function denoising algorithm is shown in Figure 1. Firstly, the CEEMDAN is used to decompose the vibration signal of the rolling bearing to obtain a series of IMF components. Then, the IMF is analyzed by DFA, and the scaling function values of each order of IMF are calculated. According to the correlation of DFA, the IMF component dominated by noise is selected, and the improved wavelet threshold function is used for denoising. Finally, the IMF component after denoising and the IMF component dominated by useful signals are reconstructed, and the reconstructed signal is the bearing vibration signal after denoising.

The verification data of the denoising algorithm come from the rolling bearing database provided by the Electrical Engineering Laboratory of Case Western Reserve University (CWRU) [35]. The validation data selected for this paper were the inner rings of a bearing with a damaging size of a 0.007-inch signal. To meet match the actual working environment, random Gaussian white noise was superimposed on the used bearing inner-ring fault signal. Figure 2 shows the time domain waveform of the bearing inner-ring fault noise-containing signal. The noise-containing signal was first decomposed using CEEMDAN to obtain the 12th-order IMF component, and then the IMF was calculated by the DFA for the scaling function value α. The calculation results are shown in Table 1. From the table, it can be seen that the scaling function value α of the first four-order IMF components was less than 0.5. Therefore, it can be determined that the main component of the first fourth-order IFM component is noise, and the main component of the remaining IMF is useful information. Signal reconstruction was performed on the first fourth-order IMF, and the reconstructed signal was denoised.

To compare the advantages of the proposed denoising algorithms, four denoising methods were used, including CEEMDAN-DFA, CEEMDAN-DFA-hard thresholding, CEEMDAN-DFA-soft thresholding and the CEEMDAN-DFA-improved wavelet thresholding. The denoising results are shown in Figure 3.

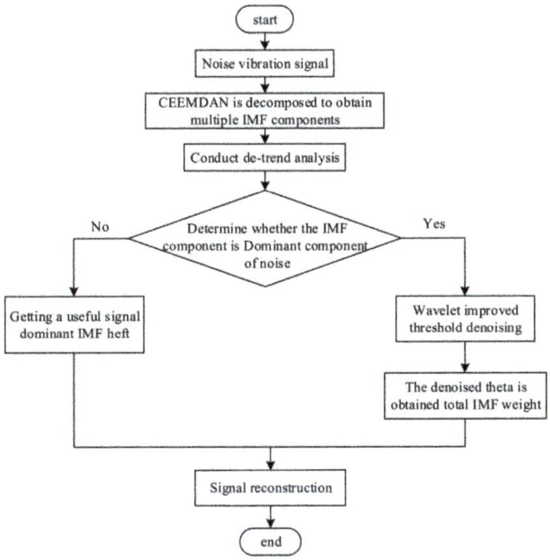

Figure 1. The operational flow of the CEEMDAN-DFA-improved wavelet threshold function algorithm.

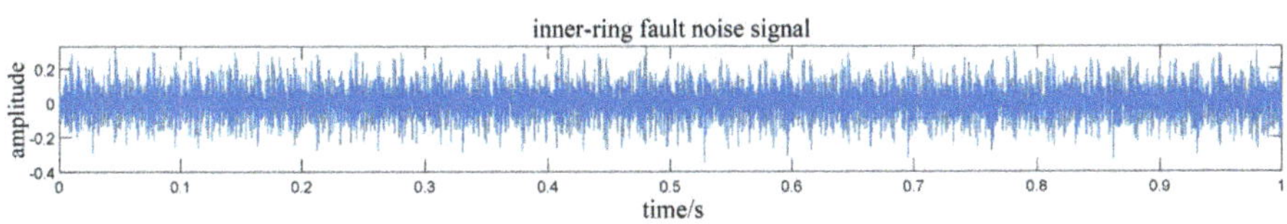

Figure 2. The time domain waveform of the inner-ring fault noise signal.

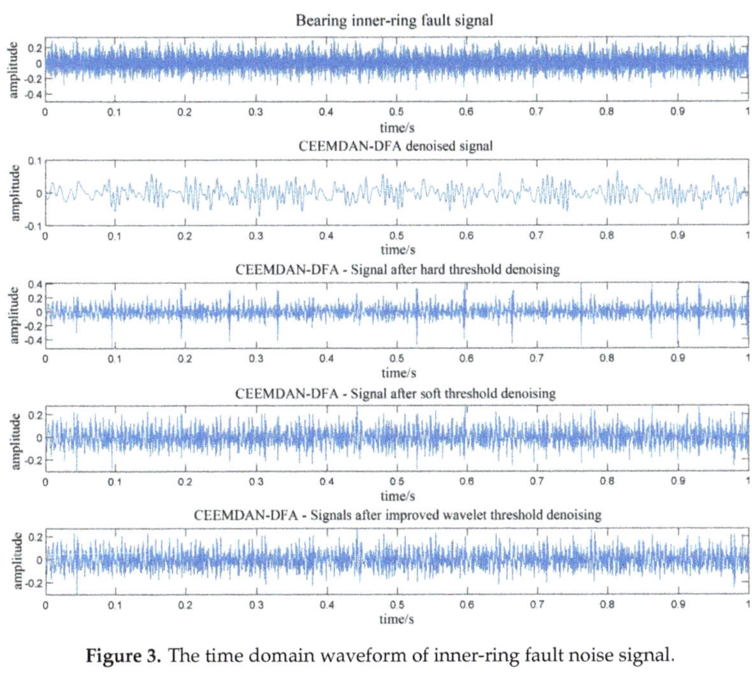

Figure 3. The time domain waveform of inner-ring fault noise signal.

Table 1. Scalar function values of each order IMF of the bearing inner ring fault signal.

IMF	**1**	**2**	**3**	**4**	**5**	**6**
α	0.4639	0.4214	0.3651	0.3124	0.5154	0.5712
IMF	**7**	**8**	**9**	**10**	**11**	**12**
α	0.6812	0.7948	0.8106	0.8637	0.8961	0.8942

The results show that both CEEMDAN-DFA and CEEMDAN-DFA-hard threshold denoising methods cause large signal distortion. The values of the signal-to-noise ratio (SNR) and root-mean-square error (RMSE) using the four denoising methods to the inner ring fault signal are shown in Table 2. The proposed CEEMDAN-DFA-improved wavelet threshold denoising method improved the SNR by 11.4% and reduced the RMSE by 16.2% compared with the CEEMDAN-DFA-wavelet soft threshold denoising method. It can be seen that the proposed denoising method in this paper has better performance than other denoising methods.

Table 2. The SNR and RMSE of the inner-ring fault denoising.

Method	**SNR**	**RMSE**
CEEMDAN-DFA	4.9134	0.12143
CEEMDAN-DFA-wavelet hard threshold function	9.7542	0.08719
CEEMDAN-DFA-wavelet soft threshold function	13.6718	0.06024
CEEMDAN-DFA-improved wavelet threshold function	15.2324	0.05047

3. Fault Feature Extraction and Identification Algorithm of the QPSO-MPE-SVM

3.1. Basic Algorithm Related to the QPSO-MPE-SVM

3.1.1. MPE Algorithm

The theoretical foundation of MPE is based on multiscale analysis and alignment entropy. MPE was used to coarsely granulate the initial time series to create a multiscale time series [35]. The calculation procedure is described as follows:

(1) The initial time series $x(i)$ is coarsely granularized to obtain the coarse-grained series y_j^τ, which is calculated as follow:

$$y_j^\tau = \frac{1}{\tau} \sum_{i=(j-1)+1}^{\tau} x_i \; 1 \leq j \leq \frac{N}{\tau} \tag{17}$$

(2) Calculate multi-scale alignment entropy based on sequence y_j^τ.

$$MPE(x, \tau, m, \lambda) = PE(y_j^\tau, m, \lambda) \tag{18}$$

The MPE calculation requires the signal length N, delay time λ, embedding dimension m, and scale factor τ. The MPE can respond very well to the changes of vibration signals, but if the parameters are not selected properly, the permutation entropy values of multiple vibration signals will be overlapped, and thus the fault characteristics cannot be extracted [36].

3.1.2. QPSO Algorithm

In order to achieve complete separation of the MPE values for different operating conditions of the bearings, the QPSO is introduced in this paper to optimize the initial parameters of the MPE. To find the global optimal solution, the calculation formula is expressed as [37]:

$$m_{best} = \frac{1}{M} \sum_{j=1}^{M} P_j \tag{19}$$

$$P = rP_j + (1 - r)P_g \tag{20}$$

$$L_j(t_1 + 1) = P \pm \varsigma \left| m_{best} - L_j(t_1) \right| \ln 1/u \tag{21}$$

where m_{best} is particle optimum average. M is race number. P_g is the global optimal solution for the particle. ς is compression expansion factor.

The fitness function has a large impact on the optimization results of QPSO when studying the trend of the data.

The MPE values of time series $x(i)$ are a formed sequence $H_p(X)$, $H_p(X)$ is described as follows:

$$H_P(X) = \{H_P(1), H_P(2), \cdots, H_P(n)\} \tag{22}$$

Skewness formula is described as follows:

$$ske = \frac{E\left[H_P(X) - H_P^m(X)\right]^3}{\left[H_P^d(X)\right]^3} \tag{23}$$

where E is the expected value. $H_p^m(X)$ and $H_p^d(X)$ are the mean and standard deviations of $H_p(X)$.

The fitness function about QPSO is described as follows:

$$F(X) = \frac{1}{ske^2 + 1} \tag{24}$$

The fitness function has a large impact on the optimization results of QPSO. The skewness value is inversely proportional to the fitness function. Therefore, by calculating the minimum value of the skewness, we can obtain the best fitness function value. Using this value to optimize the parameters of the alignment entropy, we can make the distinction between the alignment entropy values of different fault types more obvious and make it easier to carry out fault diagnosis and classification.

3.1.3. SVM Algorithm

Assume that $\{(x,y)|i = 1,2,\cdots,k\}$ is the input training set, where k is the number of training set samples.$x_i \in T^d$ is the d-dimensional feature vector. $y_i \in \{-1,1\}$ denotes the category of sample x_i.

In order to maximize the sample interval, the optimization formula was constructed and expressed as follows [38]:

$$\begin{gathered} \min_{w,b} \frac{\|w\|^2}{2} + C\sum_{i=1}^{k} \xi_i \\ s.t.\ y_i[(w_i x) + b] \geq 1 - \xi_i\ i = 1,2,\cdots,k \end{gathered} \tag{25}$$

where w is the weighting of the optimal classification surface. C denotes the penalty parameter of the deviation item ξ_i. When $C > 0$, ξ_i is the error caused when the sample points are misclassified.

The above nonlinear planning problem was transformed into a linear planning problem using the Lagrangian equation. The Lagrangian equation fuses the objective function with the constraint function and then finds the optimum value of that equation, which is the optimal classification surface. The classification decision function was obtained as:

$$f(x) = \mathrm{sgn}\left(\sum_{i=1}^{k} \alpha_i y_i K(x_i, x) + b^*\right) \tag{26}$$

where α_i is a Lagrangian multiplier. $K(x_i, x)$ represents the kernel function of the Mercer condition, which is denoted as follows:

$$K(x, y) = \exp\left(-\frac{||x - y||^2}{2g^2}\right) \quad (27)$$

where g indicates the complexity of the subspace distribution of the sample data.

3.2. The Validation of the QPSO-MPE-SVM Algorithm

The process of QPSO-MPE-SVM fault feature extraction and identification is shown in Figure 4, which is as follows.

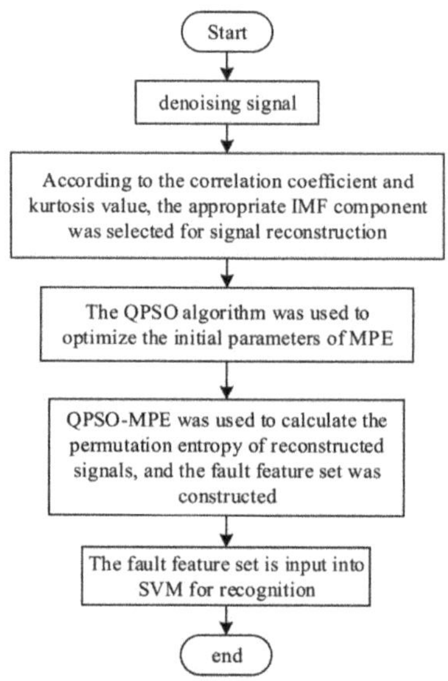

Figure 4. QPSO-MPE-SVM fault identification flow chart.

(1) The denoised vibration signal is again disintegrated by CEEMDAN, and the IMFs are selected according to the correlation coefficient and Kurtosis values of the IMF for signal reconstruction;
(2) The initial parameters of the MPE are optimized using the QPSO to obtain the better MPE parameters;
(3) The MPE values of the reconstructed signals are calculated using the optimized MPE parameters, and the MPE values with obvious differentiation are selected to construct the bearing fault feature set;
(4) The obtained MPE fault feature set is input to the SVM for fault identification.

The correlation coefficient value is proportional to the fault information content in the IMF component. The kurtosis value reflects the amount of shock information in the IMF component; the larger the kurtosis value, the more shock information in the IMF component, and the smaller the kurtosis value, the less oscillating information in the IMF component.

The validated data came from the rolling bearing database provided by the Electrical Engineering Laboratory of Case Western Reserve University (CWRU) [35]. The correlation coefficient values and Kurtosis values of each order IMF component of the bearing fault signal (normal signal, inner-ring fault signal, outer-ring signal, and rolling-body fault signal) at 1797 rmp were calculated and the results are shown in Figure 5.

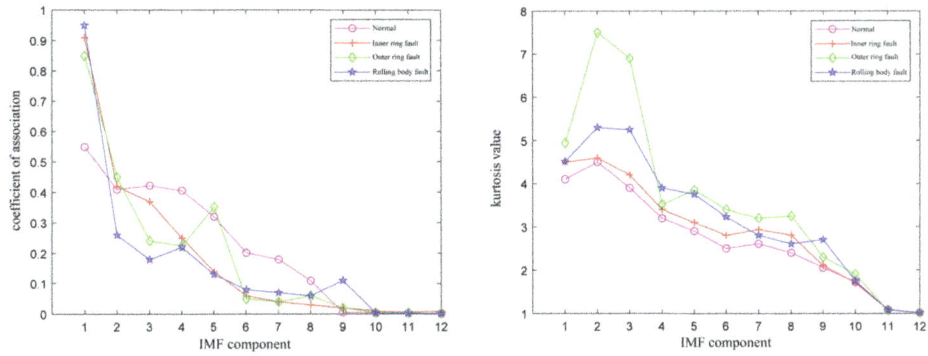

(a) (b)

Figure 5. Plots of IMF component correlation coefficients and Kurtosis values. (**a**) The correlation coefficients of IMF; (**b**) The Kurtosis values of IMF.

From Figure 5, it can be seen that the correlation coefficients of the four vibration signals were different, the correlation coefficients of the first five order IMFs of the four-fault signals were all above 10%, and Kurtosis values were all greater than 3. Therefore, the first fifth-order IMF components were selected for signal reconstruction.

To prove whether the bearing fault characteristic information can be accurately extracted from the first fifth-order IMFs, we took the inner ring fault signal as an example for research. The CEEMDAN decomposed the fault signal of the bearing inner ring and selects the first-order and fifth-order IMF for signal reconstruction. Then, the reconstructed signal was analyzed by Hilbert-Huang transform (HHT) envelope spectrum. The envelope spectrum analysis of the original signal and the reconstructed signal is shown in Figure 6.

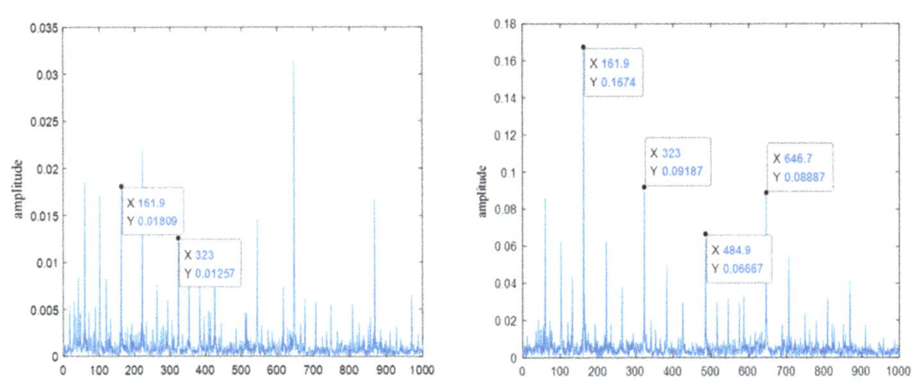

(a) (b)

Figure 6. The HHT envelope spectrum of the inner-ring fault signal. (**a**) The envelope spectrum analysis of the original signal; (**b**) The envelope spectrum analysis of the reconstructed signal.

The theoretical value of the characteristic frequency of the inner-ring fault signal is 162.1 Hz, and the actual value is 161.9 Hz. The error of no more than 10% between the theoretical frequency value and the actual frequency can be determined as the same fault. From the original signal envelope spectrum shown in Figure 6a, it can be seen that there were other interference peaks near the peak of the fault characteristic frequency (161.9 Hz), which could not effectively extract the fault characteristic frequency. The HHT envelope spectrum of the reconstructed signal shown in Figure 6b, indicating that the characteristic frequency (161.9 Hz) can be easily extracted, and it is also easy to extract the di-frequency (323 Hz) and tri-frequency (484.9 Hz) of the fault features.

In summary, it was proved that the first fifth-order IMF components can retain the fault feature information, which further verifies the effectiveness of the proposed denoising method in this paper. This scheme can realize the real-time processing of fault signals and can be looped.

3.2.1. Optimize MPE Values Using QPSO

The initial parameters of MPE were set to $N = 1024$, $\lambda = 1$, $m = 6$ and $\tau = 12$. The first fifth-order IMF components of the normal signal, inner-ring fault signal, outer-ring fault signal, and rolling-element fault signal were selected for signal reconstruction. The MPE values reconstructed signal was calculated and the results are shown in Figure 7. It can be seen that the MPE values of each signal were not distinguishable, so they could not be used as an effective feature set for fault classification.

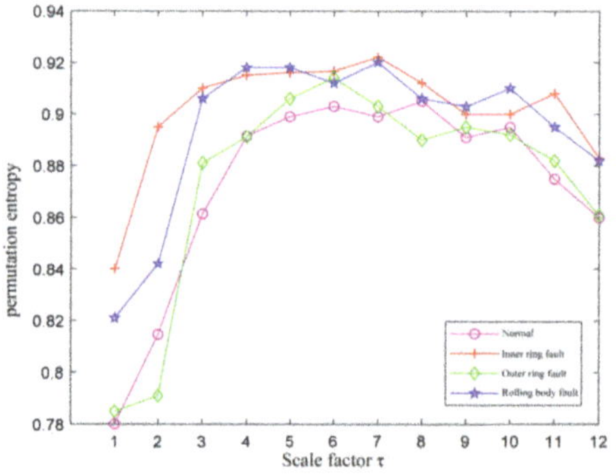

Figure 7. The MPE values without QPSO optimization.

To separate the MPE values of the four operating conditions of the bearings, the QPSO was used to optimize the initial MPE parameters. The parameters of QPSO were set to 30 for the number of races, 200 for the maximum number of iterations, and 10 and 0.2 for the maximum and minimum inertia weights, respectively. The parameters of the optimized MPE are shown in Table 3. The MPE values were calculated using the optimized parameters, and the results are shown in Figure 8. It can be seen that the MPE values of each signal were independent of each other, which can better construct the fault feature set. In this paper, the scale factors with stable and larger partition degrees were selected to construct the fault feature vector, so the selected factors were 5, 6, 7, 8, and 9.

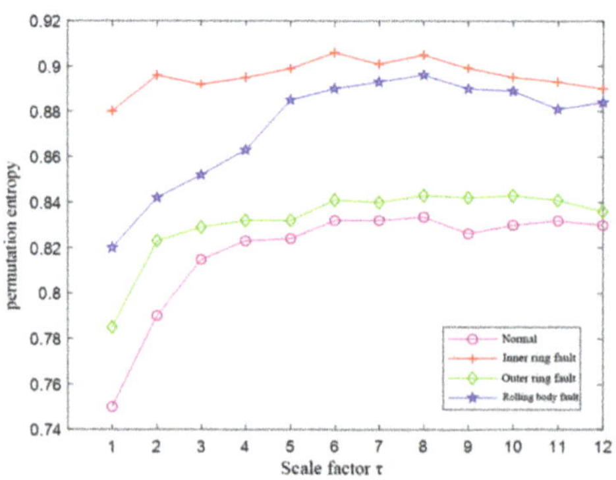

Figure 8. The MPE values optimized after QPSO.

Table 3. Parameters of MPE after QPSO optimization.

Signal	N	λ	m	τ
Normal signal	1182	1	5	12
Inner-ring fault signal	1467	2	6	14
Outer-ring fault signal	1384	3	7	13
Rolling-body fault signal	953	1	6	12

3.2.2. Fault Feature Extraction and Identification Using the QPSO-MPE-SVM

Sixty sets of vibration data were selected for each of the four operating conditions of the bearings, for a total of 240 sets. These data were divided equally into two groups, one for the training set, and the other for the test set. Two fault-identification experiments were operated. One experiment is with the initial parameters of MPE, and the fault identification results are shown in Figure 9; the other experiment was with the optimized parameters of MPE by QPSO, and the fault identification results are shown in Figure 10. The indications in the figure include 1 for the normal state, 2 for the inner-ring failure state, 3 for the outer-ring failure state, and 4 for the rolling-element failure state.

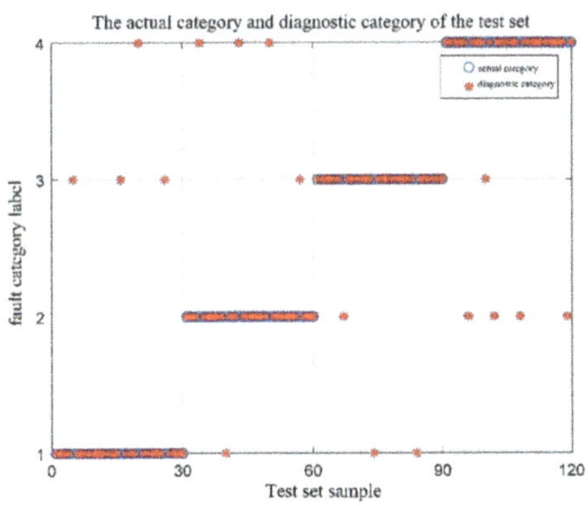

Figure 9. MPE-SVM fault identification results.

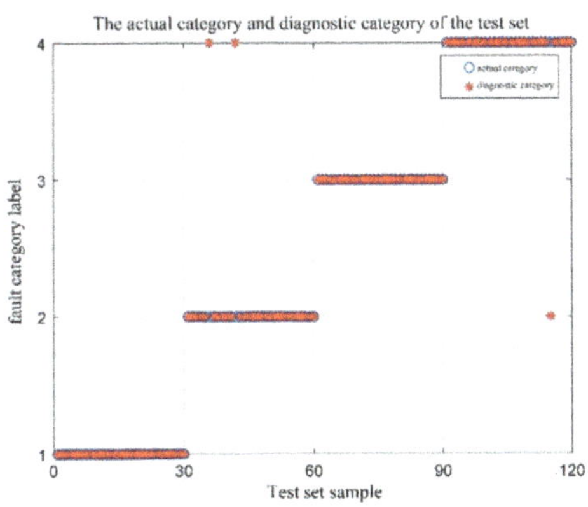

Figure 10. QPSO-MPE-SVM fault identification results.

Analysis of Figures 9 and 10 reveals that when the MPE values without optimization were directly input to the SVM for fault identification, the fault identification accuracy was 85.83%, while the fault identification accuracy could reach 97.5% after optimization by the QPSO. In order to verify the reliability of the proposed method, several simulation experiments were repeated, and the simulation results all showed that the proposed method could effectively improve the accuracy of bearing fault diagnosis. Thus, the effectiveness of the QSPO-MPE-SVM method proposed is confirmed.

4. Fault Diagnosis of Rolling Bearing of Sine Roller Screen Based on Vibration Signal

The experimental platform is shown in Figure 11. The experimental platform completes the tasks of installing sensors, driving bearings, applying radial loads and acquiring vibration signals. The test bearing was a grease-lubricated deep-groove ball bearing, and rolling bearing parameters as shown in Table 4. The working limit temperature of the bearing is 400 °C. The rotational speed is 10–800 r/min, and the limit rotational speed is 2000 r/min. The speed can be continuously adjusted, and the error is less than ±1%. The applied radial load is 120 kg, the ultimate is 500 kg, the model number of the load sensor is MZLF-2, and the sensor sensitivity is 2.0 ± 0.01 mV/V. The model number of the acceleration sensor is KH-HZD-B-2-12, and the sensitivity is 20 ± 0.05 mV/mm/s. The model number of data acquisition cards is USB3200N (32 bit, 4 channels, 20 MHz); the vibration sensor (MIC-HZB-F-2-12) collected the vibration acceleration of the bearing in real-time, and transferred the vibration signal to the data-acquisition card. The signal was finally transferred to the computer for data processing. All the experimental data were within the allowable error range.

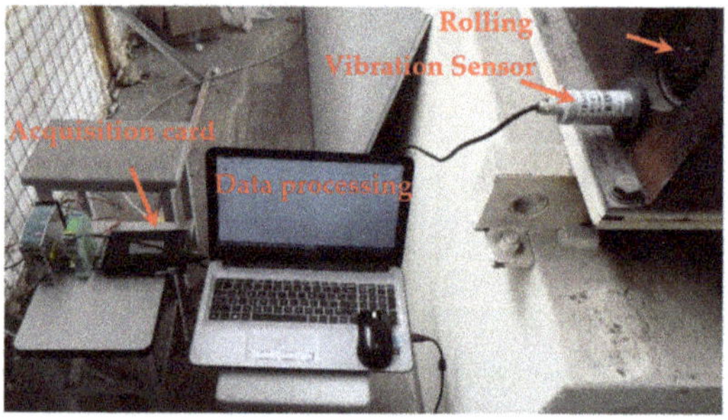

Figure 11. Sine roller screen rolling bearing fault experimental platform.

Table 4. Rolling bearing parameters.

Inner Ring Diameter D_1	Outer Ring Diameter D_2	Rolling Body Diameter d	Bearing Mid Diameter D	Number of Rolling Bodies n	Contact Angle θ
70 mm	130 mm	20.43 mm	100 mm	8	0°

4.1. Rolling Bearing Feature Frequency Calculation

The rolling bearing speed is 500 r/min. The theoretical formula of the fault frequency is expressed as:

(1) Journal rotation frequency is expressed as:

$$f_r = \frac{v}{60} \tag{28}$$

(2) Inner-ring fault characteristic frequency is expressed as:

$$f_a = \frac{nf_r}{2}\left(1 + \frac{d}{D}\cos\theta\right) \tag{29}$$

(3) Outer-ring fault characteristic frequency is expressed as:

$$f_b = \frac{nf_r}{2}\left(1 - \frac{d}{D}\cos\theta\right) \tag{30}$$

(4) Rolling-body fault characteristic frequency is expressed as:

$$f_d = \frac{Df_r}{2d}\left[1 - \left(\frac{d}{D}\right)^2\cos^2\theta\right] \tag{31}$$

where v is journal rotation speed.

The calculation results by the formulas are shown in Table 5.

Table 5. The fault feature frequency of rolling bearing.

Rotational Speed	Inner Ring Fault Frequency	Outer Ring Fault Frequency	Rolling Body Fault Frequency
500 r/min	40.143 Hz	26.523 Hz	19.545 Hz

4.2. Rolling Bearing Fault Signal Denoising and Feature Extraction

Taking the inner ring fault signal as an example, the original vibration signal and HHT envelope spectrum of the inner-ring fault are shown in Figures 12 and 13, respectively. It was denoised using CEEMDAN-DFA-improved wavelet thresholding.

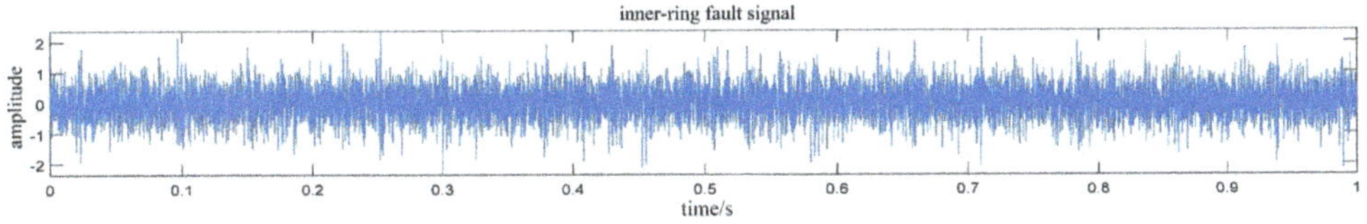

Figure 12. Time domain diagram of the original signal of the inner-ring fault.

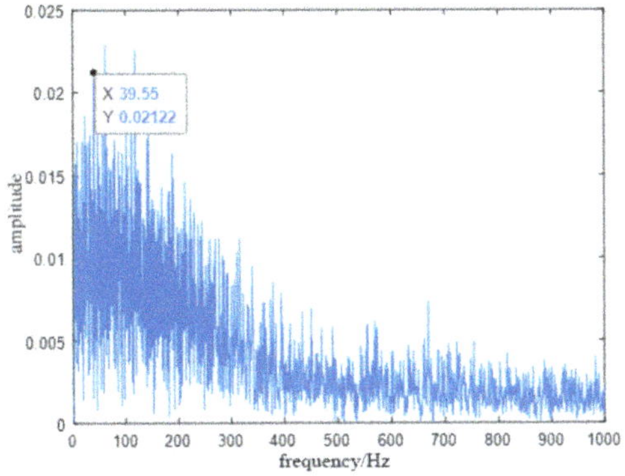

Figure 13. Original signal envelope spectrum of the inner-ring fault.

The IMFs were generated by the CEEMDAN decomposition, and the value of the scaling function α of the IMFs was calculated. α values of the first eighth-order IMFs are shown in Table 6.

Table 6. Scalar function value of each IMF component of the inner-ring signal.

IMF	IMF_1	IMF_2	IMF_3	IMF_4	IMF_5	IMF_6	IMF_7	IMF_8
α value	0.4138	0.4096	0.2961	0.4537	0.5068	0.5564	0.6224	0.6743

According to the DFA, correlation can be judged that the first fourth-order IMFs were dominated by noise, so the first four-order IMFs need to be denoised. The time-domain signal of the inner-ring fault after denoising is shown in Figure 14. The denoised signal and the remaining IMFs were reconstructed, and the reconstructed signal is the fault signal of the bearing inner ring.

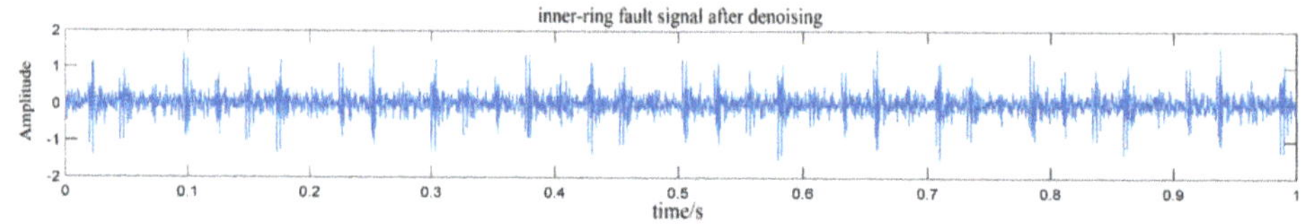

Figure 14. Time domain diagram of the inner-ring fault signal after denoising.

The denoised signal was again decomposed by the CEEMDAN, and the correlation coefficients and Kurtosis values of each IMF were calculated. The calculation results are shown in Table 7.

Table 7. Correlation coefficients and Kurtosis values of IMFs of the inner-ring signal.

IMF	IMF_1	IMF_2	IMF_3	IMF_4	IMF_5	IMF_6	IMF_7	IMF_8
Correlation coefficient	0.9214	0.4327	0.3961	0.2715	0.2049	0.0621	0.0592	0.0357
Kurtosis value	4.5217	4.9371	4.1964	3.4922	3.2614	2.5147	2.2291	2.0634

From Table 7, it can be found that the correlation coefficients of the first fifth-order IMF components were greater than 0.1, and the Kurtosis values were greater than three. Therefore, the first fifth-order IMF component signal was reconstructed, and the HHT envelope spectrum of the reconstructed signal was analyzed. The analysis result is shown in Figure 15. It can be seen that the characteristic frequency of the inner-ring fault signal was 39.55 Hz, which is close to the theoretically calculated value (40.143 Hz).

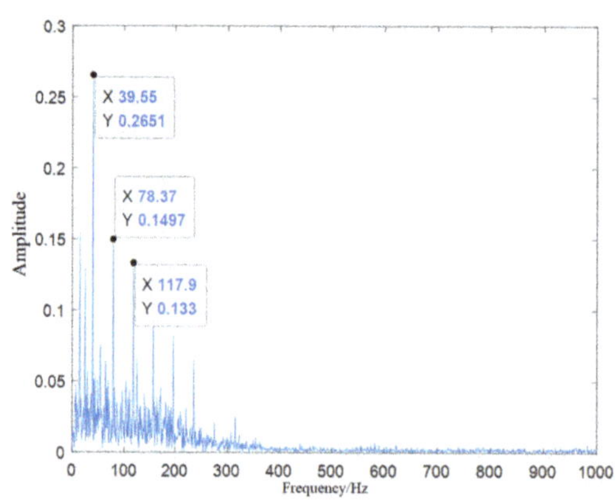

Figure 15. Envelope spectrum of the inner-ring fault signal after denoising.

The initial parameters of the MPE of the four running signals were all set as $N = 1024$, $\lambda = 1$, $m = 6$, $\tau = 12$. The MPE values were calculated using the initial parameters, and the results are shown in Figure 16a. It can be seen that the MPE values of four signals had small differences and overlapped each other, which is not conducive to the construction of the fault feature set.

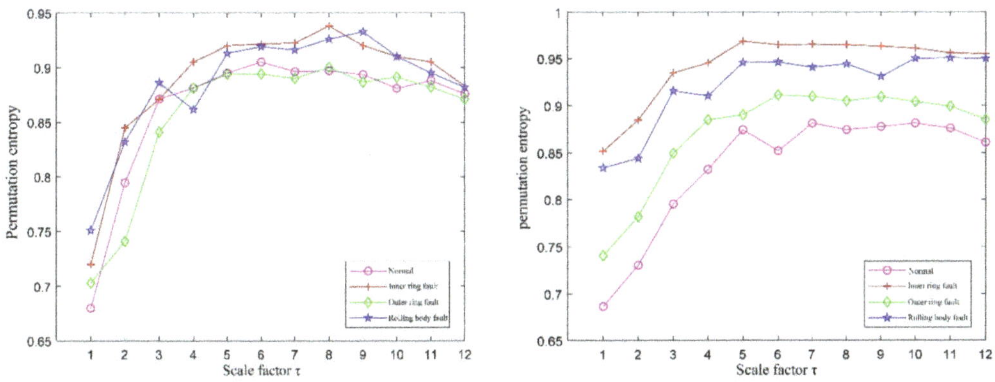

(**a**) (**b**)

Figure 16. The MPE values of the four running signals. (**a**) The MPE values using the initial parameters; (**b**) The MPE values using the optimized parameters.

The initial parameters were optimized using the QPSO, and the results are shown in Table 8.

Table 8. The optimized parameters of MPE after QPSO.

Signal	N	λ	m	τ
Normal Signal	1285	1	5	12
Inner-ring fault signal	1836	2	7	14
Outer-ring fault signal	1587	1	6	14
Rolling-body fault signal	1054	1	5	13

The MPE values of the four running signals were calculated using the optimized parameters, and the results are shown in Figure 16b. It can be seen that the MPE values of the four signals achieved separation without overlap, which is suitable for constructing fault feature sets for fault-type identification. The MPE values with scale factors (5, 6, 7, 8, 9, and 10) were selected to construct the fault feature vectors.

4.3. Rolling Bearing Failure Identification and Control Experiment

A total of 160 sets of vibration experimental data were selected from 40 sets of four running signals, of which 100 sets were used as training samples and the remaining 60 sets were used as test samples. Comparing the four test sets, the recognition results are shown in Figure 17. The fault identification result of the original signal by MPE-SVM is shown in Figure 17a. The fault identification result of QPSO-MPE-SVM to the original signal is shown in Figure 17b. The results of MPE-SVM fault identification on the denoised (CEEMDAN-DFA-improved wavelet threshold function) signal are shown in Figure 17c. The results of QPSO-MPE-SVM fault identification on the denoised (CEEMDAN-DFA-improved wavelet threshold function) signal are shown in Figure 17d.

The following conclusions can be drawn from the analysis of Figure 17.

When using the MPE-SVM method for fault identification on the original signal without denoising, there were 23 false identifications, and the identification accuracy was only 61.67%. When using the QPSO-MPE-SVM method for fault identification on the original signal without denoising, there were 17 false identifications, and the identification accuracy was 71.67%, which shows that the noise in the collected vibration signal had a large interference to the bearing fault diagnosis.

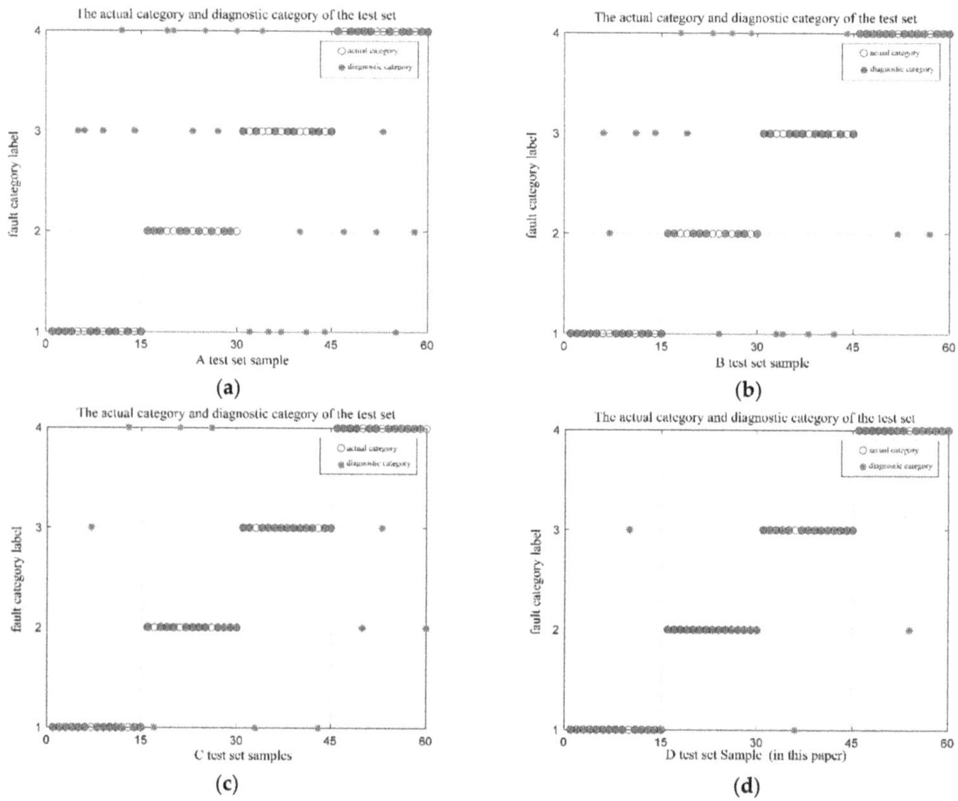

Figure 17. Four types of test sets identify results. (**a**) The results of MPE-SVM fault identification on the original signal; (**b**) the results of the QPSO-MPE-SVM fault identification on the original signal; (**c**) the results of MPE-SVM fault identification on the denoised signal; (**d**) the results of QPSO-MPE-SVM fault identification on the denoised signal.

When using the MPE-SVM method for fault identification on the denoised signal, the number of false identifications was reduced to 10, and the identification accuracy increased to 83.33%. When using the QPSO-MPE-SVM method for fault identification on the denoised signal, the number of false identifications was only three and the fault identification accuracy was 95%.

By applying the comprehensive diagnosis method proposed in this paper to the experimental platform of rolling bearings, the fault diagnosis of the actual measured fault signal of the inner ring of the bearing was achieved. The de-noising process and fault feature extraction were completed, and the fault feature set was constructed to achieve fault identification. The experimental results showed that the fault identification accuracy could be 95%. The method was verified, by several experiments, to not only have high accuracy and reliability, but also not be limited to inner-ring fault diagnosis, also being applicable to vibration signals with non-smooth and non-linear characteristics (such as outer-ring fault signals). Therefore, the method has good practical application value.

5. Conclusions

This paper proposed a CEEMDAN-DFA-improved wavelet threshold denoising method and QPSO-MPE-SVM fault feature extraction and identification method, the purpose being to realize the fault diagnosis of the rolling-bearing vibration signal.

The denoising process with the CEEMDAN-DFA-improved wavelet threshold method is as follows: Firstly, the vibration signal is decomposed into IMF by CEEMDAN, and the

DFA is performed on the IMF components. Then, the scaling function of each IMF component is calculated to select the noise-dominated IMF component. Finally, the improved wavelet threshold function is applied to denoise the noise-dominated IMFs. The de-noised IMFs and the remaining other IMFs are merged to get reconfigured signals. The validation results showed that compared with traditional denoising methods, the CEEMDAN-DFA-improved wavelet threshold function method proposed in this paper could better remove noise, effectively reduce the signal distortion, and maintain the original characteristics of the signal.

The fault feature extraction and identification with the QPSO-MPE-SVM method are as follows: Firstly, the reconfigured signal is decomposed again by CEEMDAN, and the correlation coefficients and Kurtosis values of each order IMF component are calculated; the IMF components with larger values about correlation coefficients and Kurtosis are selected for signal reconstruction, and the HHT envelope spectrum analysis is performed on the reconstructed signal to extract the fault characteristic frequencies. Then, the initial parameters of MPE are optimized with QPSO, the MPE value is calculated for the reconstructed signal, and the appropriate MPE value is selected to construct the rolling-bearing fault feature vectors. Finally, the fault feature vectors are inputted to the trained SVM for rolling-bearing fault type identification. The validation results showed that the MPE parameters were optimized by the QPSO, which makes the MPE values of the four signals achieve separation without overlap, which is more suitable for constructing fault feature sets for fault type identification.

The algorithms proposed in the paper were all validated by building an experimental platform of rolling bearings. The experimental results showed that the fault identification accuracy of rolling bearings could reach 95%; it not only has high accuracy and reliability, but also can be applicable to vibration signals with non-smooth and non-linear characteristics, having good practical application value.

Author Contributions: Conceptualization, Y.W. (Yi Wang); data curation, Y.W. (Yi Wang); formal analysis, Y.W. (Yi Wang); funding acquisition, X.C.; Investigation, C.X.; methodology, Y.W. (Yi Wang); project administration, X.C.; resources, X.C.; software, Y.W. (Yu Wang) and C.X.; supervision, X.C. and C.X.; validation, Y.W. (Yu Wang) and C.X.; writing—original draft, Y.W. (Yi Wang); writing—review and editing, Y.W. (Yu Wang), X.C. and C.X. All authors have read and agreed to the published version of the manuscript.

Funding: This research was funded by the National Natural Science Foundation of China Program (No.62073198), the Major Research Development Program of Shandong Province of China (No.2016GSF117009).

Institutional Review Board Statement: Not applicable.

Informed Consent Statement: Informed consent was obtained from all subjects involved in the study.

Data Availability Statement: https://csegroups.case.edu/bearingdatacenter/pages/12k-drive-end-bearing-fault-data accessed on 30 August 2021.

Conflicts of Interest: The authors declare no conflict of interest.

References

1. El Laithy, M.; Wang, L.; Harvey, T.J.; Vierneusel, B.; Correns, M.; Blass, T. Further understanding of rolling contact fatigue in rolling element bearings—A review. *Tribol. Int.* **2019**, *140*, 105849. [CrossRef]
2. Liu, J.; Shao, Y. Overview of dynamic modelling and analysis of rolling element bearings with localized and distributed faults. *Nonlinear Dyn.* **2018**, *93*, 1765–1798. [CrossRef]
3. Shi, W.; Yan, L.J.; Xie, X.B.; Zhou, L. Magnetotelluric noise supperssion based on CEEMDAN-DFA combined with endpoint detection technology using for oil and gas exploration. *Fresenius Environ. Bull.* **2020**, *29*, 8969–8978.
4. Huang, S.X.; Wang, X.P.; Li, C.F.; Kang, C. Data decomposition method combining permutation entropy and spectral substitution with ensemble empirical mode decomposition. *Measurement* **2019**, *139*, 438–453. [CrossRef]
5. Wang, D.; Zhao, Y.; Yi, C.; Tsui, K.-L.; Lin, J.H. Sparsity guided empirical wavelet transform for fault diagnosis of rolling element bearings. *Mech. Syst. Signal Process.* **2018**, *101*, 292–308. [CrossRef]

6. Zhang, J.; Zhang, P.L.; Hua, C.R.; Qin, P. Improved method for bearing AE signal denoising based on K-SVD algorithms. *J. Vib. Shock* **2017**, *36*, 150–156.
7. Ma, J.; Wu, J.D.; Wang, X.D. A hybrid fault diagnosis method based on singular value difference spectrum denoising and local mean decomposition for rolling bearing. *J. Low Freq. Noise Vib. Act. Control* **2018**, *37*, 928–954. [CrossRef]
8. Zeng, M.; Chen, Z. SOSO Boosting of the K-SVD denoising algorithm for enhancing fault-induced impulse responses of rolling element bearings. *IEEE Trans. Ind. Electron.* **2019**, *67*, 1282–1292. [CrossRef]
9. Abdelkader, R.; Kaddour, A.; Derouiche, Z. Enhancement of rolling bearing fault diagnosis based on improvement of empirical mode decomposition denoising method. *Int. J. Adv. Manuf. Tech.* **2018**, *97*, 3099–3117. [CrossRef]
10. Zhao, D.Z.; Li, J.Y.; Cheng, W.D.; Wang, T.Y.; Wen, W.G. Rolling element bearing instantaneous rotational frequency estimation based on EMD soft-thresholding denoising and instantaneous fault characteristic frequency. *J. Cent. South Univ.* **2016**, *23*, 1682–1689. [CrossRef]
11. Gao, S.Z.; Li, T.C.; Zhang, Y.M. Rolling bearing fault diagnosis of PSO–LSSVM based on CEEMD entropy fusion. *Trans. Can. Soc. Mech. Eng.* **2020**, *44*, 405–418. [CrossRef]
12. Tajeddini, M.A.; Aalipour, A.; Safarinejadian, B. Decision fusion method for bearing faults classification based on wavelet denoising and dempster-shafer theory. *Iran. J. Sci. Technol. Trans. Electr. Eng.* **2019**, *43*, 295–305. [CrossRef]
13. Kumar, H.; Pai, P.; Sriram, N. Classification of rolling element bearing fault using singular value. *J. Qual. Maint. Eng.* **2019**, *26*, 181–197. [CrossRef]
14. Yang, X.; Qiu, M.; Chen, L.H.; Chen, Y. Adaptive wavelet threshold function based on PSO-RWE for vibration signal denoising of rolling bearing. *J. Aerosp. Power* **2020**, *35*, 2339–2347.
15. Chegini, S.N.; Bagheri, A.; Najafi, F. Application of a new EWT-based denoising technique in bearing fault diagnosis. *Measurement* **2019**, *144*, 275–297. [CrossRef]
16. Chaabi, L.; Lemzadmi, A.; Djebala, A.; Bouhalais, M.L.; Ouelaa, N. Fault diagnosis of rolling bearings in non-stationary running conditions using improved CEEMDAN and multivariate denoising based on wavelet and principal component analyses. *Int. J. Adv. Manuf. Technol.* **2020**, *107*, 3859–3873. [CrossRef]
17. Chen, W.G.; Li, J.N.; Wang, Q.; Han, K. Fault feature extraction and diagnosis of rolling bearings based on wavelet thresholding denoising with CEEMDAN energy entropy and PSO-LSSVM. *Measurement* **2021**, *172*, 108901. [CrossRef]
18. Bie, F.F.; Zhang, S.J.; Pei, J.F.; Xu, P.Q. Study on the Method of Wavelet Threshold and CEEMDAN in Rolling Baring Recognize. *Mach. Des. Manuf.* **2020**, *2*, 68–71.
19. Zhen, D.; Zhu, J.R.; Zhang, C.; Shi, Z.Q.; Gu, F.S. Fault diagnosis of rolling bearing based on wavelet packet energy and spectral kurtosis. *J. Mach. Des.* **2021**, *38*, 23–28.
20. Han, T.; Liu, Q.N.; Zhang, L.; Tan, A.C.C. Fault feature extraction of low speed roller bearing based on teager energy operator and CEEMD. *Measurement* **2019**, *138*, 400–408. [CrossRef]
21. Li, H.; Xie, J.; Wei, W. Permutation entropy and Lyapunov exponent: Detecting and monitoring the chaotic edge of a closed planar under-actuated system. *Mech. Syst. Signal Process.* **2019**, *123*, 206–221. [CrossRef]
22. Li, J.; Shang, P.; Zhang, X. Financial time series analysis based on fractional and multiscale permutation entropy. *Commun. Nonlinear Sci. Numer. Simul.* **2019**, *78*, 104880. [CrossRef]
23. Yin, Y.; Shang, P. Multivariate weighted multiscale permutation entropy for complex time series. *Nonlinear Dyn.* **2017**, *88*, 1707–1722. [CrossRef]
24. Ye, M.Y.; Yan, X.A.; Jia, M.P. Rolling Bearing Fault Diagnosis Based on VMD-MPE and PSO-SVM. *Entropy* **2021**, *23*, 762. [CrossRef] [PubMed]
25. Du, W.; Guo, X.; Wang, Z.; Wang, J.; Yu, M.; Li, C.; Wang, G.; Wang, L.; Guo, H.; Zhou, J.; et al. A new fuzzy logic classifier based on multiscale permutation entropy and its application in bearing fault diagnosis. *Entropy* **2020**, *22*, 27. [CrossRef]
26. Mocanu, D.C.; Mocanu, E.; Stone, P.; Nguyen, P.H.; Gibescu, M.; Liotta, A. Scalable training of artificial neural networks with adaptive sparse connectivity inspired by network science. *Nat. Commun.* **2018**, *9*, 2383. [CrossRef] [PubMed]
27. Yan, X.; Liu, Y.; Ding, P.; Jia, M. Fault diagnosis of rolling-element bearing using multiscale pattern gradient spectrum entropy coupled with laplacian score. *Complexity* **2020**, *2020*, 4032628. [CrossRef]
28. Zhou, Z.; Chen, J.; Zhu, Z. Regularization incremental extreme learning machine with random reduced kernel for regression. *Neurocomputing* **2018**, *321*, 72–81. [CrossRef]
29. Wang, Z.; Yao, L.; Cai, Y.; Zhang, J. Mahalanobis semi-supervised mapping and beetle antennae search based support vector machine for wind turbine rolling bearings fault diagnosis. *Renew. Energy* **2020**, *155*, 1312–1327. [CrossRef]
30. Yan, X.; Liu, Y.; Xu, Y.; Jia, M. Multistep forecasting for diurnal wind speed based on hybrid deep learning model with improved singular spectrum decomposition. *Energy Convers. Manag.* **2020**, *225*, 113456. [CrossRef]
31. Wu, Z.; Huang, N.E. Ensemble empirical mode decomposition: A noise-assisted data analysis method. *Adv. Adapt. Data Anal.* **2009**, *1*, 1–41. [CrossRef]
32. Colominas, M.A.; Schlotthauer, G.; Torres, M.E. Improved complete ensemble EMD; a suitable tool for biomedical signal processing. *Biomed. Signal Process. Control* **2014**, *14*, 19–29. [CrossRef]
33. Yuan, Z.; Peng, T.T.; An, D.; Cristea, D.; Pop, M.A. Rolling bearing fault diagnosis based on adaptive smooth ITD and MF-DFA method. *J. Low Freq. Noise Vib. Act. Control* **2019**, *39*, 968–986. [CrossRef]

34. Song, Q.; Zhao, S.F.; Wang, M.S. On the accuracy of fault diagnosis for rolling element bearings using improved DFA and multi-sensor data fusion method. *Sensors* **2020**, *20*, 6465. [CrossRef] [PubMed]
35. Zhao, D.F.; Liu, S.L.; Gu, D.; Sun, X.; Wang, L.; Wei, Y.; Zhang, H.L. Improved multi-scale entropy and it's application in rolling bearing fault feature extraction. *Measurement* **2019**, *152*, 248–263. [CrossRef]
36. Ren, M.L.; Huang, X.D.; Zhu, X.X.; Shao, L.J. Optimized PSO algorithm based on the simplicial algorithm of fixed point theory. *Appl. Intell.* **2020**, *50*, 2009–2024. [CrossRef]
37. Wan, L.J.; Li, H.Y.; Chen, Y.W.; Li, C.Y. Rolling bearing fault prediction method based on QPSO-BP neural network and dempster–shafer evidence theory. *Energies* **2020**, *13*, 1094. [CrossRef]
38. Liu, W.Q.; Shen, J.X.; Yang, X.Q. Rolling bearing fault detection approach based on improved dispersion entropy and AFSA optimized SVM. *Int. J. Electr. Eng. Educ.* **2020**, *15*, 1–16. [CrossRef]

entropy

MDPI

Article

Sensor and Actuator Fault Diagnosis for Robot Joint Based on Deep CNN

Jinghui Pan [1,*], Lili Qu [2] and Kaixiang Peng [1]

[1] School of Automation and Electrical Engineering, University of Science and Technology Beijing, Beijing 100083, China; kaixiang@ustb.edu.cn
[2] School of Mechatronic Engineering and Automation, Foshan University, Foshan 528231, China; qulili@fosu.edu.cn
* Correspondence: panjinghuiwork@126.com

Abstract: This paper proposes a data-driven method-based fault diagnosis method using the deep convolutional neural network (DCNN). The DCNN is used to deal with sensor and actuator faults of robot joints, such as gain error, offset error, and malfunction for both sensors and actuators, and different fault types are diagnosed using the trained neural network. In order to achieve the above goal, the fused data of sensors and actuators are used, where both types of fault are described in one formulation. Then, the deep convolutional neural network is applied to learn characteristic features from the merged data to try to find discriminative information for each kind of fault. After that, the fully connected layer does prediction work based on learned features. In order to verify the effectiveness of the proposed deep convolutional neural network model, different fault diagnosis methods including support vector machine (SVM), artificial neural network (ANN), conventional neural network (CNN) using the LeNet-5 method, and long-term memory network (LTMN) are investigated and compared with DCNN method. The results show that the DCNN fault diagnosis method can realize high fault recognition accuracy while needing less model training time.

Keywords: fault diagnosis; sensor fault; actuator fault; deep convolutional neural network; robot joints

Citation: Pan, J.; Qu, L.; Peng, K. Sensor and Actuator Fault Diagnosis for Robot Joint Based on Deep CNN. *Entropy* **2021**, *23*, 751. https://doi.org/10.3390/e23060751

Academic Editor: José A. Tenreiro Machado

Received: 4 May 2021
Accepted: 8 June 2021
Published: 15 June 2021

1. Introduction

With the development of robot and control technology, various robots are widely used in industry. Different applications present specific requirements for robot systems, such as rapidity, robustness, and safety [1–3]. However, among all the indices required by applications, the controllability of the robot in fault state has become the most critical factor. Fault identification is the precondition for realizing this goal, which promotes the investigation of our research [4–6].

Robot systems cannot work without the support of different kinds of sensors and actuators. Miniaturization and multi-functionality are required for development. The rapid development of sensors, material science, and micro-electro-mechanical technology allows modern robot joint modules—such as hollow motor, servo driver, harmonic reducer, brake, and encoder—to be integrated within limited space [7]. Sensors and actuators are key components in the robot system, but their working environment is very complex, with electromagnetic interference, vibration, etc., which will affect the output of the sensors and then the actuators. Moreover, the variable load on manipulators is also a challenge for system state feedback or estimation. All of the above factors make the faults diagnosis of robot system sensors and actuators an urgent task [8].

In most robot system faults, sensor and actuator malfunction are the main causes of robot system failure. Therefore, diagnosis for the sensors and actuators is very important. In order to improve the reliability of robot joints and realize fault detection and fault-tolerant control of robot systems, researchers have been focused on fault detection and fault-tolerant control of robot joints for many years, and many practical fault diagnosis methods have been proposed. In [9], redundant sensors are used on the robot joint, and then fuzzy rules are designed to adjust the threshold of the fault signal adaptively to carry out fault diagnosis. In [10], for a six-degree-of-freedom robot joint system, low-cost MEMS magnetic, angular velocity, and gravity sensors are used to estimate the joint angle of a rotating manipulator. In [11], a discrete-time framework for fault diagnosis of robot joint sensors, force or torque sensors, and actuators is proposed. The redundant sensors are used on the robot joint, and the feedback values from redundant sensors and the estimated values calculated by two isolation observers are input into the fault decision system. The data from redundant sensors are used to provide information for a group of diagnostic observers to detect, isolate, and identify faults of joint actuators, force, or torque sensors.

However, there may be another consideration when using redundant sensors for fault diagnosis. A robot fault diagnosis system based on redundant sensors not only increases structural complexity, but also increases the hardware cost of the system. In addition, redundant sensors also increase the probability of a sensor fault when the running time of a robot system approaches the sensor's life cycle.

In order to overcome the shortcomings of using redundant sensors for fault diagnosis, observers have been widely used. There are many novel theories that could be used to design state observers for robot fault diagnosis. A robot-fault diagnosis method using fuzzy logic is proposed in [12] to evaluate residuals. Fuzzy logic applied to robot fault diagnosis does not require any redundant sensors, but it relies on the fault model of the robot system. The sliding mode method can be seen everywhere in robot fault diagnosis. Daniele uses a second-order sliding mode observer for fault detection and isolation of the rigid manipulator of the COMAU robot and uses the suboptimal second-order sliding mode algorithm to design the input law of the proposed observer, which can detect a single fault on a specific brake or a specific sensor of the manipulator [13]. Since the high order sliding mode observer can detect possible faults in specific parts of the robot system, the sliding mode method is greatly expanded [14]. The observer design methods mentioned above are just some typical representatives, actually, there are many other methods that could be used for robot fault diagnosis, such as the output feedback method [15], nonlinear disturbance observer [16], and feedback linearization disturbance observer design method [17]. As it is well known, the difficulty of observer-based robot fault diagnosis lies in the gain design process [18].

Machine learning introduces an effective solution to the above problems caused by redundant sensors and observers. Typical application methods include, but are not limited to, genetic algorithm [19], support vector machine [20], cluster analysis [21], and neural network [22]. Among them, the neural network is widely used in the field of fault diagnosis because of its superior nonlinear fitting ability. Traditional methods of fault diagnosis manually realize feature extraction, so prior knowledge about fault information is needed, which increases the difficulty of analyzing the results. Neural networks, especially deep learning methods, can learn representations and patterns hierarchically from the input data, and realize effective feature extraction, so the deep learning method has the ability to model complex working conditions and output accurate predictions. Several typical deep learning methods have been successfully applied to fault diagnosis [23–26], including autoencoders [27], deep belief networks [28], and CNN [29]. The autoencoders and feature ensemble method is applied in actuator fault diagnosis [30]. Furthermore, the one-layer autoencoder-based neural network is proven to be effective in the task of fault classification [31]. The deep belief nets model is successfully applied for fault diagnosis in actuators using vibration signals [32]. One-dimensional CNN is used to analyze the raw time series data of the actuator and proved to be successful in diagnosing fault states [33], and a new CNN architecture based on LeNet-5 is set to process the bearing data set [34].

Considering that the output of sensor and actuator are similar when faults occur, the normal neural network fault diagnosis methods cannot exactly tell the difference between them. In this paper, the DCNN is used to diagnose sensor and actuator faults of robot joints. DCNN can extract the features from the input data and realize fault classification by increasing the depth of the network. In addition, flexible selection of convolution kernel width makes it an efficient way to deal with classification problems with weak characteristics. Actually, there may be many types of sensors and actuators; our research mainly focuses on the problems of fault diagnosis in position sensors for the robot joint and torque sensors for the actuator. The robot joint is forced to move in a sinusoidal trajectory with the control of actuator, and the position sensor feeds back corresponding signals under different sensor states. Position sensor and torque sensor are separately denoted by sensors and actuators in the following main text. The main contributions of this paper are as follows.

(1) This paper gives a fused sensor and actuator fault diagnosis model, where sensor and actuator fault could be expressed in one formulation. Still, different faults could be distinguished, which contributes to our study.
(2) This paper proposes a DCNN fault diagnosis method. There are several convolution blocks in the architecture, and the depth of each kernel on different blocks varies, which helps to extract features from the time domain of input data.
(3) Experiments with different neural network fault diagnosis methods, such as SVM, ANN, CNN, LTMN, are conducted and compared with DCNN to give a comparison.

The rest of this paper is organized as follows: Section 2 introduces the basic structure of DCNN, Section 3 introduces the neural network module training method based on deep CNN, simulation experiments are conducted in Section 4 and the results are compared, the authors conclude the paper at the end.

2. Basic Structure of DCNN

Generally, CNN consists of five parts: the convolutional layer, Batch normalization, activation layer, pooling layer, and dropout layer, as shown in Figure 1. In the CNN fault diagnosis architecture, each layer plays a different role. The following part of this section will briefly introduce the function of each layer.

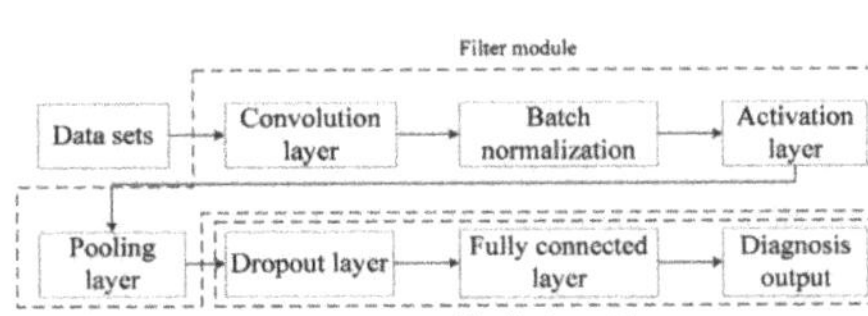

Figure 1. Diagram for CNN fault diagnosis system.

2.1. Convolution Layer

The convolution layer is one of the most important parts of CNN. It is an effective feature extraction method using a convolution operation. The expression of convolution operation in discrete format is as follows.

$$S(l,k) = X_l^k * W_l^k + b_l \tag{1}$$

where $S(l,k)$ is the output of convolution core, X_l^k is the input of convolution core, W_l^k is the core function, and b_l is the bias term. The numbers l, k are the serial number of the layer and convolution kernel. The mathematical operator $*$ in the above equation denotes the sum based on the multiplication of corresponding elements. Figure 2 gives an example of convolution operation with kernel dimension of 2×2. The input data with a dimension of 4×4 is divided into 9 subsets when the sliding step is 1, and each subset has the same dimension compared with the kernel. The multiplication of corresponding elements between kernel and subsets generates a 3×3 matrix, which is called the output of convolution core.

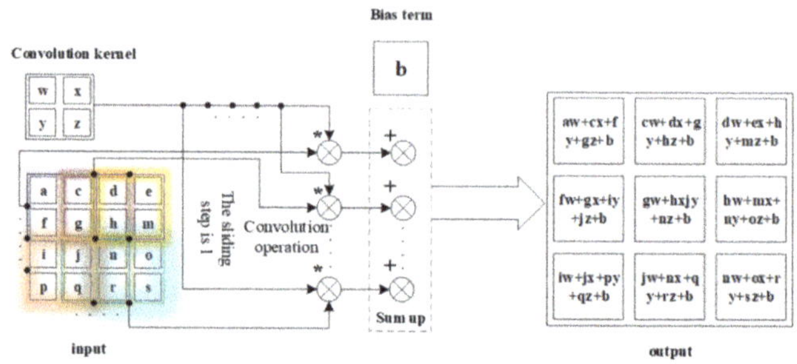

Figure 2. Diagram for convolution calculation.

2.2. Batch Normalization

Batch normalization is one of the most computing-intensive layers in CNN architecture. Meanwhile, it is also a widely used method to accelerate and stabilize the CNN training process. The main purpose of batch normalization is to force the input data sets back to the standard normal distribution with the mean value of zero and the variance of one, so that the input of the nonlinear transformation function falls into the sensitive region, which could avoid gradient loss.

The input of batch normalization comes from the output of the convolution core, and the output y_i of batch normalization is relative to data features, as shown in the following equation:

$$y_i = \gamma \hat{x}_i + \beta \tag{2}$$

where the gain γ and the bias term β are used to accelerate the convergence process. $\hat{x}_i$ is the function of mean value and standard deviation σ_β, see the following equation.

$$\hat{x}_i = \frac{1}{\sigma_\beta}(x_i - u_\beta) \tag{3}$$

$$\begin{cases} \sigma_\beta = \sqrt{\frac{1}{m}\sum\limits_{i=1}^{m}(x_i - u_\beta)^2 + \varepsilon} \\ u_\beta = \frac{1}{m}\sum\limits_{i=1}^{m} x_i \end{cases} \tag{4}$$

where ε is a very small positive value, x_i is the output of convolution operation module, m is the length of data sets.

2.3. Activation Layers

A convolutional neural network consists of stacked layers, with two basic parts on each layer; separately, they are trainable linear convolutional filters and a fixed nonlinear activation function. The activation function used in our research is ReLU [35], and its expression is as follows.

$$g(t) = \max(t, 0) = \begin{cases} 0, & t < 0 \\ t, & t > 0 \end{cases} \tag{5}$$

The ReLU function indicated by Equation (5) is a nonlinear function, where its derivative is one when $t > 0$, and zero when $t \leq 0$. The graph of the ReLU function is shown in Figure 3. The use of the ReLU function eliminates the problem of gradient vanishing compared with Sigmoid and tanh functions.

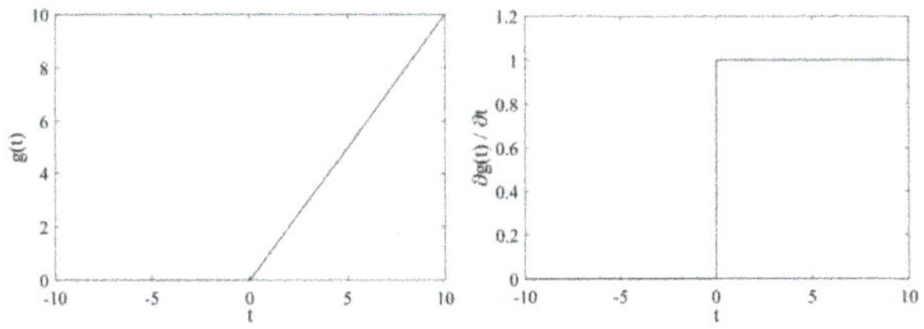

Figure 3. Graph of ReLU.

2.4. Pooling Layer

The pooling layer is used to reduce the amount of feature data needed and enhance the operational performance of the network. The main pooling methods could be classified into two categories, and they are maximum value pooling and mean value pooling methods. Figure 4 shows the basic operation of the above two pooling methods, where the input data dimension is 4×4, and the pooling kernel dimension is 2×2. The pooling kernel matrix is used to multiply with input data using the sliding step of 2. The maximum value pooling method chooses the maximum value in all four data with the same color, while the mean value pooling methods get the average value.

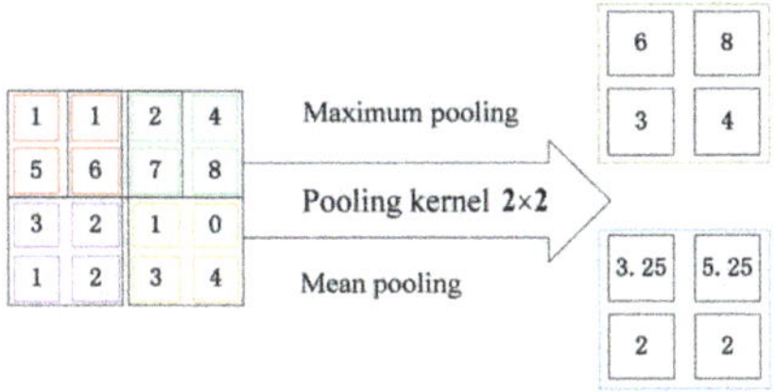

Figure 4. Operation of pooling process.

2.5. Dropout Layer

To alleviate the overfitting problem of the neural network, some neurons in the hidden layer are temporarily discarded according to a certain proportion in the training process of the neural network, and all neurons are restored when used again. When the rate of discard is fifty percent, each neuron has one half of the probability to be removed, so that the training of one neuron does not rely on another one, thus, the interaction between features is weakened, so as to alleviate the overfitting phenomenon caused by severe neural network depth [36]. The dropout process is shown in Figure 5.

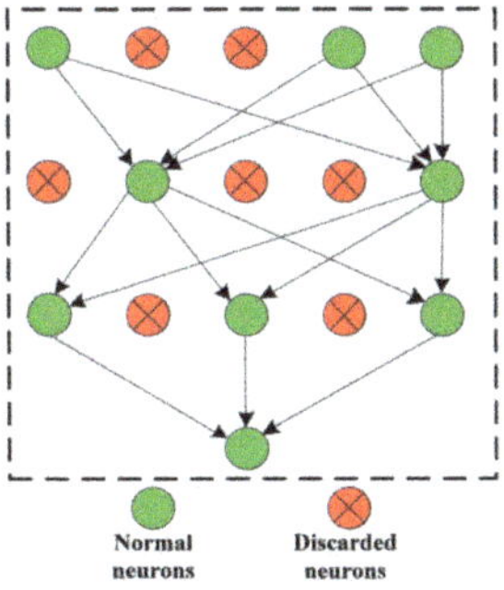

Figure 5. Process of dropout.

2.6. Fully Connected Layer

The fully connected layer is a classifier, which can map the features extracted from the filter modules to the marked data features [37]. The input of the fully connected layer is the output of the last pooling layer. Each neuron of the fully connected layer relates to all other input neurons. Then the data on each neuron is processed using the ReLU activation function, and the final output is classified through the Softmax regression method. Taking one dimensional convolution as an example, the full connection process is shown in Figure 6.

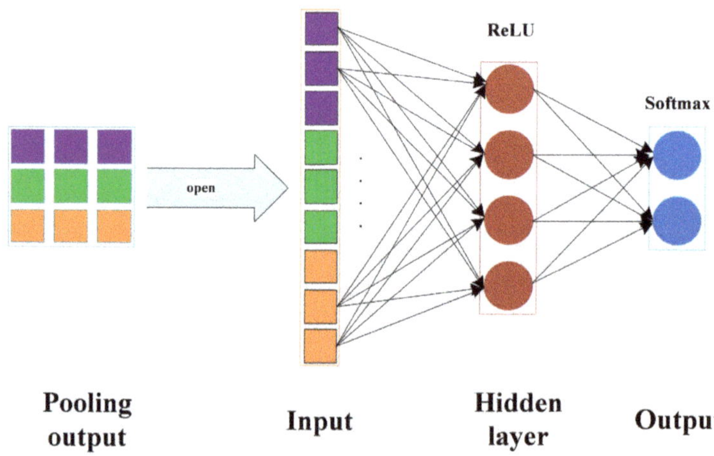

Figure 6. Full connection process.

2.7. Loss Function

There should be an evaluation method for the precision of the model during the training process. In order to make the prediction outputs of the neural network model consistent with the real values, cross-entropy function is calculated to evaluate the differences between them. The cross-entropy function is called loss function and its expression is as follows.

$$loss\_ce = -\sum_{n} P(y) \log P(\hat{y}) \tag{6}$$

where y denotes the known result, $\hat{y}$ is the output of the network, $P(y)$ is the probability distribution of known results, $P(\hat{y})$ is the probability distribution of network prediction results, and *loss_ce* is the cross-entropy function. The cross-entropy function emphasizes the difference in the probability distribution of each data class and is often used in the multi-classification problem of neural networks.

The softmax regression method makes the outputs of the neural network submit to established distribution. Assuming outputs are divided into n classes, then $(\hat{y}_1, \hat{y}_2, \ldots, \hat{y}_n)$ is obtained. When the Softmax function is used, the outputs of the neural network meet the desired probability distribution. The calculation of the loss function is shown in Figure 7.

$$Softmax(\hat{y}_i) = \frac{e^{\hat{y}_i}}{\sum_{j=1}^{n} e^{\hat{y}_j}} \tag{7}$$

$$\forall \hat{y}\ P(\hat{Y} = \hat{y}) \in [0,1] \cap \sum P(\hat{Y} = \hat{y}) = 1 \tag{8}$$

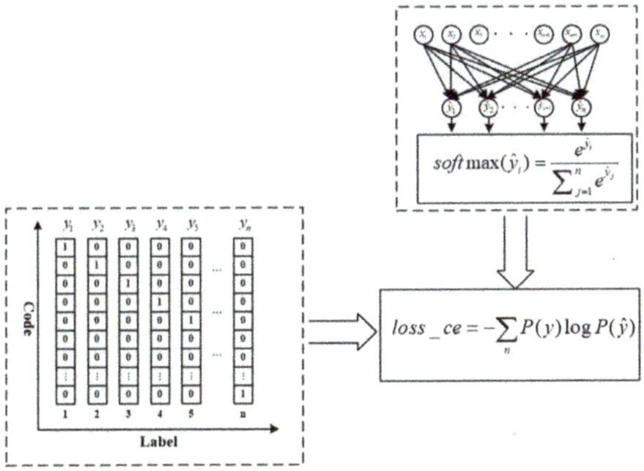

Figure 7. Calculation of loss function.

3. Sensor and Actuator Fault Diagnosis Framework Using DCNN

The proposed robot joint sensor and actuator fault diagnosis framework based on DCNN is shown in Figure 8. It is shown that the whole fault diagnosis framework can be divided into data fusion, multi-level feature extraction, dropout of neurons, full connection, and diagnosis output parts. There are six "Blocks" in the framework and each framework consists of the convolution layer, batch standardization layer, activation layer, and pooling layer.

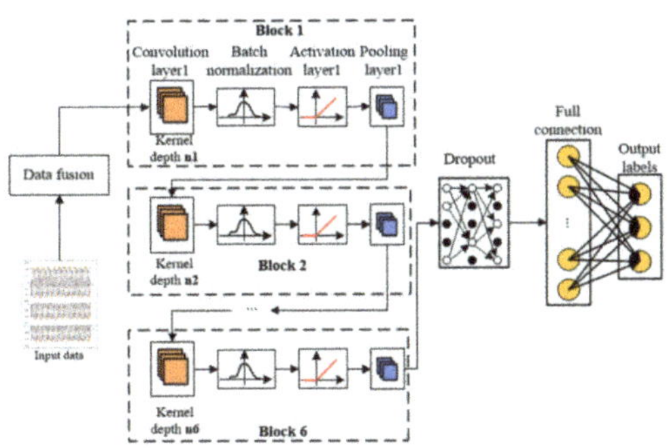

Figure 8. Fault diagnosis framework of DCNN.

The convolution layer parameters of Block1 are [32 * 1, 8, 4], where 32 * 1 represents the dimension of one-dimensional convolution kernel, convolution depth is 8, and sliding step is 4. The convolution parameters of Block2 to Block6 are [3 * 1, 16, 1], [3 * 1, 32, 1], [3 * 1, 64, 1], [3 * 1128, 1], [3 * 1128, 1]. The pooling parameters of Block1 are [2 * 1, 8, 2], where 2 * 1 represents the dimension of the pooling kernel, pooling depth is 8, and sliding step is 2. The pooled layer parameters from Block2 to Block6 are [2 * 1, 16, 2], [2 * 1, 32, 2], [2 * 1, 64, 2], [2 * 1128, 2], [2 * 1, 16, 2]. The pooled output of Block6 is used as the input of Dropout, and the rejection rate is set to 50% to slow down the over fitting of the model. The number of neurons in the full junction layer is 100, and they are connected to 10 categories.

3.1. Data Fusion

This paper aims at the fault diagnosis problem of robot sensor and actuator, so there are many kinds of input data that contain a variety of fault information. Therefore, this paper adopts a data fusion scheme to unify the sensor fault and actuator fault in one expression, so that the output of the fused model contains both sensor and actuator fault characteristics. The first thing to do is to establish the mathematical model of sensor and actuator respectively. According to the laws of mechanics, the mathematical model of the actuator is as follows.

$$S_1 : \begin{cases} \dot{x} = Ax + Bf(q, \dot{q}, \tau) \\ y = Ex \end{cases} \tag{9}$$

where,

$$A = \begin{bmatrix} 0 & 0 \\ 0 & 1 \end{bmatrix}, B = \begin{bmatrix} 0 \\ 1 \end{bmatrix}, E = \begin{bmatrix} 1 & 0 \\ 0 & 1 \end{bmatrix}, f(q, \dot{q}, \tau) = D^{-1}(q)[\tau - C(q, \dot{q})\dot{q} - G(q)],$$

$x = [x_1\ x_2]^T = [q\ \dot{q}]^T$, and $q, \dot{q} \in R^n$ are state variables of angular position and angular velocity.

From Equation (9), the torque equation of the actuator can be obtained, as follows.

$$D(q)\ddot{q} + C(q, \dot{q})\dot{q} + G(q) = \rho\tau + f_a \tag{10}$$

where τ is the vector for torque with a dimension of n; $D(q)$ and $C(q, \dot{q})$ are square matrices with the dimension of n, denoting inertia matrix and Coriolis force matrix, respectively; and $G(q) \in R^n$ is the gravity moment vector. $\rho \in [0,1]$ is the effective factor of the actuator. $\rho = 0$ means the actuator is completely broken. f_a is the bias term and its value is positively correlated with the degree of actuator damage. The actuator faults with different combinations of ρ and f_a are listed in Table 1.

Table 1. Actuator fault type.

ρ	f_a	**Fault Type**
1	Not zero	Constant deviation fault
$0 < \rho < 1$	0	Constant gain fault
0	Not zero	Actuator stuck
0	0	Actuator broken

It could be seen from Equation (9) that the output y variable of the system S_1 is the state variable x multiplied by the coefficient matrix E. Thus, the robot sensor fault can be directly expressed by the output equation, as follows.

$$S_2 : \begin{cases} \dot{x} = Ax + Bf(q, \dot{q}, \tau) \\ y = \lambda Ex + f_b \end{cases} \tag{11}$$

where $\lambda \in [0,1]$ is the effective factor of the sensor. $\lambda = 0$ means the sensor does not work anymore. f_b is the bias term and is positively correlated with the degree of sensor damage. The sensor faults with different combinations of λ and f_b are listed in Table 2.

Table 2. Sensor fault type.

λ	f_b	**Fault Type**
1	Not zero	Constant deviation fault
$0 < \lambda < 1$	0	Constant gain fault
0	Not zero	sensor stuck
0	0	sensor broken

From the robot joint sensor fault model and actuator fault model, it could be seen that both of them affect the system in a different way. However, through the model derivation and transformation, the sensor fault can be transformed into an actuator fault through the first-order filter [38], which simplifies the model. The sensor and actuator fault data are fused according to the following formula.

$$Fault = a\Delta Sensor + bActuator \tag{12}$$

where *Fault* denotes fault data set needed, $\Delta Sensor$ represents the difference between sensor output and settings, *Actuator* is the output of actuator, a and b are sensor and actuator faults coefficient respectively. Equation (12) has now unified two kinds of fault from sensor and actuator, which helps to obtain training data sets.

The robot sensor and actuator mixed faults studied in this paper are listed in Table 3. F1 to F10 will be used to represent different fault labels in the following description for ease of use.

Table 3. Investigated Sensor and actuator fault type.

Label	Fault Description
F1	Constant deviation fault of the actuator
F2	Constant gain fault of the actuator
F3	Actuator stuck
F4	Actuator broken
F5	Constant deviation fault for both sensor and actuator
F6	Constant deviation fault of the sensor
F7	Constant gain fault of the sensor
F8	Sensor stuck
F9	Sensor broken
F10	Sensor and actuator are normal

3.2. Training of Model and Diagnosis

The fault diagnosis model needs to be well trained before it is used to realize fault diagnosis. The basic training process of the fault diagnosis model based on the DCNN proposed in this paper is as follows.

Fused data containing single or mixed fault of sensor and actuator is input into Block1 (refer to Figure 8). Convolution layer1 uses kernel1 to carry out convolution operations. The result of the convolution operation is input into the batch standardization module, and the extracted data features are standardized to make the extracted feature data conform to the standard normal distribution. The activation function is then used to activate the neurons. The activation function used here is the ReLU function, and it owns the fine linear property, which overcomes the saturation effect of using Sigmoid or Tanh functions.

Finally, the activated features are input into the pooling layer. The pooling method used here is the maximum pooling method, which can extract the maximum features. The above operation completes the main steps of Block1. The output of Block1 is propagated to Block2, and the above steps conducted in Block1 are repeated until all six Blocks finished the corresponding operation.

After feature extraction is finished, the data should flow into the fully connected layer. however, in order to speed up the training process, the dropout layer is introduced to inactivate some neurons with a certain probability, and then the rest of the neurons come into the fully connected layer, and finally the fault diagnosis model gives the prediction labels.

In the training stage of the model, updating the network weight parameters is key. Figure 9 gives the flowchart of the parameters updating process, where the network outputs are evaluated using the loss function to determine the direction of parameter update. After the model is well trained, the real-time data from the robot sensor and actuator can be propagated into the fault diagnosis model to realize fault diagnosis.

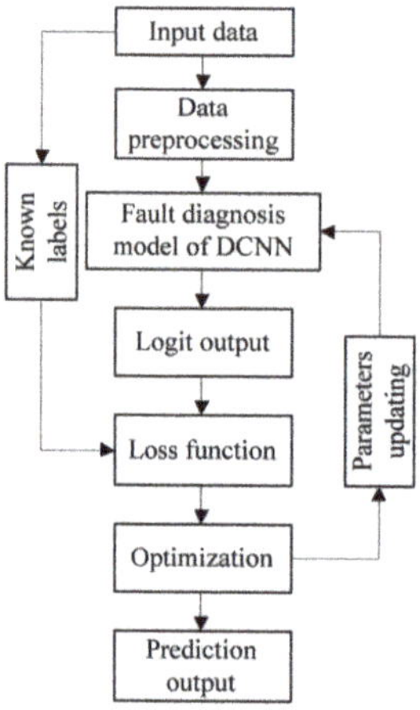

Figure 9. Process of training based on DCNN.

4. Experiment and Analysis

Data needed is based on multi-joint cooperative robot AUBO i3, with a DOF of 6, and six revolute joints with a maximum working radius of 625 mm. Our Matlab model is constructed based on the above platform. There are several position sensors and one actuator for each joint, so it is necessary to monitor their working state. Experiments using different fault diagnosis methods are conducted to reveal the effectiveness of the proposed DCNN. A PC with i7-10510U 1.8Ghz processor and 16G of RAM is used. The PyCharm software is installed, and combined with a Python3.6 interpreter. All the algorithms are implemented on Keras with Tensorflow as its backend.

The basic architecture of several experiments is shown in Figure 10, where the fused data is expanded through the data set enhancement method, and then several neural network fault diagnosis methods are investigated and their results are compared.

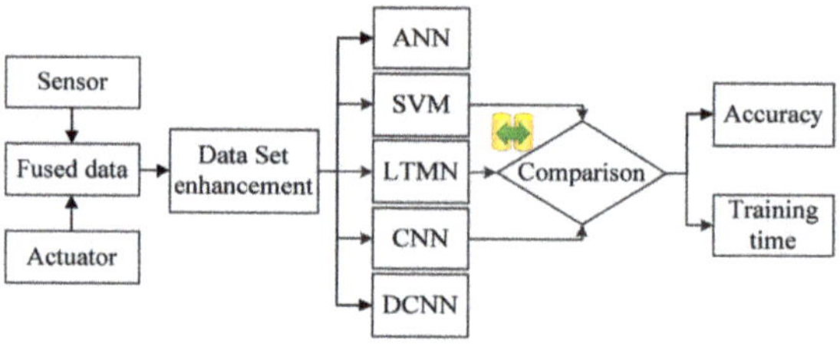

Figure 10. The basic architecture of the Comparative experiments.

4.1. Data Sets Enhancement

Considering that the amount of fault sample is not enough in a real robot joint system, in this paper, the data set enhancement method is used to expand the fault sample data sets and improve the generalization ability of the model. In order to expand the acquired robot joint fault data, a sliding sampling data set enhancement method is proposed in this paper, and the schematic diagram of data set enhancement is shown in Figure 11.

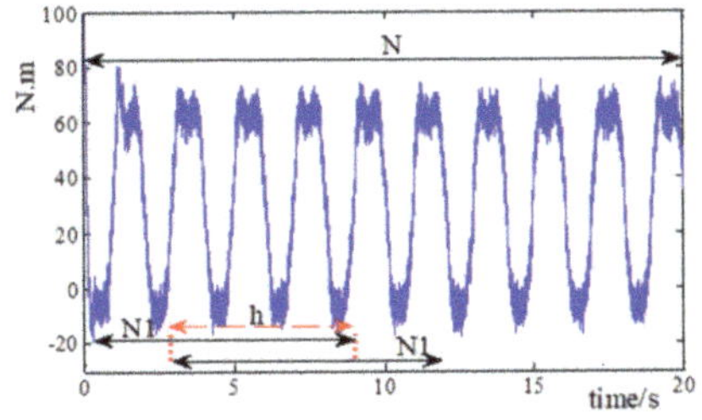

Figure 11. Diagram of data set enhancement.

The system data over a period is obtained, and a sample data segment with N_1 points is needed for single training. Assuming that the length of the data obtained is N, then the network could be trained by N/N_1 times according to the above method. In order to expand the coefficient of utilization for data, the start point of the second data set is h backwards compared to the first one, and the rest is roughly the same. The difference of samples before and after the data set enhancement method is used is shown in the following equation.

$$\frac{N-h-1}{N_1-h-1}-\frac{N}{N_1}=\frac{N-N_1}{N_1(N_1-h-1)}\geq 0 \tag{13}$$

It is obvious that when the sliding step size is small, which means h is quite large, more data samples could be obtained, which can well meet the needs of data sets in the training process. Here in our research, the sliding step selected is h = 29, and thus we could obtain plenty of data samples for model training and validation.

We have constructed a robot fault model in MATLAB/Simulink according to Equation (12), and the fault data needed in our study is obtained. In the process of data acquisition, the sampling rate is set to1000 hertz, and the sampling time is 8 s, so 8000 sample points for each kind of conditions are obtained, see Figure 12. The aforementioned data set enhancement method is adopted, so we can get 2000 samples for each kind of conditions. The 2000 samples are divided into three parts, and they are training, verification, and testing samples respectively, and the proportion of the above samples are 70%, 20%, and 10%, which would be used in the later model training and verification process. It should be noted that the input data set is one-dimensional, which is different from the conventional two-dimensional image data.

From Figure 12, it is very clear to see that some of the fault types could be easily distinguished, such as F1, F3; F2, F9. while some of them could not, such as F3, F4. When the "F3" fault happens, if the diagnosis model output is "F4", it may bring adverse effects. Thus, realizing high accuracy prediction makes sense.

4.2. Hyper Parameters of DCNN

Hyper parameters of the neural network include learning rate, regularization parameters, and iterations. Actually, these hyper parameters control the values of the weight coefficient. According to existing researches, hyper parameters in deep learning algorithms not only affect the performance of the algorithm itself, but also affect the expression ability of the trained model. This paper takes the advice of Bengio [39]. The corresponding hyper parameters are set according to whether these hyper parameters can increase or decrease the capacity of the model.

In the process of parameter updating, the exponential decay learning rate is adopted. At first, a large learning rate is used to get the optimal solution quickly, and then the learning rate is gradually reduced to keep the model stable in the later stage of training. The initial learning rate η_0 is set to 0.2, and the decay learning rate ξ is set to 0.99, so the decay rate is updated per round. The expression for decay rate is as follows.

$$\begin{cases} \eta = \eta_0 \times \xi^{(H/L)} \\ H = Epoch/Batch\_k \end{cases} \tag{14}$$

where η denotes exponential decay learning rate, H stands for the number of the current round, and L represents the turns the decay should be executed once, *Batch_k* is the number of iterations. When a complete data set passes through the neural network once and then returns, this process is called *Epoch*.

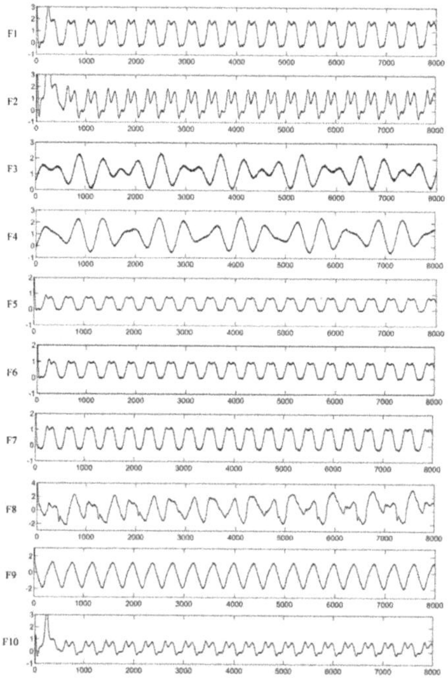

Figure 12. Data samples under different working conditions.

In order to alleviate the overfitting of the neural network, the l_2 regularization method is used in this paper. Regularization is to introduce the model complexity index into the loss function and suppress the noise in the training data set by weighting the parameters in the neural network. The expression of the loss function is as follows.

$$\begin{cases} Loss = Loss\_all + REGULARIZER \times \\ Loss(w) \\ Loss(w) = \sum_i |w_i| \end{cases} \tag{15}$$

where *Loss_all* represents the loss function of all parameters in the model, *REGULARIZER* is the regularization weight, w generally refers to the weight in the forward propagation of neural network, and *Loss(w)* is the result of l_2 regularization of parameter w.

The Adma optimal algorithm is used [40] and the procedure of weight updating is as follows.

Step 1: give the iteration step $\varepsilon = 0.001$.

Step 2: set the decay rate for matrix calculation, $\rho_1 = 0.99, \rho_2 = 0.999$.

Step 3: Determine the convergence threshold $\delta = 10^{-8}$.

Step 4: Initialize network weight θ, 1st and 2st moment variables s, r, and set s = 0, r = 0.

Step 5: Set the simulation time step to 0.0001.

Step 6: Small data set with m samples are collected from the training set, using $\{x^{(1)},x^{(2)},\ldots,x^{(\mathrm{m})}\}$. to denote it, and set corresponding goals $y^{(i)}$.

Step 7: Calculate gradient variable $g \leftarrow \frac{1}{m}\nabla_{\theta}\sum_{i} L\left(f\left(x^{(i)};\theta\right),y^{(i)}\right)$, and update biased first moment estimation $s \leftarrow \rho_1 s + (1-\rho_1)g$ as well as biased second moment estimation $r \leftarrow \rho_2 r + (1-\rho_2)g$.

Step 8: Correct the deviation of the first moment $\hat{s} \leftarrow \frac{s}{1-\rho_1^t}$ and deviation of the second moment $\hat{r} \leftarrow \frac{r}{1-\rho_2^t}$.

Step 9: Calculate incremental weight error $\Delta\theta = -\varepsilon\frac{\hat{s}}{\sqrt{\hat{r}+\delta}}$, and update it to the network weight $\theta \leftarrow \theta + \Delta\theta$.

Step 10: If the convergence threshold in Step 3 is not met, then, back to Step 6, otherwise end the iterative process.

4.3. Simulation and Results

In order to verify the feasibility and effectiveness of the DCNN used in this paper for robot joint sensor and actuator fault diagnosis, the ANN, SVM, CNN, and LTMN methods are studied for comparative analysis and verification. The diagnosis accuracy of the different network is shown in Figure 13.

Unit: %

Methods	F1	F2	F3	F4	F5	F6	F7	F8	F9
ANN	66.82	67.13	67.35	66.78	66.72	66.88	68.05	67.89	67.83
SVM	64.73	65.52	65.34	64.89	64.12	65.38	65.15	64.58	65.33
CNN	99.12	98.56	98.87	98.64	99.23	98.31	98.69	98.79	99.23
LTMN	99.89	99.91	99.89	99.90	99.87	99.91	99.88	99.93	99.89
DCNN	99.97	99.99	99.96	99.98	99.96	99.98	99.99	99.97	99.98

Fault Labels

Figure 13. Fault diagnosis accuracy on training process of the model.

As shown in Table 3, There are nine fault states for sensor and actuator. For each fault state F1 to F9, 2000 samples are obtained and then divided into training, testing, and verification data subset. The accuracy of different diagnosis methods is summarized in Figure 13. It can be seen that the accuracy of fault recognition using DCNN is significantly improved compared with ANN, SVM. The average accuracy of the DCNN in the training process is over 99%. However, it is worth noting that the lowest accuracy of the five methods appears in F5. It could be interpreted that F5 is a mixed fault of sensor and actuator, and Figure 12 shows that there is no obvious difference between the fault curves of sensor and actuator, so the fault diagnosis models cannot effectively tell the difference between them.

The confusion matrix of DCNN used on robot sensor and actuator fault diagnosis is shown in Figure 14. It could be seen that none of the fault types could be 100% recognized, and meanwhile, confusion matrix data shows that between that "Misjudgment" categories, their waveforms seem alike.

Further experiment research is conducted to compare the fault diagnosis effect of CNN, LTMN, and DCNN on robot joint sensor and actuator faults, and the accuracy and loss function figure of each diagnosis method in the training set and testing set are drawn with the help of TensorFlow.

Three of five fault diagnosis methods with an accuracy over 90% are investigated. From Figure 15, it can be seen that the LTMN and DCNN could realize a fault recognition rate of 100% on both training and testing data sets, and there is no gap between training and testing loss function, which shows the robustness of LTMN and DCNN.

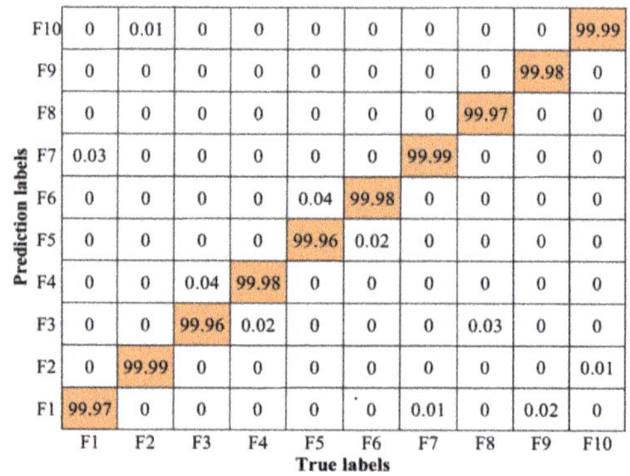

Figure 14. Confusion matrix of DCNN.

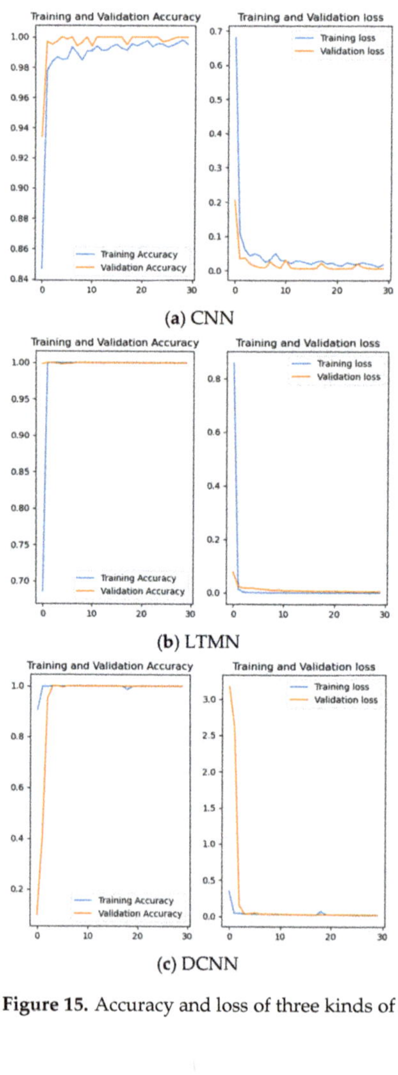

Figure 15. Accuracy and loss of three kinds of neural networks.

The model training time and accuracy are shown in Figure 16. This shows DCNN needs fewer training time but still gets the maximum diagnosis accuracy, which proves the high performance of the proposed DCNN fault diagnosis method.

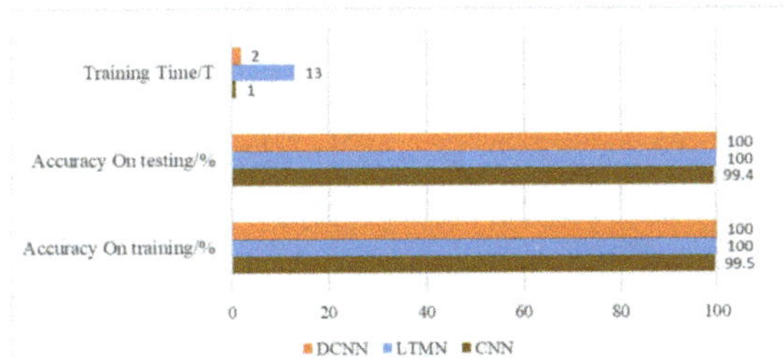

Figure 16. Accuracy and training time of three kinds of neural networks.

The initial value of the neural network is randomly given, in order to eliminate the influence of accidental factors. This paper uses the cross method to train the model 10 times, and each time the network is initialized with a random value. The accuracy of CNN, LTMN, and DCNN are shown in Figure 17.

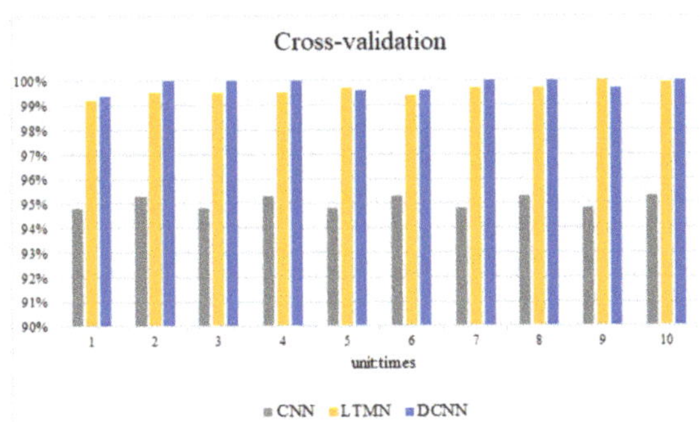

Figure 17. Cross-validation result.

As can be seen from Figure 17, the recognition accuracy of DCNN in each experiment is more than 99%, and its lowest accuracy is 99.6%. Thus, the model initial value has little effect on fault diagnosis accuracy, which also proves the robustness of DCNN.

5. Conclusions

The DCNN fault diagnosis method is used to recognize the sensor and actuator faults of the robot system. The robot sensor and actuator output data are fused. In order to increase the number of training samples, the fault data set is expanded way of the data set enhancement method, and then the fault diagnosis is carried out using a deep convolution neural network. SVM, ANN, CNN, and LTMN-based neural network fault diagnosis methods are compared with the proposed DCNN and conclusions can be drawn that DCNN can better extract the fault information from the original input data and makes a more accurate classification of the sensor and actuator fault types.

Author Contributions: Conceptualization, J.P.; methodology, J.P., L.Q. and K.P.; software, L.Q.; validation, J.P.; formal analysis, K.P.; investigation, J.P. and K.P.; resources, L.Q. and K.P.; data curation, J.P. and L.Q.; writing—original draft preparation, J.P.; writing—review and editing, J.P. and K.P.; visualization, K.P. and L.Q.; supervision, L.Q.; project administration, L.Q.; funding acquisition, L.Q. and K.P. All authors have read and agreed to the published version of the manuscript.

Funding: This research was funded by the National Key R&D Program of China, grant number 2019YFB1309900.

Data Availability Statement: The data presented in this study are available on request from the corresponding author.

Conflicts of Interest: The authors declare no conflict of interest.

References

1. Robla-Gomez, S.; Becerra, V.M.; Llata, J.R.; Gonzalez-Sarabia, E.; Perez-Oria, J. Working together: A review on safe human-robot collaboration in industrial environments. *IEEE Access* **2017**, *5*, 26754–26773. [CrossRef]
2. Zhang, S.; Wang, S.; Jing, F.; Tan, M. A sensorless hand guiding scheme based on model identification and control for industrial robot. *IEEE Trans. Ind. Inform.* **2019**, *15*, 5204–5213. [CrossRef]
3. Ceriani, N.M.; Zanchettin, A.M.; Rocco, P.; Stolt, A.; Robertsson, A. Reactive task adaptation based on hierarchical constraints classification for safe industrial robots. *IEEE/ASME Trans. Mechatron.* **2015**, *20*, 2935–2949. [CrossRef]
4. Sabry, A.H.; Nordin, F.H.; Sabry, A.H.; Ab-Kadir, M. Fault detection and diagnosis of industrial robot based on power consumption modeling. *IEEE Trans. Ind. Electron.* **2019**, *67*, 7929–7940. [CrossRef]
5. Chih-Min, C.M.L.; Boldbaatar, E.A. Fault accommodation control for a biped robot using a recurrent wavelet elman neural network. *IEEE Syst. J.* **2017**, *11*, 2882–2893.
6. Sun, X.; Jia, X. A fault diagnosis method of industrial robot rolling bearing based on data driven and random intuitive fuzzy decision. *IEEE Access* **2019**, *7*, 148764–148770. [CrossRef]
7. Kim, J. Multi-Axis Force-Torque Sensors for Measuring Zero-Moment Point in Humanoid Robots: A Review. *IEEE Sens. J.* **2020**, *20*, 1126–1141. [CrossRef]
8. Zhao, B.; Skjetne, R.; Blanke, M.; Dukan, F. Particle Filter for Fault Diagnosis and Robust Navigation of Underwater Robot. *IEEE Trans. Control. Syst. Technol.* **2014**, *22*, 2399–2407. [CrossRef]
9. Liu, W.; Chen, Y. Integrated Design of Fault Diagnosis and Fault- Tolerant Control for Spacecraft Autonomous Rendezvous. *Aerosp. Control.* **2015**, *63*, 30–39.
10. Xu, X.; Sun, Y.; Tian, X.; Zhou, L. A novel joint angle estimation method for serial manipulator using mems sensors. *IEEE Trans. Ind. Electron.* **2020**, *67*, 10610–10620. [CrossRef]
11. Caccavale, F.; Marino, A.; Muscio, G.; Pierri, F. Discrete-Time Framework for Fault Diagnosis in Robotic Manipulators. *IEEE Trans. Control. Syst. Technol.* **2013**, *21*, 1858–1873. [CrossRef]
12. Sneider, H.; Frank, P.M. Observer-based supervision and fault detection in robots using nonlinear and fuzzy logic residual evaluation. *IEEE Trans. Control. Syst. Technol.* **1996**, *4*, 274–282. [CrossRef]
13. Brambilla, D.; Capisani, L.M.; Ferrara, A.; Pisu, P. Fault detection for robot manipulators via second-order sliding modes. *IEEE Trans. Ind. Electron.* **2008**, *55*, 3954–3963. [CrossRef]
14. Brambilla, D.; Capisani, L.M.; Ferrara, A.; Pisu, P. Second order sliding mode observers for Fault Detection of robot manipulators. In Proceedings of the IEEE Conference on Decision & Control, Cancun, Mexico, 9–11 December 2008.
15. Ding, S.; Chen, W.H.; Mei, K.; Murray-Smith, D.J. Disturbance observer design for nonlinear systems represented by input–output models. *IEEE Trans. Ind. Electron.* **2020**, *67*, 1222–1232. [CrossRef]
16. Yang, X.; Wei, P.; Zhang, Y. Disturbance observer based on biologically inspired integral sliding mode control for trajectory tracking of mobile robots. *IEEE Access* **2019**, *7*, 48382–48391. [CrossRef]
17. Guo, J. Robust Tracking Control of Variable Stiffness Joint Based on Feedback Linearization and Disturbance Observer with Estimation Error Compensation. *IEEE Access* **2020**, *8*, 173732–173754. [CrossRef]
18. Chen, W.H.; Yang, J.; Guo, L.; Li, S. Disturbance-observer-based control and related methods—an overview. *IEEE Trans. Ind. Electron.* **2016**, *63*, 1083–1095. [CrossRef]
19. Yuan, Y.; Ma, S.; Wu, J.; Jia, B.; Luo, X. Fault diagnosis in gas insulated switchgear based on genetic algorithm and density- based spatial clustering of applications with noise. *IEEE Sens. J.* **2019**, *21*, 965–973. [CrossRef]
20. Borges, F.; Pinto, A.; Ribeiro, D.; Barbosa, T.; Ferreira, D. An unsupervised method based on support vector machines and higher-order statistics for mechanical faults detection. *IEEE Lat. Am. Trans.* **2020**, *18*, 1093–1101. [CrossRef]
21. Shi, J.; He, Q.; Wang, Z. Gmm clustering-based decision trees considering fault rate and cluster validity for analog circuit fault diagnosis. *IEEE Access* **2019**, *7*, 140637–140650. [CrossRef]
22. Li, C.; Xiong, J.; Zhu, X.; Zhang, Q.; Wang, S. Fault diagnosis method based on encoding time series and convolutional neural network. *IEEE Access* **2020**, *8*, 165232–165246. [CrossRef]
23. Rui, Z.; Yan, R.; Chen, Z. Deep learning and its applications to machine health monitoring. *Mech. Syst. Signal Process.* **2019**, *115*, 213–237.
24. Ding, X.; He, Q. Energy-Fluctuated Multiscale Feature Learning With Deep ConvNet for Intelligent Spindle Bearing Fault Diagnosis. *IEEE Trans. Instrum. Meas.* **2017**, *66*, 1926–1935. [CrossRef]
25. Chen, Z.; Li, W. Multisensor Feature Fusion for Bearing Fault Diagnosis Using Sparse Autoencoder and Deep Belief Network. *IEEE Trans. Instrum. Meas.* **2017**, *66*, 1693–1702. [CrossRef]
26. Wang, P.; Yan, R. Virtualization and deep recognition for system fault classification. *J. Manuf. Syst.* **2017**, *44*, 310–316. [CrossRef]

27. Vincent, P.; Larochelle, H.; Bengio, Y. Extracting and Composing Robust Features with Denoising Autoencoders. In Proceedings of the Twenty-Fifth International Conference, Helsinki, Finland, 5–9 June 2008.
28. Hinton, G.E.; Osindero, S.; The, Y.W. A fast learning algorithm for deep belief nets. *Neural Comput.* **2006**, *18*, 1527–1554. [CrossRef] [PubMed]
29. Rawat, W.; Wang, Z. Deep Convolutional Neural Networks for Image Classification: A Comprehensive Review. *Neural Comput.* **2017**, *29*, 2352–2449. [CrossRef]
30. Wang, J.; Sun, C.; Zhao, Z. Feature ensemble learning using stacked denoising autoencoders for induction motor fault diagnosis. In Proceedings of the Prognostics & System Health Management Conference, Harbin, China, 9–12 July 2017.
31. Sun, W.; Shao, S.; Zhao, R. A sparse auto-encoder-based deep neural network approach for induction motor faults classification. *Measurement* **2016**, *89*, 171–178. [CrossRef]
32. Shao, S.Y.; Sun, W.J.; Yan, R.Q. A Deep Learning Approach for Fault Diagnosis of Induction Motors in Manufacturing. *Chin. J. Mech. Eng.* **2017**, *30*, 1347–1356. [CrossRef]
33. Ince, T.; Kiranyaz, S.; Eren, L. Real-Time Motor Fault Detection by 1D Convolutional Neural Networks. *IEEE Trans. Ind. Electron.* **2016**, *63*, 7067–7075. [CrossRef]
34. Wen, L.; Li, X.; Gao, L.; Zhang, Y. A New Convolutional Neural Network-Based Data-Driven Fault Diagnosis Method. *IEEE Trans. Ind. Electron.* **2018**, *65*, 5990–5998. [CrossRef]
35. Wang, G.; Giannakis, G.B.; Chen, J. Learning relu networks on linearly separable data: Algorithm, optimality, and generalization. *IEEE Trans. Signal Process.* **2019**, *67*, 2357–2370. [CrossRef]
36. Lian, Z.; Jing, X.; Wang, X.; Hai, H.; Tan, Y.; Cui, Y. Dropconnect regularization method with sparsity constraint for neural networks. *Chin. J. Electron.* **2016**, *25*, 152–158. [CrossRef]
37. Yang, J.; Li, J. Application of deep convolution neural network. 2017 14th International Computer Conference on Wavelet, Active Media. In Proceedings of the Technology and Information Processing (ICCWAMTIP), Chengdu, China, 15–17 December 2017.
38. Zhao, B.; Li, Y.C. Active substituting decentralized fault-tolerant control for reconfigurable manipulators with multi-sensor failures. *Control. Decis.* **2014**, *29*, 226–230.
39. Bengio, Y. Practical recommendations for gradient-based training of deep architectures. *Neural Netw. Tricks Trade* **2012**, *13*, 437–478.
40. Kingma, D.; Ba, J. Adam: A method for stochastic optimization. In Proceedings of the International Conference on Learning Representations, San Diego, CA, USA, 7–9 May 2015.

entropy

MDPI

Article

X-ray Pulsar Signal Denoising Based on Variational Mode Decomposition

Qiang Chen, Yong Zhao and Lixia Yan *

School of Automation Science and Electrical Engineering, Beihang University (BUAA), Beijing 100191, China; qiangchen@buaa.edu.cn (Q.C.); zhaoyong1996@buaa.edu.cn (Y.Z.)
* Correspondence: yanlixia@buaa.edu.cn

Abstract: Pulsars, especially X-ray pulsars detectable for small-size detectors, are highly accurate natural clocks suggesting potential applications such as interplanetary navigation control. Due to various complex cosmic background noise, the original pulsar signals, namely photon sequences, observed by detectors have low signal-to-noise ratios (SNRs) that obstruct the practical uses. This paper presents the pulsar denoising strategy developed based on the variational mode decomposition (VMD) approach. It is actually the initial work of our interplanetary navigation control research. The original pulsar signals are decomposed into intrinsic mode functions (IMFs) via VMD, by which the Gaussian noise contaminating the pulsar signals can be attenuated because of the filtering effect during signal decomposition and reconstruction. Comparison experiments based on both simulation and HEASARC-archived X-ray pulsar signals are carried out to validate the effectiveness of the proposed pulsar denoising strategy.

Keywords: X-ray pulsar; signal denoising; variational mode decomposition

Citation: Chen, Q.; Zhao, Y.; Yan, L. X-ray Pulsar Signal Denoising Based on Variational Mode Decomposition. *Entropy* **2021**, *23*, 1181. https://doi.org/10.3390/e23091181

Academic Editor: Quanmin Zhu

Received: 31 July 2021
Accepted: 2 September 2021
Published: 8 September 2021

1. Introduction

Pulsars are rapidly rotating neutron stars that emit electromagnetic signals and have periods ranging from milliseconds to thousands of seconds [1,2]; see Figure 1 for an illustration.

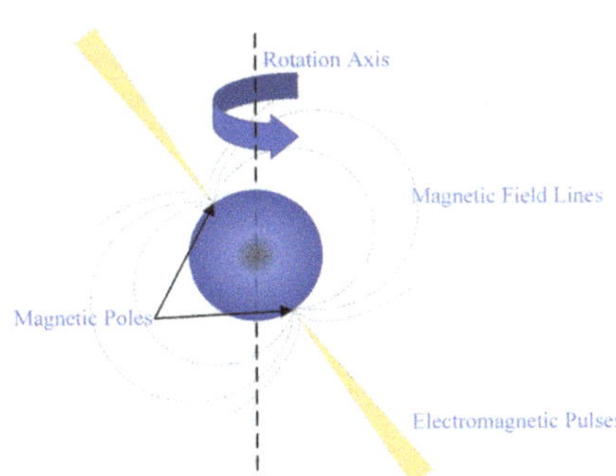

Figure 1. Illustration of the pulsar model.

The pulsar emission mechanism is complex, featuring chaotic characteristics [3,4]. Various methods have been developed to calculate the Lyapunov exponent and its complementary measures to characterize the time scales on which chaotic systems, such as pulsars, become unpredictable [5–7]. Luckily, the potential applications of pulsars, especially accurate pulsar timing features, do not rely on a comprehensive understanding of pulsar emission models. Researchers have found that the rotation periods of pulsars over long timescales are as precise as the state-of-art terrestrial atomic clock, which suggests pulsars are the perfect choice for precise timing and autonomous interplanetary navigation [8–11]. The precise timing properties of pulsars can be used for unmanned out-terrestrial vehicles, for instance, incorporating advanced robust control algorithms, such as those reported in [12,13], with the pulsar TOA sequence as well as the pulsar positioning algorithm. Nowadays, there are already over 2000 known pulsars that build a firm base for governmental and civilian use, as well as space applications not only for the military but also for astronomy exploration [14,15].

Among all types of pulsars, X-ray pulsars appear to be the most favorable ones as they are detectable by small-size detectors [1,8,9,16–19]. However, the X-ray pulsar radiation, namely the effective signals, would degenerate into photon sequences with very low intensity when it arrives at the terrestrial detector after long-distance space travel. An even more troublesome problem arises due to the complex, noisy photons, up to about nine times that of the effective photons in quantity, including the diffuse X-ray background and the cosmic X-ray background, which inherently contend with the originally observed signal [11]. These facts imply the low signal-to-noise (SNR) ratio of the pulsar signal and, hence, signify the necessity of pulsar denoising research, especially when the observation time is short.

The pulsar denoising refers to filtering the pulsar profile, generally obtained via epoch folding performed on photon sequences, that stores the time of arrivals (TOAs) of photons [1,17,20]. The Fourier filtering in the frequency domain is the original attempt to denoise pulsar profiles, which has already been proven invalid when the signal profile comprises nonsinusoidal components or nongaussian noise [16,18,21]. The modified kernel regression denoising method reported in [22] develops the second-order derivative compensation and reveals that a clear positive relationship between the TOA accuracy and the profile SNR does not exist. In [23], the wavelet analysis method using noise dispersion to determine the threshold value is applied to denoise the pulsar profile via the wavelet inverse transform. The denoising strategy based on biorthogonal lifting wavelet transforms (WT) reported in [17] studied the statistical properties of the X-ray background noise. The core idea of wavelet transform denoising is choosing an appropriate wavelet library and decomposition level in terms of finding a suitable wavelet basis and threshold function [24]. In particular, the wavelet basis plays an essential role in denoising performance as it is the base of wavelet denoising. However, at present, there is no universal wavelet basis that accommodates different pulsar signals. It has also been shown in [21] that the wavelet denoising methodology heavily relies on expertise and experience and is not compatible with an average processor. New approaches to increase the SNR of X-ray pulsar signals can be found in [25] with a machine learning method and in [26] with a recurrent neural network method, respectively. These two kinds of denoising strategies actually require high computational costs too.

Recently, empirical mode decomposition (EMD) has achieved great success in removing both white and fractional Gaussian noise by reconstructing the signal with predetermined thresholding intrinsic mode functions (IMFs) [27]. Nevertheless, the traditional EMD framework only allows for analyzing the single signal due to possible mixing phenomena when analyzing multiple signals simultaneously. This problem can be overcome by the multivariate strategy reported in [28], in which multiple pulsar signals are processed at the time while the mode mixing is avoided. The ensemble empirical mode decomposition (EEMD) synthesizes white Gaussian noise in the input signal for the later decomposition, in which the average operation can also avoid the mode mixing problem that existed in the output IMFs after EMD [29–31]. Given the limitation from the fixed threshold functions, the adaptive threshold mechanism together with the EMD denoising method shown in [32] remarkably increases the SNR of X-ray pulsar profiles. Via creating the Hausdorff measures between the probability density functions of the input signal and the mode, the filtering approach in [9] aliases and reconstructs the X-ray pulsar profiles to achieve the purpose of denoising. Though the EMD framework allows for analyzing nonstationary and nonlinear signals, it lacks a firm mathematical foundation and is sensitive to sampling noise, which as a particular result, would lead to mode mixing when it is applied on pulsar denoising processing. Additionally, one might find that eliminating the mode mixing problem would build a firm base for performing autonomous mode decomposition and signal denoising, which is essential for embedding advanced control algorithms, such as in [33], into the pulsar-based spacecraft navigation system. Regarding the possible deficiencies of EMD on pulsar signal denoising, Dragomiretskiy developed a novel signal decomposition technique called variational mode decomposition (VMD) that nonrecursively decompose the signal into several quasi-orthogonal IMFs [34]. The VMD has already been applied in many areas such as seismic wave analysis [35], pipeline leakage detection [36] and vibration estimation of rotor-stator machinery [37]. For VMD-based applications, it is important to determine an accurate number of modes that have an essential effect on denoising performance [38]. The detrended fluctuation analysis (DFA) appears to be a favorable choice for this issue as it is a systematic scaling analysis approach to evaluate long-term dependency for nonstationary signal series [35,36]. In our previous work [39], a brief VMD-based denoising algorithm was developed for an X-ray pulsar profile after epoch folding without applying DFA to determine the mode number. To the authors' best knowledge, no literature reported had shown a detailed discussion about applying the VMD framework on denoising X-ray pulsar signals.

Motivated by the discussion above, this brief undertakes further endeavors on the research of denoising designs of X-ray pulsar signals so as to increase the SNR of the pulsar profiles and pave the way for potential applications. More precisely, a VMD-based denoising strategy is developed. We first perform the epoch folding method on the faint X-ray pulsar signals and obtain the noisy pulsar profile, followed by applying DFA to obtain the number of IMFs, after which the reconstruction of the pulsar signals achieves the denoising goal.

The contribution of this paper includes presenting the denoising algorithm for X-ray pulsar signals based on the VMD method. Compared with those in [18,20,40], the prior knowledge of pulsar profiles is unnecessary for the proposed VMD-based pulsar denoising strategies. In comparison with wavelet analysis [17,23,24], the denoising algorithm in this paper removes the reliance on choosing perfect basis and threshold functions. The VMD-based design in this paper also allows for processing the original pulsar signals that contain nonstationary background noise derived from many sources without leading to mode mixing problems, like that of the EMD-based analysis in [9,28,32].

The rest of the paper is organized as follows. Section 2 introduces basic knowledge of the VMD framework. Section 3 involves the VMD-based denoising design for the contaminated faint X-ray pulsar signals. Section 4 presents the comparison experiments with simulation and measured data. Section 5 concludes this work briefly.

2. Preliminaries

Variational Mode Decomposition

Variational Mode Decomposition (VMD) is a novel framework for signal processing that outperforms the traditional decomposition approaches, allowing for a systematic analysis of nonstationary and nonlinear signals [34]. It is built upon classical theories such as the Wiener filter, Hilbert transformation, frequency mixing, and decomposition method. Compared with the traditional empirical mode decomposition (EMD) technique, the VMD theoretically eliminates the potential problem of mode mixing by decomposing the signal into a sum of IMFs with the limited center frequency and bandwidth calculated analytically. It is worth noting that the reconstruction of the VMD-based decomposed signals, namely summing the obtained IMFs, would attenuate noise and increase SNR.

The VMD-based decomposition performed on the input signal f intends to achieve

$$\min_{\{h_k\},\{\omega_k\}} \left\{ \sum_{k=1}^{K} \left\| \frac{\partial}{\partial t} \left[\left(\delta(t) + j\frac{1}{\pi t} \right) * h_k(t) \right] e^{-j\omega t} \right\|_2^2 \right\}, \tag{1}$$

so that the f can be reconstructed by the sum $\sum_{k=1}^{K} h_k = f$, where $\{h_k\}$ and $\{\omega_k\}$ denote the sets of all modes and center frequencies, respectively; $\delta(t)$ stands for Dirac function and notation $*$ denotes convolution. For definitions and discussions about the mode, refer to [34,41]. For the sake of completeness and simplicity, the VMD algorithm renders the constraint problem (1) into an unconstrained one via introducing a quadratic penalty term α and Lagrangian multipliers $\lambda(t)$, resulting in the following augmented Lagrangian.

$$\begin{aligned}\mathcal{L}(\{h_k\},\{\omega_k\},\lambda) = \alpha \sum_{k=1}^{K} \left\| \frac{\partial}{\partial t}\left[\left(\delta(t)+\frac{j}{\pi t}\right) * h_k(t)\right] e^{-j\omega_k t} \right\|_2^2 \\ + \left\| f(t) - \sum_{k=1}^{K} h_k(t) \right\|_2^2 \\ + \left\langle \lambda(t), f(t) - \sum_{k=1}^{K} h_k(t) \right\rangle.\end{aligned} \tag{2}$$

It then converts the original minimization problem (1) into the saddle point problem of (2). Iterative sub-optimizations in terms of the alternate direction multiplier method (ADMM) can be used to obtain the saddle point of (2) and the optimal solution of (1). Let ω_k and $h_{i\neq k}$ denote the most recently updated values, and the minimization problem for h_k^{n+1} becomes

$$\begin{aligned}h_k^{n+1} = \arg\min_{h_k \in X} \{\alpha \left\| \frac{\partial}{\partial t}\left[\left(\delta(t)+\frac{j}{\pi t}\right) * h_k(t)\right] e^{-j\omega_k t} \right\|_2^2 \\ + \left\| f(t) - \sum_{t} h_i(t) + \frac{\lambda(t)}{2} \right\|_2^2 \}.\end{aligned} \tag{3}$$

By applying the Parseval/Plancherel formula for Fourier transform together with Hermitian properties under the L_2 norm, we can find the optimal solution in the spectrum domain as follows,

$$\hat{h}_k^{n+1}(\omega) = \frac{\hat{f}(\omega) - \sum_{i\neq k} \hat{h}_i(\omega) + \frac{\hat{\lambda}(\omega)}{2}}{1 + 2\alpha(\omega - \omega_k)^2}. \tag{4}$$

Implementing the same technical principles above, we obtain the center frequency in the form of

$$\omega_k^{n+1} = \frac{\int_0^\infty \omega \left| \hat{h}_k(\omega) \right|^2 d\omega}{\int_0^\infty \left| \hat{h}_k(\omega) \right|^2 d\omega}. \tag{5}$$

The specific calculation steps of variational mode decomposition, intuitively including denoising functionality, can be summarized as follows.

1. Initialization: $h_k^1, \hat{\omega}_k^1, \hat{\lambda}^1, n \leftarrow 0$;
2. $n \leftarrow n+1$;
3. Update $\hat{h}_k$ and ω_k via (4) and (5), respectively, where $k = 1, 2, \ldots, k, \forall\omega \geq 0$;
4. Use $\hat{\lambda}^{n+1}(\omega) = \hat{\lambda}^n(\omega) + \tau[\hat{f}(\omega) - \sum_k \hat{h}_k^{n+1}(\omega)]$ and update $\hat{\lambda}$, $\forall\omega \geq 0$;
5. Stop the iteration until $\sum_{k=1}^{K} \frac{\left\| \hat{h}_k^{n+1} - \hat{h}_k^n \right\|_2^2}{\left\| \hat{h}_k^n \right\|} < \varepsilon$ for a chosen criterion ε, otherwise return to step 2.

Actually, one should calculate an appropriate number of mode K before applying VMD to decompose the input signals. For this sake, the DFA approach is applicable to determine K as it can characterize different components contained in the signal and provides us with a threshold functionality for the calculation of K [21,35,36].

Remark 1. *Adopting the same simulation conditions as our previous work [42], we perform here a comparison experiment of EMD and VMD on decomposing the analog signal defined by*

$$f(t) = \begin{cases} \cos(20\pi t), \forall t \in [0, 0.5], \\ \cos(20\pi t) + \cos(6\pi t) + \cos(120\pi t), \\ \qquad \forall t \in (0.5, 0.8], \\ \cos(20\pi t), \forall t \in (0.8, 1]. \end{cases} \quad (6)$$

The simulation results are drawn in Figure 2. As shown in Figure 2a, the EMD decomposes $f(t)$ into four intrinsic mode functions and residue, leading to background signals mixing with frequencies 10 and 60 Hz. This mixing phenomenon distorts the intrinsic mode functions obtained later, lowering the signal decomposition performance. Comparing the results in Figure 2b with that in Figure 2a, the VMD features its superiority by accurately decomposing $f(t)$ into three types of signals with different time frames and frequencies.

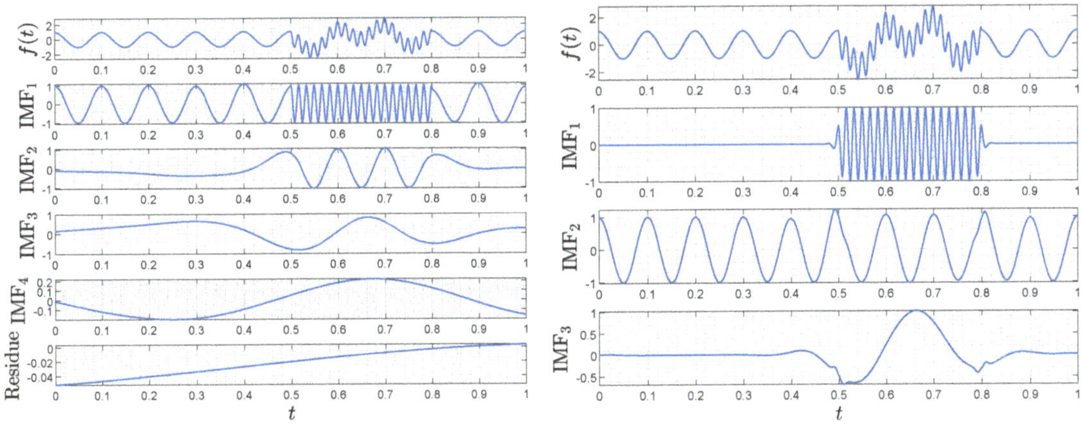

(**a**) EMD-based decomposition result for $f(t)$. (**b**) VMD-based decomposition result for $f(t)$.

Figure 2. Decomposing results of $f(t)$ via EMD and VMD.

3. VMD-Based Denoising Design for X-ray Pulsar Signals

The X-ray pulsar signal observed by detectors contains the time of arrival and the number of photons. It is contaminated by various types of electromagnetic noise. Denoising X-ray pulsar signals appear essential for characterizing different pulsars and building a base for accurate timing in space. In this section, we present a VMD-based denoising method of X-ray pulsars step by step. First, we apply the epoch folding to obtain the noisy pulsar profile. Second, we use the DFA approach to calculate the number of modes of the pulsar profile. Third, we apply the VMD method to decompose and reconstruct the pulsar profile, which achieves the purpose of denoising.

3.1. X-ray Pulsar Profile

Epoch folding is applied in this subsection to obtain the pulsar profile. Sort the TOAs of photons into,

$$t'_0 \le t'_1 \le t'_2 \le \ldots \le t'_{N-1}, \quad (7)$$

where $t'_i, i \in Z$ denotes the time of arrival of the $i+1$-th photon. In (7), the symbol "$\le$" suggests that numerous photons would run into the detector area simultaneously at a certain observing window.

For eliminating the effects of earth-observatory motions and interstellar medium during the pulsar signal transmitting path, we convert the TOAs of photons from observed time t_{obs} to the solar system barycenter (SSB) as follows,

$$t_{\text{SSB}} = t_{\text{obs}} + \Delta_{\text{E}} + \Delta_{\text{R}} + \Delta_{\text{S}}, \tag{8}$$

where Δ_{E} denotes Einstein delay, Δ_{R} is the Roemer delay and Δ_{S} represents the Shapiro delay [43]. Many tools can be applied to finish the time conversion (8). For example, one can synthesis t_{obs} with the orbit parameters and evoke the barycorr function provided by High Energy Astrophysics Science Archive Research Center (HEASARC) to complete the time conversion automatically.

Denote the TOA sequences in SSB by

$$t_0 \leq t_1 \leq t_2 \leq \ldots \leq t_{N-1}, \tag{9}$$

Supposing that the pulsar period is T_0, the phase of (9) in the normalized time frame $[0, 1)$ with respect to the initial time instant can be calculated as

$$\varphi_i = \frac{t_i - t_0}{T_0} \text{mod } 1. \tag{10}$$

Let us averagely scatter each period into m bins and compute the number of photons N_i in each bin. Construct

$$\chi^2 = \sum_{i=1}^{m} \frac{(N_i - \bar{N})^2}{\bar{N}}, \bar{N} = \frac{N}{m}. \tag{11}$$

where N is the total number of photons, $\bar{N}$ stands for the mean value of each bin. When we only perform period estimation, the variable χ^2 in (11) satisfies the χ^2-distribution with $m-1$ degrees of freedom. Different pulsar periods vary from phases and photon quantities in each bin, which, as a result, shows that the obtained χ^2-distributions are different. A period with errors would make the estimated phases inaccurate, simultaneously reducing N_i, $\bar{N}$ and χ^2. In contrast with that, an accurate estimated period suggests that the estimated phases are reliable, and N_i and $\bar{N}$ will differ a lot from each other, which increases χ^2. Therefore, achieving the best estimation of period T relates to finding the maximum of χ^2. Let

$$T = \text{arg}_{\text{T} \in [\text{T}_{\min}, \text{T}_{\max}]} \max\{\text{Ø}^2\}. \tag{12}$$

The estimated phase of the arrival time of every photon is assigned to a certain bin according to the pulsar period, resulting in the estimated pulsar profile. The relationship between the phase and the number of photons in each bin is drawn. For instance, the horizontal axis of the pulsar profile is the phase with multiple bins, while the vertical axis denotes the number of photons. Theoretically speaking, the longer the observing time frame and the more the photon quantity, the more accurate the pulsar profile would become [17,20].

3.2. *Denoise of Pulse Profile Based on VMD*

Without causing any confusion, let f, defined below, be the pulse profile,

$$f = x + d, \tag{13}$$

where x is the original pulsar signal, and d denotes noise.

The number of modes K should be determined before performing VMD on f. The K plays an essential role in VMD decomposition, or in the view of this brief, the wrong choice of K would lower the denoising performance. For these considerations, we apply the detrended fluctuation analysis (DFA) method to determine K. The DFA is a favorable scaling tool generally utilized to analyze nonstationary signals, by which the scaling exponent α depicts how tough the signal performs, and a large α means that the volatility is small [21]. The K can then be calculated according to α.

Given any signal $\{u(i), i = 1, 2, \ldots, N\}$, let us apply DFA to calculate α by the steps below.

1. Calculate the sum

$$y(k) = \sum_{i=1}^{k} u(i) - k\bar{u}, k = 1, 2, \ldots, N, \tag{14}$$

where $\bar{u} = \frac{1}{N} \sum\limits_{i=1}^{N} u(i)$.

2. Divide the sequence $y(k)$ into $N_n = (N/n)$ nonoverlap length-of-n pieces. As for each local trend, one can apply l-order polynomial to fit $y_n(k)$. For example, let $l = 2$ and define

$$y_n(k) = a_n k^2 + b_n k + c_n, \tag{15}$$

where a_n, b_n and c_n denote constants.

3. Define the root-mean-square (RMS) function by

$$F(n) = \sqrt{\frac{1}{N} \sum_{k=1}^{N} [y(k) - y_n(k)]^2}. \tag{16}$$

4. Finally, calculate the scaling exponent α by the least square regression approach as follows,

$$\ln(F(n)) = \alpha \ln(n) + C. \tag{17}$$

The relationship between K and α can be understood in the sense that the quantity of all scaling exponents $\alpha_{1:K}$ under the constraint $\alpha_{1:K} \geq \theta$ equals J, where $\theta = \alpha_\theta = 0.25$ denotes the threshold and $\alpha_\theta = 0.5$ for the white Gaussian noise. Without losing generality, we suppose that the noise of the pulsar signal received by probes satisfies the Gaussian distribution.

The variable J is determined by α_θ via,

$$J = \begin{cases} 1, & \alpha_0 \leq 0.8 \\ 2, & 0.8 < \alpha_0 \leq 1.0 \\ 3, & 1.0 < \alpha_0 \leq 1.5 \\ 4, & \alpha_0 > 1.5 \end{cases} \tag{18}$$

Then, we utilize the obtained K to decompose the signal f via VMD, after which we reconstruct the nominal functions and achieve denoising; see (19).

$$\hat{x} = \sum_k h_k, k = \{k | \alpha_k \geq \theta\}. \tag{19}$$

The flow chart in Figure 3 depicts the VMD denoising process on X-ray pulsar signals. The VMD denoising applied on pulsar signals consists of reconstructing the pulsar profile via summing intrinsic mode functions after decomposition. This process actually attenuates the noise and would remarkably increase the SNR of the pulsar profile. As can be seen, the presented X-ray pulsar denoising strategy does not require any prior knowledge, such as the known pulsar profiles required by the denoising algorithms in [18,20,40].

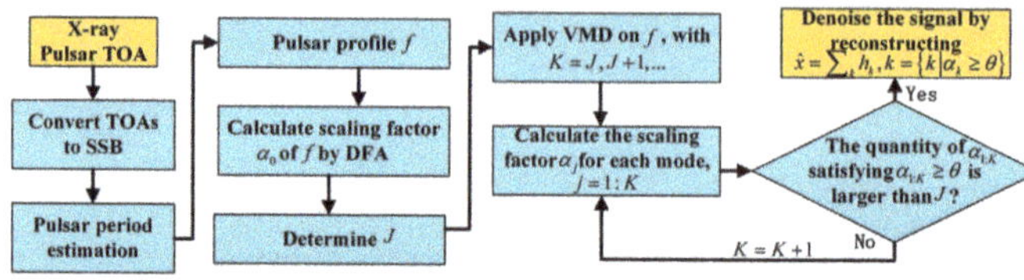

Figure 3. The denoising process of VMD on X-ray pulsar signals.

Remark 2. *The VMD approach provides a general framework for signal processing. In addition to X-ray pulsar denoising designs, various VMD-based applications, such as seismic wave analysis [35], pipeline leakage testing [36] and machinery vibration research [37], have achieved significant attention.*

4. Experimental Analysis

This section presents the experimental study focusing on validating the effectiveness of the proposed X-ray pulsar denoising strategies. Both simulated and measured data of the X-ray pulsar are considered. For performance evaluation purposes, various measures, including the Signal to Noise Ratio (SNR), Root-Mean-Square Error (RMSE), and Pearson's correlation coefficient (PCC), are considered as follows.

$$\mathrm{SNR} = 10\log\frac{\sum\limits_{i=1}^{M} x(i)^2}{\sum\limits_{i=1}^{M}[x(i)-\hat{x}(i)]^2}, \mathrm{RMSE} = \sqrt{\frac{1}{\mathrm{M}}\sum_{\mathrm{i}=1}^{\mathrm{M}}\frac{[\mathrm{x(i)}-\hat{\mathrm{x}}(\mathrm{i})]^2}{\mathrm{x(i)}^2}},$$

$$\mathrm{PCC} = \frac{M\sum\limits_{i=1}^{M} x(i)\hat{x}(i) - \sum\limits_{i=1}^{M} x(i)\sum\limits_{i=1}^{M}\hat{x}(i)}{\sqrt{M\sum\limits_{i=1}^{M} x(i)^2 - \left(\sum\limits_{i=1}^{M} x(i)\right)^2}\sqrt{M\sum\limits_{i=1}^{M}\hat{x}(i)^2 - \left(\sum\limits_{i=1}^{M}\hat{x}(i)\right)^2}},$$

where $x(i)$ denotes the nominal signal, $\hat{x}(i)$ represents the estimated denoised signal, and M stands for the signal length. A higher SNR value or a lower RMSE value leads to better denoising performance. The PCC in the range $[-1,1]$ is generally utilized to address the linear relativity of paired variables. The larger PCC would suggest better denoising performance.

4.1. Experiments of Simulation Data

For pulsar simulation purposes, we applied the Psrsigsim pulsar signal simulator to construct a virtual reference X-ray pulsar profile and associated noisy pulsar profile. The Psrsigsim is a python toolbox developed for simulating pulsar signals based on the known and open databases. Parameters including pulsar profile type, interstellar medium, and telescope information should be set before propagating the virtual pulsar signals. In our settings, we saved the pulsar profile data in the portable '%.csv' form and processed the data via Matlab. Figure 4 depicts the overall production of propagating data and processing.

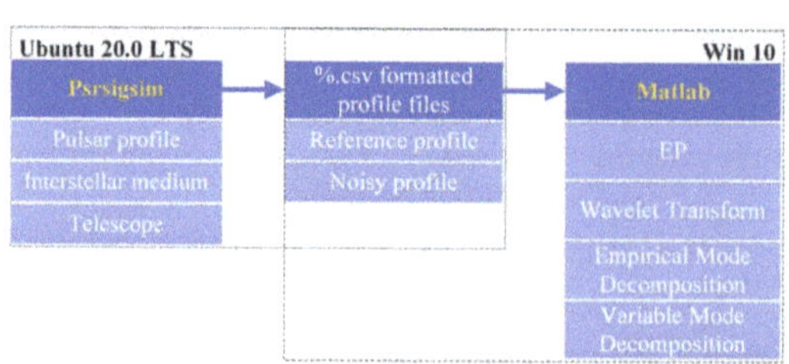

Figure 4. Experiment flow chart of Psrsigsim-propagated pulsar data.

We adopted the default pulsar profile 'J1713 + 07474' provided by Psrsigsim as the reference profile, by which the noisy pulsar profile is constructed via the 'pulsar.make_pulse' function of the Psrsigsim class with a 2000-s observing time. As a result, two million pulsar events in terms of phases were obtained and taken as noisy pulsar signals. The reference pulsar profile is plotted in Figure 5a, while the noisy profiles in the first 1000 samples are drawn in Figure 5b. The vertical coordinate unit is omitted for the convenient processing of simulation data. Two hundred thousand noisy samples were adopted to perform filtering techniques that include epoch folding (EP), wavelet transform (WT), empirical mode decomposition (EMD), and variational mode decomposition (VMD). It is worth noting that the pulsar profile data after EP builds the base for the latter three approaches, or say, the latter three ones use the folding profile as the input source to perform filtering.

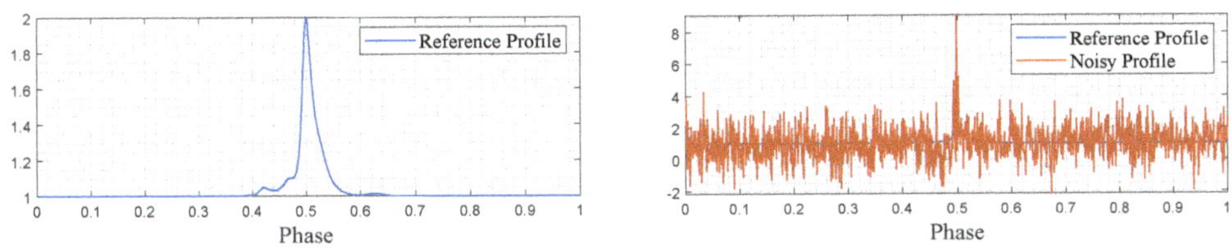

(a) Reference pulsar profile obtained from Psrsigsim.

(b) Noisy pulsar profile for experimental use.

Figure 5. The reference and noisy pulsar profile of simulation data.

For EP filtering, we divided the adopted 200,000 noisy data into 1000 bins per the whole period, following the routine of (7)–(11), and depicted the folding results in Figure 6a. For WT filtering, the Matlab widenoise function was applied to denoise the pulsar profile after EP, and we decomposed the profile signal into three layers, after which the coefficients with high frequencies were eliminated manually. The reconstructed filtered profile after WT is shown in Figure 6b. For EMD filtering, the Matlab emd function was directly applied to process the EP-folded profile data sequence, and eight IMFs were obtained in which the last seven ones were summed together to reconstruct the profile signal, as plotted in Figure 6c. Finally, we implemented the proposed VMD-based denoising strategy to denoise the EP-folded profile data and plotted the results in Figure 6d. For VMD decomposition, we determined the mode number $K = 3$ via the DFA, as shown in (13)–(19), after which the VMD function shown in [34] is evoked.

As shown in Figure 5b, the noisy pulsar profile lacks basic profile shape features, which are important to characterize the pulsar property. Figure 6a–d demonstrates the denoising performance of the EP, WT, EMD, and VMD, respectively. Compared with the other three methods, it can be found that our VMD-based denoising strategy obtains better denoising results, as the denoised profile features a more fluent shape and fewer disturbances. For supporting this viewpoint, the SNRs, RMSEs, and PCCs of the original noisy profile and the denoised profile samples are shown in Table 1.

Table 1. Denoising performance with simulation data.

Method	ORI	EP	WT	EMD	VMD
SNR	−0.1145	22.8210	25.9410	27.2745	**29.8821**
RMSE	1.0121	0.0727	0.0508	0.0435	**0.0313**
PCC	0.1554	0.8619	0.9251	0.9441	**0.9668**

In Table 1, the term 'ORI' refers to the original noisy profile samples. As can be seen, all denoising methods chosen for comparison dramatically increase the SNR values compared with the original profiles of negative SNR, which means that the pulsar signal intensity is less than 1/10 of the noise. In particular, our VMD-based denoising strategy outperforms the compared methods as it gained the highest SNR, smallest RMSE, and largest PCC among the others.

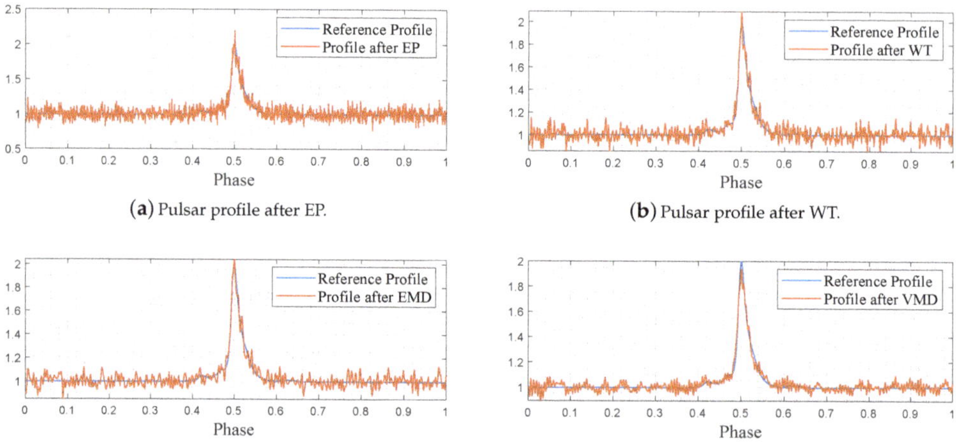

(**a**) Pulsar profile after EP. (**b**) Pulsar profile after WT. (**c**) Pulsar profile after EMD. (**d**) Pulsar profile after VMD.

Figure 6. Denoising results of simulation data via different methods.

4.2. Experiments of HEASARC-Archived Data

The real pulsar TOA data are implemented in this subsection to validate the proposed pulsar denoising strategy. We chose the ni1011010301.cl file to perform denoising experiments from the XAMIN database. The chosen '%.cl' file includes the TOAs of Crab Pulsar photons observed and archived by the NICER mission in standard FITS format. The ni1011010301.cl file was created on 20 November 2017 after a 99-min continuous observation of the Crab X-ray Pulsar and includes 1,680,000 observed photons.

Because the motions of earth and observatory, as well as the interstellar medium, would distort the pulsar signals, we needed to convert the recorded TOAs to the solar system barycenter (SSB). More precisely, the Web Hera, an online HEASARC HEAsoft toolbox set, was applied to evoke the barycorr function that converts the TOA information in '%.cl' files into TOA in SSB, in which the associated orbit information should be set simultaneously. Later on, we performed denoising analysis via EP, WT, EMD, and VMD, respectively. The overall process is depicted in the Figure 7.

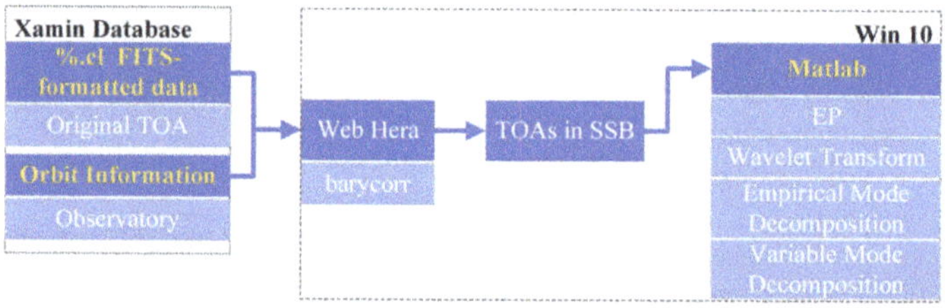

Figure 7. Experiment flow chart of HEASARC-archived pulsar data.

According to (12), we estimated the pulsar period by the Chi-square period estimation method. The estimated pulsar period is 0.033742389697 s. As the SNR value becomes large due to an increase in time frames used for folding, we view the pulse profile created by all 1,680,000 observed data as the reference signal while viewing the pulse profile created by the former 40,000 photons as a disturbance signal for experimental use. The reference profile is then obtained by multiplying the nominal signal by the ratio of noisy photon quantity and all photons, in which the phase part is divided into 100 bins. The reference and noisy profiles are depicted in Figure 8a,b.

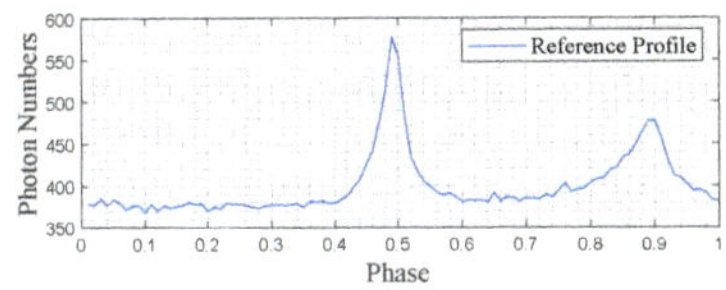

(**a**)

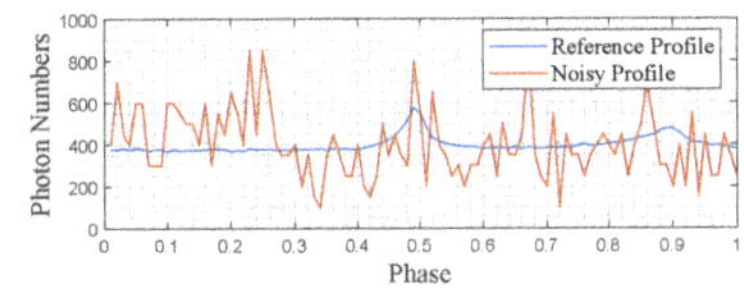

(**b**)

Figure 8. The reference and noisy pulsar profile archived pulsar data. (**a**) Reference pulsar profile obtained via folding all photons; (**b**) Noisy pulsar profile.

The EP method was firstly used to obtain the X-ray pulsar profile, and the results are shown in Figure 9a. Note that the obtained profile in Figure 9a includes 400 times of folding because the total photon quantity in use is 40,000, and each period is divided into 100 bins. For WT filtering, we followed the same routine as the previous subsection and manually removed the high-frequency part, and the results are depicted in Figure 9b. For EMD filtering, the EP-folded profile data were decomposed into six modes, and we summed the later five ones to reconstruct the profile. The EMD-based denoising result is shown in Figure 9c. For VMD filtering, we applied the method of (13)–(18) to calculate the number of modes and obtain $K = 3$, based on which we performed VMD decomposition and signal reconstruction. The VMD-based denoising results are shown in Figure 9d.

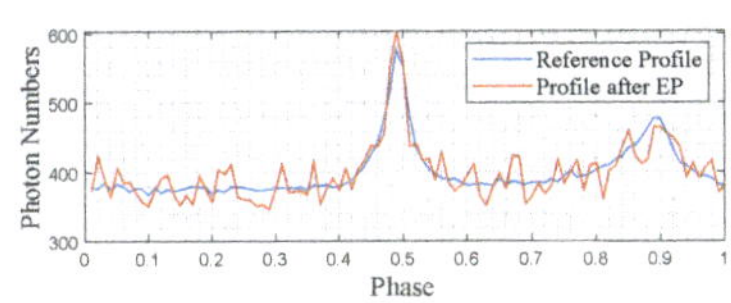

(**a**) Pulsar profile after EP.

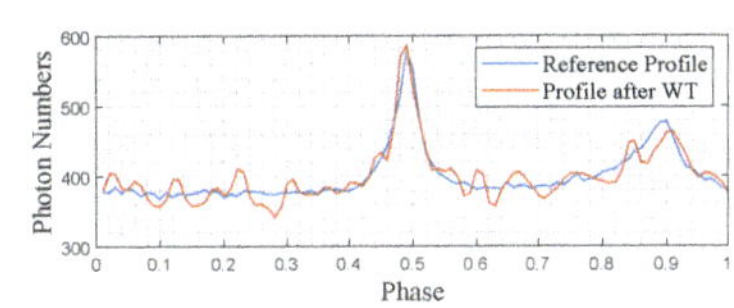

(**b**) Pulsar profile after WT.

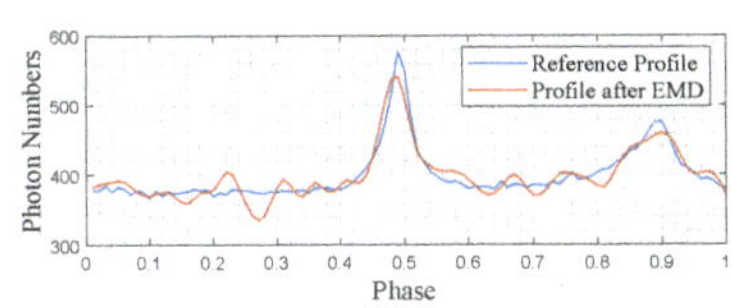

(**c**) Pulsar profile after EMD.

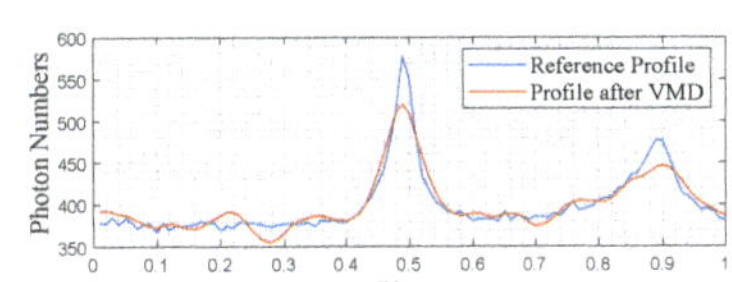

(**d**) Pulsar profile after VMD.

Figure 9. Denoising results of archived pulsar data via different methods.

It is observed from Figure 9a–d that the adopted approaches achieve denoising at different levels. Figure 9c depicts the results after EP, in which obvious saw signals exist, even though the pulse shape matches that of the reference. It is hard to distinguish the best denoising algorithm among WT, EMD, and VMD from the plotted figures. We then numerically summarized the SNRs, RMSEs, and PCCs of the denoised signals in the following table for better comparisons.

The performance indexes in Table 2 indicate that all chosen experimental approaches have realized denoising purposes, and the SNRs are obviously increased compared with the original noisy signal. In particular, the proposed VMD-based denoising algorithm outperforms the other three approaches when inputting virtual Psrsigsim-propagated pulsar data. The obtained experimental results validate the effectiveness of the proposed VMD-based X-ray pulsar denoising design.

Table 2. Denoising performance with measured data.

Method	Ori	EP	WT	EMD	VMD
SNR	7.8000	25.9117	27.9890	29.1157	**30.3373**
RMSE	0.4132	0.0506	0.0391	0.0339	**0.0270**
PCC	0.0462	0.8748	0.9138	0.9277	**0.9480**

5. Conclusions

This paper has made further efforts on X-ray pulsar denoising techniques. The VMD approach incorporated with EP and DFA is facilitated to derive a novel pulsar denoising algorithm that features filtering capacity while eliminating the mode mixing problem after EMD decomposition. Experimental results with both simulation and HEASARC-archived pulsar data demonstrate that the proposed denoising algorithm outperforms the EP-, WT- and EMD-based denoising approaches. In the future, the authors will apply the proposed denoising algorithm on pulsar identification and pulsar-based navigation simulation of orbital satellites.

Author Contributions: Conceptualization, Q.C. and Y.Z.; methodology, Q.C., Y.Z. and L.Y.; software, Y.Z. and L.Y.; validation, Y.Z. and L.Y.; formal analysis, Q.C., Y.Z. and L.Y.; investigation, Q.C., Y.Z. and L.Y.; resources, Q.C., Y.Z. and L.Y.; data curation, Y.Z. and L.Y.; writing—original draft preparation, L.Y.; writing—review and editing, L.Y.; visualization, L.Y.; supervision, L.Y.; project administration, L.Y.; funding acquisition, Q.C. All authors have read and agreed to the published version of the manuscript.

Funding: This work was supported by National Key R&D Program of China (No.2017YFB0503304).

Data Availability Statement: The HEASARC-archived %.cl file is available at https://heasarc.gsfc.nasa.gov (accessed on 20 November 2017).

Conflicts of Interest: The authors declare no conflict of interest.

References

1. Xue, M.; Li, X.; Fu, L.; Liu, X.; Sun, H.; Shen, L. Denoising of X-ray pulsar observed profile in the undecimated wavelet domain. *Acta Astronaut.* **2016**, *118*, 1–10. [CrossRef]
2. Buhler, R.; Blandford, R. The surprising crab pulsar and its nebula: A review. *Rep. Prog. Phys.* **2014**, *77*, 066901. [CrossRef]
3. DeLaney, T.; Weatherall, J. Model for deterministic chaos in pulsar radio signals and search for attractors in the Crab and Vela pulsars. *Astrophys. J.* **1999**, *519*, 291. [CrossRef]
4. Hankins, T.; Kern, J.; Weatherall, J.; Elike, J. Nanosecond radio bursts from strong plasma turbulence in the Crab pulsar. *Nature* **2003**, *422*, 141–143. [CrossRef] [PubMed]
5. Schembri, F.; Sapuppo, F.; Bucolo, M. Experimental classification of nonlinear dynamics in microfluidic bubbles' flow. *Nonlinear Dyn.* **2012**, *67*, 2807–2819. [CrossRef]
6. Bucolo, M.; Grazia, F.D.; Sapuppo, F.; Virzi, M.C. A new approach for nonlinear time series characterization. In Proceedings of the 2008 16th Mediterranean Conference on Control and Automation, Ajaccio, France, 25–27 June 2008.
7. Bonasera, A.; Bucolo, M.; Fortuna, L.; Frasca, M.; Rizzo, A. A new characterization of chaotic dynamics: The d∞ parameter. *Nonlinear Phenom. Complex Syst.* **2003**, *6*, 779–786.
8. Shuai, P.; Chen, S.; Wu, Y.; Zhang, C.; Li, M. Navigation principles using X-ray pulsars. *J. Astronaut.* **2007**, *28*, 1538–1543.
9. Wang, L.; Zhang, S.; Lu, F. Pulsar signal denoising method based on empirical mode decomposition and independent component analysis. In Proceedings of the 2018 Chinese Automation Congress (CAC), Xi'an, China, 30 November–2 December 2018; pp. 3218–3221.
10. Buist, P.J.; Engelen, S.; Noroozi, A.; Sundaramoorthy, P.; Verhagen, S.; Verhoeven, C. Overview of pulsar navigation: Past, present and future trends. *Navigation* **2011**, *58*, 153–164. [CrossRef]
11. Sun, J.; Xu, L.; Wang, T. New denoising method for pulsar signal. *J. Xidian Univ.* **2010**.
12. Jia, Y. General solution to diagonal model matching control of multiple-output-delay systems and its applications in adaptive scheme. *Prog. Nat. Sci.* **2009**, *19*, 79–90. [CrossRef]
13. Jia, Y. Robust control with decoupling performance for steering and traction of 4WS vehicles under velocity-varying motion. *IEEE Trans. Control. Syst. Technol.* **2000**, *8*, 554–569.
14. Lan, S.; Xu, G.; Zhang, J. Interspacecraft ranging for formation flying based on correlation of pulsars. *Syst. Eng. Electron.* **2010**, *3*, 650–654.
15. Dahal, P. Review of pulsar timing array for gravitational wave research. *J. Astrophys. Astron.* **2020**, *41*, 8. [CrossRef]

16. Xie, Q.; Xu, L.; Zhang, H. X-ray pulsar signal detection using photon interarrival time. *J. Syst. Eng. Electron.* **2013**, *24*, 899–905. [CrossRef]
17. Xue, M.; Li, X.; Liu, Y.; Fang, H.; Sun, H.; Shen, L. Denoising of X-ray pulsar observed profile using biorthogonal lifting wavelet transform. *J. Syst. Eng. Electron.* **2016**, *27*, 514–523. [CrossRef]
18. Liang, H.; Zhan, Y. A fast detection algorithm for the X-ray pulsar signal. *Math. Probl. Eng.* **2017**, *2017*, 9607821. [CrossRef]
19. Mitchell, J.; Winternitz, L.; Hassouneh, M.; Price, S.; Semper, S.; Yu, W.; Ray, P.; Wolf, M.; Kerr, M.; Wood, K. Sextant X-ray pulsar navigation demonstration: Initial on-orbit results. In Proceedings of the American Astronautical Society 41st Annual Guidance and Control Conference, Breckenridge, CO, USA, 1–7 February 2018.
20. Su, Z.; Wang, Y.; Xu, L.-P.; Luo, N. A new pulsar integrated pulse profile recognition algorithm. *J. Astronaut.* **2010**, *31*, 1563–1568.
21. Li, F.; Zhang, B.; Verma, S.; Marfurt, K. Seismic signal denoising using thresholded variational mode decomposition. *Explor. Geophys.* **2018**, *49*, 450–461. [CrossRef]
22. Song, J.; Qu, J.; Xu, G. Modified kernel regression method for the denoising of X-ray pulsar profiles. *Adv. Space Res.* **2018**, *62*, 683–691. [CrossRef]
23. Garvanov, I.; Iyinbor, R.; Garvanova, M.; Geshev, N. Denoising of pulsar signal using wavelet transform. In Proceedings of the 2019 16th Conference on Electrical Machines, Drives and Power Systems (ELMA), Varna, Bulgaria, 6–8 June 2019; pp. 1–4.
24. You, S.; Wang, H.; He, Y.; Xu, Q.; Feng, L. Frequency domain design method of wavelet basis based on pulsar signal. *J. Navig.* **2020**, *73*, 1223–1236. [CrossRef]
25. Singh, A.; Pathak, K. A machine learning-based approach towards the improvement of snr of pulsar signals. *arXiv* **2020**, arXiv:2011.14388.
26. Jiang, Y.; Jin, J.; Yu, Y.; Hu, S.; Wang, L.; Zhao, H. Denoising method of pulsar photon signal based on recurrent neural network. In Proceedings of the 2019 IEEE International Conference on Unmanned Systems (ICUS), Beijing, China, 17–19 October 2019; pp. 661–665.
27. Huang, N.; Shen, Z.; Long, S.R.; Wu, M.C.; Shih, H.; Zheng, Q.; Yen, N.; Tung, C.; Liu, H. The empirical mode decomposition and the hilbert spectrum for nonlinear and non-stationary time series analysis. *Proc. R. Soc. Lond.* **1998**, *454*, 903–995. [CrossRef]
28. Jin, J.; Ma, X.; Li, X.; Shen, Y.; Huang, L.; He, L. Pulsar signal de-noising method based on multivariate empirical mode decomposition. In Proceedings of the 2015 IEEE International Instrumentation and Measurement Technology Conference (I2MTC) Proceedings, Pisa, Italy, 11–14 May 2015; pp. 46–51.
29. Caesarendra, W.; Kosasih, P.; Tieu, A.; Moodie, C.; Choi, B. Condition monitoring of naturally damaged slow speed slewing bearing based on ensemble empirical mode decomposition. *J. Mech. Sci. Technol.* **2013**, *27*, 2253–2262. [CrossRef]
30. Wu, Z.; Huang, N. Ensemble empirical mode decomposition: A noise-assisted data analysis method. *Adv. Adapt. Data Anal.* **2009**, *1*, 1–41. [CrossRef]
31. Wang, L.; Shao, Y. Fault feature extraction of rotating machinery using a reweighted complete ensemble empirical mode decomposition with adaptive noise and demodulation analysis. *Mech. Syst. Signal Process.* **2020**, *138*, 106545. [CrossRef]
32. Li, X.; Jin, J.; Wang, M.; Liu, Y.; Shen, Y. X-ray pulsar signal denoising based on emd with adaptive thresholding. In Proceedings of the 2014 IEEE International Instrumentation and Measurement Technology Conference (I2MTC) Proceedings, Montevideo, Uruguay, 12–15 May 2014; pp. 977–982.
33. Jia, Y. Alternative proofs for improved LMI representations for the analysis and the design of continuous-time systems with polytopic type uncertainty: A predictive approach. *IEEE Trans. Autom. Control* **2003**, *48*, 1413–1416.
34. Dragomiretskiy, K.; Zosso, D. Variational mode decomposition. *IEEE Trans. Signal Process.* **2013**, *62*, 531–544. [CrossRef]
35. Liu, Y.; Yang, G.; Li, M.; Yin, H. Variational mode decomposition denoising combined the detrended fluctuation analysis. *Signal Process.* **2016**, *125*, 349–364. [CrossRef]
36. Lu, J.; Yue, J.; Zhu, L.; Li, G. Variational mode decomposition denoising combined with improved bhattacharyya distance. *Measurement* **2020**, *151*, 107283. [CrossRef]
37. Wang, Y.; Markert, R.; Xiang, J.; Zheng, W. Research on variational mode decomposition and its application in detecting rub-impact fault of the rotor system. *Mech. Syst. Signal Process.* **2015**, *60*, 243–251. [CrossRef]
38. Lian, J.; Liu, Z.; Wang, H.; Dong, X. Adaptive variational mode decomposition method for signal processing based on mode characteristic. *Mech. Syst. Signal Process.* **2018**, *107*, 53–77. [CrossRef]
39. Zhao, Y.; Jia, Y.; Chen, Q. Denoising of X-Ray Pulsar Signal Based on Variational Mode Decomposition. In *Chinese Intelligent Systems Conference*; Springer: Singapore, 2020; pp. 419–427.
40. Garvanov, I.; Garvanova, M.; Kabakchiev, C. Pulsar signal detection and recognition. In Proceedings of the Eighth International Conference on Telecommunications and Remote Sensing, Rhodes, Greece, 16–17 September 2019; pp. 30–34.
41. El, S.; Alexandre, R.; Boudraa, A. Analysis of intrinsic mode functions: A pde approach. *IEEE Signal Process. Lett.* **2009**, *17*, 398–401.
42. Zhao, Y. On the Denoising of X-ray Pulsar Signals. Bachelor's Thesis, Beihang University, Beijing, China, 2020.
43. Shen, L. Research on the X-ray Pulsar Navigation. Bachelor's Thesis, Xidian University, Xi'an, China, 2017.

entropy

Article

Causality-Network-Based Critical Hazard Identification for Railway Accident Prevention: Complex Network-Based Model Development and Comparison

Qian Li [1], Zhe Zhang [2,*] and Fei Peng [3]

[1] Integrated Transportation Research Center, China Academy of Transportation Science, Beijing 100029, China; ecliqian@126.com
[2] State Key Laboratory of Rail Traffic Control and Safety, Beijing Jiaotong University, Beijing 100044, China
[3] School of Automotive Engineering, Beijing Polytechnic, Beijing 100176, China; pengfei@bpi.edu.cn
* Correspondence: zhangzhe@bjtu.edu.cn

Abstract: This study investigates a critical hazard identification method for railway accident prevention. A new accident causation network is proposed to model the interaction between hazards and accidents. To realize consistency between the most likely and shortest causation paths in terms of hazards to accidents, a method for measuring the length between adjacent nodes is proposed, and the most-likely causation path problem is first transformed to the shortest causation path problem. To identify critical hazard factors that should be alleviated for accident prevention, a novel critical hazard identification model is proposed based on a controllability analysis of hazards. Five critical hazard identification methods are proposed to select critical hazard nodes in an accident causality network. A comparison of results shows that the combination of an integer programming-based critical hazard identification method and the proposed weighted direction accident causality network considering length has the best performance in terms of accident prevention.

Keywords: railway accident prevention; critical hazard identification; accident causality network; integer programming

Citation: Li, Q.; Zhang, Z.; Peng, F. Causality-Network-Based Critical Hazard Identification for Railway Accident Prevention: Complex Network-Based Model Development and Comparison. *Entropy* **2021**, *23*, 864. https://doi.org/10.3390/e23070864

Academic Editors: Quanmin Zhu, Giuseppe Fusco, Jing Na, Weicun Zhang and Ahmad Taher Azar

Received: 25 May 2021
Accepted: 3 July 2021
Published: 6 July 2021

1. Introduction

1.1. Background

Railway transportation has become the main transportation mode, with the advantages of high speed and low cost. However, railway accidents often interrupt railway transportation processes. Therefore, railway companies emphasize hazard control and emergency management to improve the safety and efficiency of railway transportation. Hazard control is used to alleviate critical or frequent hazard factors and can be considered an accident prevention measure. Emergency management addresses accidents and reduces the negative effects of railway accidents after their occurrence.

This study focuses on identifying critical hazards, which is the core aspect of hazard control. Using accident analysis methods, railway safety managers can investigate the causes of accidents from the aspects of humans, organizations, the environment, and technology. With the increasing number of accident investigation reports, more railway accident causes or hazards can be determined. Typically, some hazards more significantly contribute to accidents than others. These hazards or critical hazards should be identified to support the risk management of railway systems. In the following section, we review the related literature and discuss the contributions of this study.

1.2. Related Studies

Experience or data-based risk analysis methods are often used to determine the causes of accidents. The data are typically obtained from accident reports or railway experts. Based on accident data or experience, many accident causation models have been developed to determine hazards that should be controlled or alleviated. Some researchers have classified accident causation models into various groups based on their attention preference toward accidents [1–3]. The first group is the domino theory-based accident causation model. The domino model, which was developed by Heintich, considers an accident as a result of several sequential hazard events [4]. Because the domino model simplifies the control of human behavior in accident causation, researchers improved the model by focusing more on management failures, and the modified domino models are defined as management-based accident causation models [5,6]. The third group is human error models, where the occurrence of accidents is attributed to human hazards or errors such as unsafe incorrect responses and improper activities [7]. For example, the human performance railway index operational index has been proposed to predict the probability of human failure in railway operations. Human error analysis, human error reduction, and human factor analysis and classification system (HFACS) methods have been developed to analyze human errors in railway operations [8–11]. A Poisson regression method was used to determine the relationship between driver personality and driving safety [12]. Many factors affect the safe operation of railway systems in addition to operators, including train drivers, signalers, and controllers. Therefore, the systemic accident models proposed by Hollnagel may be more suitable for railway accident causation analysis [13,14]. An accident causation model for the railway industry has been proposed to investigate the contributions of human failure, technical failure, and external intrusion to final accidents [15]. A system theoretic accident model and process-based model has been proposed to investigate the various causes of railway accidents, including human and management failures [16,17]. Furthermore, the HFACS-RA (Human Factors Analysis and Classification System-railway Accidents) method was proposed to identify and analyze human and organizational factors that affect railway accidents [18]. An integrated evolutionary model called the scenario-risk-accident chain ontology (SRAC) is proposed to determine the risk relevance of railway systems from accident reports [19]. To support the safety management of railway systems, a quantitative causal analysis was proposed to identify the most important factors that contribute to the risk of passenger train accidents [20]. The associated rules have been derived by mining train accident data [21].

Various equipment, machines, and human resources are interrelated in railway systems, which increase the complexity of railway accidents [22,23]. Therefore, network-based accident causation models have been developed to analyze the causes of accidents and model interdependent hazards in recent years [24]. The nodes in the network represent hazard events or accidents, and the links indicate the relationship between hazards and accidents. A directed network has been formulated to analyze the causes of accidents or railway operational accidents [25,26]. The increasing number of accident reports has enabled the weight between hazards to be measured; hence, a directed weighted accident causation network has been proposed to investigate the accident causation complexity [27]. Three hazard control strategies have been compared using the tailored accident causality network (ACN) [28]; however, the hazard control model has not been formulated to identify the optimal hazards to be removed. Some entropy-based methods have been proposed to determine the critical nodes in real-world complex networks [29–31] such as power networks [32], biological networks [33], and transportation networks [34,35]; however, there is no method to identify critical hazard nodes in ACNs from the perspective of both global optimization and accident prevention. Therefore, a hazard identification model was developed in this study to select optimal hazards to be removed for railway accident prevention.

The selected critical hazards should block the path from hazards to accidents or lengthen the distance from hazards to accidents to prevent accidents. However, the length of the edges cannot be appropriately obtained; hence, the distance from hazards to accidents cannot be measured, although the proposed ACN model can facilitate the understanding of railway accidents. Herein, we propose a novel accident causation network model inspired by a Bayesian network model [36]; the model enables us to easily measure the weights and lengths of the links. Consequently, the shortest path from hazards to accidents can be obtained, and a critical hazard identification (CHI) model can be formulated. The best CHI model was selected by comparing the performances of different models. The proposed integer programming method performs the best because it can solve the CHI problem from a global perspective.

1.3. Contributions and Organization

The proposed model contributes to research pertaining to complex network-based accident analysis methods in three aspects:

(1) We propose an improved ACN model (WLDACN) to model the relationship between hazards and accidents. The shortest distances from hazards to accidents can be obtained based on the proposed new length metrics of edges in WLDACN, which is proven to be superior to other previous methods.

(2) Given the proposed length metrics of WLDACN, the network efficiency is used to represent the difficulty of hazards causing accidents. Therefore, the accident prevention problem is transferred to successfully minimize the WLDACN efficiency.

(3) To support the hazard management of railway systems, we propose a high centrality adaptive and integer programming method to identify critical hazards that greatly contribute to railway accidents. A heuristic algorithm is proposed to solve the integer programming model. The comparison results show that the integer programming method can help prevent accidents better than other models.

The remainder of this paper is organized as follows. Section 2 introduces the ACN construction method and analyzes the objective of hazard control and the formulation of five CHI models. Section 3 presents a real-world case study to verify the effectiveness of the proposed ACN and CHI models. Finally, the conclusions and directions for future research are presented in Section 4.

2. Problem Description and Formulation

2.1. ACN Model

The first step in our study was to construct a causality network from accident investigation reports, which contain reports of many events. We focus on the description of the accident process because it contains the immediate cause and causal factors of accidents. The causal factors of accidents are typically defined as hazards. Several hazards can be extracted, and the causal relationship between hazards can be described using various causality connectors [37]. For example, we can identify two textual causality connectors such as "because" and "due to" in the sentence "the incident occurred because the driver of the first tram did not stop at the platform or stop signal due to a temporary loss of awareness." Therefore, the hazard event "loss of awareness" causes the hazard event "tram did not stop at the platform." We can obtain one hazard pair $\langle i, j \rangle$ if hazard event i causes hazard event j. If hazards i and j are denoted by two nodes, then an arrow from hazard node i to j can be added to represent their causal relationship. After manually extracting all hazard events and accidents from the set of accident investigation reports, we can construct an ACN for railway systems. Each node in a network corresponds to a hazard event or accident. Each edge is associated with a cause–effect pair and directed from the cause (hazard event) to the effect (hazard or accident). Table 1 defines the notations in this paper to formulate the proposed model.

2.1.1. Edge Weight Metrics

We can obtain the same hazard pairs in different accident reports. Therefore, the ACN is a weighted direct accident causality network (WDACN), as shown in Figure 1a. Let

S denote the set of nodes in the WDACN. In the WDACN, the weight $w(i,j)$ of the edge from node i to node j is equal to the number of hazard pairs $\langle i,j \rangle$ in the reports n_{ij}, i.e.,:

$$w(i,j) \ = n_{ij} \tag{1}$$

Table 1. Notations and abbreviations.

Variable	Description
Abbreviations	
CHI	critical hazard indentification
ACN	accident causality network
WDACN	weighted direct accident causality network
DACN	direct accident causality network
SCP	shortest causation path
MPCP	most probable causation path
WLDACN	directed ACN with weights and length metrics
HDA	high degree adaptive
HBA	high betweenness adaptive
HCA	high closeness adaptive
HPA	high pagerank adaptive
ASCP	average SCP
IPM	Integer Programming method
Notations	
h	hazard node
a	accident node
$w(i,j)$	the weight of the edge from node i to j
$F_k(h,a)$	The frequencies of the causation route k from hazard node h to accident node a
$S_k^{h,a}$	the set of points on causation route k
$p(i,j)$	the normalized weight of the edge from node i to j
$P_k(h,a)$	the active probability of causation route k
$l(i,j)$	the length of the edge from node i to j
E	the ACN efficiency
D_h	degree centrality of hazard node h
B_h	betweenness centrality of hazard node h
C_h	closeness centrality of hazard node h
R_h	PageRank of hazard node h

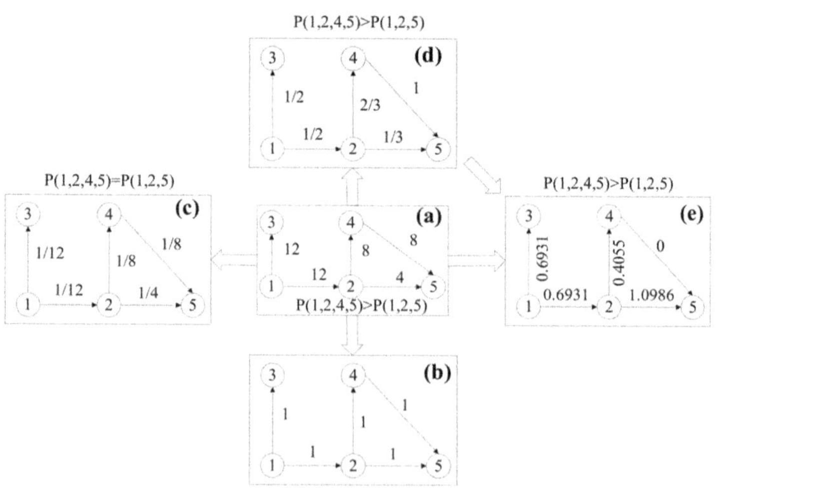

Figure 1. Accident causation network. (**a**) WDACN, (**b**) DACN (**c**) ACN after reversing the weight of the edge, (**d**) Normalized WDACN, (**e**) WLDACN.

The causation routes are defined as the set of links from the source hazard node to the accident node. There may be more than one route from source hazard node h to accident node a, and the same route may appear several times. The frequencies of the causation route k from hazard node h to accident node a $F_k(h,a)$ can be defined as $F_k(h,a) \; = \min(w(i,j)), i,j \in S_k^{h,a}$, where $S_k^{h,a}$ is the set of points on causation route k.

As shown in Figure 1a, the causation route $(1 \to 2 \to 5)$ has a lower frequency than the causation route $(1 \to 2 \to 4 \to 5)$. Therefore, the causation path $(1 \to 2 \to 5)$ has a lower active probability than the causation path $(1 \to 2 \to 4 \to 5)$. In this study, the causation route with the highest active probability is defined as the most likely causation path (MPCP).

The shortest causation path (SCP) is often used to represent the MPCP from one node to another in a complex network. To analyze the interaction between hazards from the perspective of complex networks, the WDACN was simplified as a direct accident causation network (DACN) [28], as shown in Figure 1b. However, the frequency difference of the hazards cannot be observed in the DACN. As shown in Figure 1c, the MPCP from a to e is $(1 \to 2 \to 5)$, which is contrary to the evidence from the WDACN; hence, this method cannot be used to model the interaction between hazard factors and accidents.

Another method is to preserve the frequency information by reversing the weight of the edge [27], as shown in Figure 1c. However, the SCP still cannot indicate the MPCP precisely because paths $(1 \to 2 \to 5)$ and $(1 \to 2 \to 4 \to 5)$ have the same distance (i.e., $1/3$), as shown in Figure 1c. Therefore, the SCP resulting from these two methods is not consistent with the MPCP observed from the WDACN.

2.1.2. Edge Length Metrics

To realize the consistency between SCP and MPCP, we propose a new method to measure the length of the edges. First, the weight of the edge (i,j) in the WDACN can be normalized as follows:

$$p(i,j) = \frac{w(i,j)}{D_i^{out}}, \tag{2}$$

where D_i^{out} is the out-degree of node i. As shown in Figure 1d, the normalized weight $p(i,j)$ can be interpreted as the active probability of hazard node j for active node i. Let H and A denote the sets of hazard accident nodes. Let $S_k^{h,a}$ denote the set of points on causation route k from hazard node h to accident node a. Therefore, the active probability of causation route k can be expressed as:

$$P_k(h,a) = \prod_{i,j \in S_k^{h,a}} p(i,j) \tag{3}$$

$P_k(h,a)$ reflects the conditional probability of causation path k given the occurrence of source hazard h. The MPCP from hazard node h to accident node a can be obtained by solving the following equation:

$$MPCP(h,a) \; = \text{argmax}(P_k(h,a)), k = 1,2,3\ldots K \tag{4}$$

In order to transform the MPCP problem into an SCP problem, the transformation function should satisfy two conditions: (1) it is a monotone decreasing function of probability $p(i,j)$; (2) the uncertainty of route k can be represented as the sum of uncertainties of each edge. Therefore, logarithmic function is selected as our transformation function. The natural logarithm of Equation (3) can be used as follows:

$$-\ln P_k(h,a) = -\sum_{i,j \in S_k^{h,a}} \ln(p(i,j)) \tag{5}$$

Equation (5) can denote the distance of route k which is the sum of length of edges. Therefore, the length $l(i,j)$ of edge (i,j) can be described as follows:

$$l(i,j) = -\ln(p(i,j)) \tag{6}$$

Figure 1e shows the length of edges. The MPCP problem can be transformed into an SCP problem. The length of causation path r from hazard node h to accident node a can be expressed as:

$$L_k(h,a) = -\ln P_k(h,a) = l_k(h,a), \tag{7}$$

where $l_k(h,a) = \sum_{i,j\in S_k^{h,a}} l(i,j)$. Therefore, the SCP can be obtained by solving the following problem:

$$SCP(h,a) = \text{argmin}(l_k(h,a)), k = 1,2,3\ldots K \tag{8}$$

The SCP can be obtained using Dijkstra algorithm and used to measure the difficulty of a hazard factor in causing an accident. The probability of hazard node h causing accident a is greater if $SCP(h,a)$ is shorter. Finally, edges (i,j) in the ACN exhibit three attributes: weight $w(i,j)$, active probability $p(i,j)$, and length $l(i,j)$. Because the proposed ACN contains weights and length metrics and is a directed network, the proposed network model can be named the WLDACN.

2.2. CHI Method Development

2.2.1. Objective of CHI

The objective of CHI is to prevent accidents by reducing or eliminating hazards. The ACN efficiency is used as an indicator to reflect the difficulty of hazards causing accidents.

$$E = \sum_{h\in H, a\in A} \frac{1}{SCP(h,a)} \tag{9}$$

A lower network efficiency corresponds to a longer SCP from hazards to accidents. Therefore, we should eliminate hazard–accident interactions by removing the nodes in the ACN. We assume that all hazards can be controlled or removed via technological development and management improvement. For example, equipment failures can be eliminated by improving the maintenance strategy, human factors can be controlled via safety training, and environmental hazards can be prevented by monitoring the operation surroundings. Additionally, the cost increases with the number of removed hazard nodes. Therefore, the maximum number of removed hazard nodes in the network should be defined.

Some hazard factors such as wind, snow, and rain cannot be controlled because they originate from the natural environment instead of the railway system. Therefore, we should discuss the controllability of hazard nodes. Hazard factors can be controlled if their causes can be determined, so only hazard nodes with parent nodes can be alleviated.

2.2.2. High Centrality Adaptive Methods

The high centrality adaptive method is typically used to reduce the network efficiency by removing hazard nodes that have the highest centrality. Four methods can be used to measure the centrality of nodes in a complex network: based on the node degree, node betweenness, node closeness, and PageRank [38]. Therefore, high degree adaptive (HDA), high betweenness adaptive (HBA), high closeness adaptive (HCA), and high pagerank adaptive (HPA) methods are used to remove critical hazard nodes in a WLDACN.

The HDA method removes the node with the highest degree centrality (DC) in each iteration. The HDA method recomputes the DCs of the remaining nodes after node removal. The DC of hazard node h can be expressed as:

$$D_h = D_h^{in} + D_h^{out} = \sum_{i\neq h} w(i,h) + \sum_{j\neq h} w(h,j) \tag{10}$$

Based on Equation (10), a greater DC of hazard node h results in more causal relationships between hazard node h and other nodes. If a node with a higher DC is eliminated, then more causal relationships can be eliminated.

HBA sequentially removes the node with the highest betweenness centrality (BC) and recomputes the BC for the remaining nodes in each iteration. The BC of hazard node h can be expressed as:

$$B_h = \sum_{\substack{i \in H, j \in A \\ i \neq h, j \neq h}} \frac{\rho_{ij}(h)}{\rho_{ij}}, \tag{11}$$

where ρ_{ij} is the number of shortest paths from hazard node i to accident node j, and $\rho_{ij}(h)$ is the number of shortest paths from hazard node i to accident node j that pass through hazard node h.

Based on the definition of BC, a greater value of the BC of hazard node h results in more SCPs from hazards to accidents to pass hazard h. If a node with a higher BC is eliminated, then more SCPs from hazards to accidents can be removed.

The HCA method removes the node with the highest closeness centrality (CC) and updates the CC for the remaining nodes in each iteration. CC describes the proximity of a hazard node to all other nodes in the ACN. It is calculated as the reciprocal of the average distances from one hazard node to all accident nodes as follows:

$$C_h = \frac{N_A}{\sum_{a \in A} SCP(h,a)}, \tag{12}$$

where N_A is the number of accident nodes. Based on Equation (12), a greater C_h results in fewer steps from hazard node h to accidents. If the node with the higher CC is eliminated, then the average SCP increases from hazards to accidents.

The HPA method deletes the node with the highest PageRank centrality and subsequently recomputes PageRank for the remaining nodes in each iteration. The PageRank of hazard node h can be expressed as:

$$R_h = \sum_{j \neq h} p(j,h)p(j) \tag{13}$$

Based on Equation (13), a greater value of R_h results in a greater possibility of hazard node h being activated by other hazards. If the nodes with higher PageRank values are eliminated, then the propagation probability from hazards to accidents decreases.

In fact, the hazard link, including hazard h in the source dataset, should be changed if hazard h is deleted. Assuming that we extract one hazard link $\langle i, j, h, a \rangle$ from one accident report, if hazard h has been removed by our CHI methods, then accident a cannot occur; hence, the hazard link should be deleted.

2.2.3. Integer Programming Method

High centrality adaptive methods delete the node with the highest centrality at each iteration, which is a local optimal solution. Therefore, a model should be developed to solve the global optimal solution. From an economical and safety perspective, the CHI problem can be described as follows: To identify the nodes to be removed to reduce the network efficiency, an integer programming model for the CHI problem is formulated, i.e.,:

$$\begin{gathered} Min\ \ E \\ \begin{cases} \sum_i x_i \leq M, \forall i & (a) \\ x_i \in \{0,1\}, \forall i & (b), \end{cases} \end{gathered} \tag{14}$$

where x_i is a decision variable for removing hazard node i. $x_i = 1$ indicates that hazard node i is removed. Equation (14a) states that a maximum of M nodes can be removed.

Because the objective function of the CHI model cannot be transformed into a linear form, a heuristic algorithm is proposed to solve the integer programming model. The procedures of the algorithm can be described as follows:

Step 1: Based on the controllability of hazards, randomly select M controllable hazard nodes to be deleted from the accident causation network.

Step 2: For each hazard node in a circular order, maintain the other $M-1$ hazard nodes and identify the optimal solution for the selected hazard node. The solution for the selected hazard node is updated before determining the optimal hazard node in the WLDACN to minimize the network efficiency.

Step 3: Terminate the algorithm when M consecutive solutions for a hazard node do not reduce the network efficiency; otherwise, return to Step 2.

2.3. Model Performance Comparison

The objective of the CHI is to increase the overall average degree of difficulty in causing accidents. To compare the performances of CHI methods from a local perspective, the average SCP (ASCP) from all hazards to each accident type and the ASCP from each hazard type to accidents ($ASCP(H_t)$) were used as indicators to evaluate the proposed CHI methods. The ASCP from all hazards to accident type a, $ASCP(a)$, can be formulated as follows:

$$ASCP(a) = \frac{\sum_{h \in H} SCP(h,a)}{N_H}, \tag{15}$$

where N_H is the number of hazard nodes. If $ASCP(a)$ decreases, then the overall probability of all hazards causing accident a increases; otherwise, the overall probability of all hazards causing accident a will decrease.

Hazards can be classified into different types. Each hazard type is composed of various hazard factors. The ASCP from hazard type H_t to accident a can be formulated as follows:

$$ASCP(H_t,a) = \frac{\sum_{h \in H_t} SCP(h,a)}{N_{H_t}}, \tag{16}$$

where N_{H_t} is the number of hazard nodes that belong to hazard type H_t. Similarly, the ASCP from hazard type H_t to all accidents $ASCP(H_t)$ can be formulated as:

$$ASCP(H_t) = \frac{\sum_{a \in A} ASCP(H_t,a)}{N_A} \tag{17}$$

If $ASCP(H_t)$ decreases, then the overall probability of hazard type H_t causing accidents increases; otherwise, the overall probability of hazard type H_t causing accidents decreases.

3. Case Studies

The proposed model was applied to the hazard management of railway systems. All computational experiments were conducted on a PC with a 2.8-GHz CPU operating the Windows 10 operating system.

3.1. Data Description

The data in this study were obtained from RAIB (https://www.gov.uk/government/publications/raib-investigation-reports-and-bulletins (accessed on 1 September 2020)). We obtained 240 accident investigation reports from 2012 to 2020. The hazards were classified into four types: human (H), equipment and machine (EM), environment (E), and management (M). We extracted 20 H-type hazards (H01-H2O), 53 EM-type hazards (EM01–EM53), 12 E-type hazards (E01–E12), and six M-type hazards (M01–M06). Eighteen types of accidents were obtained, as graphically illustrated in Figure 2.

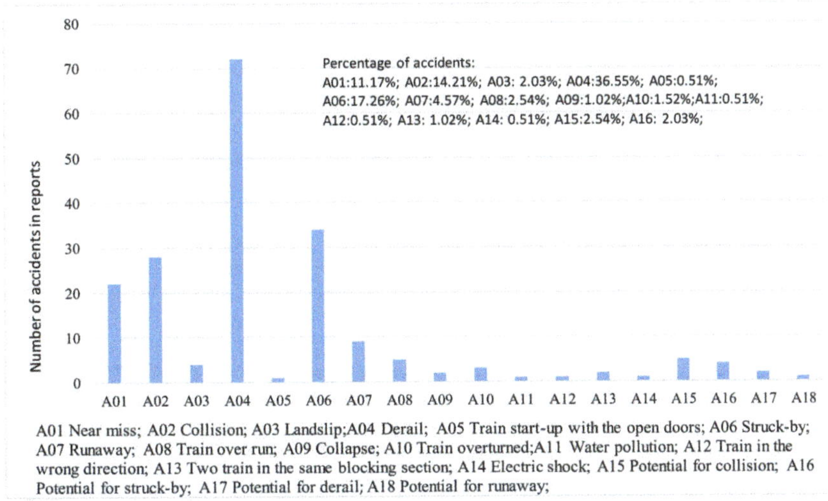

Figure 2. Collection of accidents from RAIB.

3.2. ACN Construction and Analysis

Figures 3 and 4 show the WLDACN based on the interaction between hazard factors and accidents. The weight was computed using Equation (1) and depicted near the edges in the network, as shown in Figure 3. The length of each edge was computed using Equation (6) and depicted near the edges of the network, as shown in Figure 4.

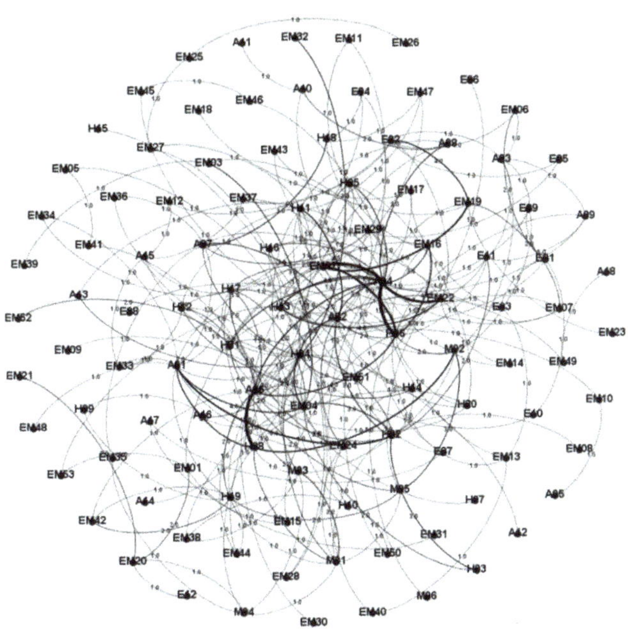

Figure 3. ACN with edge weight metrics.

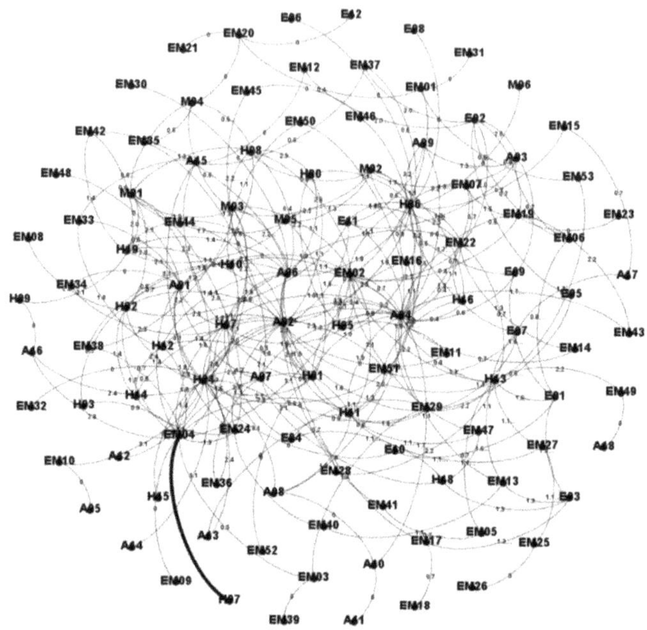

Figure 4. ACN with edge length metrics.

The SCP is an important feature of the WLDACN because it reflects the overall probability of hazards that cause accidents. Table 2 shows the distance from four types of hazards to 18 types of accidents. As shown in Table 2, the derailment accident (A04) had the shortest ASCP (4.44), which indicates that the most likely accident that results from hazards is derailment. Among the four types of hazards causing derailment, the H-type hazard had the shortest path length, which indicates that human failure can much more easily cause derailment. The H-type hazard can much more easily cause 15 types of accidents than the other types of hazards. The EM- and E-type hazards can cause derailment (A04), as indicated by the distance values of 3.54 and 5.62, respectively. M-type hazards were more likely to cause accidents "struck-by (A06)," due to their distance value of 4.78.

Table 2. Distances from 4 types of hazards to 18 types of accidents.

Accident Type	**A01**	**A02**	**A03**	**A04**	**A05**	**A06**	**A07**	**A08**	**A09**
H-type hazard	4.19	4.35	–	3.42	5.73	3.50	4.89	3.71	5.98
EM-type hazard	4.54	5.16	598	3.54	5.98	5.47	5.07	3.83	7.04
E-type hazard	5.80	5.98	5.98	5.62	9.03	6.90	5.98	5.91	7.37
M-type hazard	5.88	6.09	–	5.19	8.34	4.78	5.58	5.47	8.47
All types	5.10	5.39	5.98	4.44	7.27	5.16	5.38	4.73	7.22
Accident type	A10	A11	A12	A13	A14	A15	A16	A17	A18
H-type hazard	3.71	3.71	5.73	5.04	5.73	5.73	5.04	–	5.76
EM-type hazard	3.83	3.82	7.58	6.89	7.58	5.29	5.92	5.98	5.98
E-type hazard	5.91	5.91	9.03	8.34	9.03	5.98	8.34	5.98	8.06
M-type hazard	5.47	5.47	8.34	7.64	8.34	8.29	7.37	–	9.21
All types	4.73	4.73	7.67	6.98	7.67	6.32	6.67	5.98	7.25

Figure 5 shows the ASCP for each hazard type for accidents. As shown in Figure 5, EM-type hazards had the shortest distance to accidents (5.69) among the four types of hazards, which indicates that the most likely cause of accidents is equipment or machine failure. The second most likely cause of accidents is human-type hazards.

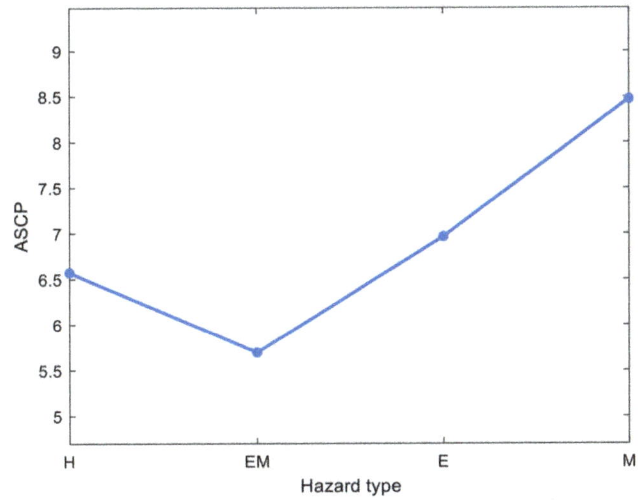

Figure 5. ASCP from each hazard type to accidents.

Figure 6 shows the ASCP from each H-type hazard node and E-type hazard node to accidents. As shown in Figure 6, the distances from different hazard nodes to accidents were different. H06 (rail line inspector did not identify problems in a timely manner) and H08 (track worker negligence) had shorter ASCPs to accidents than the other H-type hazard nodes. Therefore, hazards H06 and H08 can more easily cause railway accidents than the other H-type hazard nodes. E02 (water hazard) had a shorter ASCP to accidents than other E-type hazard nodes. Therefore, hazard E02 can more easily cause railway accidents than the other E-type hazard nodes.

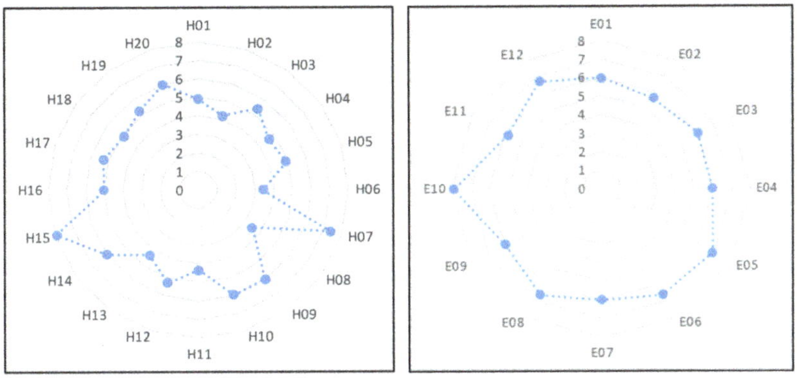

Figure 6. ASCP from each H-type hazard node and E-type hazard node to accidents.

Railway safety managers should delete as many hazard nodes as possible through hazard control efforts. However, different ASCPs from each hazard node to accidents indicate different contributions of various hazards to railway accidents. Therefore, the CHI model should be used to obtain the critical hazard nodes.

3.3. CHI Model Application and Comparison

Based on the obtained WLDACN and hazard controllability analyses, we obtained 64 controllable hazard nodes. Herein, five CHI methods have been proposed to find the critical hazard nodes. Ten nodes were applied to compare these CHI methods, including the

HBA, HCA, HDA, HPA, and IPM. The comparison results are shown in Figure 7. Among the four high centrality adaptive strategies, the HBA strategy is much more effective in preventing accidents than the other three methods because it enables more SCPs from hazards to accidents to be removed. It is difficult to distinguish the effectiveness of the HDA and HPA methods because the network efficiency depends on the number of removed nodes. The HCA strategy performed the worst among the four strategies. Therefore, we suggest using the HBA method to identify critical hazards in railway systems among high centrality adaptive strategies. However, the proposed IPM performed better than all high centrality adaptive strategies, as shown in Figure 7. The high centrality adaptive method iteratively removes the nodes with the highest centrality; therefore, it only solves one critical hazard node at each step. However, the IPM obtains the critical hazard nodes from a global perspective. Consequently, the proposed IPM more effectively performed hazard management and accident prevention.

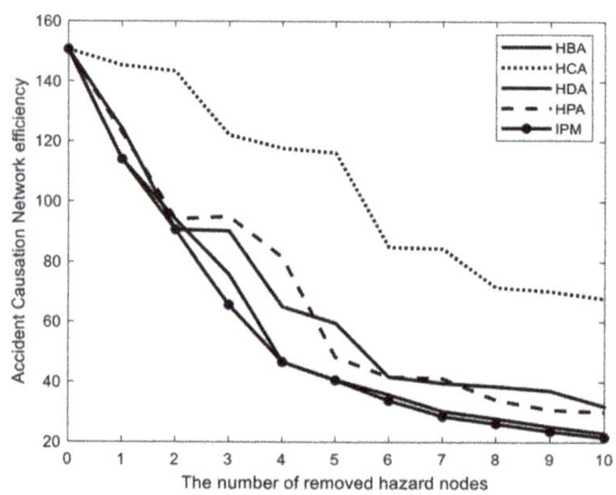

Figure 7. Comparison of different CHI methods.

Additionally, the ASCP was used to compare the performances of five CHI methods. Figure 8 shows the ASCP from all hazards for each accident type after removing 10 hazard nodes. We assume that the ASCP is 20 if accidents cannot be caused by hazards. As shown in Figure 8, some accidents cannot be caused by hazards after we remove 10 hazard nodes. Nine types of accidents, including A05, A12, A13, and A14, cannot be caused by hazards when the HBA method is adopted; eight types cannot be caused by hazards when the HCA method is used; nine types cannot be caused by hazards when the HDA method is used; eight types of accidents, including A05, A08, and A14, cannot be caused by hazards when the HPA method is adopted; nine types of accidents, including A05, A08-A14, A16, and A18, cannot be caused by hazards when the IPM method is adopted. Therefore, HBA and IPM better prevent accidents than the other three methods.

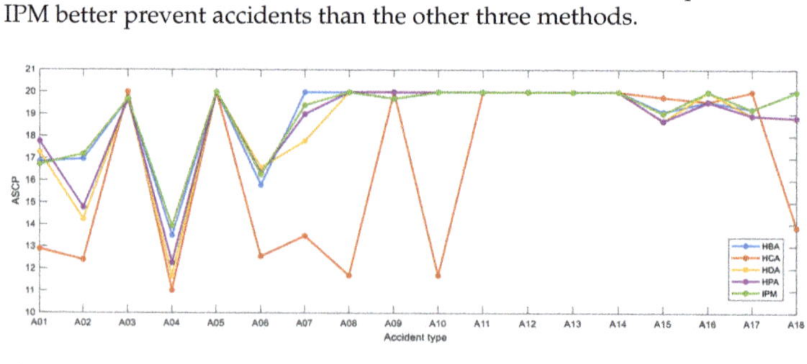

Figure 8. ASCP from all hazards to each accident type.

Based on Equation (17), Figure 9 shows the ASCP from each hazard type to accidents after we removed 10 hazard nodes. As shown in Figure 9, the IPM and HBA methods performed better than the other three methods because IPM and HBA increased the ASCP of EM-, E-, and M-type more than the HPA, HCA, and HDA methods. The ASCP of H-type hazards can be lengthened by the HDA method to a higher level. However, HAD has a worse overall performance than the IPM and HBA methods, which implies that the proposed IPM and HBA methods significantly reduce the accident-causing probability of each hazard type. Therefore, the IPM and HBA methods can be selected as CHI models to identify the critical hazard nodes. The IPM can reduce the probability of H-, EM-, and E-type hazard-related accidents to a lower level than the HBA. The HBA method can reduce the probability of M-type hazard-related accidents to a lower level than the IPM method. However, the overall length of the ASCP resulting from the IPM was longer than that from the HBA method. Therefore, the proposed IPM is suggested as the CHI method for accident prevention in railway systems.

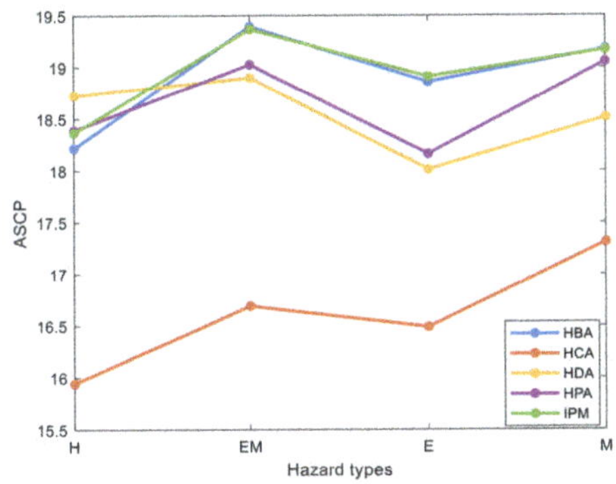

Figure 9. ASCP from each hazard type to accidents.

The hazard nodes to be alleviated were H02 (driver emergency brake failure), H04 (driver's operation mistake), H06 (rail line inspector did not identify problems in a timely manner), H12 (level crossing watchman's mistake), H13 (train maintainer's inadequate maintenance), EM02 (damaged track), EM04 (false signal displayed), EM24 (equipment failure signal), EM29 (train braking system failure), and M01 (poor safety education for workers). The hazard types included 1 M-, 5 H-, and 4 EM-type hazards. Although we obtained 20 H-type hazards from the accident reports, which only constituted 22% of all hazard nodes, the critical H-type hazards still constituted 50% of the total critical hazard nodes, which implies that human failure was the main cause of accidents. Based on the obtained critical hazard nodes and their parent nodes, hazard control measures can be implemented. For example, the top two causes of hazard event H02 were the conductor's mistake (H03) and false signal displayed (EM04). For hazard node H03, we could not identify the causes from the accident reports. Therefore, railway safety managers should employ professional or experienced conductors to reduce these mistakes. Additionally, hazard node EM04 is a critical hazard node, which is mainly caused by EM03 (power supply failure). Therefore, the power supply departments of railway companies should strive to maintain a stable power supply.

The top 10 critical hazard nodes did not include E-type hazard nodes because the railway system is more robust to environmental changes than other transportation modes [39]. If the controllability of hazards is not analyzed, then E03 (wind) will constitute one of the top 10 critical hazard nodes. However, the wind cannot be alleviated because it is a natural hazard; as such, railway companies strive to protect railway systems from wind damage by technology improvement. In summary, the controllability of hazard factors must be analyzed to obtain feasible hazard control measures.

3.4. ACN Model Comparison

The proposed WLDACN model uses a different length metric method from other network-based accident causation models such as DACN and WDACN. To validate the efficiency of the WLDACN, we compared the performances of the five CHI methods under three accident causation models. The comparison results are listed in Table 3.

Table 3. Difficulty of hazards causing accidents under different ACN models and CHI methods. (Bold represent the optimal values).

Models	HBA	HCA	HDA	HPA	IPM
DACN	27.3342	72.3560	32.4598	31.8879	27.1856
WDACN	25.8917	70.3325	31.8904	30.0392	25.3462
WLDACN	**22.9560**	**67.7097**	**31.8904**	**30.0392**	**21.6170**

All three accident causation models performed the best when they were implemented using the HBA and IPM methods. However, the proposed IPM performed slightly better than the HBA method, although different accident causation models were used. Because WDACN and DACN obtain the edge length by reversing or disregarding the weights of edges, as described in Section 2, the proposed WLDACN performed the best among the three accident causation models. The WDACN and WLDACN demonstrated identical performance when implemented using the HDA and HPA methods because they used the same method to compute the weights of the edges. In general, combining the IPM and WLDACN yielded the best accident prevention results.

3.5. Limitation of the Method

The railway company may use facilities with various levels of reliability or employ staff with different professional abilities, which may lead to uncertain accidents. Therefore, the causality network obtained from different railway companies may be different. The obtained results only represent the dataset in this paper. Although the causality network based critical hazard identification method can help safety managers understand the causes of railway accidents to avoid similar occurrences in the future, the number of accident reports determines the efficiency of the method. In fact, the railway industry has developed for a long time, and we believe that the accumulated accident data are so large that the method can be used to suggest critical hazards.

4. Conclusions

Accident causation models enable us to effectively understand the causes of railway accidents. In previous ACN models, the effects of the frequency of hazard events or accidents are typically simplified; hence, the MPCP from hazards to accidents is inconsistent with the obtained SCP. Therefore, we proposed a new ACN model, where the MPCP problem is transformed into an SCP problem. The nodes in the network can be classified as hazard or accident nodes. As such, we can use the distance from hazards to accidents to measure the difficulty of various hazards causing accidents; consequently, critical hazards can be identified.

The causes of railway accidents should be understood to prevent accidents through hazard control. Therefore, we proposed CHI models to select critical hazards. The controllability of hazards, which is often disregarded, was discussed to determine the variables in CHI models. In this study, five CHI models were proposed to identify the critical hazard nodes in the ACN. To compare the performances of the models, three indices that can measure the efficiency of accident prevention were proposed. Comparison results from the case study indicated that the proposed IPM and WLDACN models performed better than the other models.

Because the results from the proposed model were based on real-world data, they offer useful insights into the hazard management of railway systems. The proposed model can suggest critical hazard factors that should be controlled. Safety managers can select hazard control options based on the critical causes of accidents.

In future studies, more accident data will be used to validate the proposed CHI model. In our model, the results of each type of accident, such as damage, injury, or death, were not considered due to inadequate data. Therefore, more data regarding accident damage should be obtained to measure the severity of accidents, and the weight of distance from hazards to each type of accident can be measured. Furthermore, the management cost of each hazard can be differentiated in future models.

Author Contributions: Conceptualization, Z.Z. and Q.L.; methodology, Z.Z.; software, F.P.; validation, Q.L.; writing—original draft preparation, Z.Z.; project administration, Z.Z. All authors have read and agreed to the published version of the manuscript.

Funding: This research was funded by the National Key R&D Program of China, Grant/Award Number: 2018YFB1601200, National Natural Science Foundation of China under Grant 71801009, State Key Laboratory of Rail Traffic Control and Safety (Contract No.: RCS2021ZT004), Key Program of Beijing Polytechnic, Grant/Award Number: 2021Z018-KXZ.

Institutional Review Board Statement: Not applicable.

Informed Consent Statement: Not applicable.

Data Availability Statement: The data presented in this study are openly available in https://www.gov.uk/government/publications/raib-investigation-reports-and-bulletins (accessed date 6 July 2021).

Conflicts of Interest: The authors declare no conflict of interest.

References

1. Katsakiori, P.; Sakellaropoulos, G.; Manatakis, E. Towards an evaluation of accident investigation methods in terms of their alignment with accident causation models. *Saf. Sci.* **2009**, *47*, 1007–1015. [CrossRef]
2. Hollnagel, E. *Barriers and Accident Prevention*; Routledge: London, UK, 2016.
3. Hosseinian, S.S.; Torghabeh, Z.J. Major theories of construction accident causation models: A literature review. *Int. J. Adv. Eng. Technol.* **2012**, *4*, 53.
4. Heinrich, H.W. *Industrial Accident Prevention. A Scientific Approach*, 2nd ed.; McGraw-Hill Book Company, Inc.: New York, NY, USA; London, UK, 1941.
5. Weaver, D. Symptoms of operational error. *Prof. Saf.* **1971**, *104*, 39–42.
6. Bird, F.E.; Cecchi, F.; Tilche, A.; Mata-Alvarez, J. *Management Guide to Loss Control*; Institute Press: Durham, NC, USA, 1974.
7. Hughes, P.; Ferrett, E. *Introduction to Health and Safety in Construction*; Routledge: London, UK, 2012.
8. Kyriakidis, M.; Majumdar, A.; Grote, G.; Ochieng, W.Y. Development and assessment of taxonomy for performance-shaping factors for railway operations. *Transp. Res. Rec.* **2012**, *2289*, 145–153. [CrossRef]
9. San Kim, D.; Baek, D.H.; Yoon, W.C. Development and evaluation of a computer-aided system for analyzing human error in railway operations. *Reliab. Eng. Syst. Saf.* **2010**, *95*, 87–98.
10. Baysari, M.T.; McIntosh, A.S.; Wilson, J.R. Understanding the human factors contribution to railway accidents and incidents in Australia. *Accid. Anal. Prev.* **2008**, *40*, 1750–1757. [CrossRef]
11. Madigan, R.; Golightly, D.; Madders, R. Application of Human Factors Analysis and Classification System (HFACS) to UK rail safety of the line incidents. *Accid. Anal. Prev.* **2016**, *97*, 122–131. [CrossRef] [PubMed]
12. Guo, M.; Wei, W.; Liao, G.; Chu, F. The impact of personality on driving safety among Chinese high-speed railway drivers. *Accid. Anal. Prev.* **2016**, *92*, 9–14. [CrossRef] [PubMed]
13. Hollnagel, E. Understanding accidents-from root causes to performance variability. In Proceedings of the IEEE 7th Conference on Human Factors and Power Plants, Scottsdale, Arizona, 15–19 September 2002; pp. 1–4.
14. Underwood, P.; Waterson, P. Systemic accident analysis: Examining the gap between research and practice. *Accid. Anal. Prev.* **2013**, *55*, 154–164. [CrossRef]
15. San Kim, D.; Yoon, W.C. An accident causation model for the railway industry: Application of the model to 80 rail accident investigation reports from the UK. *Saf. Sci.* **2013**, *60*, 57–68.
16. Ouyang, M.; Hong, L.; Yu, M.H.; Fei, Q. STAMP-based analysis on the railway accident and accident spreading: Taking the China–Jiaoji railway accident for example. *Saf. Sci.* **2010**, *48*, 544–555. [CrossRef]
17. Li, C.; Tang, T.; Chatzimichailidou, M.M.; Jun, G.T.; Waterson, P. A hybrid human and organisational analysis method for railway accidents based on STAMP-HFACS and human information processing. *Appl. Ergon.* **2019**, *79*, 122–142. [CrossRef]
18. Zhan, Q.; Zheng, W.; Zhao, B. A hybrid human and organizational analysis method for railway accidents based on HFACS-Railway Accidents (HFACS-RAs). *Saf. Sci.* **2017**, *91*, 232–250. [CrossRef]

19. Cao, T.; Mu, W.; Gou, J.; Peng, L. A study of risk relevance reasoning based on a context ontology of railway accidents. *Risk Anal.* **2020**, *40*, 1589–1611. [CrossRef]
20. Lin, C.Y.; Rapik Saat, M.; Barkan, C.P.L. Quantitative causal analysis of mainline passenger train accidents in the United States. *Proc. Inst. Mech. Eng. Part. F J. Rail Rapid Transit.* **2020**, *234*, 869–884. [CrossRef]
21. Chen, D.; Xu, C.; Ni, S. Data mining on Chinese train accidents to derive associated rules. *Proc. Inst. Mech. Eng. Part. F J. Rail Rapid Transit.* **2017**, *231*, 239–252. [CrossRef]
22. Read, G.J.M.; Naweed, A.; Salmon, P.M. Complexity on the rails: A systems-based approach to understanding safety management in rail transport. *Reliab. Eng. Syst. Saf.* **2019**, *188*, 352–365. [CrossRef]
23. An, M.; Huang, S.; Baker, C.J. Railway risk assessment-the fuzzy reasoning approach and fuzzy analytic hierarchy process approaches: A case study of shunting at waterloo depot. *Proc. Inst. Mech. Eng. Part. F J. Rail Rapid Transit.* **2007**, *221*, 365–383. [CrossRef]
24. Guo, S.; Zhou, X.; Tang, B.; Gong, P. Exploring the behavioral risk chains of accidents using complex network theory in the construction industry. *Phys. A Stat. Mech. Appl.* **2020**, *560*, 125012. [CrossRef]
25. Xin, M.; Ke-Ping, L.; Zi-Yan, L.; Jin, Z. Analyzing the causation of a railway accident based on a complex network. *Chin. Phys. B* **2013**, *23*, 028904.
26. Li, Q.; Song, L.; List, G.F.; Deng, Y.; Zhou, Z.; Liu, P. A new approach to understand metro operation safety by exploring metro operation hazard network (MOHN). *Saf. Sci.* **2017**, *93*, 50–61. [CrossRef]
27. Zhou, J.; Xu, W.; Guo, X.; Ding, J. A method for modeling and analysis of directed weighted accident causation network (DWACN). *Phys. A Stat. Mech. Appl.* **2015**, *437*, 263–277. [CrossRef]
28. Liu, J.; Schmid, F.; Zheng, W.; Zhu, J. Understanding railway operational accidents using network theory. *Reliab. Eng. Syst. Saf.* **2019**, *189*, 218–231. [CrossRef]
29. Ai, X. Node importance ranking of complex networks with entropy variation. *Entropy* **2017**, *19*, 303. [CrossRef]
30. Qiao, T.; Shan, W.; Zhou, C. How to identify the most powerful node in complex networks? A novel entropy centrality approach. *Entropy* **2017**, *19*, 614. [CrossRef]
31. Li, Y.; Cai, W.; Li, Y.; Du, X. Key node ranking in complex networks: A novel entropy and mutual information-based approach. *Entropy* **2020**, *22*, 52. [CrossRef]
32. Nguyen, D.T.; Shen, Y.; Thai, M.T. Detecting critical nodes in interdependent power networks for vulnerability assessment. *IEEE Trans. Smart Grid* **2013**, *4*, 151–159. [CrossRef]
33. Han, C.; Sun, X.; Yang, Y.; Che, Y.; Qin, Y. Brain complex network characteristic analysis of fatigue during simulated driving based on electroencephalogram signals. *Entropy* **2019**, *21*, 353. [CrossRef] [PubMed]
34. Liu, K.; Wang, M.; Cao, Y.; Zhu, W.; Wu, J.; Yan, X. A Comprehensive Risk Analysis of Transportation Networks Affected by Rainfall-Induced Multihazards. *Risk Anal.* **2018**, *38*, 1618–1633. [CrossRef] [PubMed]
35. Valério, D.; Lopes, A.M.; Tenreiro Machado, J.A. Entropy analysis of a railway network's complexity. *Entropy* **2016**, *18*, 388. [CrossRef]
36. Xu, H.; Zhang, Y.; Li, H.; Skitmore, M.; Yang, J. Safety risks in rail stations: An interactive approach. *J. Rail Transp. Plan. Manag.* **2019**, *11*, 100148. [CrossRef]
37. Xuelan, F.; Kennedy, G. Expressing causation in written English. *Relc J.* **1992**, *23*, 62–80. [CrossRef]
38. Fan, C.; Zeng, L.; Sun, Y.; Liu, Y.Y. Finding key players in complex networks through deep reinforcement learning. *Nat. Mach. Intell.* **2020**, *2*, 317–324. [CrossRef]
39. He, Z.; Navneet, K.; van Dam, W.; Van Mieghem, P. Robustness assessment of multimodal freight transport networks. *Reliab. Eng. Syst. Saf.* **2021**, *207*, 107315. [CrossRef]

entropy

Article

Estimation of Feeding Composition of Industrial Process Based on Data Reconciliation

Yusi Luan, Mengxuan Jiang, Zhenxiang Feng and Bei Sun *

School of Automation, Central South University, Changsha 410083, China; 184611053@csu.edu.cn (Y.L.); jiangmengxuan@csu.edu.cn (M.J.); fengzxcsu@csu.edu.cn (Z.F.)
* Correspondence: sunbei@csu.edu.cn

Abstract: For an industrial process, the estimation of feeding composition is important for analyzing production status and making control decisions. However, random errors or even gross ones inevitably contaminate the actual measurements. Feeding composition is conventionally obtained via discrete and low-rate artificial testing. To address these problems, a feeding composition estimation approach based on data reconciliation procedure is developed. To improve the variable accuracy, a novel robust M-estimator is first proposed. Then, an iterative robust hierarchical data reconciliation and estimation strategy is applied to estimate the feeding composition. The feasibility and effectiveness of the estimation approach are verified on a fluidized bed roaster. The proposed M-estimator showed better overall performance.

Keywords: data reconciliation; robust estimator; gross error detection; feeding composition

Citation: Luan, Y.; Jiang, M.; Feng, Z.; Sun, B. Estimation of Feeding Composition of Industrial Process Based on Data Reconciliation. *Entropy* **2021**, *23*, 473. https://doi.org/10.3390/e23040473

Academic Editor: Quan Min Zhu

Received: 1 March 2021
Accepted: 13 April 2021
Published: 16 April 2021

1. Introduction

Complete, accurate and reliable data measurements are important for process modeling, model identification, real-time online optimization and process control. However, the measured values of variables in the actual measurement process are inevitably contaminated by random errors or even gross errors. Therefore, the process measurements will deviate from the real values. In addition, the feeding composition, which is important in making control decisions, cannot be measured online due to the high economic cost and technical limitations. Therefore, it is necessary to process the measurements to guarantee the reliability and observability of the system. Based on this, the estimation of feeding composition can be better performed. Data reconciliation is an effective approach for this aim [1]. The principle of data reconciliation is to minimize the sum of square deviations between the coordinated value of variables and their measurements under the condition of satisfying material balances, heat balances and boundary constraints of process variables. From the mathematical point of view, data reconciliation is a constrained optimization problem [2], which has been widely used in many important procedures such as in-line process monitoring [3,4], enhanced control strategy performance [5,6], parameter estimation [7], soft sensor applications [8], quality control [9] and industrial optimization [10,11] among others [3–11].

Traditional data reconciliation typically assumes that measurements contain only random errors with an average value of zero and normal distributions. Nonetheless, the measurements will be contaminated by gross errors. If gross errors are not processed, the accuracy of data reconciliation results will be reduced. The gross error detection is therefore particularly important for data reconciliation. There are currently three main methods for handling measurements with gross errors. The first strategy is sequential gross error detection and data reconciliation. A variety of statistical tests, including the global test [12], the measurement test [13], the nodal test [14], generalized likelihood ratio test [15] and principal component analysis test [16], are used for gross error detection. Following the detection and elimination of gross errors, the traditional data reconciliation is carried out. The second strategy is simultaneous gross error detection and data reconciliation. Yu et al. [17] proposed a support vector regression method for recursive simultaneous data reconciliation and gross error detection. Zhang et al. [18] used a novel particle filter (PF) algorithm based on the measurement test (MT) to solve the dynamic simultaneous gross error detection and data reconciliation. Yuan et al. [19] established a new hierarchical Bayesian framework, which can simultaneously estimate the real value of process variables and obtain the magnitudes of gross errors. The last strategy is robust data reconciliation. The estimations close to the true value can be obtained by reducing the influence of gross errors on data reconciliation results. Tjoa and Biegler [20] constructed an M-estimator based on mixed distribution, which takes both random errors and gross errors into account and minimized the M-estimator as an objective function. Johnson and Kramer [21] presented a new robust method for data reconciliation based on a probability bootstrapping theory. Arora and Biegler [22] evaluated the Fair and Hampel estimators in stationary and dynamical constrained problems. Hampel's three-part estimator presented the best performance. Wang and Romagnoli [23] designed a partially adaptive estimator based on a generalized T distribution and a fully adaptive estimator based on density estimation by studying the robustness of a set of existing estimators. Özyurt and Pike [24] analyzed the performance of a number of robust estimators such as Hampel, Cauchy and Fair. Ragot et al. [25] adopted a cost function that has been improved based on the contaminated probability density distribution model for the design of robust estimator. Prata et al. [26] applied the proposed strategy based on the particle swarm optimization (PSO) and the robust Welsch estimator to real-time online detection of Bulk polymerization of propylene. Zhang et al. [27] constructed a robust estimator by using the quasi-weighted least squares. Through the comparative analysis of various methods like Welsch, quasi-weighted least squares and comentropy M-estimators, the feasibility and effectiveness of robust estimators in simultaneous gross error detection and data reconciliation were demonstrated [28]. Alighardashi et al. [29] proposed a maximum likelihood framework for simultaneous data reconciliation and gross error detection for steady-state data. Xie et al. [30] utilized a novel robust estimator to improve the robustness of data reconciliation. Forty-eight different robust M-estimators can be founded in a recent review study, including Fair, Cauchy, Biweight, Jin, Welsch estimator and Xie [31].

The key to estimating feeding composition based on data reconciliation is the robust estimator and the design of the estimation strategy. An excellent robust estimator is embodied in two aspects: On the one hand, the objective function is bounded, and the bound value is small; on the other hand, the influence function can converge quickly and converge to zero. The objective functions of Cauchy [32] and Fair [33] estimators increase with the increase in standard measurement residuals, indicating that the error will greatly affect the reconciled results. Besides, the convergence speed of their influence functions is very slow, and the bound values are large, which shows that the two estimators cannot suppress the influence of gross errors on reconciled results. The objective function of Welsch [26] estimator is limited, but the bound value is also large, and the influence function does not converge rapidly. Although the influence function of Xie [30] estimator may converge to zero and the convergence speed is improved in comparison to the estimators above, it still cannot achieve the fast standard. Based on the robust estimation theory, a novel robust estimator is proposed to enhance the performance of the robust estimators. The objective function is bounded, the bound value is smaller, and the influence function converges toward zero faster. At the same time, aiming at the problem that the change of feeding composition caused by the long test period of industrial processes is unknown, an iterative robust hierarchical data reconciliation and estimation strategy based on the heat balance is proposed. In order to further verify the robustness and effectiveness of the proposed robust estimator and estimation of feeding composition strategy, a series of comparative experiments are carried out through two numerical examples and the real data from a fluidized bed roaster for zinc smelting.

After initial remarks, motivation aspects and applications of the data reconciliation issue, the paper is organized as follows. Section 1 presents preliminary approaches to traditional data reconciliation and robust estimator. The novel robust estimator is proposed, and two numerical examples are used to demonstrate the effectiveness of the proposed estimator in Section 2. Section 3 provides an iterative robust hierarchical data reconciliation and estimation strategy to estimate feeding composition. The feeding composition of fluidized bed roaster is then estimated in Section 4. Finally, Section 5 provides the conclusions.

2. Preliminaries

2.1. Data Reconciliation

The purpose of data reconciliation is to obtain the reconciled data by processing the raw measurements. The reconciled data not only satisfy the constraints of the process model but are also closer to the true value. There are three main assumptions for the system using data reconciliation: The system is in a steady state, the measurement errors follow a normal distribution with zero mean, and each measured variable is independent of each other. Taking into account the presence of measurement errors, the measurement model may be expressed as follows:

$$x = x^* + \varepsilon \tag{1}$$

where x denotes the vector of raw process measurements, x^* denotes the vector of true values of the process variables and ε is the random measurement errors.

Based on the principle of data reconciliation, the general steady-state data reconciliation problem can be stated as a form of solving the weighted least squares solution satisfying the process model and boundary constraints:

$$\begin{aligned} &\min\ (x-\widetilde{x})^T \textstyle\sum^{-1}(x-\widetilde{x}) \\ &s.t.\ \ F(\widetilde{x},u)=0 \\ &\qquad\ G(\widetilde{x},u)\leq 0 \end{aligned} \tag{2}$$

where $\widetilde{x}$ is the vector of reconciled data, $\sum$ is the diagonal covariance matrix of measurement errors, u is the vector of unmeasured variables; F represents the process model, which is used as equality constraints in the optimization problem, and G denotes the inequality constraints indicating variable boundaries.

2.2. M-Estimator

In the classical weighted least square of data reconciliation, Equation (2), it is usually assumed that the process measurements contain only random errors. However, in the actual measurement process, the measured values may contain gross errors. The presence of gross errors has a serious effect on the conventional data reconciliation, resulting in the propagation of gross errors. This will cause the reconciled data to not satisfy the process model and deviate from the true value of variables seriously. As a result, the robust estimator is used to account for gross errors in measurements.

Currently, there are numerous approaches for robust data reconciliation, the majority of which are based on the theory of M-estimator. The function of the measurement residuals is constructed as the objective function of the robust estimator. By reducing the weight of the measurements with gross errors, it can prevent gross errors smearing other measurements. Hence, the robust estimator can well reconcile the data with gross errors. The problem of robust data reconciliation can be expressed as follows:

$$\begin{aligned} &\min \sum_{i=1}^{n} \rho(\xi_i) = \min \sum_{i=1}^{n} \rho\left(\tfrac{x_i-\widetilde{x}_i}{\sigma_i}\right) \\ &s.t.\ \ F(\widetilde{x},u)=0 \\ &\qquad \widetilde{x}_{i\min} \leq \widetilde{x}_i \leq \widetilde{x}_{i\max},\ \mathrm{i}=1,2,\ldots,n \\ &\qquad u_{l\min} \leq u_l \leq u_{l\max},\ l=1,2,\ldots,N-n \end{aligned} \tag{3}$$

where ρ is the robust estimator, $\xi_i = (x_i - \widetilde{x}_i)/\sigma_i$ is the standardized residual for the ith measured variable, x_i, $\widetilde{x}_i$ and u_l are, respectively, the measured data, the reconciled data for the ith measured variable and the estimated value for the jth unmeasured variable, $\widetilde{x}_{i\min}$, $\widetilde{x}_{i\max}$, $u_{l\min}$ and $u_{l\max}$ are, respectively, the lower limit, the upper limit for the ith reconciled variable, the lower limit, the upper limit for the lth unmeasured variable. N, n, $N-n$ denote the total number of variables, the total number of measured variables and the total number of variables not measured, respectively. σ_i represents the standard deviation for variable measurement errors.

Most M-estimators are not based on a clearly defined probability distribution known in advance. Most of them are based on a simple mathematical structure. The mathematical structure of M-estimator, i.e., ρ, has some general characteristics as follows:

- ρ is is continuous;
- $\rho(\xi) = \rho(-\xi)$;
- $\rho(\xi) \geq 0$ and ρ is integrable
- $\rho(\xi_1) \leq \rho(\xi_2)$, for $|\xi_i| < |\xi_j|$;
- $\rho(0) = 0$.

As a crucial index in assessing the robustness of the M-estimator, the influence function (IF) [34] is directly proportional to the derivative of the objective function of the estimator. The influence function value refers to the effect of different deviation on the estimator. In general, the function may be defined as the first derivative of the objective function to standardized residuals, which can be expressed as follows:

$$\mathrm{IF}(\xi) = \frac{d\rho(\xi)}{d\xi} \tag{4}$$

Some general characteristics of IF [35] are:

- IF is limited;
- IF is continuous or piecewise continuous;
- $\mathrm{IF}(-\xi) = -\mathrm{IF}(\xi)$;
- IF is nearly linear near the origin ($\mathrm{IF}(\xi) \approx k \cdot \xi, k \neq 0$, for small ξ), but this characteristic is not necessary;
- the rejection point of IF (the point where IF is zero) is finite to suppress large deviations.

On the basis of a robust M-estimation theory, the influence function needs to be continuous and bounded. When ξ is small, the influence function is proportional to ξ; when ξ is infinite, the influence function can converge to a constant, which indicates that the robust estimator can suppress the effect of gross errors in the reconciled results. The influence function of the weighted least squares estimator is the standardized residual ξ, which indicates that the influence function will increase with the measurement errors. It can be seen that the weighted least squares estimator is not robust because its influence function is boundless, and the reconciled results can be easily affected by gross errors.

3. Data Reconciliation Based on a Novel Robust Estimator

3.1. A Novel Robust Estimator

The most robust estimators are constructed according to the following requirements: The objective function of the estimator is bounded, and the bound value is small; the influence function can rapidly converge to a constant. From the observation of the mathematical structure of 48 different robust estimators in the literature [31], it has been shown that the objective functions of these M-estimators comprise limitless function, limited function, non-piecewise function, piecewise function, multi parameters and a single parameter. The performance of the M-estimator whose objective function is limited, non-piecewise and has single parameter is analyzed. It is found that the M-estimator with a relatively rapid convergence rate basically contains the mathematical expression with e as the base number. The derivative of exponential function based on e is equal to itself. Moreover, $e^{-\xi}$ decreases with increasing ξ and tends towards 0. When ξ tends towards 0, $e^{-\xi}$ tends towards 1.

Therefore, the objective function can be limited if the structure of $1 - e^{-\xi}$ is included in. For instance, the Welsch estimator and the objective function is presented in Equation (5). In order to improve the robustness of the M-estimator, Xie divided $1 - e^{-\xi}$ with $1 + e^{-\xi}$. By the decline of $1 + e^{-\xi}$, the overall rise speed of the objective function is improved. Furthermore, the converging speed of the influence function is accelerated. The objective function of Xie estimator is illustrated under Equation (6).

Welsch:

$$\rho(\xi_i) = \frac{c_w^2}{2}\left(1 - \exp\left(-\left(\frac{\xi_i}{c_w}\right)^2\right)\right) \tag{5}$$

Xie:

$$\rho(\xi_i) = \frac{1 - \exp\left(-\left(\frac{\xi_i}{c_x}\right)^2\right)}{1 + \exp\left(-\left(\frac{\xi_i}{c_x}\right)^2\right)} \tag{6}$$

where c_w and c_x are, respectively, tuning parameters of Welsch and Xie estimators.

As discussed above, in order to limit the effect of gross errors on the reconciled results, a novel robust estimator is created based on the general characteristics of the M-estimator and influence function. The objective function of the novel robust estimator is defined as follows:

$$\rho(\xi_i) = \frac{c_p^2\left[1 - \exp\left(-(\xi_i/c_p)^2\right)\right]}{4\left[1 + \exp\left(-(\xi_i/c_p)^4\right)\right]} \tag{7}$$

where c_p is the tuning parameter; the definition of ξ_i is the same as above. By increasing the index of ξ, the downward velocity of $1 + e^{-\xi}$ is increased. Then, the objective function increases faster, and the influence function converts to 0 faster. At the same time, in order to satisfy the characteristic that the influence function is almost linear near the origin, the objective function is multiplied by $c_p^2/4$. The influence function of the novel robust estimator function expressed in Equation (7) is as follows:

$$\mathrm{IF}(\xi_i) = \frac{d\rho(\xi_i)}{d\xi_i} = \frac{\xi_i \exp\left(-\left(\frac{\xi_i}{c_p}\right)^2\right)\left[1 + \exp\left(-\left(\frac{\xi_i}{c_p}\right)^4\right) + \frac{2\xi_i^2}{c_p^2}\left(1 - \exp\left(-\left(\frac{\xi_i}{c_p}\right)^2\right)\right)\right]}{2\left[1 + \exp\left(-\left(\frac{\xi_i}{c_p}\right)^4\right)\right]^2} \tag{8}$$

It can be seen from Equations (7) and (8) that when the standardized residual is small, the objective function of the novel robust estimator tends to be gradually zero, and the influence function is proportional to the standardized residual. It shows that the weighted least squares method can be used to reconcile the data where measurements contain only random errors. When the standardized residual is significant, the objective function of the novel robust estimator tends to be constant little by little, and the influence function quickly converges to zero. It is demonstrated that the novel robust estimator can adjust the weight of the estimator based on the standardized residual size when the measurements contain gross errors. The larger the gross errors, the lower the weights, and the greater the effects of restraining the gross errors. Therefore, the proposed estimator is more robust.

3.2. The Tuning Parameter of the Novel Robust Estimator

The use of robust estimators in data regression typically involves the inverse relationship between relative efficiency and robustness [36]. Relative efficiency refers to the fit quality of the estimated value reconciled by M-estimator, when errors follow another distribution relative to the reference distribution (usually assumed to be Normal distribution). The so-called robustness refers to the performance of the M-estimator under various nonnormal error distributions. The more robust an M-estimator is, the less its relative efficiency is. The relationship between robustness and relative efficiency is addressed by the tuning parameters. When comparing the performance of different M-estimators, the level of relative efficiency and the reference distribution should be consistent. In general, the tuning parameters for each M-estimator are obtained at the relative efficiency level of 95%. Then the robustness of distinct M-estimators is judged.

The mathematical definition of relative efficiency is illustrated in Equation (9):

$$E_{ff}[\mathrm{IF}(\xi), f(\xi)] = \frac{V_f[\mathrm{IF}_f(\xi), f(\xi)]}{V[\mathrm{IF}(\xi), f(\xi)]} \tag{9}$$

where V_f is the asymptotic variance of the reference estimator, V is the asymptotic variance of the specified estimator, $f(\xi)$ is the error reference probability distribution, $\mathrm{IF}_f(\xi)$ is the influence function of the reference estimator and $\mathrm{IF}(\xi)$ is the influence function of the specified estimator. The asymptotic variance V is defined as follows [37]:

$$V[\mathrm{IF}(\xi), f(\xi)] = \frac{\int_{-\infty}^{+\infty} \mathrm{IF}^2(\xi) f(\xi) d\xi}{\left[\int_{-\infty}^{+\infty} \mathrm{IF}'(\xi) f(\xi) d\xi\right]^2} \tag{10}$$

where $\mathrm{IF}'(\xi)$ is the derivative of the influence function $\mathrm{IF}(\xi)$. Since $\mathrm{IF}'(\xi)$ may be discontinuous, Equation (10) can be further expressed by Equation (11):

$$V[\mathrm{IF}(\xi), f(\xi)] = \frac{\int_{-\infty}^{+\infty} \mathrm{IF}^2(\xi) f(\xi) d\xi}{\left[-\int_{-\infty}^{+\infty} \mathrm{IF}(\xi) f'(\xi) d\xi\right]^2} = \frac{\int_{-\infty}^{+\infty} \mathrm{IF}^2(\xi) f(\xi) d\xi}{\left[-\int_{-\infty}^{+\infty} \mathrm{IF}(\xi) f'(\xi) d\xi\right]^2} \tag{11}$$

where $f'(\xi)$ is the derivative of $f(\xi)$.

The tuning parameter of the proposed estimator at the relative efficiency level of 95% in respect to the Normal distribution is calculated as follows:

(1) The Normal distribution is selected as the reference distribution, and its probabilistic distribution and the first derivative are respectively expressed as follows:

$$f(\xi) = \frac{1}{\sigma\sqrt{2\pi}} \exp\left(-\frac{\xi^2}{2}\right) \tag{12}$$

$$f'(\xi) = -\xi \frac{1}{\sigma\sqrt{2\pi}} \exp\left(-\frac{\xi^2}{2}\right) = -\xi f(\xi) \tag{13}$$

(2) The least square estimator is selected as the reference estimator, and its influence function is expressed as follows:

$$\mathrm{IF}_f(\xi) = \frac{d\rho(\xi)}{d\xi} = \frac{d}{d\xi}\left(\frac{\xi^2}{2}\right) = \xi \tag{14}$$

(3) The influence function of the proposed estimator is shown in Equation (8). The expressions of V_f and V can be obtained by using Equation (11):

$$V_f = \frac{2 \cdot \int_0^{+\infty} \mathrm{IF}_f^2(\xi) f(\xi) d\xi}{\left(2 \cdot \int_0^{+\infty} \mathrm{IF}_f(\xi) \xi f(\xi) d\xi\right)^2} \tag{15}$$

$$V = \frac{2 \cdot \int_0^{+\infty} \mathrm{IF}^2(\xi) f(\xi) d\xi}{\left(2 \cdot \int_0^{+\infty} \mathrm{IF}(\xi) \xi f(\xi) d\xi\right)^2} \tag{16}$$

(4) The expression of relative efficiency can be obtained from Equation (9). Making the expression equal to 0.95, a univariate equation on the tuning parameter c_p can be

constructed, as shown in Equation (17). By numerical calculation, the tuning parameter at the relative efficiency level of 95% in respect to the Normal distribution is 1.5424.

$$\frac{V_f}{V} = \frac{\frac{2\int_0^{+\infty} \xi^2 \cdot \frac{1}{\sigma\sqrt{2\pi}} \exp(-(\xi^2/2))d\xi}{\left[2\int_0^{+\infty} \xi^2 \cdot \frac{1}{\sigma\sqrt{2\pi}} \exp(-(\xi^2/2))d\xi\right]^2}}{\frac{2\int_0^{+\infty} \frac{\xi^2 \exp(-2(\xi/c_p)^2)[1+\exp(-(\xi/c_p)^4)+2\xi^2(1-\exp(-(\xi/c_p)^2))/c_p^2]^2}{4[1+\exp(-(\xi/c_p)^4]^4} \cdot \frac{1}{\sigma\sqrt{2\pi}} \exp(-(\xi^2/2))d\xi}{\left[2\int_0^{+\infty} \frac{\xi^2 \exp(-(\xi/c_p)^2)[1+\exp(-(\xi/c_p)^4)+2\xi^2(1-\exp(-(\xi/c_p)^2))/c_p^2]}{2[1+\exp(-(\xi/c_p)^4]^2} \cdot \frac{1}{\sigma\sqrt{2\pi}} \exp(-(\xi^2/2))d\xi\right]^2}} = 0.95 \quad (17)$$

To more accurately reflect the competitive relationship between the relative efficiency and robustness of the M-estimator, the comparative images of the objective function and influence function of the proposed, Xie and Welsch estimators are shown in Figure 1. Relative efficiency levels are 90%, 95%, 98% and 99%, respectively. The tuning parameters of the three estimators at different relative efficiency levels are presented in Table 1. Table 1 and Figure 1 show that the higher the relative efficiency level, the higher the values of the tuning parameters for the three estimators. The higher the tuning parameter, the slower the convergence speed of the influence function to 0 and the weaker the robustness. Hence, when comparing the robustness of distinct estimators, the comparative analysis has to be performed at the same level of relative efficiency.

Table 1. Tuning parameters of three estimators at distinct relative efficiency levels.

	M-Estimator	Tuning Parameter
1	Welsch	$c_w = 2.3828 \quad E_{ff} = 90\%$ $c_w = 2.9846 \quad E_{ff} = 95\%$ $c_w = 3.9077 \quad E_{ff} = 98\%$ $c_w = 4.7343 \quad E_{ff} = 99\%$
2	Xie	$c_x = 1.6705 \quad E_{ff} = 90\%$ $c_x = 1.9597 \quad E_{ff} = 95\%$ $c_x = 2.3409 \quad E_{ff} = 98\%$ $c_x = 2.6359 \quad E_{ff} = 99\%$
3	Proposed	$c_p = 1.3082 \quad E_{ff} = 90\%$ $c_p = 1.5424 \quad E_{ff} = 95\%$ $c_p = 2.0942 \quad E_{ff} = 98\%$ $c_p = 2.6060 \quad E_{ff} = 99\%$

In order to further analyze the effectiveness of the proposed estimator, the objective function and influence function of the proposed estimator are compared with the other four robust estimators. The four types of robust estimators are Fair, Cauchy, Welsch and Xie. The objective functions of Fair and Cauchy are shown in Equations (18) and (19):

Fair:

$$\rho(\xi_i) = c_f^2\left(\frac{|\xi_i|}{c_f} - \ln\left(1 + \frac{|\xi_i|}{c_f}\right)\right) \quad (18)$$

Cauchy:

$$\rho(\xi_i) = c_c^2 \ln\left(1 + \frac{\xi_i^2}{c_c^2}\right) \quad (19)$$

where c_f and c_c are, respectively, tuning parameters of Fair and Cauchy estimators. The tuning parameters of the five M-estimators under the level of 95% relative efficiency are $c_f = 1.3998$, $c_w = 2.9846$, $c_c = 2.3849$, $c_x = 1.9597$ and $c_p = 1.5424$. The results of the comparison are illustrated in Figures 2 and 3.

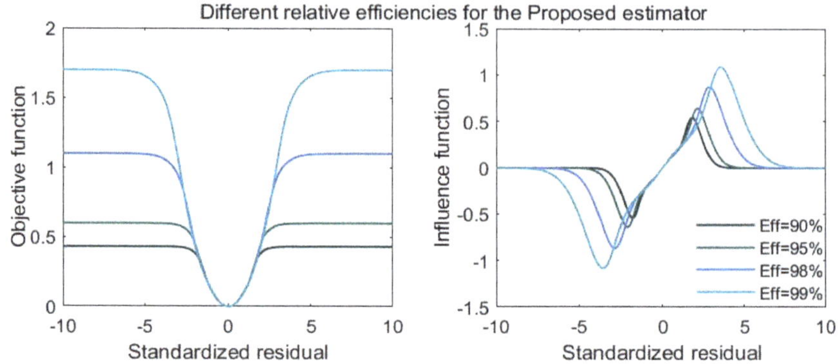

(**a**) Proposed

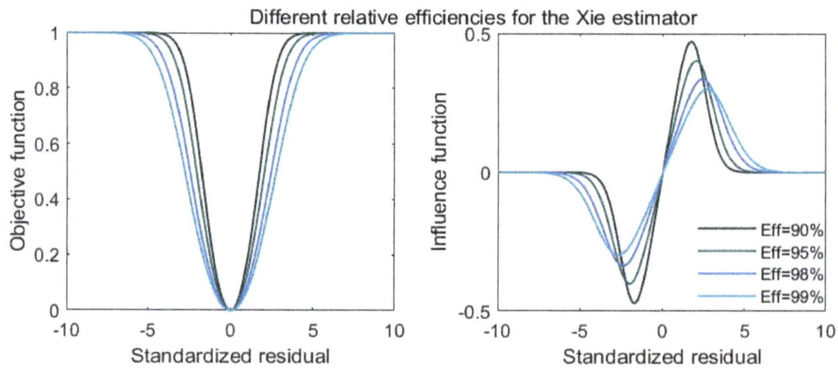

(**b**) Xie

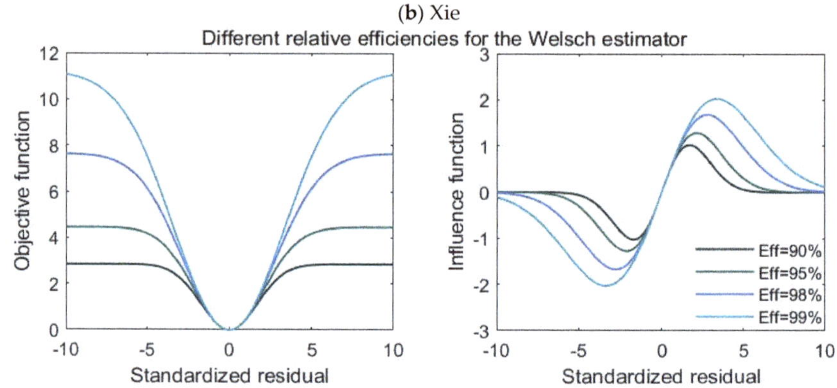

(**c**) Welsch.

Figure 1. Characteristicfunctions of three estimators at distinct relative efficiency levels.

As can be seen from Figures 2 and 3, the objective function and influence function of the proposed M-estimator satisfy their general characteristics. When the standardized residual is low, the objective and influence functions of several robust estimators are all close, which indicates that low measurement errors have little effect on these estimators. When the standardized residual is large, the objective function of the Fair and Cauchy estimator increases at a higher rate. Their influence functions do not converge rapidly and eventually converge to a nonzero value. It shows that both Fair and Cauchy estimators are sensitive to gross errors. Although the objective function of Welsch does not diverge, it tends to be a greater constant than Xie and the novel robust estimator. The objective functions of the novel robust estimator and Xie estimator are less than those of the other three estimators in the case of gross errors in measurements. Furthermore, the proposed M-estimator tends to be a constant lower than Xie. It can be seen further from Figure 2 that when there are gross errors in measurements, the influence function of the proposed M-estimator decreases much faster than that of Xie. When the standardized residual exceeds 4, the influence function of the proposed M-estimator converges to 0. Nevertheless, the influence function of Xie converges to 0 only when the standardized residual is above 5. Hence one can see that the convergence speed of influence function of the novel robust estimator is faster than that of Xie, and the effect of suppressing gross errors is excellent. The proposed method is therefore more robust and efficient than other estimators.

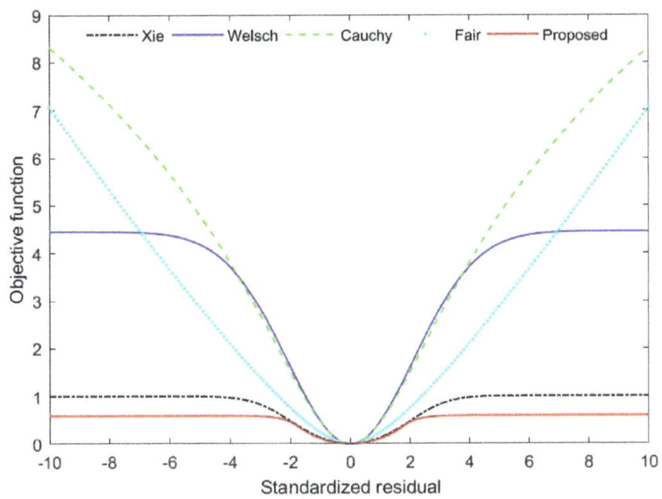

Figure 2. Objective functions for Xie, Welsch, Cauchy, Fair and Proposed estimator.

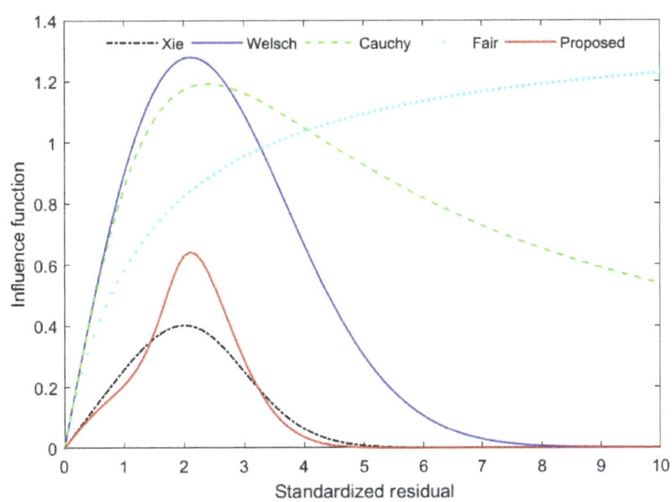

Figure 3. Influence functions for Xie, Welsch, Cauchy, Fair and Proposed estimator.

Next, two numerical examples are used to further verify the effectiveness of the proposed robust estimator, and the proposed robust estimator is compared with several estimators. The data reconciliation procedure can be interpreted mathematically as an optimization problem. In this paper, the state transition algorithm (STA) is applied as the optimization method for data reconciliation problems [38]. The local search operator, the global search operator and the heuristic search operator are adopted as the state transformation operators. This can prevent the solution process from falling into the local optimum and reduce the search time for the global optimum.

3.3. Linear Case

In this part, the measurement network [26] as shown in Figure 4 is adopted. The network is a linear structure consisting of four nodes and seven streams. As can be seen from Figure 4, the seven streams meet the following material balance

$$\begin{aligned} &x_1 - x_2 + x_4 + x_6 = 0 \\ &x_2 - x_3 = 0 \\ &x_3 - x_4 - x_5 = 0 \\ &x_5 - x_6 - x_7 = 0 \end{aligned} \tag{20}$$

where $x = [x_1, x_2, \ldots, x_7]^T$ is set of stream variables. In this linear example, all variables are assumed to be measured, and the true value of each variable is X = [5, 15, 15, 5, 10, 5, 5]. The true value of each variable is added with some random errors to get the corresponding measurements. Let the standard deviation of random errors be 2.5% of the true value of each stream. The diagonal matrix of the variance is as follows:

$$\sum = 0.01 \times diag(1.562, 14.062, 14.062, 1.562, 6.250, 1.562, 1.562) \tag{21}$$

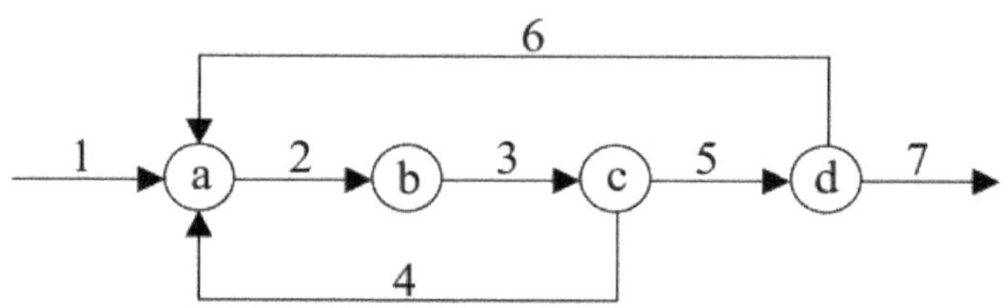

Figure 4. Diagram of the measurement network.

For a precise analysis of the performance of each robust estimator, SSE (sum of squares due to error), TER (total error reduction) and RER (relative error reduction) [27] are used as indicators to assess the precision of the reconciled results. The definitions included are:

$$\text{SSE} = \sum_{i=1}^{n} (\tilde{x}_i - x_i^*)^2 + \sum_{j=1}^{m} (\tilde{u}_j - u_j^*)^2 \tag{22}$$

$$\text{TER} = \frac{\sqrt{\sum_{i=1}^{n} \left(\frac{x_i - x_i^*}{\sigma_i}\right)^2} - \sqrt{\sum_{i=1}^{n} \left(\frac{\tilde{x}_i - x_i^*}{\sigma_i}\right)^2}}{\sqrt{\sum_{i=1}^{n} \left(\frac{x_i - x_i^*}{\sigma_i}\right)^2}} \tag{23}$$

$$\text{RER} = \frac{\sum_{i=1}^{n} \left(\frac{|x_i^* - x_i|}{x_i^*} - \frac{|x_i^* - \tilde{x}_i|}{x_i^*}\right)}{\sum_{i=1}^{n} \frac{|x_i^* - x_i|}{x_i^*}} \tag{24}$$

where $\tilde{x}_i$ represents the reconciled data for the ith measurable variable; x_i^* represents the true value for the ith measurable variable; x_i represents the measured data for the ith measurable variable; $\tilde{u}_j$ is the estimated value of the jth unmeasurable variable; u_j^* is the true value of the jth unmeasurable variable; σ_i denotes the standard deviation for the ith measurable variable; n is the total number of measurable variables; m is the total number of unmeasurable variables. The smaller the SSE, the closer the reconciled results are to the true value. The larger the TER, the more accurate the reconciled results. The larger RER shows that the reconciled results are less affected by gross errors.

3.3.1. There Are Two Gross Errors in the Measurement Variables

Streams 2 and 5 are selected to add gross errors in the magnitude of 2 and 1, respectively, and other streams contain only random errors. The results of the data reconciliation and indicators for several methods are presented in Table 2. For convenience, "Proposed" is used to represent the proposed estimator in this table and subsequent tables.

Table 2. Reconciled results of different methods with two gross errors.

Stream	True	Original Meas.	Meas. with Gross Error	Proposed	Xie	Welsch	Cauchy	Fair
x_1	5	4.995	4.995	5.0111	5.0152	5.0346	5.0519	5.1777
x_2	15	14.91	16.91	15.0420	15.0540	15.1202	15.1970	15.6413
x_3	15	15.01	15.01	15.0420	15.0540	15.1202	15.1970	15.6413
x_4	5	5.002	5.002	4.9987	4.9985	5.0053	5.0230	5.1795
x_5	10	9.98	10.98	10.0433	10.0555	10.1149	10.1740	10.4618
x_6	5	5.019	5.019	5.0322	5.0403	5.0803	5.1221	5.2841
x_7	5	5.014	5.014	5.0111	5.0152	5.0346	5.0519	5.1777
SSE	–	–	–	0.0067	0.0110	0.0510	0.1287	1.2117
TER	–	–	–	0.9424	0.9263	0.8457	0.7606	0.2953
RER	–	–	–	0.9101	0.8838	0.7501	0.6008	−0.2626

Table 2 shows that the SSE of the proposed robust estimator (Proposed) is much lower than that of Fair with 1.2117 and Cauchy with 0.1287 and is less than that of Xie and Welsch estimator. The TER of the novel robust estimator is much larger than that of Fair with 0.2953 and Cauchy with 0.7606, and higher than that of Xie and Welsch estimator. In addition, the RER of Proposed is also higher than that of the other four robust estimators. The results show that in the case of two gross errors, the reconciled results of the novel robust estimator are closer to the true value, and the proposed estimator performs better than other methods.

3.3.2. There Are Three Gross Errors in the Measurement Variables

Streams 2, 5 and 7 are selected to add gross errors with the magnitudes of 2, 1 and −1.5, respectively, and other streams contain only random errors. The results of the data reconciliation and indicators for several methods are presented in Table 3.

Table 3. Reconciled results of different methods with three gross errors.

Stream	True	Original Meas.	Meas. with Gross Error	Proposed	Xie	Welsch	Cauchy	Fair
x_1	5	4.995	4.995	5.0080	5.0164	5.0574	5.0323	4.3533
x_2	15	14.91	16.91	15.0390	15.0552	15.1425	15.1802	15.1833
x_3	15	15.01	15.01	15.0390	15.0552	15.1425	15.1802	15.1833
x_4	5	5.002	5.002	4.9991	4.9984	5.0037	5.0244	5.3066
x_5	10	9.98	10.98	10.0399	10.0568	10.1388	10.1558	9.8767
x_6	5	5.019	5.019	5.0320	5.0404	5.0814	5.1235	5.5233
x_7	5	5.014	3.514	5.0080	5.0164	5.0574	5.0323	4.3533
SSE	–	–	–	0.0058	0.0115	0.0731	0.1071	1.2866
TER	–	–	–	0.9743	0.9642	0.9111	0.8954	0.3474
RER	–	–	–	0.9641	0.9470	0.8621	0.8446	0.1268

Table 3 shows that the SSE of the proposed robust estimator (Proposed) is much lower than that of Fair with 1.2866 and Cauchy with 0.1071 and is less than that of Xie and Welsch estimator. The TER of the novel robust estimator is much larger than that of Fair with 0.3474, and higher than that of Xie, Welsch and Cauchy estimator. In addition, the RER of Proposed is also higher than that of the other four robust estimators. The results show that in the case of three gross errors, the reconciled results of the novel robust estimator are closer to the true value, and the proposed estimator performs better than other methods.

In order to better illustrate the effectiveness of the proposed estimator, the linear numerical example is divided into three groups for comparative experiments. Each group comprises 100 data samples. In the first group, there is gross error in stream 2 (x_2). In group two, there are gross errors in stream 2 (x_2) and stream 5 (x_5). The third group contains gross errors in stream 2 (x_2), stream 5 (x_5) and stream 7 (x_7). In each group of experiments, the selected variables are added with gross errors ranging from 3 times to 9 times the standard deviation. The measurement test (MT) is used to detect gross errors. Two criteria are selected to assess the performance of different methods, including the observed power (*OP*) and the Average Number of Type I errors (*ATVI*) [2]. *OP* and *ATVI* are defined as follows:

$$OP = \frac{\text{The number of gross errors detected correctly}}{\text{The total number of gross errors}} \tag{25}$$

$$ATVI = \frac{\text{The number of gross errors detected mistakenly}}{\text{The total number of simulations}} \tag{26}$$

OP indicates the number of gross errors correctly detected. *ATVI* indicates the number of gross errors mistakenly detected. The larger value of the *OP* is, the stronger ability of the estimator to identify gross errors. The lower value of the *ATVI* is, the less frequently gross errors are detected incorrectly. The statistical graph of the process in which gross errors are detected by distinct methods in each group is shown in Figure 5. The abscissa of each graph corresponds to seven stream variables. The y-axis is the number of times each variable is detected to contain gross errors. The *OP* and *ATVI* of the proposed estimator, Xie, Welsch, Cauchy and Fair are obtained by statistical analysis of the process diagram, as illustrated in Figure 6.

As depicted in Figure 6, when one variable, two variables and three variables contain gross errors, the *OP* obtained by the proposed estimator is higher than that obtained by other estimators. The result shows that the proposed estimator has a higher probability of identifying gross errors correctly. In addition, the *AVTI* obtained by the proposed estimator is lower than that obtained by other methods. It can also be seen from Figure 5 that the proposed estimator has a lower number of gross errors mistakenly detected than other estimators. It is shown that gross errors have less influence on the data reconciliation based on the Proposed estimator.

3.4. Nonlinear Case

A nonlinear numerical example [22] is used in this part. Within the nonlinear system, there are five measurable variables, three unmeasurable variables and six nonlinear constraint equations, which are described as Equation (27):

$$\begin{cases} 0.5x_1^2 - 0.7x_2 + x_3u_1 + x_2^2u_1u_2 + 2x_3u_3^2 - 255.8 = 0 \\ x_1 - 2x_2 + 3x_1x_3 - 2x_2u_1 - x_2u_2u_3 + 111.2 = 0 \\ x_3u_1 - x_1 + 3x_2 + x_1u_2 - x_3\sqrt{u_3} - 33.57 = 0 \\ x_4 - x_1 - x_3^2 + u_2 + 3u_3 = 0 \\ x_5 - 2x_3u_2u_3 = 0 \\ 2x_1 + x_2x_3u_1 + u_2 - u_3 - 126.6 = 0 \end{cases} \tag{27}$$

where $x_i (i = 1,2,\ldots,5)$ is the measured variable and $u_j (j = 1,2,3)$ is the unmeasured variable. Measurements of the measured variables are obtained from the true value added, some random errors and gross errors. The true value, the measured value, the standard deviation of the measured variables and the true value of the unmeasured variables are given in Table 4. Amongst these, x_2 , x_3 and x_5 contain gross errors. The results of the data reconciliation and indicators for several methods are presented in Table 4.

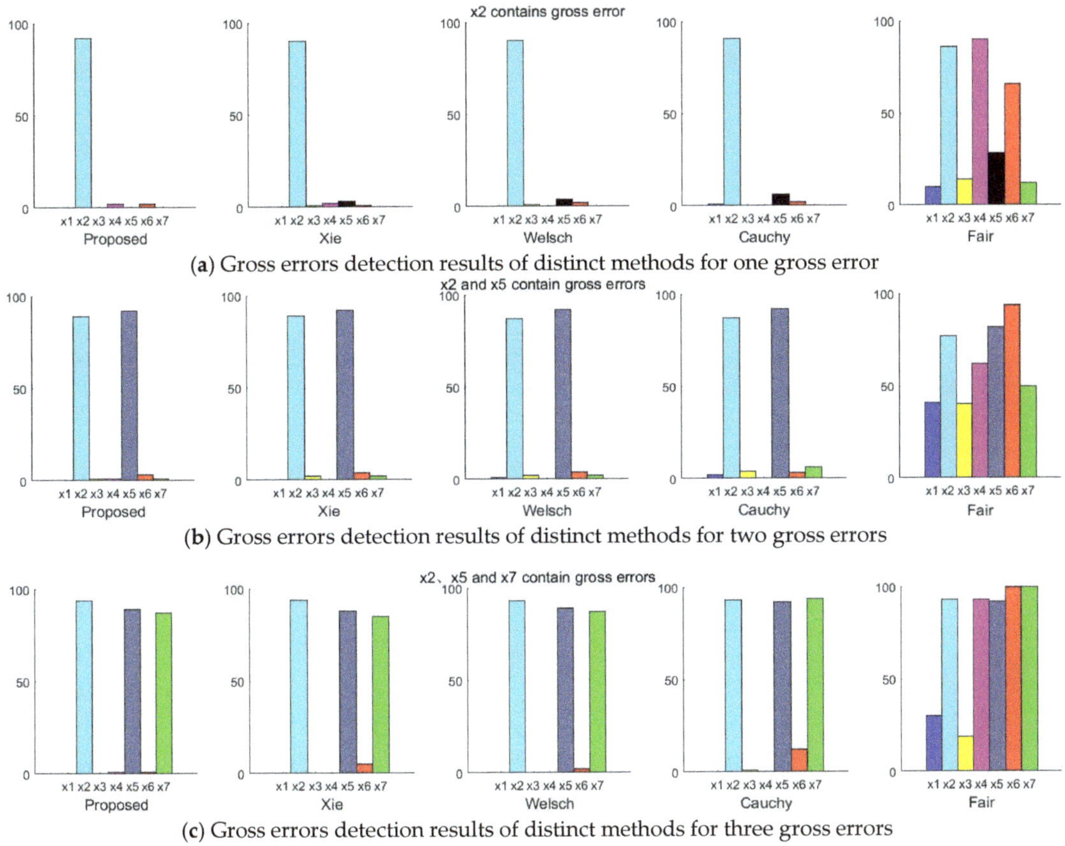

Figure 5. Statistical diagram of gross errors detection process for distinct methods.

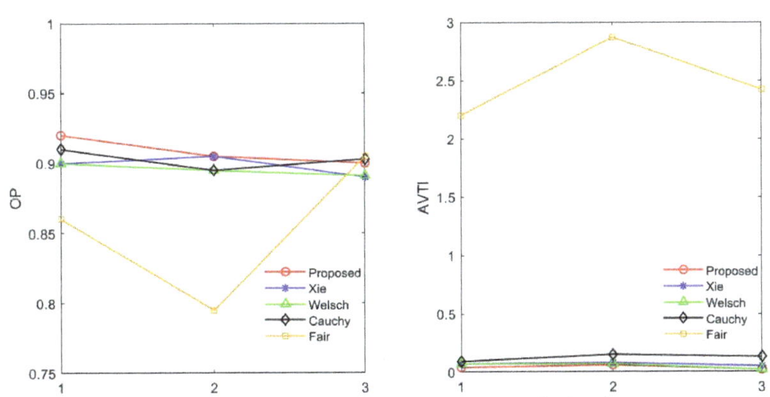

Figure 6. Results of *OP* and *AVTI* with different methods.

Table 4. Reconciled results of different methods for nonlinear constraints.

Stream	Standard Deviation	True	Meas.	Proposed	Xie	Welsch	Cauchy	Fair
x_1	0.5	4.5124	4.5360	4.5378	4.5280	4.5379	4.5796	4.4727
x_2	0.6	5.5819	5.9070	5.5754	5.5655	5.5331	5.5360	5.6514
x_3	0.2	1.9260	1.8074	1.9221	1.9223	1.9200	1.9153	1.9321
x_4	0.2	1.4560	1.4653	1.4653	1.4842	1.4924	1.5096	1.4655
x_5	0.5	4.8545	4.5491	4.8156	4.8010	4.8083	4.7882	4.7870
u_1	–	11.070	–	11.0988	11.1178	11.1939	11.2079	10.9076
u_2	–	0.6147	–	0.6143	0.6160	0.6187	0.6168	0.6104
u_3	–	2.0504	–	2.0469	2.0349	2.0317	2.0345	2.0372
SSE	–	–	–	0.0031	0.0067	0.0222	0.0333	0.0377
TER	–	–	–	0.8950	0.8192	0.7751	0.6631	0.7994
RER	–	–	–	0.8805	0.8008	0.7324	0.5930	0.7693

Table 4 shows that the SSE calculated by the proposed robust estimator is significantly less than that calculated by Fair, Cauchy and Welsch estimator, and it is less than that calculated by Xie estimator. Furthermore, the TER and RER of the proposed estimator are larger than those of the other four estimators. The results show tha for the nonlinear case, the reconciled results of the new robust estimator are also near the true value. Consequently, the proposed robust estimator can effectively suppress the effect of gross errors on the reconciled results and has greater robustness and excellent effectiveness.

4. Feeding Composition Estimation Based on Iterative Data Reconciliation

In order to address the problem caused by the unknown change in feeding composition, an iterative robust hierarchical data reconciliation and estimation strategy is proposed. The method is applied to a fluidized bed roaster for zinc smelting. Firstly, the industrial process of fluidized bed roaster is introduced. At the same time, difficulties in reconciling the fluidized bed roaster are discussed in Section 3.1. Then, on the basis of the establishment of the mechanism balance model of the fluidized bed roaster, specific solutions are presented in accordance with the existing difficulties.

4.1. Industrial Process Description

The fluidized bed roaster is a type of thermal equipment that applies fluidization technology to make materials desulphurized by oxidation roasting in the metallurgical industry. The roasting process of the fluidized bed roaster is a gas-solid reaction procedure. By sending a great deal of air into the oven from the bottom, a very intense exothermic reaction occurs in the material layer of the sulfur ore under the agitation of the air. Oxygen combines with sulfide to form sulfur dioxide and precious metals are converted into oxides or sulphates. The structure diagram for the fluidized bed roaster is shown in Figure 7.

In the fluidized roasting process, accurate and reliable measurements are the basis of process control, functional analysis and production management. However, in the real process, due to inaccurate sensors, equipment leaks and measurement bias, real-time measurements are inevitably affected by random errors and gross errors. That has a serious effect on the accuracy of measurements. In addition, some important variables, such as smoke volume and calcine quality, cannot be measured due to limitations in measurement technology and the environment.

The ZnS content of the feed during the roasting of the zinc concentrate will affect the indices of operation, such as the resistance to roasting, the temperature of the boiling layer and the content in the gaseous effluents. Then, the technical-economic index of the level of zinc soluble in calcine is affected, which field staff can judge from according to roasting. The compositions of mixed zinc concentrate and calcine are sampled only twice a day, once in the morning and once at night. Then the percentage of Zn, S and other elements in the mixed zinc concentrate and that of soluble zinc and soluble sulfur in the calcine are detected by the laboratory. Nevertheless, the sampling frequency of the variables measured in real time, like feed volume and blast volume, is 5 seconds, which is much greater than that of the laboratory data. Hence, due to the long test cycle of the mixed zinc concentrate composition in the fluidized bed roaster, the laboratory data cannot reflect the change in the real-time feeding composition.

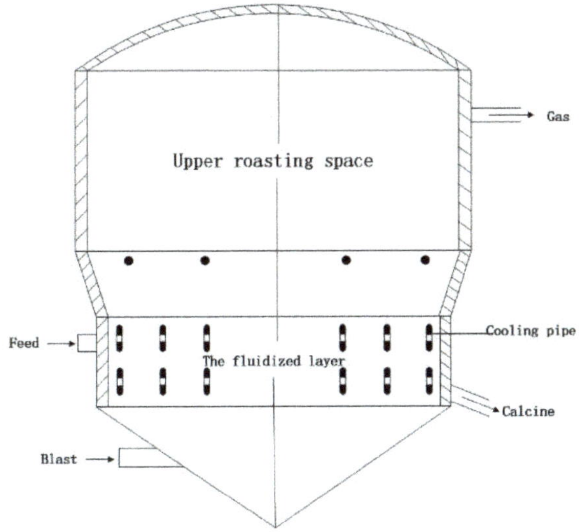

Figure 7. Structure diagram of fluidized bed roaster.

Concerning the problems of measurements contaminated by gross errors, irredundant process data and a long test cycle of the feeding composition, an iterative robust hierarchical data reconciliation and estimation strategy for feeding composition is proposed.

4.2. Iterative Robust Hierarchical Data Reconciliation and Composition Estimation Framework

Figure 8 illustrates the framework of iterative robust hierarchical data reconciliation and estimation strategy for feeding composition. The major steps are as follows:

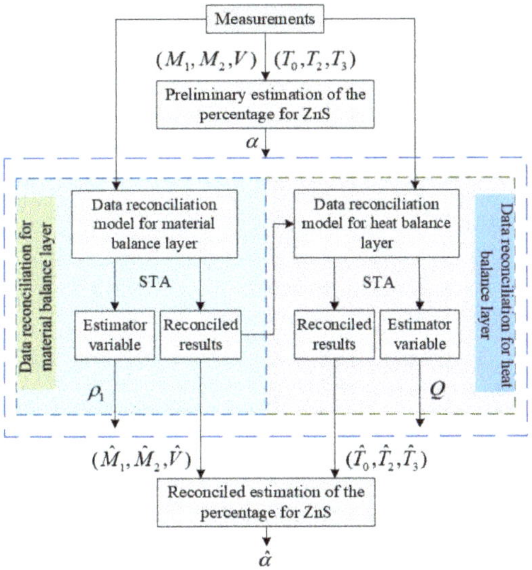

Figure 8. The framework of iterative robust hierarchical data reconciliation and composition estimation.

Step 1: Establish the steady-state mechanism model of the fluidized bed roaster The establishment of a reasonable process model is the basis for data reconciliation. However, the actual industrial process is complex and needs to be simplified accordingly. The following assumptions are put forward to establish a credible and comprehensible mechanism model:

(1) The only exothermal component of the mixed zinc concentrate is ZnS.

(2) Since the specific heat of calcine and soot is close and the mass ratio is close to 1:1, soot can be classified as calcine in the calculation.

Therefore, the material and heat balance at steady-state are expressed by Equations (28) and (29):

Material balance:

$$\frac{16}{97}\alpha(M_1+M_2)1000+V_0(\rho_0-\rho_1)=0 \tag{28}$$

Heat balance:

$$\begin{aligned}&1000(M_1+M_2)T_1cp_1+V_0T_0cp_2+\tfrac{443508}{97}\alpha(M_1+M_2)1000\\&=1000(1-\tfrac{16}{97}\alpha)(M_1+M_2)T_3cp_3+V_0T_2cp_0+Q\end{aligned} \tag{29}$$

where α denotes the percentage of ZnS in zinc concentrate; M_1 and M_2 represent the first and second feeding quantity, respectively; V_0 is the blast velocity; T_0, T_1, T_2 and T_3 denote the temperature variables for blast, zinc concentrate, gas and calcine, respectively; cp_0, cp_1, cp_2 and cp_3 are specific heat for gas, zinc concentrate, blast and calcine, respectively; Q describe the heat loss; ρ_0 and ρ_1 are, respectively, the density of air and gas. Among these variables, M_1, M_2, V_0, T_0, T_2 and T_3 are measured variables. ρ_1, Q and α are unmeasured variables. The other parameters are all fixed values.

Step 2: Preliminary estimation of α based on measurements and heat balance. Estimation of α can be considered as an optimization problem. Combined with the heat balance and measurements, the ZnS estimation model before data reconciliation could be expressed as follows:

$$\begin{aligned}\min Y_1=&(1000(M_1+M_2)T_1cp_1+V_0T_0cp_2+\tfrac{443{,}508}{97}\alpha(M_1+M_2)1000\\&-1000(1-\tfrac{16}{97}\alpha)(M_1+M_2)T_3cp_3-V_0T_2cp_0-Q)^2\end{aligned} \tag{30}$$

$$s.t.\quad \alpha_{\min}\le\alpha\le\alpha_{\max} \tag{31}$$

where Y_1 is the objective of the optimization problem; $\alpha_{\min}$ and $\alpha_{\max}$ are the top and bottom bounds of α, respectively. In this case, Q can be roughly considered as a fixed value, which can be obtained by expert knowledge. The preliminary estimation of α will be used to construct the hierarchical model in Step 3.

Step 3: Robust hierarchical data reconciliation. First, the robust reconciliation model of material balance layer is established to obtain the reconciled values of the first and second feeding quantity and blast velocity. Then, the reconciled values of the material balance layer are used as precise values to construct the robust reconciliation model of heat balance layer. Finally, the reconciled results of the temperature for the blast, gas and calcine can be obtained. Therefore, the hierarchical data reconciliation model based on the novel robust estimation is expressed as follows:

Material balance layer:

$$\begin{aligned}\min f_1&=\sum_{i=1}^{n}\rho(x_i,\hat{x}_i)\\&=\sum_{i=1}^{n}(1+((x_i-\hat{x}_i)^2/\sigma_i^2c_p^{14}-1)\exp(-((x_i-\hat{x}_i)/\sigma_ic_p)^2))\\&=1+((M_1-\hat{M}_1)^2/\sigma_{M_1}^2c_p^{14}-1)\exp(-((M_1-\hat{M}_1)/\sigma_{M_1}c_p)^2)+\\&1+((M_2-\hat{M}_2)^2/\sigma_{M_2}^2c_p^{14}-1)\exp(-((M_2-\hat{M}_2)/\sigma_{M_2}c_p)^2)+\\&1+((V_0-\hat{V}_0)^2/\sigma_{V_0}^2c_p^{14}-1)\exp(-((V_0-\hat{V}_0)/\sigma_{V_0}c_p)^2)+\\&1+(\|\rho_1-\tilde{\rho}_1\|^2/c_p^{14}-1)\exp(-(\|\rho_1-\tilde{\rho}_1\|/c_p)^2)\end{aligned} \tag{32}$$

$$s.t.\begin{cases} \frac{16}{97}\alpha(\hat{M}_1+\hat{M}_2)+\hat{V}_0(\rho_0-\rho_1)=0 \\ \hat{M}_{1\min}\le\hat{M}_1\le\hat{M}_{1\max} \\ \hat{M}_{2\min}\le\hat{M}_2\le\hat{M}_{2\max} \\ \hat{V}_{0\min}\le\hat{V}_0\le\hat{V}_{0\max} \\ \rho_{1\min}\le\rho_1\le\rho_{1\max} \end{cases} \tag{33}$$

where f_1 is the objective function of the robust estimator; x_i and $\hat{x}_i$ represent the measurement and reconciled data for the ith measured variable respectively; $\hat{M}_{1\min}$, $\hat{M}_{2\min}$, $\hat{V}_{0\min}$ and $\rho_{1\min}$ denote the bottom bounds of the reconciled results for the first and second feeding quantity, blast velocity and the density of the air, respectively. $\hat{M}_{1\max}$, $\hat{M}_{2\max}$, $\hat{V}_{0\max}$ and $\rho_{1\max}$ denote the top bounds of the reconciled results for the first and second feeding quantity, blast velocity and the density of the air, respectively.

Heat balance layer:

$$\begin{aligned} \min f_2 &= \sum_{i=1}^{n}\rho(x_i,\hat{x}_i) \\ &= \sum_{i=1}^{n}(1+((x_i-\hat{x}_i)^2/\sigma_i^2c_p^{14}-1)\exp(-((x_i-\hat{x}_i)/\sigma_ic_p)^2)) \\ &= 1+((T_0-\hat{T}_0)^2/\sigma_{T_0}^2c_p^{14}-1)\exp(-((T_0-\hat{T}_0)/\sigma_{T_0}c_p)^2)+ \\ &\quad 1+((T_2-\hat{T}_2)^2/\sigma_{T_2}^2c_p^{14}-1)\exp(-((T_2-\hat{T}_2)/\sigma_{T2}c_p)^2)+ \\ &\quad 1+((T_3-\hat{T}_3)^2/\sigma_{T_3}^2c_p^{14}-1)\exp(-((T_3-\hat{T}_3)/\sigma_{T_3}c_p)^2)+ \\ &\quad 1+(\|Q-\tilde{Q}\|^2/c_p^{14}-1)\exp(-(\|Q-\tilde{Q}\|/c_p)^2) \end{aligned} \tag{34}$$

$$s.t.\begin{cases} 1000(\hat{M}_1+\hat{M}_2)T_1cp_1+\hat{V}_0\hat{T}_0cp_2+\frac{443508}{97}\alpha(\hat{M}_1+\hat{M}_2)1000 \\ =1000(1-\frac{16}{97}\alpha)(\hat{M}_1+\hat{M}_2)\hat{T}_3cp_3+\hat{V}_0\hat{T}_2cp_0+Q \\ \hat{T}_{0\min}\le\hat{T}_0\le\hat{T}_{0\max} \\ \hat{T}_{2\min}\le\hat{T}_2\le\hat{T}_{2\max} \\ \hat{T}_{3\min}\le\hat{T}_3\le\hat{T}_{3\max} \\ Q_{\min}\le Q\le Q_{\max} \end{cases} \tag{35}$$

where f_2 is the objective function of the robust estimator; $\hat{T}_{0\min}$, $\hat{T}_{2\min}$, $\hat{T}_{3\min}$ and $Q_{\min}$ denote the bottom bounds of the reconciled results for the temperature of the blast, gas, calcine and heat loss, respectively; $\hat{T}_{0\max}$, $\hat{T}_{2\max}$, $\hat{T}_{3\max}$ and $Q_{\max}$ denote the top bounds of the reconciled results for the temperature of the blast, gas, calcine and heat loss, respectively.

Step 4: Reconciled estimation of $\hat{\alpha}$ based on reconciled results and heat balance. The reconciled results obtained in Step 3 are substituted for the objective function of the optimization problem based on heat balance. Then the estimated value of the percentage for ZnS after reconciliation could be obtained. The ZnS estimation model after data reconciliation can be represented as follows:

$$\begin{aligned} \min Y_2 = (&1000(\hat{M}_1+\hat{M}_2)T_1cp_1+\hat{V}_0\hat{T}_0cp_2+\tfrac{443,508}{97}\hat{\alpha}(\hat{M}_1+\hat{M}_2)1000 \\ &-1000(1-\tfrac{16}{97}\hat{\alpha})(\hat{M}_1+\hat{M}_2)\hat{T}_3cp_3-\hat{V}_0\hat{T}_2cp_0-Q)^2 \end{aligned} \tag{36}$$

$$s.t.\hat{\alpha}_{\min}\le\hat{\alpha}\le\hat{\alpha}_{\max} \tag{37}$$

where Y_2 is the objective of the optimization problem; $\hat{\alpha}$ represents the estimated value of percentage for ZnS after reconciliation; $\hat{\alpha}_{\min}$ and $\hat{\alpha}_{\max}$ denote the bottom and top bounds of $\hat{\alpha}$ respectively. Through the iterative process of the above steps, greater accuracy and a smaller range of ZnS fluctuation in the feed can be obtained. Therefore, staff can better understand changes of the feeding composition and make adjustments as required.

5. Results from Real Industrial Data

To further demonstrate the effectiveness of the proposed method, 100 steady-state samples are collected from the real industry of a fluidized bed roaster. In addition, the data reconciliation results of the proposed estimator are compared with those of Xie estimator.

The results of the comparison between the two methods are presented in Figures 9–14, which respectively present the reconciled results for the first feeding quantity, the second feeding quantity, blast velocity, blast temperature, gas temperature and calcine temperature. From Figures 9–14, it is known that the reconciled results by the novel robust estimator have a smaller range of fluctuation than that obtained by Xie estimator. It shows that whatever the uncertainty of the fluctuation of the measurements and the number of gross errors they contain, the data reconciliation approach based on the novel robust estimator is highly robust.

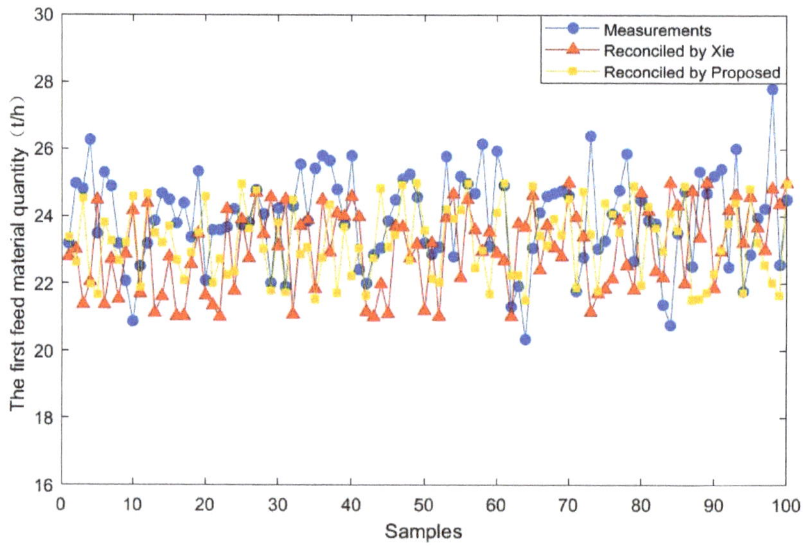

Figure 9. Reconciled results for the first feed material quantity.

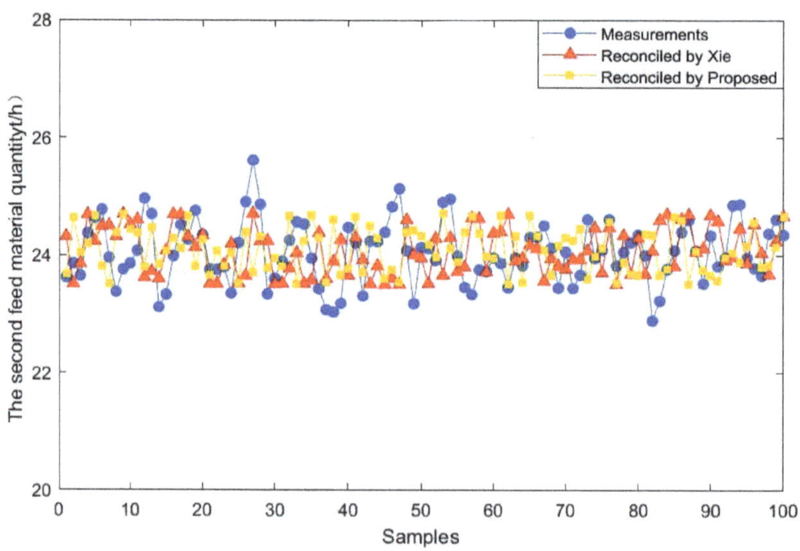

Figure 10. Reconciled results for the second feed material quantity.

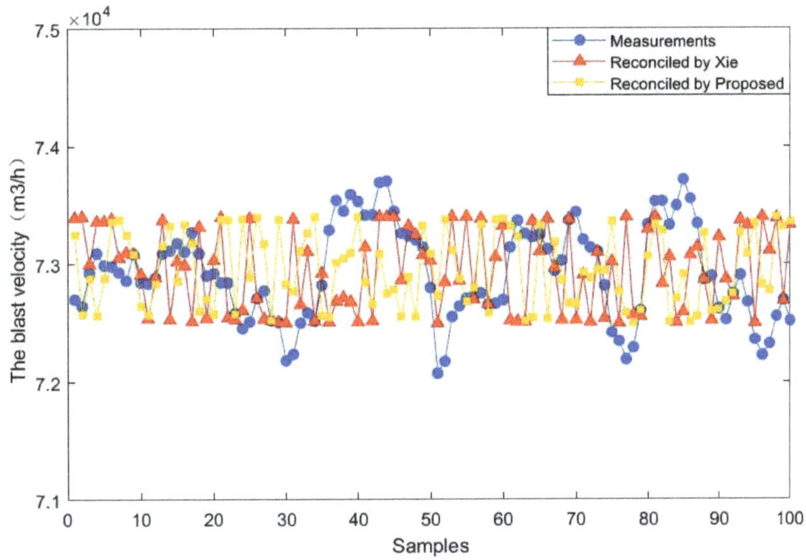

Figure 11. Reconciled results for the blast velocity.

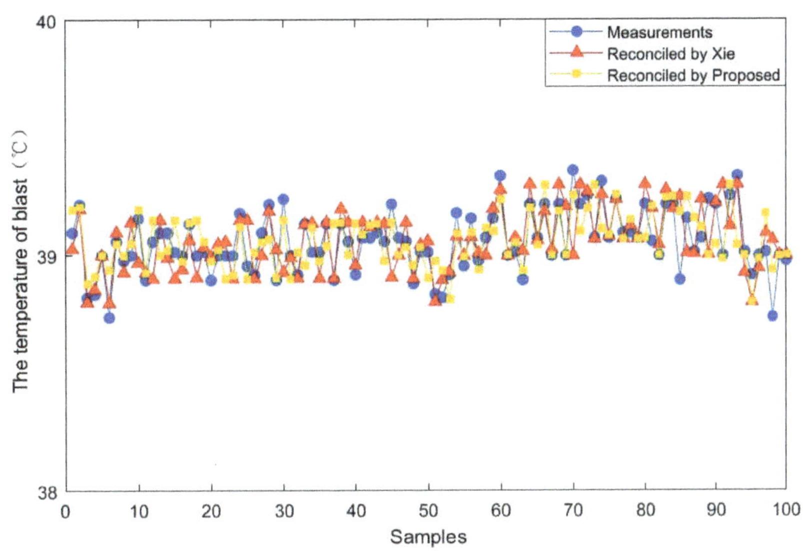

Figure 12. Reconciled results for the temperature of blast.

In addition, the standard deviations of the reconciled results and measurements for six variables obtained from the two methods are calculated, and the results of the comparison are presented in Figure 15. Figure 15 shows that, for the first feeding quantity, the seconding feed quantity, the blast velocity and the temperature of blast, the standard deviation of the reconciled results based on the novel robust estimator is less than Xie. For the temperature of gas and calcine, the standard deviations of reconciled results for the two methods are closer, because the measurement fluctuation ranges for both two variables are all very small. The results demonstrate that the data reconciliation based on the novel robust estimator provides better reconciliation results for variables with a large fluctuation range.

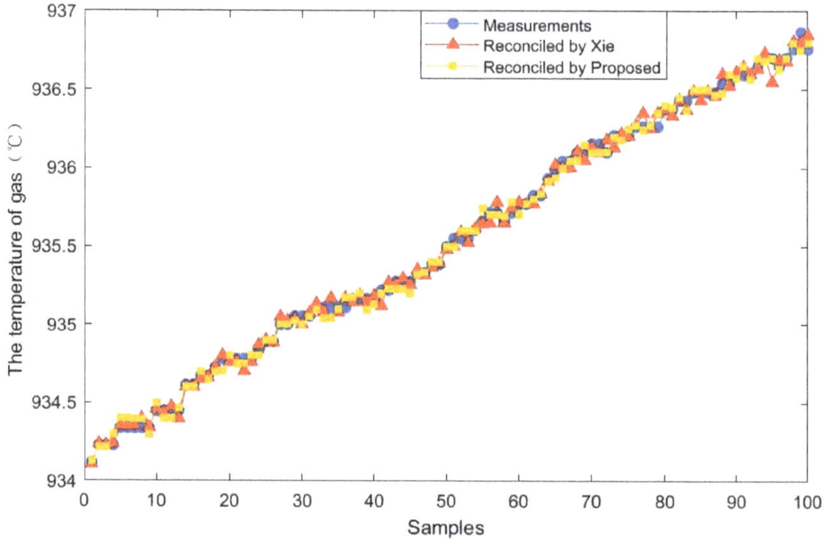

Figure 13. Reconciled results for the temperature of gas.

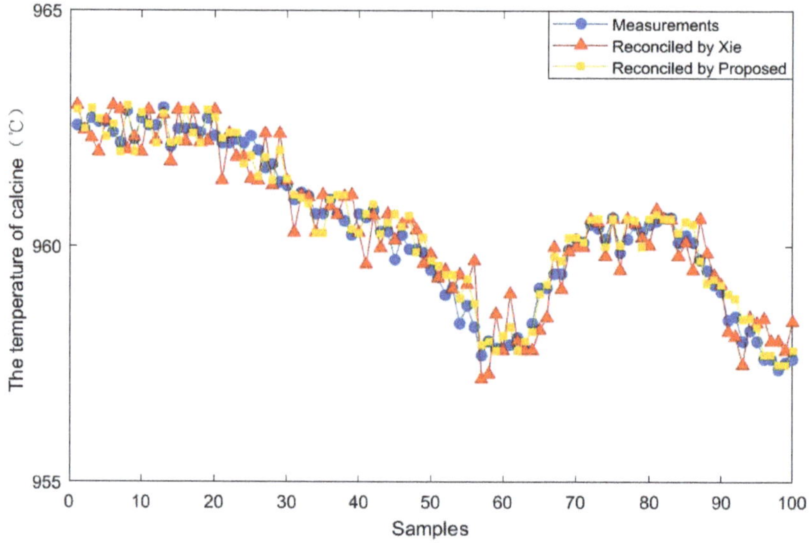

Figure 14. Reconciled results for the temperature of calcine.

For the change in feeding composition, the above 100 steady-state samples are also collected to compare the percentage of ZnS before and after data reconciliation. Experimental results are shown in Figures 16 and 17. It is evident that the ZnS fluctuation range obtained by the proposed method is smaller than that before data reconciliation. It illustrates that the percentage of ZnS estimated by the reconciled results is more accurate and could better reflect the change of feeding composition over a period of time.

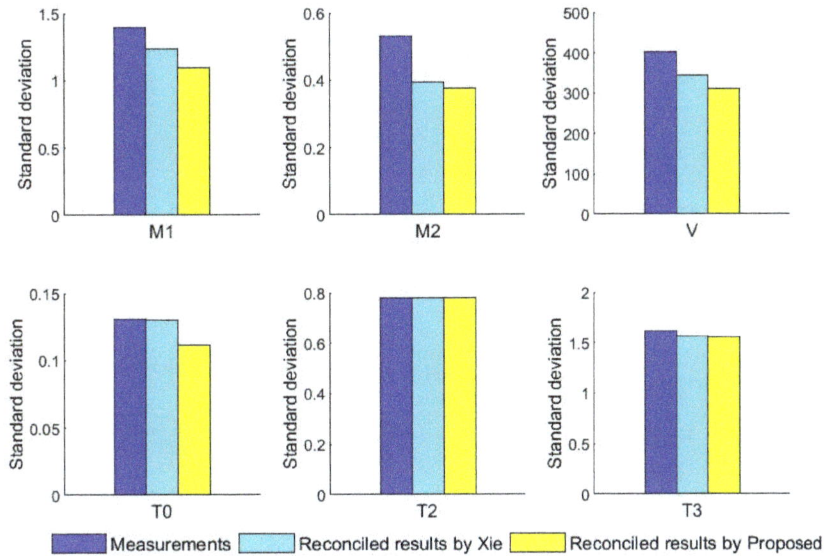

Figure 15. The standard deviation of reconciled results and measurements.

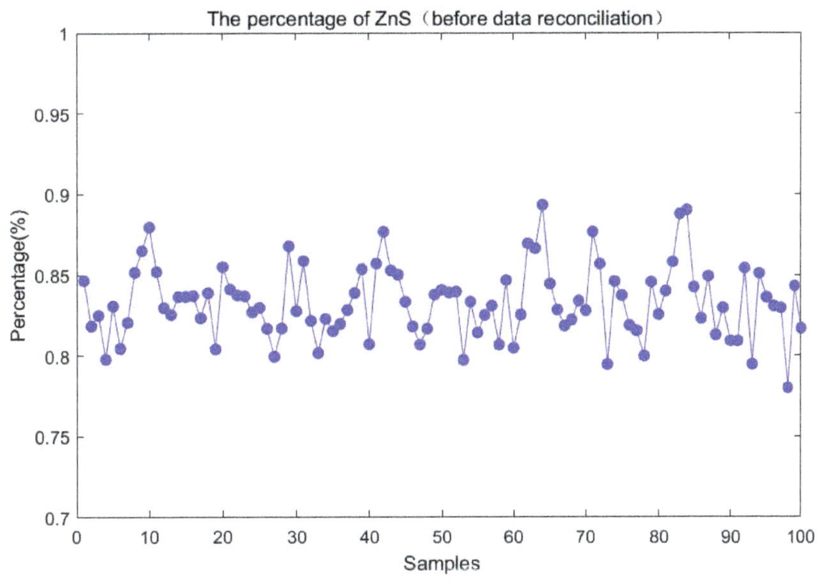

Figure 16. The percentage of ZnS before data reconciliation.

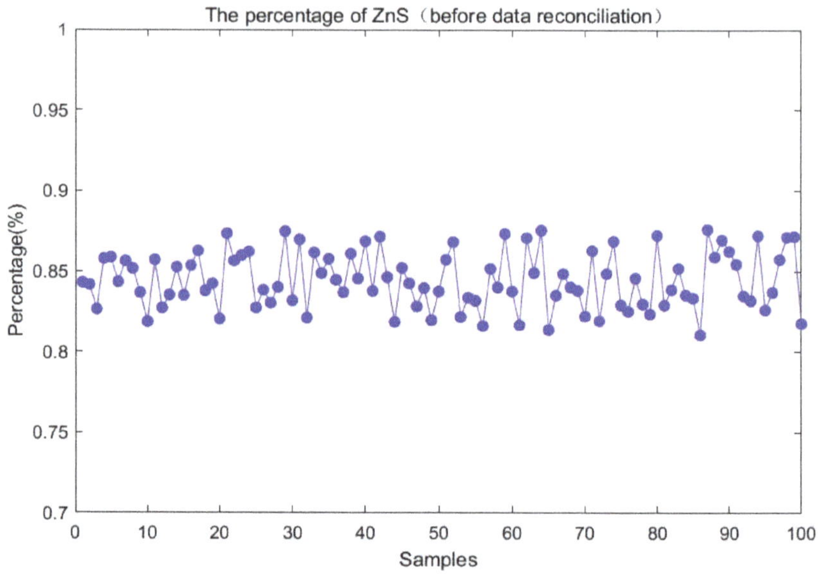

Figure 17. The percentage of ZnS after data reconciliation.

6. Conclusions

The feeding composition in an industrial process can better reflect the state of production. In this paper, estimation of feeding composition based on robust data reconciliation is proposed to account for the unknown change in feeding composition. According to the robust estimation theory, a novel robust estimator is presented to handle the measurements with random errors and gross errors. In comparison with the objective function and influence function of other robust estimators, the proposed estimator is proved to be more robust. Besides, the linear and nonlinear numerical examples are used to further verify the effectiveness of the estimator. Then, an iterative robust hierarchical data reconciliation and estimation of feeding composition strategy is proposed and used for a fluidized bed roaster. The reconciled and estimated results show that the proposed strategy is beneficial for the estimation of feeding composition in the actual industrial process.

Author Contributions: Conceptualization, Y.L. and B.S.; methodology, Y.L.; software, Y.L.; validation, Y.L.; formal analysis, Y.L.; investigation, Y.L., Z.F. and M.J.; resources, Y.L.; data curation, Y.L. and Z.F.; writing—original draft preparation, Y.L.; writing—review and editing, Y.L. and B.S.; visualization, Y.L.; supervision, Y.L.; project administration, Y.L.; funding acquisition, B.S. All authors have read and agreed to the published version of the manuscript.

Funding: This research was funded by the National Natural Science Foundation of China (61973321) and the Natural Science Foundation of Hunan Province (Grant No. 2019JJ50823).

Data Availability Statement: Publicly available datasets were analyzed in this study. It can be found here: https://open.library.ubc.ca/cIRcle/collections/ubctheses/831/items/1.0078522 (accessed on 1 February 2021).

Conflicts of Interest: The authors declare no conflict of interest.

References

1. Kuehn, D.R.; Davidson, H. Computer control II. Mathematics of control. *Chem. Eng. Prog.* **1961**, 57, 44–47.
2. Narasimhan, S.; Jordache, C. *Data Reconciliation and Gross Error Detection: An Intelligent Use of Process Data*; Elsevier: Amsterdam, The Netherlands, 1999.

3. Prata, D.M.; Lima, E.L.; Pinto, J.C. In-Line Monitoring of Bulk Polypropylene Reactors Based on Data Reconciliation Procedures. *Macromol. Symp.* **2008**, *271*, 26–37. [CrossRef]
4. Prata, D.M.; Lima, E.L.; Pinto, J.C. Nonlinear Dynamic Data Reconciliation in Real Time in Actual Processes. *Comput. Aided Chem. Eng.* **2009**, *27*, 47–54.
5. Abu-El-Zeet, Z.H.; Roberts, P.D.; Becerra, V.M. Enhancing model predictive control using dynamic data reconciliation. *Alche J.* **2002**, *48*, 324–333. [CrossRef]
6. Sun, B.; Yang, C.; Wang, Y.; Gui, W.; Craig, I.; Olivier, L. A comprehensive hybrid first principles/machine learning modeling framework for complex industrial processes. *J. Process Control* **2020**, *86*, 30–43. [CrossRef]
7. Faber, R.; Arellano-Garcia, H.; Li, P.; Wozny, G. An optimization framework for parameter estimation of large-scale systems. *Chem. Eng. Process.* **2007**, *46*, 1085–1095. [CrossRef]
8. Schladt, M.; Hu, B. Soft Sensor Based on Nonlinear Steady-State Data Reconciliation in the Process Industry. *Chem. Eng. Process.* **2007**, *46*, 1107–1115. [CrossRef]
9. Su, Q.; Bommireddy, Y.; Shah, Y.; Ganesh, S.; Moreno, M.; Liu, J.; Gonzalez, M.; Yasdanpanah, N.; O'Connor, T.; Reklaitis, G.V.; et al. Data reconciliation in the Quality-byDesign (QbD) implementation of pharmaceutical continuous tablet manufacturing. *Int. J. Pharm.* **2019**, *563*, 259–272. [CrossRef]
10. Chiari, M.; Bussani, G.; Grotolli, M.G.; Pierucci, S. On-line Data Reconciliation and Optimisation: Refinery Applications. *Comput. Chem. Eng.* **1997**, *21*, s1185–s1190. [CrossRef]
11. Lee, M.H.; Lee, S.J.; Han, C.; Chang, K.S.; Kim, S.H.; Sang, H.Y. Hierarchical on-line data reconciliation and optimization for an industrial utility plant. *Comput. Chem. Eng.* **1998**, *22*, s247–s254. [CrossRef]
12. Almásy, G.A.; Sztanó, T. Checking and correction of measurements on the basis of linear system model. *Probl. Control Inf. Theory* **1975**, *4*, 57–69.
13. Mah, R.S.H.; Tamhane, A.C. Detection of gross errors in process data. *Aiche J.* **2010**, *28*, 828–830. [CrossRef]
14. Mah, R.S.; Stanley, G.M.; Downing, D.M. Reconciliation and rectification of process flow and inventory data. *Ind. Eng. Chem. Process Des. Dev.* **1976**, *15*, 175–183. [CrossRef]
15. Narasimhan, S.; Mah, R.S.H. Generalized likelihood ratio method for gross error identification. *Aiche J.* **1987**, *33*, 1514–1521. [CrossRef]
16. Tong, H.; Crowe, C.M. Detection of gross erros in data reconciliation by principal component analysis. *Aiche J.* **2010**, *41*, 1712–1722. [CrossRef]
17. Yu, M.; Hong-Ye, S.U.; Jian, C.H.U. A support vector regression approach for recursive simultaneous data reconciliation and gross error detection in nonlinear dynamical systems. *Acta Autom. Sin.* **2009**, *35*, 707–716.
18. Zhang, Z.; Chen, J. Simultaneous data reconciliation and gross error detection for dynamic systems using particle filter and measurement test. *Comput. Chem. Eng.* **2014**, *69*, 66–74. [CrossRef]
19. Yuan, Y.; Khatibisepehr, S.; Huang, B.; Li, Z. Bayesian method for simultaneous gross error detection and data reconciliation. *Aiche J.* **2015**, *61*, 3232–3248. [CrossRef]
20. Tjoa, I.B.; Biegler, L.T. Simultaneous strategies for data reconciliation and gross error detection of nonlinear systems. *Comput. Chem. Eng.* **1991**, *15*, 679–690. [CrossRef]
21. Johnson, L.P.M.; Kramer, M.A. Maximum likelihood data rectification: Steady-state systems. *Alche J.* **1995**, *41*, 2415–2426. [CrossRef]
22. Arora, N.; Biegler, L.T. Redescending estimators for data reconciliation and parameter estimation. *Comput. Chem. Eng.* **2001**, *25*, 1585–1599. [CrossRef]
23. Wang, D.; Romagnoli, J.A. A Framework for Robust Data Reconciliation Based on a Generalized Objective Function. *Ind. Eng. Chem. Res.* **2003**, *42*, 3075–3084. [CrossRef]
24. Özyurt, D.B.; Pike, R.W. Theory and practice of simultaneous data reconciliation and gross error detection for chemical processes. *Comput. Chem. Eng.* **2004**, *28*, 381–402. [CrossRef]
25. Ragot, J.; Chadli, M.; Maquin, D. Mass balance equilibration: A robust approach using contaminated distribution. *Aiche J.* **2005**, *51*, 1569–1575. [CrossRef]
26. Prata, D.M.; Schwaab, M.; Lima, E.L.; Pinto, J.C. Simultaneous robust data reconciliation and gross error detection through particle swarm optimization for an industrial polypropylene reactor. *Chem. Eng. Sci.* **2010**, *65*, 4943–4954. [CrossRef]
27. Zhang, Z.; Shao, Z.; Chen, X.; Wang, K.; Qian, J. Quasi-weighted least squares estimator for data reconciliation. *Comput. Chem. Eng.* **2010**, *34*, 154–162. [CrossRef]
28. Llanos, C.E.; Sanchéz, M.C.; Maronna, R.A. Robust estimators for data reconciliation. *Ind. Eng. Chem. Res.* **2015**, *54*, 5096–5105. [CrossRef]
29. Alighardashi, H.; Magbool Jan, N.; Huang, B. Expectation Maximization Approach for Simultaneous Gross Error Detection and Data Reconciliation Using Gaussian Mixture Distribution. *Ind. Eng. Chem. Res.* **2017**, *56*, 14530–14544. [CrossRef]
30. Xie, S.; Yang, C.; Yuan, X.; Wang, X.; Xie, Y. A novel robust data reconciliation method for industrial processes. *Control Eng. Pract.* **2019**, *83*, 203–212. [CrossRef]
31. de Menezes, D.Q.F.; Prata, D.M.; Secchi, A.R.; Pinto, J.C. A review on robust M-estimators for regression analysis. *Comput. Chem. Eng.* **2021**, *147*, 107254. [CrossRef]
32. Rey, W.J.J. *Introduction to Robust and Quasi-Robust Statistical Methods*; Springer Science & Business Media: Berlin, Germany, 2012.

33. Fair, R.C. On the robust estimation of econometric models. *Ann. Econ. Soc. Meas.* **1974**, *3*, 667.
34. Jin, S.Y.; Li, X.W.; Huang, Z.J.; Meng, L. A new target function for robust data reconciliation. *Ind. Eng. Chem. Res.* **2012**, *51*, 10220–10224. [CrossRef]
35. Hoaglin, D.C.; Mosteller, F.; Tukey, J.W. *Understanding Robust and Exploratory Data Analysis*; Wiley: New York, NY, USA, 1983; Volume 3.
36. Albuquerque, J.S.; Biegler, L.T. Data reconciliation and gross-error detection for dynamic systems. *Aiche J.* **1996**, *42*, 2841–2856. [CrossRef]
37. Shevlyakov, G.; Morgenthaler, S.; Shurygin, A. Redescending m-estimators. *J. Stat. Plann. Inference* **2008**, *138*, 2906–2917. [CrossRef]
38. Zhou, X.J.; Yang, C.H.; Gui, W.H. State transition algorithm. *J. Ind. Manag. Optim.* **2012**, *8*, 1039–1056. [CrossRef]

entropy

MDPI

Article

Constrained Parameter Estimation for a Mechanistic Kinetic Model of Cobalt–Hydrogen Electrochemical Competition during a Cobalt Removal Process

Yiting Liang, Yuanhua Zhang and Yonggang Li *

School of Automation, Central South University, Changsha 410083, China; liang_yiting@csu.edu.cn (Y.L.); 184612192@csu.edu.cn (Y.Z.)
* Correspondence: liyonggang@csu.edu.cn

Abstract: A mechanistic kinetic model of cobalt–hydrogen electrochemical competition for the cobalt removal process in zinc hydrometallurgical was proposed. In addition, to overcome the parameter estimation difficulties arising from the model nonlinearities and the lack of information on the possible value ranges of parameters to be estimated, a constrained guided parameter estimation scheme was derived based on model equations and experimental data. The proposed model and the parameter estimation scheme have two advantages: (i) The model reflected for the first time the mechanism of the electrochemical competition between cobalt and hydrogen ions in the process of cobalt removal in zinc hydrometallurgy; (ii) The proposed constrained parameter estimation scheme did not depend on the information of the possible value ranges of parameters to be estimated; (iii) the constraint conditions provided in that scheme directly linked the experimental phenomenon metrics to the model parameters thereby providing deeper insights into the model parameters for model users. Numerical experiments showed that the proposed constrained parameter estimation algorithm significantly improved the estimation efficiency. Meanwhile, the proposed cobalt–hydrogen electrochemical competition model allowed for accurate simulation of the impact of hydrogen ions on cobalt removal rate as well as simulation of the trend of hydrogen ion concentration, which would be helpful for the actual cobalt removal process in zinc hydrometallurgy.

Keywords: cobalt removal process; mechanistic kinetic model; parameter estimation; constrained parameter estimation

Citation: Liang, Y.; Zhang, Y.; Li, Y. Constrained Parameter Estimation for a Mechanistic Kinetic Model of Cobalt–Hydrogen Electrochemical Competition during a Cobalt Removal Process. *Entropy* **2021**, *23*, 387. https://doi.org/10.3390/e23040387

Academic Editor: Jing Na

Received: 22 February 2021
Accepted: 21 March 2021
Published: 24 March 2021

Current zinc production is mainly based on zinc pyrometallurgy and zinc hydrometallurgy with the latter accounting for more than 80% of zinc production capacity [1]. Zinc hydrometallurgy consists of four processes: calcination, leaching, purification, and electrolysis. In the purification process, zinc dust is added for displacement deposition of impurity elements such as cobalt, nickel, and copper in the leaching solution thereby preventing co-deposition of impurity ions with zinc during the electrolysis process and the consequent decrease in electrolysis efficiency. The purification process involves the removal of copper, cobalt, and cadmium ions; cobalt ions are the most difficult to remove thereby making the cobalt removal step the most crucial.

The main reaction in the cobalt removal step is generally the electrochemical reaction of zinc–copper microcells [2–4]. Exposed zinc on the microcell surface undergoes oxidation to become zinc ions while cobalt ions obtain electrons and are reduced to elemental cobalt accompanied by reduction of hydrogen ions to hydrogen molecules. Low pH conditions (pH = 3–5) are usually adopted for cobalt removal to avoid zinc ions from forming basic zinc sulfate, which would deactivate the zinc dust by absorbing it onto its surface [5]. However, hydrogen ions have greater reactivity than cobalt ions on the zinc surface thereby making high-concentration hydrogen ions compete with cobalt ions for electrons result in a significant drop in the removal rate of cobalt [6]. To circumvent this competition, it is necessary to add copper salts, antimony salts, or arsenic salts as a catalyst to reduce the competitive reduction of hydrogen ions.

Despite such catalyst application, zinc dust in an amount twice the theoretically required amount should be used to decrease the concentration of cobalt ions to a proper level. The excess consumption of zinc dust arises from the competitive reduction of hydrogen ions. Given the large consumption of zinc dust in the purification process, how to model and estimate the cobalt removal process to achieve model-based optimal control of the cobalt removal process is an important research topic [7–10].

To model cobalt removal in the purification process of zinc hydrometallurgy, Wang [7] developed a first-order reaction kinetic model of cobalt removal assuming that the cobalt removal rate was proportional to the surface area of the zinc dust. They further used this model to develop a cobalt removal process model for continuous stirred-tank reactors (CSTRs) in series. Sun et al. [8] used the electrochemical mechanism of microcells in the cobalt removal process and treated the cathodic and anodic reactions in the cobalt removal process as independent processes under mixed potential control. They proposed a first-order electrochemical kinetic model based on mixed potential and seed concentration for the first time. Sun [9] proposed a modeling framework for strongly nonlinear processes based on multi-mechanism model synthesis, which nicely simulated the removal process of cobalt by arsenic salts. Li et al. [10] modeled cobalt removal via several interacting CSTR (ICSTR) systems with multiple time delays based on a mechanism of sharing the surface area of zinc dust in view of the competitive reduction of multiple ions in the cobalt removal process as well as the use of multiple stirred reactors in series. However, there remain multiple reactions in the zinc sulfate solution during the cobalt removal process: (1) competitive reduction of hydrogen ions and cobalt ions on the microcell surface; and (2) zinc ion hydrolysis that compensates for the loss of hydrogen ions due to reductive hydrogen evolution thereby slowing the rise of pH and affecting the cobalt removal rate. However, the cobalt removal process models proposed by Wang [7] and Sun [8,9] did not consider the competitive inhibitory effect of hydrogen ions on cobalt displacement, and the competitive reaction model was based on a mechanism of sharing the surface area of zinc dust [10]; it ignored the electrochemical nature of the competitive reaction. In addition, the impact of zinc hydrolysis on the concentration of hydrogen ions was ignored when modeling hydrogen reaction rates. These shortcomings are unfavorable for accurate assessment of the impact of pH fluctuation on zinc dust consumption and cobalt ion concentration during the cobalt removal process thereby adversely affecting optimal control of the cobalt removal process. Therefore, it is necessary to develop a kinetic model that not only incorporates the underlying electrochemical mechanism of the cobalt–hydrogen competitive reaction in the cobalt removal process but also reveals the trend of hydrogen ion concentration.

In addition to the difficulty in laying out the model structure, parameter estimates for the cobalt removal kinetic model also face many other challenges. The electrochemical reaction of various ions in the cobalt removal process and the coupling of hydrogen evolution with zinc ion hydrolysis are highly nonlinear processes. Parameter estimates for such strongly nonlinear dynamic systems usually occur through two classes of approaches.

The first class of approach estimate directly identifies the dynamic equations based on the time series of full-state values rather than solving the dynamic system [11,12]. For example, the SINDy algorithm [11] is a sparse estimation algorithm for nonlinear systems and estimates the first-order or even higher-order derivatives of the state according to available data. It then substitutes the derivatives into the state equation to estimate parameters. This approach is advantageous in that it avoids the strong nonlinearity arising from the integration of the dynamic system, but the disadvantage is that it requires high-frequency sampling of full-state data. For systems whose measurement values are partially missing or whose states are not all measurable, the state distribution at each time point can be recursively estimated through filtering and expectation-maximization followed by model parameter estimation through a state recursion equation [12,13]. However, this algorithm ensures convergence only when the initial estimates of parameters are relatively accurate.

The second class of approach is to solve the dynamic system and estimate model parameters by minimizing the differences between predicted outputs and measurements, which are usually unconstrained or simple constraints, like the upper and lower bounds of the parameters to be estimated. Different optimization algorithms may be involved in this approach—namely gradient-based optimization algorithms such as a sequential quadratic programming algorithm) versus heuristic optimization algorithms. Chai et al. [14] proposed a parameter estimation scheme for nonlinear time-delay dynamic systems that adopted a variational method to derive a formula for rapid gradient calculation greatly improving parameter estimation efficiency. However, gradient-based parameter estimation algorithms can only find locally optimal parameters and are not suitable for estimating model parameters of strongly nonlinear dynamic systems. Therefore, heuristic optimization algorithms such as particle swarm optimization (PSO) algorithms [15], state transition algorithms [16], differential evolution algorithms [17] and bee colony algorithm [18] have been widely used in parameter estimation of nonlinear systems. For example, Zhang [19] used a PSO algorithm to estimate parameters for a non-linear model of copper removal in the purification process of zinc hydrometallurgy. Deng et al. [20] used a state transition algorithm to estimate model parameters for the electrolysis process of zinc hydrometallurgy. In addition to heuristic optimization algorithms, additional prior information was incorporated into the estimation framework to narrow the parameter search space thus improving estimation efficiency and accuracy. Liu et al. [21] constructed a parameter optimization framework for a model of microbial batch fermentation that incorporated continuous state constraints according to the process mechanism and adopted an improved differential evolution algorithm to solve the estimation optimization problem. The continuous state constraints provide additional prior information for model parameter estimation to improve the reliability of estimation results. Improved differential evolution algorithms are more likely than gradient-based optimization algorithms to find globally optimal solutions, but they still use constraint penalty functions, which leads to a low estimation efficiency.

High-frequency sampling of state data is difficult for the cobalt removal process. Meanwhile, the kinetic and thermodynamic parameters in chemical databases only apply to specific experimental conditions (e.g., temperature, solution composition, and reactive electrode material) and model structures, which makes it difficult to accurately estimate the value range of a given parameter using chemical databases. These limitations prevent the algorithms above from being directly used in estimating the parameters of a given mechanistic kinetic model of the cobalt removal process. In the case of an unknown range of parameter values, a wide range of parameter searching is required to improve the solution accuracy. However, the non-linear dynamic system is solved each time that the objective function is calculated making the wide-range search time-consuming and inefficient and thereby limiting the estimation accuracy.

To overcome the obstacles to kinetic model development and parameter estimation of the cobalt removal process, firstly, this study developed a mechanistic kinetic model of cobalt–hydrogen electrochemical competition during the cobalt removal process. This model was composed of three models for the cobalt removal process: an electrochemical rate model, a mixed potential model of zinc dust, cobalt ions, and hydrogen ions, and a hydrolysis rate model of zinc ions. These models were all constructed in accordance with the principles of microcells and electrochemical kinetics. Secondly, this study proposed a constrained estimation scheme for model parameters. Specifically, mechanistic equations were simplified according to experimental data, and were then used to obtain approximate estimation equations for experimental metrics. Next, value ranges of experimental metrics were estimated according to experimental data to impose constraints on parameter estimation and narrow the parameter search space followed by transforming such constrained parameter estimation problem to an easy-to-solve box-constrained optimization problem through quasi-linear mapping of the parameters to be estimated. That is, the estimation problem of a given parameter with a difficult-to-estimate value range was transformed to the estimation problem of an experimental metric whose value range would be easily estimated from available experimental data. Meanwhile, the constraint conditions provided a relational expression directly linking the experimental metrics to the model parameters thereby providing deeper insights into the model parameters for model users.

The proposed parameter estimation scheme has three advantages. First of all, compared with the conventional unconstrained identification algorithm, the proposed constrained parameter estimation scheme does not depend on the information about the possible range of the parameters to be estimated. It obtains the searching space of parameters by building constraints of parameters based on a reasonable approximation of the experimental metrics of which rough ranges are easy to be obtained from available experimental data. As a result, the time consumption for searching optimal parameter value is dramatically reduced. Secondly, compared to the identification algorithms based on high-frequency sampling data, such as SINDy or filtering-based parameter estimation methods, it requires only low-frequency sampling experimental data. Thirdly, the constraint conditions provided by the scheme directly link the experimental phenomenon metrics to the model parameters, which provide a sense of intuition about how to adjust the parameters while the phenomenon changes in actual processes for model users.

The rest of this paper is organized as follows. In Section 1, a mechanistic kinetic model of cobalt–hydrogen electrochemical competition in the cobalt removal process is developed. In Section 2, a state-space model for cobalt removal in batch-type reactors is established based on the kinetic model and the principle of conservation of matter followed by introducing constraints into the model to construct a constrained parameter estimation problem and then transforming the problem into an easy-to-solve box-constrained optimization problem. In Section 3, the accuracy and efficiency of the proposed constrained parameter estimation algorithm in estimating model parameters based on experimental data are verified versus a conventional algorithm.

1. Cobalt–Hydrogen Electrochemical Competition Model for the Cobalt Removal Process

1.1. Mechanism Analysis of Cobalt Removal Process

During zinc hydrometallurgy, the zinc ore is calcined and dissolved to form a zinc sulfate solution that is purified and sent for electrolysis to obtain pure zinc products. However, the zinc sulfate solution usually contains copper, cobalt, nickel, and other impurity ions. Of these, cobalt ions are the most difficult to remove, and excessive cobalt ions will cause a serious decrease in the efficiency of zinc electrolysis. Therefore, the cobalt removal step is a core step in the purification process of zinc metallurgy. The cobalt removal step is generally considered to involve electrochemical reactions on the surface of microcells as illustrated in Figure 1 [2–4]. Hydrogen ions have a high reactivity on the zinc and cobalt surfaces allowing hydrogen ions to compete with cobalt ions for electrons and thereby inhibit the displacement of cobalt. It is often necessary to add catalysts such as copper and antimony salts in the cobalt removal process to suppress the competitive hydrogen evolution on the zinc dust surface.

During the process of catalytic removal of cobalt by copper and antimony salts, the first reaction is the displacement of copper and antimony, which are deposited on the zinc dust surface to form active microcells. The displacement deposition of copper and antimony is very fast (complete within 5 min as reported by Lew [4]). Generally, the copper-antimony deposition is complete before cobalt begins to be deposited with cobalt deposition and hydrogen evolution on the microcells as well as zinc ion hydrolysis being the main reactions afterwards. The displacement of copper does not occur at the same time as that of copper and antimony; the competitive inhibitory effect of hydrogen ions on cobalt removal was the main focus of this study—we only investigated the electrochemical reactions of cobalt and hydrogen ions as well as zinc ion hydrolysis generating hydrogen ions. Three reversible half-cell reactions occur on the zinc-based microcell (Figure 1):

$$Zn^{2+} + 2e^{-} \leftrightarrow Zn \tag{1}$$

$$Co^{2+} + 2e^- \leftrightarrow Co \quad (2)$$

$$2H^+ + 2e^- \leftrightarrow H_2 \quad (3)$$

Zinc ion hydrolysis occurs in the main solution as shown by the following formula:

$$Zn^{2+} + 2H_2O \leftrightarrow Zn(OH)_2 + 2H^+ \quad (4)$$

In the actual process of cobalt removal, the net direction of half-cell Reaction (1) is to the left providing the electrons required for the evolution of hydrogen and displacement deposition of cobalt. Cobalt ions are displaced and deposited through a half-cell Reaction (2). Hydron evolution occurs through half-cell Reaction (3) in which hydrogen ions compete with cobalt ions for electrons reducing the cobalt removal rate. Zinc ion hydrolysis replenishes hydrogen ions as they are reduced and evolve as hydrogen gas exerting a buffering effect and indirectly affecting the cobalt removal rate.

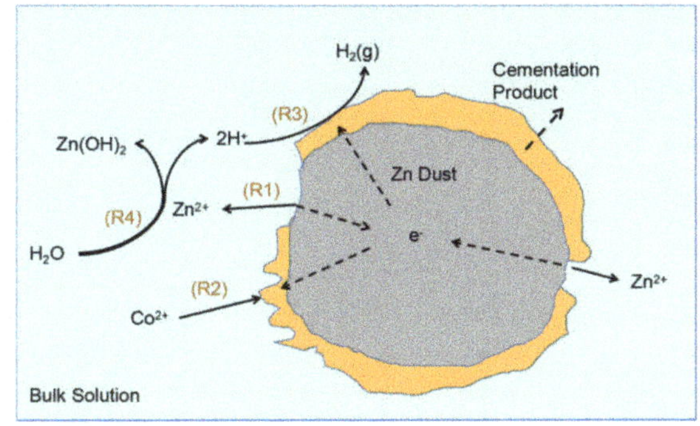

Figure 1. The micro-battery reaction mechanism of the cobalt removal process.

1.2. Reaction Kinetics of Cobalt Removal

Almost all the data published [2–6] showed that the activation energy of cobalt removal reaction was higher than 40 kJ/mol, so it must be a chemically controlled process rather than a diffusion-controlled one. The electrochemical reaction for cobalt removal occurs on the microcell surface and is driven by the potential difference between the microcell surface and the solution. Given the very small diameters of the microcells and the high electrical conductivity of zinc and copper as the major microcell electrodes, we assumed that the potential at different areas on the microcell surface is identical, which is referred to as mixed potential and denoted as $E(t)$ in V. According to the principles of electrochemical kinetics—assuming that the total reactive surface area of microcells is proportional to the elemental zinc concentration—one can express the reverse rate of half-cell Reaction (1):

$$r_1(c(t), E(t)) = k_1 c_{Zn,0}(c_{Zn^{2+}}(t) exp(-\frac{(2-\beta_1)FE(t)}{RT}) - b_1 \frac{c_{Zn}(t)}{c_{Zn,0}} exp(\frac{\beta_1 FE(t)}{RT})) \quad (5)$$

where $c(t)$ represents the reactant concentration vector, $c(t) = [c_{Zn^{2+}}(t), c_{Zn}(t), c_{Co^{2+}}(t), c_{Co}(t), c_{H^+}(t), c_{Zn(OH)_2}(t)]^T$; $c_M(t)$, M $=$ Zn^{2+}, Zn, Co^{2+}, Co, H^+, $Zn(OH)_2$ is the concentration of reactant M in mol/L; $r_i, i = 1, \ldots, 4$ is the rate of reaction (i), which is a function of $c(t)$ and $E(t)$; $c_{Zn,0}$ is the initial concentration of elemental zinc in mol/L; k_1 is the rate constant of Reaction (1); b_1 is the equilibrium constant of Reaction (1); β_1 is the anode transfer coefficient of Reaction (1); F is the Faraday constant (96,485 C/mol); R is the gas constant (8.314); and T is the reaction temperature (346 K in this study).

Assuming that cobalt displacement is determined by the electrochemical reaction and that the number of electrons involved is 2, the reaction rate of half-cell Reaction (2) can be expressed as:

$$r_2(c(t),E(t)) = k_2 c_{Zn,0}(c_{Co^{2+}}(t)exp(-\frac{\alpha_2 E(t)F}{RT}) - b_2\frac{c_{Co}(t)}{c_{Zn,0}}exp(\frac{(2-\alpha_2)E(t)F}{RT})) \quad (6)$$

where k_2 is the rate constant of Reaction (2), b_2 is the equilibrium constant of Reaction (2), and α_2 is the cathode transfer coefficient of Reaction (2).

Given that the reverse of hydrogen evolution Reaction (3) is usually weak and can be ignored, only the forward reaction is considered in this study. Accordingly, the rate of hydrogen evolution reaction is:

$$r_3(c(t),E(t)) = k_3 c_{Zn,0} c_{H^+}(t)exp(-\frac{\alpha_3 FE(t)}{RT}) \quad (7)$$

where k_3 is the rate constant of Reaction (3), b_3 is the equilibrium constant of Reaction (3), and α_3 is the cathode transfer coefficient of Reaction (3).

According to the principle of an electric double layer (EDL), the rate of change of mixed potential of a microcell is proportional to the total electrochemical current on the microcell as follows:

$$\dot{E}(t) = C^{-1}F\sum_{i=1}^{3} 2r_i(c(t),E(t)) \quad (8)$$

where C is the capacitance of the EDL. The rate of hydrolysis Reaction (4) of zinc ions is:

$$r_4(c(t),E(t)) = k_4(c_{Zn^{2+}}(t) - b_4 c_{Zn(OH)_2}(t)c_{H^+}(t)^2) \quad (9)$$

where k_4 is the rate constant of Reaction (4), and b_4 is the equilibrium constant of Reaction (4).

Formulas (5)–(9) constitute a kinetic model of cobalt–hydrogen electrochemical competition in the cobalt removal process, which contain many unknown parameters and whose estimation algorithms will be elaborated in the next section.

2. Constrained Parameter Estimation Algorithms for The Cobalt–Hydrogen Electrochemical Competition Model

For the cobalt–hydrogen electrochemical competition model, the value of hydrolysis rate constant k_4 was set to 1×10^8 in this study considering that its fluctuation within a certain range would have little impact on modeled concentration of cobalt and hydrogen ions. In addition, the charge and discharge time of a microcell was assumed to be less than the reaction characteristic time, i.e., FC^{-1} was relatively large. In such a scenario, the fluctuation of FC^{-1} within a certain range would have little impact on model results and accordingly its value was set to 1×10^4. The important parameter vector to be estimated to be:

$$p^{(0)} = [\beta_1,\alpha_2,\alpha_3,k_1,k_2,k_3,b_1,b_2,b_4] \quad (10)$$

In view of the great order of magnitudes in the fluctuation of reaction rate constants and equilibrium constants, the last six constants in $p^{(0)}$ were logarithm-transformed thereby leading to the following estimates of the parameter vector:

$$p^{(1)} = [\beta_1,\alpha_2,\alpha_3,ln(k_1),ln(k_2),ln(k_3),ln(b_1),ln(b_2),ln(b_4)] \quad (11)$$

Lew and colleagues [4] performed experimental tests under 20 conditions to explore the impacts of copper and antimony contents, pH, zinc dust dosage, and zinc dust particle size on the efficiency of batch removal of cobalt under catalysis by copper and antimony. They obtained abundant kinetic data on the cobalt removal process. Experimental data under five conditions regarding the impact of pH on cobalt removal rate in that study were taken as source data to estimate the parameters of the present mechanistic kinetic model. These experimental data were the monitored values of cobalt ion concentration at constant pH values of 3.0, 3.6, 4.0, and 4.4 as well as the monitored values of cobalt ion concentration and pH under an uncontrolled pH condition (Page 62, Figure 4-31; Page 47, Figure 4-10; page 42, Figure 4-4). Since the cobalt removal tests by Lew [4] were performed in batch reactors, a state-space model will be developed in Section 2.1 based on the cobalt–hydrogen competition model developed in Section 1 to describe the batch removal process of cobalt. Section 2.2 will construct a parameter estimation problem based on the state-space model and address the difficulty in solving the problem. Section 2.3 proposes constraint conditions to make the above parameter estimation problem become a constrained parameter estimation problem. Section 2.4 further transforms the problem into an easy-to-solve box-constrained optimization problem.

2.1. State Space Model for the Batch Removal Process of Cobalt

For a given half-cell reaction, its state function is constructed in terms of the cumulative concentration change of the reactant as follows:

$$\dot{x}_i(t) = r_i(c(t), E(t)), i = 1, \ldots, 4 \tag{12}$$

where t represents time in min, and $x_i(t)$ the cumulative concentration changes of the reactant of reaction (i) in mol/L. The material balance equation in the batch reactor is:

$$c(t) = c_0 + \Lambda x(t) \tag{13}$$

where c_0 is the vector of the initial concentration of reactants added in batch:

$$c_0 = \left[c_{Zn^{2+},0}, c_{Zn,0}, c_{Co^{2+},0}, c_{Co,0}, c_{H^+,0}, c_{Zn(OH)_2,0}\right]^T$$

Here, $c_{M,0}$ (M = Zn^{2+}, Zn, Co^{2+}, Co, H^+, $Zn(OH)_2$) is the initial concentration of reactant M in mol/L, and Λ is a stoichiometric coefficient matrix where column i represents the stoichiometric coefficients of all the six reactants for reaction (i), i = 1, 2, 3, and 4, based on Equations (1)–(4):

$$\Lambda = \begin{bmatrix} -1 & 0 & 0 & -1 \\ 1 & 0 & 0 & 0 \\ 0 & -1 & 0 & 0 \\ 0 & 1 & 0 & 0 \\ 0 & 0 & -2 & 2 \\ 0 & 0 & 0 & 1 \end{bmatrix} \tag{14}$$

According to Equations (8) and (12), the mixed potential in the batch reactor is:

$$E(t) = E_0 + FC^{-1}\begin{bmatrix} 2 & 2 & 2 \end{bmatrix} x(t) \tag{15}$$

where E_0 is the initial mixed potential in V. Considering that the experimental data used for model parameter estimation in this study included cobalt ion concentration and pH, the model output is:

$$y(t) = \begin{bmatrix} c_{Co^{2+}}(t) \\ c_{H^+}(t) \end{bmatrix} = \begin{bmatrix} 0 & 0 & 1 & 0 & 0 & 0 \\ 0 & 0 & 0 & 0 & 1 & 0 \end{bmatrix} c_0 + \begin{bmatrix} 0 & -1 & 0 & 0 \\ 0 & 0 & -2 & 2 \end{bmatrix} x(t) \tag{16}$$

Equations (12)–(16) constitute a state-space model of the batch removal process of cobalt where the state vector is x and the output is y. For the convenience of description, function G is introduced to represent the relationship between the parameters, the initial concentration of reactants, and the model output in the state-space model (13) of the batch removal process of cobalt as follows:

$$y(t) = G(t|p, c_0) \tag{17}$$

Since the experimental data used for model parameter estimation in this study also included the experimental data of batch removal of cobalt under constant-pH conditions, it is necessary to develop a state-space model for such removal processes under constant-pH conditions. Accordingly, the hydrogen ion concentration is set to a constant value in the state space model Equation (13) followed by solving the equations to obtain the model output. Specifically, Equations (13) and (16) are replaced by Equations (18) and (19), respectively:

$$c(t) = c_0 + \Lambda^* x(t) \tag{18}$$

$$y(t) = \begin{bmatrix} c_{Co^{2+}}(t) \\ c_{H^+}(t) \end{bmatrix} = \begin{bmatrix} 0 & 0 & 1 & 0 & 0 & 0 \\ 0 & 0 & 0 & 0 & 1 & 0 \end{bmatrix} c_0 + \begin{bmatrix} 0 & -1 & 0 & 0 \\ 0 & 0 & 0 & 0 \end{bmatrix} x \tag{19}$$

In Equation (18), one has:

$$\Lambda^* = \begin{bmatrix} -1 & 0 & 0 & -1 \\ 1 & 0 & 0 & 0 \\ 0 & -1 & 0 & 0 \\ 0 & 1 & 0 & 0 \\ 0 & 0 & 0 & 0 \\ 0 & 0 & 0 & 1 \end{bmatrix} \tag{20}$$

To simplify subsequent expressions, the relationship between the parameters obtained based on Equations (12), (15), (18), (19) and (20), the initial concentration of the reactants, and the model output is expressed as:

$$y(t) = G^*(t|p, c_0) \tag{21}$$

2.2. Parameter Estimation Problem of Cobalt–Hydrogen Electrochemical Competition System

For electrochemical rate Equations (5)–(7), the cathode and anode reaction transfer coefficients are confined in the following ranges:

$$\beta_1 \in [0.7, 1.5], \alpha_2 \in [0.2, 0.8], \alpha_3 \in [0.2, 0.8]$$

In order to identify the model parameters, experimental data published in Lew [4], a doctoral thesis on the mechanism of the copper-antimony activated cobalt removal from zinc sulphate solutions, were used in this paper. In Lew [4], the cobalt removal experiments were carried out in a batch stirred reactor, as shown in Figure 2. The volume of the reactor is 4 L, and the stirring speed is 700 RPM. The main component of the reaction solution was $ZnSO_4$, and a certain amount of copper, antimony and cobalt ions were added one time in advance according to the design of each experiment. Zinc powder was added at one time at the initial stage of the reaction. During the experiments, the cobalt ion concentration was measured by sampling using a 50-mL pipette at intervals of 5, 15, 30, 45, 60 and 90 min, and the pH values are monitored online. The data from five experiments are selected from Lew [4] for model identification: experiment with natural pH (initial pH = 3.7), controlled pH (pH = 3.0), controlled pH (pH = 3.6), controlled pH (pH = 4.0), and controlled pH (pH = 4.4). In general, the experimental setup of the data used by this study was performed under the following conditions:

1. 4 g/L loading with −100 + 140 mesh (126 to 174 micron) zinc dust;
2. The batch reactor was heated in a thermostated water bath and maintained to within 1 °C of the target value of temperature 73 °C;
3. Natural pH: the pH was allowed to come to its own equilibrium from a starting pH of 3.7;
4. Controlled pH: the pH was controlled to be within 0.2 units of the target value using a pH controller;
5. 26 mg/L Co;
6. 46 mg/L Cu, 1.5 mg/L Sb were added as catalysts.

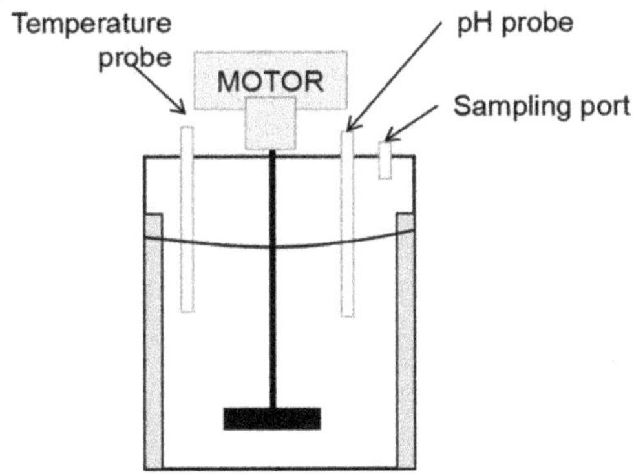

Figure 2. The batch reactor for cobalt removal experiments of Lew [4].

Accordingly, the following parameter estimation problem is proposed:

$$\begin{array}{l} \min\limits_{p^{(1)}} \sum\limits_{j=1}^{5} \sum\limits_{i=1}^{n_{\mathrm{Co},j}} \left(\frac{1}{n_{\mathrm{Co},j}} \left| ln\left(\frac{c^{(j)}_{\mathrm{Co}^{2+},0}}{\hat{c}^{(j)}_{\mathrm{Co}^{2+}}\left(t^{(j)}_{i,\mathrm{Co}^{2+}}\right)}\right) - ln\left(\frac{c^{(j)}_{\mathrm{Co}^{2+},0}}{c^{(j)}_{\mathrm{Co}^{2+},i}}\right) \right| \right) \\ + \frac{\lambda_{\mathrm{pH}}}{n_{\mathrm{pH}}} \sum\limits_{l=1}^{n_{\mathrm{pH}}} \left| -log_{10}\left(\hat{c}^{(1)}_{H^+}\left(t^{(1)}_{l,H^+}\right)\right) - \mathrm{pH}_l \right| \\ \text{s.t.} \begin{bmatrix} \hat{c}^{(j)}_{\mathrm{Co}^{2+}}(t) \\ \hat{c}^{(j)}_{H^+}(t) \end{bmatrix} = \begin{cases} G\left(t \middle| p^{(1)}, c_0^{(j)}\right), \text{if } j = 1, \\ G^*\left(t \middle| p^{(1)}, c_0^{(j)}\right), \text{if } j > 1, \end{cases} \\ \beta_1 \in [0.7, 1.5], \alpha_2 \in [0.2, 0.8], \alpha_3 \in [0.2, 0.8]. \end{array} \tag{22}$$

where j = 1,2,3,4,5 represent natural pH, controlled pH (pH = 3.0), controlled pH (pH = 3.6), controlled pH (pH = 4.0), and controlled pH (pH = 4.4), respectively. For the natural pH condition, cobalt concentration data (Page 47, Figure 4-10) and pH values (page 42, Figure 4-4) were collected while for the other four controlled-pH conditions (j = 2,3,4,5) only cobalt concentration data were collected (Page 62, Figure 4-31). $c_0^{(j)}$ represents the initial concentration in test j; $\hat{c}^{(j)}_{\mathrm{Co}^{2+}}(t)$ and $\hat{c}^{(j)}_{H^+}(t)$ represent the model-predicted concentration of cobalt ion and hydrogen ion in test j, respectively; $c^{(j)}_{\mathrm{Co}^{2+},i}$, and pH_l represent the i-th measurement value of cobalt ion concentration and the l-th measurement value of pH in test j, respectively; $t^{(j)}_{i,\mathrm{Co}^{2+}}$ and $t^{(j)}_{l,H^+}$ represent the measurement time of cobalt ion concentration and pH in test j, respectively; λ_{pH} is the weight coefficient of pH; n_{pH} is the number of collected pH values in test 1; and $n_{\mathrm{Co},j}$ is the number of collected concentration values of cobalt ions in test j.

The model parameters can be estimated by solving Equation (22), which is very difficult to solve. The nonlinearity of mechanistic kinetic Equations (5)–(9) makes Equation (22) a non-convex optimization problem, which is difficult to directly solve via gradient-based optimization algorithms (such as sequential quadratic programming algorithms and interior-point algorithms); however, it is usually solvable by heuristic optimization algorithms only when the ranges of parameter values are known. However, since the kinetic and thermodynamic parameters in the standard chemical databases apply to specific experimental conditions, such as temperature, solution composition, electrode matrix composition, and kinetic models, it is difficult to find corresponding values from chemical databases for the present model parameters (i.e., $k_1, k_2, k_3, b_1, b_2, b_4$) thereby making it impossible to estimate the reasonable value ranges of the parameters. To ensure that the optimal parameters are found, it is necessary to give a large parameter search range, but a large range search will result in low optimization efficiency. A more serious issue is that there is coupling between the competitive electrochemical reactions, the hydrogen evolution, and the zinc ion hydrolysis in the cobalt–hydrogen electrochemical competition mechanism model, which forms a highly nonlinear process. As a result, there would be large differences in reaction characteristic time between reactions (1)–(4) when model parameters are randomly set to a value, thereby causing the state-space model of batch reactor to achieve a very stiff dynamic system. A trial solution of such a system is time-consuming and results in low optimization search efficiency.

The key to overcoming the limitations above is to accurately determine the parameter value ranges. Although it is impossible to directly obtain the value ranges of the model parameters, it is still possible to experimentally estimate a reasonable value range using some experimental data measured under certain conditions such as the mixed potential within a certain period of time, the apparent first-order reaction rate constant, the equilibrium concentration, and the change of the rate of solid deposition or gas evolution at a certain time. Provided that an approximate relationship between these experimental data and the model parameters is obtained, it would be possible to further impose constraints on the parameters to rapidly determine a reasonable range of parameter values and improve the solution efficiency.

In the next sub-section, assumptions are made based on experimental data to simplify the mechanistic equations so that they can be partially yet directly solved. The solutions are directly used to establish approximate expressions linking the model parameters to some experimental metrics. Next, constraints are introduced for the model parameters by specifying value ranges of the experimental metrics based on experimental data, thereby leading to a constrained parameter estimation scheme.

2.3. *Constrained Parameter Estimation Algorithm*

2.3.1. Constraints on Electrochemical Parameters

In this sub-section, the cobalt removal experimental data at pH 3 reported by Lew [4] (Section 4, Page 60–62) will be analyzed in view of the test conditions so as to introduce proper consumptions for simplifying the reaction kinetic models. Next, the mixed potential, the apparent first-order reaction rate constant, and equilibrium cobalt concentration were estimated as experimental metrics according to the simplified kinetic equations to impose constraints on relevant electrochemical parameters of cobalt and hydrogen. The mixed-potential change rate Equation (8) is simplified. Based on the assumption that the charge and discharge of the EDL capacitor are fast (i.e., FC^{-1} is sufficiently large), the charge and discharge time is ignored thereby allowing Equation (8) to be approximated by a balance equation:

$$\sum_{i=1}^{3} r_i(c(t), E(t)) = 0 \tag{23}$$

As indicated by the above equation, the total number of electrons lost through oxidation on a microcell is equal to the total number of electrons obtained through reduction. According to Lew [4] (Page 61, Table 4-2), the first-order reaction rate constant of cobalt was about 489×10^{-6} sec^{-1} pH 3. The first-order reaction rate constant of hydrogen was estimated here to be 767×10^{-5} sec^{-1} based on the early-stage portion (0–5 min) of the hydrogen evolution curve reported by Lew [4] (Page 70, Figure 4-42) with zero initial concentration of zinc ions. This was about 15 times that of cobalt. In the test at pH 3, the hydrogen ion concentration was approximately 2.26-times the cobalt ion concentration (26 mg/L). Considering that two electrons are consumed to displace a single cobalt ion while one electron is required for the reduction of a single hydrogen ion, the current of hydrogen evolution Reaction (3) is approximately $15 \times 2.26/2$ that of cobalt reduction (2). The concentration of cobalt ions became smaller with the passage of time. However, the concentration of hydrogen ions remained nearly constant under constant-pH conditions, which makes the current of Reaction (3) increasingly large relative to that of Reaction (2). Consequently, the current of Reaction (2) would have little impact on the mixed potential, which was largely affected by Reaction (3) and zinc dissolution Reaction (1). That is, the coefficient r_2 in Equation (23) can be ignored. In addition, the reverse rate of zinc dissolution Reaction (1) is assumed to be ignorable when estimating the mixed potential thereby allowing Equation (23) to be further simplified:

$$k_1 c_{Zn,0}(-b_1 \frac{c_{Zn}}{c_{Zn,0}} exp(\frac{\beta_1 FE}{RT})) + k_3 c_{Zn,0} c_{H^+} exp(-\frac{\alpha_3 FE}{RT}) = 0 \tag{24}$$

The mixed potential can be estimated from Equation (24):

$$E_{est} = RTF^{-1}(\beta_1 + \alpha_3)^{-1}(ln(k_3 c_{Zn,0} c_{H^+}) - ln(b_1 k_1 c_{Zn})) \tag{25}$$

The experimental data of Lew [4] (Page 61, Figure 4-29) showed that the potentiometer reading was approximately −0.6 V for sludge at pH 3 ($c_{H^+} = 10^{-3}$ mol/L); thus, we assumed that $E_{est} \in [-0.7, -0.5]$. In addition, the amount of zinc dust added during the cobalt removal reaction is excessive (more than 20 times the amount of zinc dust theoretically required for displacement of copper, antimony, and cobalt ions). Thus, we assumed that the content of zinc dust is roughly constant, namely $c_{Zn} \approx c_{Zn,0}$. Substitution of this expression in Equation (25) leads to the following constraint:

$$-(\beta_1 + \alpha_3)^{-1}(ln(b_1 k_1 k_3^{-1}) + 3ln(10)) \in [-0.7F(RT)^{-1}, -0.5F(RT)^{-1}] \tag{26}$$

Considering that the reverse rate of the displacement reaction of cobalt ions at the early stage is negligible compared to the forward rate, the displacement reaction rate of cobalt ions can be estimated using mixed-potential estimation Equation (25) as follows:

$$r_2 \approx k_2 c_{Zn,0} c_{Co^{2+}} (b_1 k_1 c_{Zn})^{\frac{\alpha_2}{\beta_1+\alpha_3}} (k_3 c_{Zn,0} c_{H^+})^{-\frac{\alpha_2}{\beta_1+\alpha_3}} \tag{27}$$

Assuming that the content of zinc dust is roughly constant, namely $c_{Zn} \approx c_{Zn,0}$, the apparent first-order reaction rate constant of cobalt displacement reaction can be calculated as follows:

$$\widetilde{k}_2 \approx k_2 c_{Zn,0} (b_1 k_1)^{\frac{\alpha_2}{\beta_1+\alpha_3}} (k_3 c_{H^+})^{-\frac{\alpha_2}{\beta_1+\alpha_3}} \tag{28}$$

The experimental data of Lew [4] (Page 61, Table 4-2) showed that at pH 3, the apparent first-order reaction rate constant of cobalt displacement reaction was 0.02934 min^{-1}. Considering the model simplification error, one may let $\widetilde{k}_2 \in [0.015\,\text{min}^{-1}, 0.060\,\text{min}^{-1}]$. Based on Equation (28), the following constraint is obtained:

$$\begin{gathered}\alpha_2(\beta_1 + \alpha_3)^{-1}(ln(k_1) + ln(b_1) - ln(k_3) + 3ln(10)) + ln(k_2) + ln(c_{Zn,0}) \\ \in [ln(0.015), ln(0.060)]\end{gathered} \tag{29}$$

The rate of hydrogen evolution reaction is:

$$r_3 \approx k_3 c_{Zn,0} c_{H^+} (b_1 k_1 c_{Zn})^{\frac{\alpha_3}{\beta_1+\alpha_3}} (k_3 c_{Zn,0} c_{H^+})^{-\frac{\alpha_3}{\beta_1+\alpha_3}} \tag{30}$$

Accordingly, the apparent first-order rate constant of hydrogen evolution reaction is:

$$\widetilde{k}_3 \approx k_3 c_{Zn,0} (b_1 k_1 c_{Zn})^{\frac{\alpha_3}{\beta_1+\alpha_3}} (k_3 c_{Zn,0} c_{H^+})^{-\frac{\alpha_3}{\beta_1+\alpha_3}} \tag{31}$$

As shown by the earlier analysis of the experimental data of Lew [4] in this sub-section, the rate constant of hydrogen evolution reaction was about 15 that of cobalt reduction at pH = 3. In view of the estimation error, the following constraint assumption is introduced for parameter estimation: At pH 3, the apparent pseudo-first-order rate of hydrogen evolution reaction is between 5–50 times the apparent first-order rate of cobalt reduction, namely $\widetilde{k}_3\left(\widetilde{k}_2\right)^{-1} \in [5,50]$. Taking logarithms of both sides based on Equations (31) and (28) while assuming that $c_{\text{Zn}} \approx c_{\text{Zn},0}$ yields the following constraint:

$$(\alpha_3 - \alpha_2)\left(\beta_1 + \alpha_3\right)^{-1}\left(ln(b_1k_1k_3^{-1}) + 3ln(10)\right) + ln(k_3k_2^{-1}) \in [ln(5), ln(10)] \quad (32)$$

Substituting mixed-potential estimation expression (25) into the reaction rate Equation (6) of cobalt displacement Reaction (2) and letting $r_2 = 0$, the equilibrium concentration of cobalt can be calculated as:

$$c_{\text{Co}^{2+},\text{eq}} = \frac{b_2(k_3c_{\text{Zn},0}c_{H^+})^{\frac{2}{\beta_1+\alpha_3}}(b_1k_1c_{\text{Zn}})^{-\frac{2}{\beta_1+\alpha_3}}}{b_2(k_3c_{\text{Zn},0}c_{H^+})^{\frac{2}{\beta_1+\alpha_3}}(b_1k_1c_{\text{Zn}})^{-\frac{2}{\beta_1+\alpha_3}} + c_{\text{Zn},0}} c_{\text{Co}^{2+},0} \quad (33)$$

According to the experimental data of Lew [4] (Page 47, Figure 4-10), $c_{\text{Co}^{2+},\text{eq}} \leq 0.1c_{\text{Co}^{2+},0}$. Accordingly, the first term in the denominator of the above equation can be ignored. Moreover, assuming $c_{\text{Zn}} \approx c_{\text{Zn},0}$:

$$c_{\text{Co}^{2+},\text{eq}} \approx b_2(k_3c_{H^+})^{\frac{2}{\beta_1+\alpha_3}}(b_1k_1)^{-\frac{2}{\beta_1+\alpha_3}} c_{\text{Zn},0}{}^{-1}c_{\text{Co}^{2+},0} \quad (34)$$

According to the experimental data of Lew [4] (Page 47, Figure 4-10), the logarithmic ratio ($ln(c_{\text{Co}^{2+},0}c_{\text{Co}^{2+},\text{eq}}^{-1})$) of the equilibrium to the initial concentration of cobalt ions was approximately 4.2 at pH 3. It is assumed that $ln(c_{\text{Co}^{2+},0}c_{\text{Co}^{2+},\text{eq}}^{-1}) \in [3,8]$ considering the model simplification error. Accordingly, the following constraint can be imposed:

$$2(\beta_1 + \alpha_3)^{-1}\left(ln(k_1b_1k_3^{-1}) + 3ln(10)\right) - ln(b_2) + ln(c_{\text{Zn},0}) \in [3,8] \quad (35)$$

Moreover, the forward rate of zinc dissolution Reaction (1) considered in the mixed-potential equation is assumed to be less than the reverse rate thereby leading to the following constraint:

$$c_{\text{Zn}^{2+}}exp\left(-\frac{(2-\beta_1)FE_{\text{est}}}{RT}\right) \leq \frac{1}{2}b_1\frac{c_{\text{Zn}}}{c_{\text{Zn},0}}exp\left(\frac{\beta_1FE_{\text{est}}}{RT}\right) \quad (36)$$

Mixed-potential estimation Equation (16) is substituted into the above equation while assuming $c_{\text{Zn}} \approx c_{\text{Zn},0}$. Meanwhile, given that the initial concentration of zinc ions (151 g/L) is far greater than the initial concentration of elementary zinc (4 g/L), it is further assumed that $c_{\text{Zn}^{2+}} \approx c_{\text{Zn}^{2+},0}$, i.e., the concentration of zinc ions remains constant. Accordingly, the following constraint can be imposed:

$$2(\beta_1 + \alpha_3)^{-1}\left(ln(b_1k_1k_3^{-1}) + 3ln(10)\right) - ln(b_1) + ln(c_{\text{Zn}^{2+},0}) \leq -ln(2) \quad (37)$$

2.3.2. Constraints on Hydrolysis Parameters

The hydrolysis Reaction (4) affects the rate of cobalt displacement reaction mainly by releasing hydrogen ions and consequently alleviating the decreasing trend of hydrogen ion concentration. Hereinafter, the simplified model is analyzed to derive an equation to estimate the attenuation coefficient for the apparent rate of hydron ion consumption in the presence of zinc ion hydrolysis. Next, the estimation equation is used to derive a constraint on hydrolysis parameter b_4. The hydrolysis Reaction (4) generally has large forward and reverse rate constants and thereby can be treated as a quasi-equilibrium reaction namely $r_4 = 0$. Therefore:

$$c_{\text{Zn}^{2+}} - b_4c_{\text{Zn(OH)}_2}c_{H^+}^2 = 0 \quad (38)$$

Taking the total differential of the above equation:

$$\delta c_{Zn^{2+}} - b_4 c_{H^+}^2 \delta c_{Zn(OH)_2} - 2b_4 c_{Zn(OH)_2} c_{H^+} \delta c_{H^+} = 0 \tag{39}$$

Taking the total differential of state Equation (12) and material balance Equation (13):

$$\begin{cases} \delta x_i(t) = r_i \delta t, i = 1,2,3 \\ \delta c_{Zn^{2+}} = -\delta x_1(t) - \delta x_4(t) \\ \delta c_{H^+} = -2\delta x_3(t) + 2\delta x_4(t) \\ \delta c_{Zn(OH)_2} = \delta x_4(t) \end{cases} \tag{40}$$

Substitution of Equation (40) into Equation (39) leads to:

$$\frac{\delta x_4(t)}{\delta t} = \frac{4b_4 c_{Zn(OH)_2} c_{H^+}}{\left(1 + b_4 c_{H^+}^2 + 4b_4 c_{Zn(OH)_2} c_{H^+}\right)} r_3 - \frac{1}{\left(1 + b_4 c_{H^+}^2 + 4b_4 c_{Zn(OH)_2} c_{H^+}\right)} r_1 \tag{41}$$

A combination of Equation (41) with Equation (23) of equilibrium mixed potential:

$$\frac{\delta x_4(t)}{\delta t} = \frac{4b_4 c_{Zn(OH)_2} c_{H^+} + 1}{\left(1 + b_4 c_{H^+}^2 + 4b_4 c_{Zn(OH)_2} c_{H^+}\right)} r_3 + \frac{1}{\left(1 + b_4 c_{H^+}^2 + 4b_4 c_{Zn(OH)_2} c_{H^+}\right)} r_2 \tag{42}$$

According to Equations (40) and (42), the rate of change of hydrogen ion concentration with time is:

$$\begin{aligned} \dot{c}_{H^+} &= \frac{\delta c_{H^+}}{\delta t} = \frac{-2\delta x_3(t) + 2\delta x_4(t)}{\delta t} \\ &= -\frac{2b_4 c_{H^+}^2}{\left(1 + b_4 c_{H^+}^2 + 4b_4 c_{Zn(OH)_2} c_{H^+}\right)} r_3 + \frac{2}{\left(1 + b_4 c_{H^+}^2 + 4b_4 c_{Zn(OH)_2} c_{H^+}\right)} r_2 \end{aligned} \tag{43}$$

The second term in equation (43) incorporates the impact of cobalt removal rate r_2 on the rate of change of hydrogen ion concentration. Close examination of the experimental data of Lew [4] (Page 58, Figure 4-27) reveals that the pH evolution curve is only slightly affected by cobalt ion concentration thereby allowing one to ignore the second term of the above equation. In addition, rearrangement of hydrolysis equilibrium Equation (38) gives:

$$c_{Zn(OH)_2} = b_4^{-1} c_{Zn^{2+}} c_{H^+}^{-2} \tag{44}$$

Substitution of Equation (44) into Equation (43) leads to Equation (45):

$$\dot{c}_{H^+} = -2b_4 c_{H^+}^2 \left(1 + b_4 c_{H^+}^2 + 4c_{Zn^{2+}} c_{H^+}^{-1}\right)^{-1} r_3 \tag{45}$$

As shown above, zinc ion hydrolysis (4) alleviates the declining trend of hydrogen ion concentration. Therefore, the attenuation coefficient for the apparent consumption rate of hydrogen ions relative to the rate of hydrogen ion reduction (3) is less than one:

$$\lambda_{H^+} = b_4 c_{H^+}^2 \left(\left(1 + b_4 c_{H^+}^2 + 4c_{Zn^{2+}} c_{H^+}^{-1}\right)\right)^{-1} \tag{46}$$

The t-pH curve generated under the condition of zero initial concentration of zinc ions by Lew [4] (Page 70, Figure 4-42) showed that the rising slope of pH was much greater when the initial concentration of zinc ions was low than when it was high in the early stage (0–5 min) (Page 42, Figure 4-4). This trend suggests that the presence of zinc ions has a significant impact on the apparent rate of hydrogen evolution. Therefore, the attenuation coefficient λ_{H^+} should be much less than 1 in the presence of zinc ions. Moreover, the third term of the denominator, which is a zinc ion-related term, is the largest one among the three terms: $4c_{Zn^{2+}}c_{H^+}^{-1}$. Accordingly, the attenuation coefficient can be approximated as:

$$\lambda_{H^+} \approx b_4 c_{H^+}^2 \left(4c_{Zn^{2+}}c_{H^+}^{-1}\right)^{-1} \tag{47}$$

The experimental data of Lew [4] (Page 42, Figure 4-4) showed that when the pH was above 4.4, the slope of the pH curve was much smaller than that in the initial stage. It is assumed here that $\lambda_{H^+}|_{\mathrm{pH}=4.4} \in [0.01, 0.1]$ and $c_{Zn^{2+}} \approx c_{Zn^{2+},0}$. Taking the logarithm of λ_{H^+}, one has the following constraint on hydrolysis rate constant b_4:

$$ln(b_4 c_{Zn^{2+},0}^{-1}) \in [11.2 \times ln(10) + ln(4), 12.2 \times ln(10) + ln(4)] \tag{48}$$

In this section, to reduce the searching space of parameters, the constraints on parameters are derived based on the approximation of experimental phenomenon metrics. However, the approximation formula of experimental phenomenon metrics can also be useful in parameter adjustment and optimizing. For example, Equation (28) for apparent first-order reaction rate of cobalt removal, can be used for determining the rough range of the required zinc loading and pH setting value under the constraints about cobalt removal rates. Equations (28), (31), (34), (47) also provide some intuitive information about the relation between the model parameters and process dynamical behavior, and thus would be useful for adjustment of model parameters while the dynamic changes in practical cobalt removal process.

2.3.3. Constrained Parameter Estimation Problem

Introduction of constraints (Equations (26), (29), (32), (35), (37) and (48)) into the parameter estimation problem (22) leads to a constrained parameter estimation problem:

$$\begin{aligned}
&\min_{p^{(1)}} \sum_{j=1}^{5} \sum_{i=1}^{n_{\mathrm{Co},j}} \left(\frac{1}{n_{\mathrm{Co},j}} \left| ln\left(\frac{c_{\mathrm{Co}^{2+},0}^{(j)}}{\hat{c}_{\mathrm{Co}^{2+}}^{(j)}\left(t_{i,\mathrm{Co}^{2+}}^{(j)}\right)}\right) - ln\left(\frac{c_{\mathrm{Co}^{2+},0}^{(j)}}{c_{\mathrm{Co}^{2+},i}^{(j)}}\right) \right| \right) + \frac{\lambda_{\mathrm{pH}}}{n_{\mathrm{pH}}} \sum_{l=1}^{n_{\mathrm{PH}}} \left| -log_{10}\left(\hat{c}_{H^+}^{(1)}\left(t_{l,H^+}^{(1)}\right)\right) - \mathrm{pH}_l \right| \\
&\text{s.t.} \begin{bmatrix} \hat{c}_{\mathrm{Co}^{2+}}^{(j)}(t) \\ \hat{c}_{H^+}^{(j)}(t) \end{bmatrix} = \begin{cases} G(t|p^{(1)}, c_0^{(j)}), \text{if } j = 1, \\ G^*(t|p^{(1)}, c_0^{(j)}), \text{if } j > 1, \end{cases} \\
&\beta_1 \in [0.7, 1.5], \alpha_2 \in [0.2, 0.8], \alpha_3 \in [0.2, 0.8], \\
&(26), (29), (32), (35), (37), (48).
\end{aligned} \tag{49}$$

Equation (49) contains nonlinear constraints (Equations (26), (29), (32), (35), (37) and (48)). Gradient-based constrained optimization algorithms such as the sequential quadratic programming algorithm and the interior-point algorithm can only handle convex constrained optimization or quasi-convex constrained optimization problems while Equation (49) is a non-convex constrained optimization problem and cannot be solved using such algorithms. Heuristic optimization algorithms are effective in solving non-convex, box-constrained optimization problems; i.e., each decision variable is bounded by two constant values as the upper and lower limits. However, because nonlinear constraints generally come in different forms and are difficult to handle in a unified manner, parameter optimization would be far less satisfactory under nonlinear constraints than under box constraints. A more serious problem is that most intelligent constrained optimization methods require sampling outside the feasible region such as various heuristic optimization algorithms with penalty functions.

In Equation (49), however, most of the parameters outside the feasible region are rigid parameters, which will cause the state-space model to be unsolvable and thus seriously affect the optimization efficiency. Therefore, the constraints contained in Equation (49) will be analyzed in the next sub-section to transform Equation (49) into an easy-to-solve box-constrained optimization problem so that a general heuristic optimization algorithm can be used to iteratively optimize the parameters within the feasible region of the constraints to achieve higher solution efficiency and accuracy.

2.4. Parameter Transformation before Estimation

Constraints (Equations (26), (29), (32), (35), (37) and (48)) will be linear constraints in which are linearly combined if are ignored. Optimization problems with linear constraints can be transformed to a box-constrained optimization problem through linear mapping of the parameters to be estimated. Let:

$$p^{(1)} = [, ln(k_1), ln(k_2), ln(k_3), ln(b_1), ln(b_2), ln(b_4)]^T \tag{50}$$

It is obvious that the following equation holds:

$$p^{(1)} = \left[\beta_1, \alpha_2, \alpha_3, (P^{(1,2)})^T\right] \tag{51}$$

Equation (49) is now re-formulated as:

$$\begin{aligned}
&\min_{\beta_1,\alpha_2,\alpha_3,p^{(1,2)}} \sum_{j=1}^{5} \sum_{i=1}^{n_{\mathrm{Co},j}} \left(\frac{1}{n_{\mathrm{Co},j}} \left| ln\left(\frac{c^{(j)}_{\mathrm{Co}^{2+},0}}{\hat{c}^{(j)}_{\mathrm{Co}^{2+}}\left(t^{(j)}_{i,\mathrm{Co}^{2+}}\right)}\right) - ln\left(\frac{c^{(j)}_{\mathrm{Co}^{2+},0}}{c^{(j)}_{\mathrm{Co}^{2+},i}}\right) \right| \right) + \frac{\lambda_{\mathrm{pH}}}{n_{\mathrm{pH}}} \sum_{l=1}^{n_{\mathrm{PH}}} \left| -log_{10}\left(\hat{c}^{(1)}_{H^+}\left(t^{(1)}_{l,H^+}\right)\right) - \mathrm{pH}_l \right| \\
&\text{s.t.} \begin{bmatrix} \hat{c}^{(j)}_{\mathrm{Co}^{2+}}(t) \\ \hat{c}^{(j)}_{H^+}(t) \end{bmatrix} = \begin{cases} G(t|p^{(1)}, c_0^{(j)}), \text{if } j = 1, \\ G^*(t|p^{(1)}, c_0^{(j)}), \text{if } j > 1, \end{cases} \\
&\beta_1 \in [0.7, 1.5], \alpha_2 \in [0.2, 0.8], \alpha_3 \in [0.2, 0.8], \\
&A(\beta_1, \alpha_2, \alpha_3)p^{(1,2)} + B(\beta_1, \alpha_2, \alpha_3) \in [z_{min}, z_{max}]
\end{aligned} \tag{52}$$

where

$$A(\beta_1, \alpha_2, \alpha_3) = \begin{bmatrix} -(\beta_1+\alpha_3)^{-1} & 0 & (\beta_1+\alpha_3)^{-1} & -(\beta_1+\alpha_3)^{-1} & 0 & 0 \\ \alpha_2(\beta_1+\alpha_3)^{-1} & 1 & -\alpha_2(\beta_1+\alpha_3)^{-1} & \alpha_2(\beta_1+\alpha_3)^{-1} & 0 & 0 \\ (\alpha_3-\alpha_2)(\beta_1+\alpha_3)^{-1} & -1 & (\beta_1+\alpha_2)(\beta_1+\alpha_3)^{-1} & (\alpha_3-\alpha_2)(\beta_1+\alpha_3)^{-1} & 0 & 0 \\ 2(\beta_1+\alpha_3)^{-1} & 0 & -2(\beta_1+\alpha_3)^{-1} & 2(\beta_1+\alpha_3)^{-1} & -1 & 0 \\ 2(\beta_1+\alpha_3)^{-1} & 0 & -2(\beta_1+\alpha_3)^{-1} & 2(\beta_1+\alpha_3)^{-1}-1 & 0 & 0 \\ 0 & 0 & 0 & 0 & 0 & 1 \end{bmatrix} \tag{53}$$

$$B(\beta_1, \alpha_2, \alpha_3) = \begin{bmatrix} -\frac{ln(10^3)}{\beta_1+\alpha_3} \\ \frac{ln(10^3)\alpha_2}{\beta_1+\alpha_3} + ln(c_{\mathrm{Zn},0}) \\ \frac{ln(10^3)(\alpha_3-\alpha_2)}{\beta_1+\alpha_3} \\ \frac{ln(10^6)}{\beta_1+\alpha_3} + ln(c_{\mathrm{Zn},0}) \\ \frac{ln(10^6)}{\beta_1+\alpha_3} + ln(c_{\mathrm{Zn}^{2+},0}) \\ -ln(c_{\mathrm{Zn}^{2+},0}) \end{bmatrix} \tag{54}$$

$$z_{min} = [-0.7\frac{F}{RT}, \ln(0.015), \ln(5), 3, -\infty, 11.2 \times \ln(10) + \ln(4)]^T \tag{55}$$

$$z_{max} = \left[-0.5\,\frac{F}{RT},\ ln(0.060), ln(10), 8, -ln(2),, 12.2 \times ln(10) + ln(4)\right]^T \tag{56}$$

The parameters to be estimated in Equation (52) are transformed through quasi-linear mapping:

$$z = A(\beta_1, \alpha_2, \alpha_3)p^{(1,2)} + B(\beta_1, \alpha_2, \alpha_3) \tag{57}$$

As indicated by the structure of $A(\beta_1, \alpha_2, \alpha_3)$, this matrix is an invertible matrix. Accordingly:

$$p^{(1,2)} = A(\beta_1, \alpha_2, \alpha_3)^{-1}(z - B(\beta_1, \alpha_2, \alpha_3)) \tag{58}$$

Equation (52) now can be transformed to problem Equation (59) as follows:

$$\begin{aligned}
&\min_{\beta_1,\alpha_2,\alpha_3,z} \sum_{j=1}^{5}\sum_{i=1}^{n_{\mathrm{Co},j}}\left(\frac{1}{n_{\mathrm{Co},j}}\left| ln(\frac{c^{(j)}_{\mathrm{Co}^{2+},0}}{\hat{c}^{(j)}_{\mathrm{Co}^{2+}}\left(t^{(j)}_{i,\mathrm{Co}^{2+}}\right)}) - ln(\frac{c^{(j)}_{\mathrm{Co}^{2+},0}}{c^{(j)}_{\mathrm{Co}^{2+},i}})\right|\right) + \frac{\lambda_{\mathrm{pH}}}{n_{\mathrm{pH}}}\sum_{l=1}^{n_{\mathrm{PH}}}\left|-log_{10}(\hat{c}^{(1)}_{H^+}\left(t^{(1)}_{l,H^+}\right)) - \mathrm{pH}_l\right| \\
&\text{s.t.}\begin{bmatrix} \hat{c}^{(j)}_{\mathrm{Co}^{2+}}(t) \\ \hat{c}^{(j)}_{H^+}(t) \end{bmatrix} = \begin{cases} G(t\,|\,A\left(\beta_1,\alpha_2,\alpha_3\right)^{-1}(z - B(\beta_1,\alpha_2,\alpha_3)), c^{(j)}_0\right), \text{if } j = 1, \\ G^*(t\,|\,A\left(\beta_1,\alpha_2,\alpha_3\right)^{-1}(z - B(\beta_1,\alpha_2,\alpha_3)), c^{(j)}_0\right), \text{if } j\rangle 1, \end{cases} \\
&\beta_1 \in [0.7, 1.5], \alpha_2 \in [0.2, 0.8], \alpha_3 \in [0.2, 0.8], z \in [z_{min}, z_{max}]
\end{aligned} \tag{59}$$

Equation (59) is a pure box-constrained optimization problem and such a problem can be solved using the general PSO algorithm [14]. Meantime, as indicated by the constraint construction process in Sections 2.3.1 and 2.3.2, some parameters of z are actually linearly related to some experimental metrics (e.g., mixed potential, apparent first-order reaction rate constant of cobalt, equilibrium concentration); z_{min} and z_{max} are the transformed lower and upper limits of the experimental metrics, respectively. Therefore, the proper solution of Equation (59) is an estimation of value ranges for experimentally measurable metrics that are transformed from kinetic model parameters whose value ranges are difficult to estimate directly.

In summary, the proposed parameter estimation algorithm for cobalt–hydron electrochemical competition is as follows (Algorithm 1):

Algorithm 1: Constrained parameter estimation algorithm (CPEA)

The inputs include the following: monitored cobalt ion concentration, monitored pH $c^{(j)}_{\mathrm{Co}^{2+},i}, j = 1, \ldots, 5, i = 1, \ldots, n_{\mathrm{Co},j}$, $\mathrm{pH}_l, l = 1, \ldots, n_{\mathrm{pH}}$ initial concentration vector $c^{(j)}_0$ initial mixed potential E_0 weight coefficient λ_{pH} Npop (the population size of the PSO algorithm), and MaxIters (the maximum number of iterations).

Output: Cobalt–hydrogen electrochemical competition model parameters $p^{(1)}$

Steps:

(1) Use the PSO algorithm to solve Equation (59) for $\beta_1, \alpha_2, \alpha_3, z$

(2) Calculate $p^{(1,2)}$ according to Equation (58)

(3) Calculate $p^{(1)}$ according to Equation (51)

---

3. Model and Algorithm Verification

The solution of Equation (59) requires one to know the initial concentration of the reactants and the initial mixed potential in each set. The initial concentration vector of test (j) is as follows:

$$c^{(j)}_0 = \left[2.310, 6.12 \times 10^{-2}, 0.44 \times 10^{-3}, 0, 10^{-pH_{0,j}}, 2.310 b_4^{-1}\left(c^{(j)}_{H^+,0}\right)^{-2}\right]^T \mathrm{mol/L} \tag{60}$$

where pH_0 is the initial pH vector:

$$pH_0 = [3.7\ 3.0\ 3.6\ 4.0\ 4.4]^T \tag{61}$$

The fifth term of Equation (60) represents the initial $Zn(OH)_2$ concentration, which can be calculated from the equilibrium Equation (38): $c_{Zn(OH)_2,0} = c_{Zn^{2+},0} b_4^{-1} c^{-2}_{H^+,0}$. The initial mixed potential can be estimated by solving the equilibrium mixed-potential Equation (23) as follows:

$$E_0 = \arg_E\left(\sum_{i=1}^{3} 2r_i(c_0, E) = 0\right) \tag{62}$$

The weight coefficient λ_{pH} is set to 50. The PSO algorithm is used to optimally solve Equation (59) with Npop set to 50 and MaxIters set to 50. The state-space model (13) is solved using the stiff differential equation solver ode15s in Matlab 2016a. During the optimization process, some randomly generated parameters may make the state-space model very stiff. To avoid wasting too much time on the solution of strongly stiff problems, the time step of the ode15s solver is set to 1×10^{-8}.

Table 1 lists the model parameters optimized by the constrained parameter estimation algorithm. Meanwhile, to evaluate the impact of constraints on the solution, the original parameter estimation problem PRO-1 without constraints is also solved by using the PSO algorithm directly but under two setting conditions—namely Npop = 50 and MaxIters = 50 versus Npop = 500 and MaxIters = 50. The PSO algorithm requires a pre-specified range of parameter values, but Equation (22) only specifies value ranges for β_1, α_2, and α_3. Therefore, the search range of $p^{(1,2)}$ is specified for the original parameter estimation problem using the results of the constrained parameter estimation:

$$p^{(1,2)} \in \left[p_{\text{best}}^{(1,2)} - 10, p_{\text{best}}^{(1,2)} + 10\right] \tag{63}$$

Table 1. Results of constraints guide identification.

Parameter	Value	Parameter	Value	Parameter	Value
β_1	1.004	$ln(k_1)$	−11.00	$ln(b_1)$	38.09
α_2	0.299	$ln(k_2)$	−8.298	$ln(b_2)$	30.03
α_3	0.603	$ln(k_3)$	−13.87	$ln(b_4)$	27.53

Here, $p_{\text{best}}^{(1,2)}$ is the result of constrained parameter estimation with its component values listed in Table 1.

The curves of the best fitness values versus the number of iterations are compared between the proposed constrained parameter estimation algorithm and the original parameter estimation algorithm(OPEA) (Figure 3), and the two algorithms are also compared in terms of estimation accuracy and optimization time (Table 2). Table 2 shows that the original parameter estimation algorithm achieved a relative fitting error of 74.1% for the cobalt concentration curve under the setting condition of Npop = 50 and MaxIters = 50; this failed to find a meaningful solution. Only when the population size Npop increased to 500 could a meaningful solution be found, but the spent time substantially increased by nearly a factor of 8 versus the constrained parameter estimation algorithm. Meanwhile, the mean relative fitting error for the logarithmic concentration of cobalt ions was also larger for the conventional PSO algorithm. These observations indicated that the constrained parameter estimation algorithm significantly improved the estimation efficiency. Moreover, as shown by the curves in Figure 3, the original PSO algorithm with Npop = 50 failed to converge to a good solution. Only when the population size was increased to 500 could the conventional PSO algorithm achieve a parameter estimation accuracy close to that of the constrained PSO algorithm. However, the former algorithm still gave a slightly higher best fitness value than the latter algorithm indicating that constrained parameter estimation is more accurate. It was also reflected on the mean relative errors for fitting the curves of cobalt ion concentrations (Table 2): the MSE from the CPEA, 7.142%, was lower than that of the OPEA,8.775%.

Table 2. Performances of parameter-identification algorithms.

No.	$\ln(\frac{c_{Co^{2+},0}}{c_{Co^{2+}}})$		pH		Ts (min)
	MSE	Rsq	MSE	Rsq	
1	7.142%	0.982	0.729%	0.860	6
2	74.1%	−0.11	0.732%	0.859	7
3	8.775%	0.975	0.721%	0.863	56

1. Constrained parameter estimation algorithm (CPEA) with Npop = 50 and MaxIter s = 50; 2. Original parameter estimation algorithm (OPEA) with Npop = 50 and MaxIters = 50; 3. Original parameter estimation algorithm (OPEA) with Npop = 500 and MaxIters = 50; MSE: mean relative error; Rsq: goodness-of-fit; Ts, solution time.

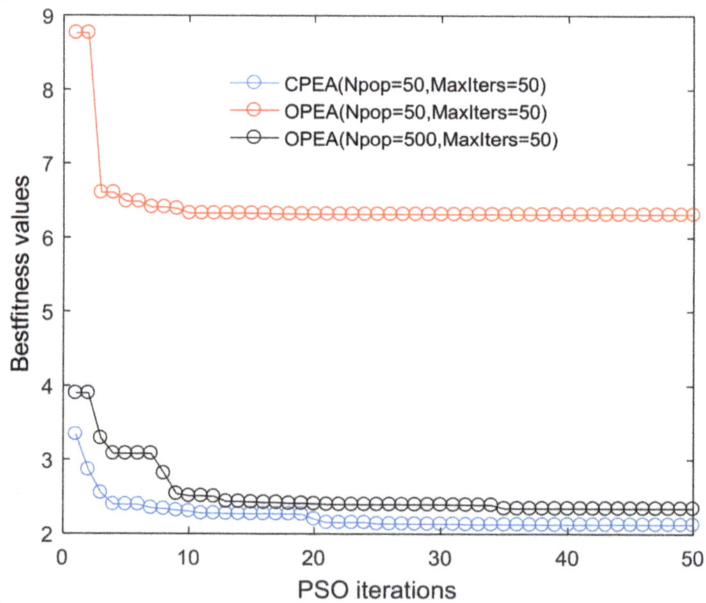

Figure 3. The iteration curves of identification algorithms. CPEA: constrained parameter estimation algorithm, OPEA: original parameter estimation algorithm.

The simulation was performed based on the results of constrained parameter estimation. Model-fit values versus experimentally measured values of cobalt ion concentration were plotted in Figures 4 and 5. The model accurately fitted the curves of cobalt ion concentration versus time under different pH conditions with a mean relative fitting error of 7.142%. The cobalt displacement rate was the lowest at pH 3. The model accurately predicted the inhibitory effect of the competitive hydrogen evolution on cobalt removal when the pH was too high. The model was accurate in predicting the cobalt removal rate at pH 3 and pH 3.6. However, the model had a large error in predicting the cobalt removal rate at pH 4.4, which may be related to the "passivation" effect in the actual cobalt removal process. In other words, $Zn(OH)_2$ is formed and covers the zinc dust surface at high pH, which causes the zinc dust to partially lose reactivity thereby decreasing the cobalt removal rate. The proposed model failed to consider this process and thereby overestimated the cobalt removal rate. The cobalt displacement Reaction (2) gradually reached equilibrium with time as the decrease in cobalt ion concentration gradually leveled off (Figure 4); this trend was well fit by the model.

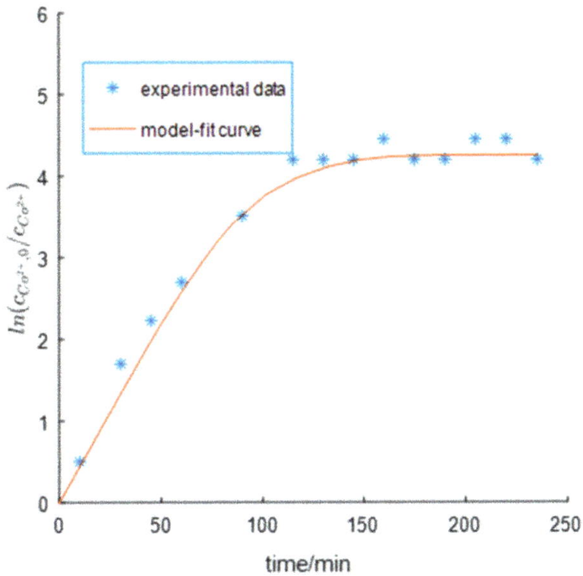

Figure 4. Comparing of the model-fit and experimental [14] dynamical cobalt concentration curves (experimental data come from Lew [4], Page 47, Figure 4-10) with uncontrolled pH started from 3.7; experimental time: 250 min.

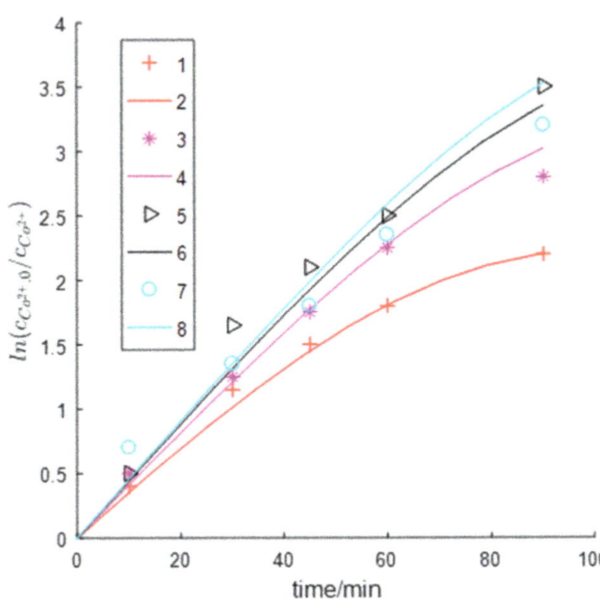

Figure 5. Comparing model-fit and experimental dynamical cobalt concentration curves (experimental data from Lew [4], Page 62, Figure 4-31). The pH is controlled to be a constant value. Symbols 1, 3, 5, and 7 represent experimental data at pH 3, pH 3.6, pH 4, and pH 4.4, respectively; lines 2, 4, 6, and 8 represent fitted lines at pH 3, pH 4, pH 4, and pH 4.4, respectively.

The curve of monitored pH versus time was well fitted by the proposed model (Figure 6) with a mean relative fitting error of 0.729% and a goodness-of-fit of 0.86. In the initial stage, the pH rapidly increased due to massive hydrogen evolution. The addition of zinc dust promoted the forward reaction of zinc hydrolysis (4), which replenished hydrogen ions. As a result, the apparent consumption rate of hydrogen ions was far less than the actual hydrogen evolution rate, which led to a dramatic decrease in the rate of change of pH with time thereby leading to a monotonic and concave-down curve.

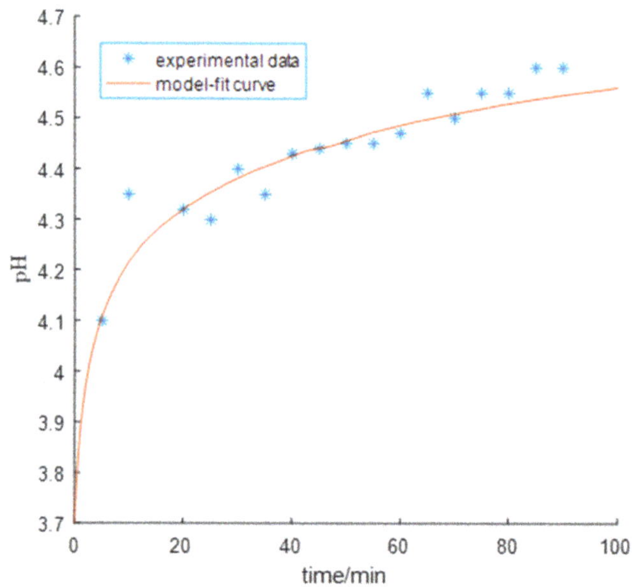

Figure 6. Comparing the model-fit and experimental [14] dynamical pH curves (experimental data come from Lew [4], Page 42, Figure 4-4) with uncontrolled pH started from 3.7.

4. Conclusions

(1) For the first time, a mechanistic kinetic model of cobalt–hydrogen electrochemical competition was proposed for the cobalt removal process in zinc hydrometallurgy by exploring the mechanism of the impact of hydrogen ions on the cobalt removal rate and the mechanism of zinc ion hydrolysis. The model parameters were estimated based on experimental data. The estimation results showed that the model accurately simulated the competitive inhibitory effect of hydrogen ions on cobalt removal and the trend of pH in the cobalt removal process. It shows a goodness-of-fit value of 0.982 on cobalt concentration values and that of 0.860 on pH values.

(2) The difficulty in using heuristic optimization algorithms to estimate the parameters of reaction kinetic models with little information about the parameter value ranges was addressed. A new approach for estimating the parameters of strongly nonlinear and chemical reaction mechanism models was proposed. Specifically, parameter estimation constraints were first constructed based on an estimation of the value ranges of experimental metrics to narrow the search space. Next, the nonlinear constraints were transformed to simple box constraints through quasi-linear mapping, and a box-constrained parameter estimation scheme was constructed for the cobalt–hydrogen electrochemical competition model. Parameter estimation tests showed that the constrained parameter estimation algorithm was far more efficient than the conventional parameter estimation algorithm. Specifically, the constrained parameter estimation algorithm reduced the solution time by 90 percent. Furthermore, the

proposed algorithm had a lower MSE on fitting the cobalt ion concentration curves than the conventional parameter estimation algorithm (7.142% vs. 8.775%).

(3) The non-linearity of the electrochemical and hydrolysis reactions involved in the cobalt removal process as well as the high coupling between multiple reactions made the model incomprehensible. To overcome this limitation, we derived estimation equations for the value ranges of relevant experimental metrics—mixed potential, apparent first-order reaction rate of cobalt removal, equilibrium concentration, and attenuation coefficient for the apparent consumption rate of hydrogen ions—based on model simplification assumptions in the constraint construction process. These equations linked the experimental phenomenon metrics to the model parameters providing deeper insights into the model parameters adjustment for model users and controller design [22,23].

(4) As stated above, the model proposed in this paper can accurately predict the change of cobalt ion concentration and the rate of hydrogen replacement reaction during the cobalt removal process, so it is helpful for optimizing the zinc powder and acid usage in the practical cobalt removal process of zinc hydrometallurgy, although this may require for adding extra equations about the relationship between dosages and the zinc powder concentration or pH. In addition, the identification framework proposed significantly reduces the time required for parameter identification of the model, which is also very helpful for improving the efficiency and accuracy of the modelling in the optimal control implementation.

(5) The model does not consider the "passivation" effect of zinc ion, as a result, it has a large error in predicting the cobalt removal rate while the value of controlled pH is larger than 4.4. Therefore, the available pH value of the model should be lower than 4.4. In addition, the model does not consider the catalytic effect of copper and antimony, so it is not suitable for the process with large fluctuations in copper and antimony concentrations.

Author Contributions: Conceptualization, Y.L. (Yiting Liang) and Y.L. (Yonggang Li); methodology, Y.L. (Yiting Liang); software, Y.Z.; validation, Y.L. (Yiting Liang); formal analysis, Y.L. (Yiting Liang); investigation, Y.L. (Yiting Liang); resources, Y.L. (Yonggang Li); data curation, Y.Z.; writing—original draft preparation, Y.L. (Yiting Liang); writing—review and editing, Y.L. (Yonggang Li); visualization, Y.L. (Yiting Liang); supervision, Y.L. (Yonggang Li); project administration, Y.L. (Yonggang Li); funding acquisition, Y.L. (Yonggang Li). All authors have read and agreed to the published version of the manuscript.

Funding: This work was supported by the National Key R&D Program of China(2019YFB1704703), the Natural Science Foundation of Hunan Province (Grant No. 2019JJ50823).

Data Availability Statement: Publicly available datasets were analyzed in this study, reference number [4]. It can be found here: https://open.library.ubc.ca/cIRcle/collections/ubctheses/831/items/1.0078522.

Conflicts of Interest: The authors declare no conflict of interest.

References

1. Davey, T.R.A.; Willis, G.M. Review of developments in extractive metallurgy of lead, zinc, and tin in 1978. *JOM* **1979**, *31*, 1–17. [CrossRef]
2. Nelson, A.; Wang, W.; Houlachi, G.P.D.G. The removal of cobalt from zinc electrolyte by cementation: A critical review. *Miner. Process. Extract. Metall. Rev.* **2000**, *20*, 325–356. [CrossRef]
3. Pas, V.V.D.; Dreisinger, D.B. A fundamental study of cobalt cementation by zinc dust in the presence of copper and antimony additives. *Hydrometallurgy* **1996**, *43*, 187–205.
4. Lew, R.W. The Removal of Cobalt from Zinc Sulphate Electrolytes Using the Copper-Antimony Process. Ph.D. Thesis, University of British Columbia, Vancouver, BC, Canada, 1994.
5. Karlsson, T.; Cao, Y.; Colombus, Y.; Steenari, B.-M. Investigation of the kinetics and the morphology of cementation products formed during purification of a synthetic zinc sulfate electrolyte. *Hydrometallurgy* **2018**, *181*, 169–179. [CrossRef]
6. Tozawa, K.; Nishimura, T.; Akahori, M.; Malaga, M.A. Comparison between purification processes for zinc leach solutions with arsenic and antimony trioxides. *Hydrometallurgy* **1992**, *30*, 445–461. [CrossRef]

7. Wang, L.Y.; Gui, W.H.; Teo, K.L.; Loxton, R.; Yang, C.H. Time delayed optimal control problems with multiple characteristic time points: Computation and industrial applications. *J. Ind. Manag.* **2009**, *5*, 705–718.
8. Sun, B.; Gui, W.; Wu, T.; Wang, Y.; Yang, C. An integrated prediction model of cobalt ion concentration based on oxidation-reduction potential. *Hydrometallurgy* **2013**, *140*, 102–110. [CrossRef]
9. Li, Y.G.; Gui, W.H.; Teo, K.L.; Zhu, H.Q.; Chai, Q.Q. Optimal control for zinc solution purification based on interacting CSTR models. *J. Process. Control.* **2012**, *22*, 1878–1889. [CrossRef]
10. <named-content content-type="background:white">Sun, B.; Yang, C.; Wang, Y.; Gui, W.; Craig, I.; Olivier, L. A comprehensive hybrid first principles/machine learning modeling framework for complex industrial processes. *J. Process. Control* **2020**, *86*, 30–43.
11. Mangan, N.M.; Brunton, S.L.; Proctor, J.L.; Kutz, J.N. Inferring Biological Networks by Sparse Identification of Nonlinear Dynamics. *IEEE Trans. Mol. Biol. Multi-Scale Commun.* **2016**, *2*, 52–63. [CrossRef]
12. Ghahramani, Z.; Roweis, S.T. Learning Nonlinear Dynamical Systems using an EM Algorithm. *Adv. Neural Inf. Process. Syst.* **1999**, *11*, 431–437.
13. Nüsken, N.; Reich, S.; Rozdeba, P.J. State and Parameter Estimation from Observed Signal Increments. *Entropy* **2019**, *21*, 505. [CrossRef] [PubMed]
14. Chai, Q.; Loxton, R.; Teo, K.L.; Yang, C. A unified parameter identification method for nonlinear time-delay systems. *J. Ind. Manag. Optim.* **2013**, *9*, 471–486. [CrossRef]
15. Eberhart, R.; Kennedy, J. A new optimizer using particle swarm theory. In Proceedings of the Sixth International Symposium on Micro Machine & Human Science, Nagoya, Japan, 6 August 2002.
16. <named-content content-type="background:white">Zhou, X.; Yang, C.; Gui, W. A Statistical Study on Parameter Selection of Operators in Continuous State Transition Algorithm. *IEEE Trans. Cybern.* **2019**, *49*, 3722–3730.
17. Storn, R. Differential evolution-a simple and efficient heuristic for global optimization over continuous space. *J. Glob. Optim.* **1997**, *11*, 341–359. [CrossRef]
18. <named-content content-type="background:white">Hu, W.; Yu, Y.; Wang, S. Parameters Estimation of Uncertain Fractional-Order Chaotic Systems via a Modified Artificial Bee Colony Algorithm. *Entropy* **2015**, *17*, 692–709.
19. Zhang, B.; Yang, C.; Zhu, H.; Li, Y.; Gui, W. Kinetic Modeling and Parameter Estimation for Competing Reactions in Copper Removal Process from Zinc Sulfate Solution. *Ind. Eng. Chem. Res.* **2013**, *52*, 17074–17086. [CrossRef]
20. Yang, C.; Deng, S.; Li, Y.; Zhu, H.; Li, F. Optimal Control for Zinc Electrowinning Process with Current Switching. *IEEE Access* **2017**, *5*, 24688–24697. [CrossRef]
21. Liu, C. Modelling and parameter identification for a nonlinear time-delay system in microbial batch fermentation. *Appl. Math. Model.* **2013**, *37*, 6899–6908. [CrossRef]
22. Wang, S.; Na, J.; Chen, Q. Adaptive Predefined Performance Sliding Mode Control of Motor Driving Systems with Disturbances. *IEEE Trans. Energy Convers.* **2020**, *1*. [CrossRef]
23. Chen, Q.; Yu, X.; Sun, M.; Wu, C.; Fu, Z. Adaptive Repetitive Learning Control of PMSM Servo Systems with Bounded Nonparametric Uncertainties: Theory and Experiments. *IEEE Trans. Ind. Electron.* **2020**, *1*. [CrossRef]

entropy

MDPI

Article

Distributed Deep Fusion Predictor for a Multi-Sensor System Based on Causality Entropy

Xue-Bo Jin [1,2,*], Xing-Hong Yu [1,2], Ting-Li Su [1,2,*], Dan-Ni Yang [3], Yu-Ting Bai [1,2], Jian-Lei Kong [1,2] and Li Wang [1,2,*]

[1] Artificial Intelligence College, Beijing Technology and Business University, Beijing 10048, China; yuxinghong@st.btbu.edu.cn (X.-H.Y.); baiyuting@btbu.edu.cn (Y.-T.B.); kongjianlei@btbu.edu.cn (J.-L.K.)
[2] China Light Industry Key Laboratory of Industrial Internet and Big Data Beijing Technology and Business University, Beijing 10048, China
[3] Electrical and Information Engineering College, Tianjin University, Tianjin 300072, China; 18693803229@163.com
* Correspondence: jinxuebo@btbu.edu.cn (X.-B.J.); sutingli@btbu.edu.cn (T.-L.S.); wangli@th.btbu.edu.cn (W.L.)

Abstract: Trend prediction based on sensor data in a multi-sensor system is an important topic. As the number of sensors increases, we can measure and store more and more data. However, the increase in data has not effectively improved prediction performance. This paper focuses on this problem and presents a distributed predictor that can overcome unrelated data and sensor noise: First, we define the causality entropy to calculate the measurement's causality. Then, the series causality coefficient (SCC) is proposed to select the high causal measurement as the input data. To overcome the traditional deep learning network's over-fitting to the sensor noise, the Bayesian method is used to obtain the weight distribution characteristics of the sub-predictor network. A multi-layer perceptron (MLP) is constructed as the fusion layer to fuse the results from different sub-predictors. The experiments were implemented to verify the effectiveness of the proposed method by meteorological data from Beijing. The results show that the proposed predictor can effectively model the multi-sensor system's big measurement data to improve prediction performance.

Keywords: series causality analysis; Bayesian LSTM; multi-sensor system; meteorological data; big measurement data; deep fusion predictor

Citation: Jin, X.-B.; Yu, X.-H.; Su, T.-L.; Yang, D.-N.; Bai, Y.-T.; Kong, J.-L.; Wang, L. Distributed Deep Fusion Predictor for a Multi-Sensor System Based on Causality Entropy. *Entropy* **2021**, *23*, 219. https://doi.org/10.3390/e23020219

Academic Editor: Quan Min Zhu, Giuseppe Fusco, Jing Na, Weicun Zhang and Ahmad Taher Azar

Received: 11 January 2021
Accepted: 7 February 2021
Published: 11 February 2021

Publisher's Note: MDPI stays neutral with regard to jurisdictional claims in published maps and institutional affiliations.

1. Introduction

Measurements have been obtained and saved in many multi-sensor systems, such as mobile robots [1], unmanned aerial vehicles (UAVs) [2,3], smart agriculture [4,5], air quality monitoring systems [6,7], etc. It is very meaningful to analyze these data and understand and predict the information in the sensor system [8], for example the analysis and prediction of meteorological elements in precision agriculture or environmental management systems [9]. Furthermore, in terms of environmental governance, the prediction for air pollution sources such as PM2.5 has played an important role [10–13].

Recently, more measurements have been collected with the development of sensor technology. Therefore, in a multi-sensor system, big data analysis has become a new research area. These data have two characteristics: noisy and numerous [14]. For example, the collected and saved meteorological data are big data and include many variables, such as temperature, wind, rainfall, humidity, etc. Further, they are related to each other [15]. However, the correlation between each type of variable is different: some of them have a strong correlation, but some have a low correlation.

In general, more data can provide more information. For big data, deep learning can extract hidden information to make more accurate predictions [16]. The recent research has proven that the recurrent neural network (RNN) and its improved version are widely used in regression prediction problems with better nonlinear modeling ability compared with the classical regression method.

We can find that the network has become larger and more complex due to the massive amount of data. However, because of the amount of data, the network's training time is getting longer. To make matters worse, the increase of the input data does not improve the prediction performance; on the contrary, it decreases.

This paper focuses on how to use this big noisy data in the multi-sensor system to efficiently improve prediction performance. This paper mainly aims at multi-sensor systems, proposes a causal entropy method for feature selection, and constructs a distributed forward multi-step prediction framework based on Bayesian deep learning theory. In this way, the dimensionality reduction of high-dimensional data feature selection is realized, and the problem of data noise affecting deep network training is initially overcome. The rest is organized as follows: Section 2 summarizes current prediction models and describes the main contribution of this paper. Section 3 proposes a distributed deep learning network predictor, and Section 4 describes the experiments and results to verify the performance of our predictor. We draw conclusions in Section 5.

2. Related Works

2.1. The Methods for Prediction

Prediction is to analyze historical data and obtain the trend of the future. With the development of computer storage technology and sensor technology, the prediction of measurement data in a multi-sensor system has been widely used in many fields. It has become a hot topic of research. The traditional prediction methods require prior knowledge of the data, such as exponential smoothing [17], moving average (MA) [18], auto-regression (AR) [19], auto-regressive integrated moving average (ARIMA) [20], etc. In practical systems, the traditional prediction methods cannot obtain a high accuracy prediction result due to the system's complexity.

For nonlinear input data, shallow machine learning methods obtain model parameters through training, such as support vector machines (SVMs) [21], the echo state network (ESO) [22], Boltzmann machines (BMs) [23], shallow artificial neural networks (ANNs) [24], generalized regression neural networks (GRNNs) [25], etc., which avoids the requirement of prior knowledge of the data. However, because of their simple structure, they cannot process large amounts of data.

With the development of the depth of the neural network, the hidden information in the massive data can be extracted to make more accurate predictions. The recurrent neural network (RNN) [26] and its improved versions, such as long short-term memory (LSTM) [27], etc., is widely used for regression prediction problems, demonstrating its superior nonlinear modeling capabilities. For example, a gated recurrent unit (GRU) network [28] and Bi-LSTM [29] were proposed to improve LSTM. Furthermore, researchers [30,31] have combined the one-dimensional convolutional neural network (CNN) with LSTM to predict the time series data. System identification is the theory, and the methods of establishing the mathematical models of dynamical systems [32–36] and some identification approaches can be used to establish the prediction models and soft sensor models [37–42] for various application problems.

2.2. The Method to Calculate Causality and Correlation

Undoubtedly, deep neural networks are currently the best solution to the big data prediction problem of multi-sensor systems. However, we found that the network's ability to predict does not increase as the amount of input variable increases. On the contrary, sometimes, the larger the amount of input measurement data from multi-sensor system, the worse performance the prediction obtains. This is contrary to what we have always believed: one advantage of the deep learning network is that it has comprehensive and robust learning capabilities for big data.

We believe one of the reasons is that the data contain too much low-relevance information; the increase in the amount of data leads to a decrease in the ratio of useful information. The weights of training for the neural network and the diluted information make the network's convergence more difficult, so the prediction performance cannot be developed, but is even reduced. Therefore, we think the data with a high correlation and strong causality with the target variable should be selected as the network's input data, rather than just increasing the number.

Then, we describe the correlation degree method and discuss a causal correlation method for measuring variables suitable for multi-sensor systems to measure big data. The Pearson correlation coefficient (PCC) [43] and Spearman correlation coefficient [44,45] have been used for such a problem. The former can be used to find the linear relationship between the two variables. For the data, the features are continuous and conform to a positive distribution; the linear relationship between the two variables can be mined by the PCC [46]. Jing et al. [47] selected the characteristic sub-sequence by PCC to improve the prediction accuracy when forecasting photo-voltaic power output. Lin et al. [48] built a hybrid model framework using the stacking scheme of integrated learning by PCC between different models. As for the prediction problem, PCC requires a known prediction target variable, so it cannot be applied for predicting.

Spearman's correlation coefficient is mainly used to solve problems related to sequential data. It applies to two variables with an order relationship. Another kind of correlation analysis method is called the Kendall correlation coefficient [49], which is suitable for sequenced variables or evenly spaced data that do not satisfy the assumption of a normal distribution. This method is usually calculated for a piece of sequence data, and it cannot obtain an effective correlation between the input and output for large amounts of data.

Contreras-Reyes et al. [50] used the frequency-domain Granger-causality method to test the statistical significance of causality between two time series and determine the direction of causality on the drivers of pelagic species' biological indicators. Since the Granger causality coefficient is used in two stationary time series, its application is limited. Podobnik et al. [51] proposed a detrending cross-correlation analysis method to explore the correlation between two non-stationary series. It shows that effectively measuring the correlation between two variables can help analyze the change characteristics of one of the variables.

The current methods to calculate the correlation and causality rely on the predicted result and cannot be applied to the prediction problem of multi-sensor systems.

2.3. The Bayesian Deep Learning Network

The big data measured by sensors contain noise, which is another reason for the degradation of prediction performance based on deep learning networks. Traditional neural network training obtains the fixed weights and biases, which are easily disturbed by noise [52,53]. On the one hand, the noise makes it difficult to converge the network, that is the loss of the network is larger. On the contrary, if the noise is also learned as a certain value until a small loss value is obtained, it will cause the problem of overfitting [54].

Suppose we use the data distribution to train the network and obtain weights and deviations to express the input data's distribution characteristics. In that case, the problem of overfitting will be avoided. Based on the distribution characteristics of the weights and deviations, the obtained neural network is a group. The output is also a group of prediction outputs with distribution characteristics, improving the prediction results' reliability. Based on this research question, the Bayesian deep learning network came into being [55]. Through Monte Carlo sampling, the Bayesian deep learning network trains the network several times and takes the average of all losses, then uses it for backpropagation to obtain the distribution of weights and deviations [56].

The Bayesian method has been used in many application systems, such as indoor tracking [57], robot systems [58], etc. The Bayesian deep learning network has been applied in modeling with noisy data, and some results have been obtained. For example,

Li et al. [59] integrated uncertainties by defining the Bayesian deep learning framework, in which a sequential Bayesian boosting algorithm is used to improve the estimation accuracy. Another example is [60], where a Bayesian framework was proposed to model the valence predictions.

2.4. Innovation

Aiming at the problem of improving the prediction performance based on the huge amount measurement data in a multi-sensor system, this paper provides a distributed deep prediction network. To solve the contradiction between data volume and performance and the influence of data noise on prediction performance, the innovation of this paper lies in the following:

(1) A series causality entropy method is developed to select the related input data for the neural network. Compared with the PCC [48], Spearman correlation [45], and the Kendall correlation coefficient method [61], the method does not depend on the prediction results and is suitable for prediction problems based on measurement data in multi-sensor systems.

(2) A distributed prediction framework is proposed, in which Bayesian training is used to suppress the noise impact of the data, and the prediction based on the selected input data is fused by a nonlinear fusion network. Compared with the classical LSTM [27], GRU [28], CNN-LSTM [11], the conv-LSTM [30] predictor, etc., the proposed method outperforms them in its prediction performance.

3. Distributed Deep Fusion Predictor

3.1. Series Causality Entropy

In a multi-sensor system, we can set up multiple sensors to obtain a variety of measurement data. For example, in the system given in Figure 1, we use four sensors to obtain four types of measurement data, and they will be used as candidate input data for the deep network. The prediction task is for Measurement 1, and we will predict its future trend.

Firstly, we will consider the method to select the input data for the networks. Obviously, the principle of selecting data is to select those measurement data that are most causal for the future trend of Measurement 1. As for the prediction problem, it can be defined as the series causality between the historical data and the future data.

We give the following definition about the causality entropy to calculate the measurement's causality between two data named X and Y:

$$CE(X,Y) = \frac{1}{N-1}\sum_{i=1}^{N}\left(\frac{X(i)-\overline{X}}{\sigma_X}\right) \times \log\left(\frac{Y(i)-\overline{Y}}{\sigma_Y}\right) \tag{1}$$

where $\overline{X}$ and $\overline{Y}$ are the mean of $X(i)$ and $Y(i)$, $i = 1, 2, \ldots\ldots, N$, respectively, and σ_X and σ_Y are the standard deviation of $X(i)$ and $Y(i)$. We can see that $CE(X,Y)$ can be positive or negative. When it is a positive number, it indicates that the two data are positively supporting. Otherwise, it is negatively supporting.

This calculating correlation method cannot be directly applied to obtain the correlation of prediction problems. Because the measured data and their prediction are considered in the prediction problem, therefore, we cannot calculate CE when the prediction is not yet available. Secondly, since the step length of the measurement data I is different from the predicted step length J, the number of data points N in Equation (1) cannot be used.

Therefore, we propose the following series causality coefficient (SCC) for the measurement in the multi-sensor system. Suppose the measured data are represented by $X_m(i)$, where $m = 1, 2, \ldots\ldots, M$ is the sensor number to obtain the measurement and $i = 1, 2, \ldots\ldots, I$ is the step number of historical data used for prediction. The target data to be predicted are represented by $Y(j)$, where $j = 1, 2, \ldots\ldots, J$ is the step number of prediction. We revised the method to calculate coefficient Equation (1) as the following.

$$S_m = \frac{1}{K-1}\sum_{i=1,j=1}^{K}\left|\frac{X_m(i) - \overline{X_m}}{\sigma_{X_m}}\right| \times \log\left|\frac{Y(j) - \overline{Y}}{\sigma_Y}\right| \tag{2}$$

where $m = 1, 2, \ldots\ldots M$ is the sensor number to obtain the measurement, $K = min(I, J)$, $\overline{X_n}$ and $\overline{Y}$ are the mean of $Xm(i)$ and $Y(i)$, $i = 1, 2, \ldots\ldots K$, respectively, and σ_{X_m} and σ_Y are the standard deviation of $\overline{X_n}$ and $\overline{Y}$. We can find that Equation (2) still has the prediction $Y(i)$, which is unknown data. To eliminate $Y(i)$ in Equation (2), we modify Equation (2) by normalization. The normalized SCC of each measurement can be obtained by the following.

$$SCC_m^* = \frac{S_m}{S_1 + S_2 + \ldots + S_M} = \frac{\sum_{i=1}^{K}\left|\frac{X_m(i) - \overline{X_m}}{\sigma_{X_m}}\right|}{\sum_{m=1}^{M}\sum_{i=1}^{K}\left|\frac{X_m(i) - \overline{X_m}}{\sigma_{X_m}}\right|} \tag{3}$$

From Equation (3), we can conclude that the value of SCC is between zero and one; the larger the SCC, the higher the causality is. For example, when the value is zero, it means that the feature is not useful for predicting the target variable. We can see that the SCC given by Equation (3) omits the calculation process for the prediction $Y(i)$.

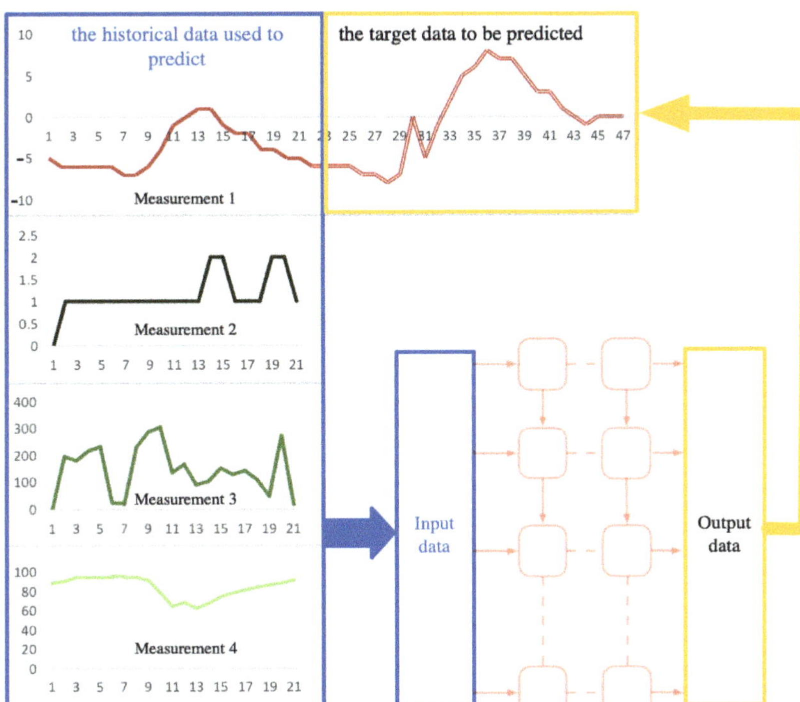

Figure 1. Relationship between the target variables to be predicted and the input variables.

We give the following examples to illustrate the SCC obtained by Equation (3). Meteorological data are used, including temperature, wind direction, wind force, rainfall, and humidity, which are used to predict the future temperature. We have five measurements, so according to Equation (3), M is five. We set $K = 24$, then SCC_m can be obtained. X is set to five meteorological elements separately, and Y is the future temperature to be predicted. The result is shown in Table 1. To clearly illustrate the difference of SCC_m, we visualize them as Figure 2.

Table 1. The order of the SCC between variables to be predicted.

Measurement	Series Causality Coefficient (SCC)
Temperature	0.3673
Humidity	0.3259
Wind force	0.175
Wind direction	0.1318
Rainfall	0

It can be seen from Table 1 and Figure 2 that the causality between historical temperature data and their future prediction is the largest, which is 0.3673. Next is humidity. We get the SSC as 0.3259 between historical humidity and future temperature. Compared with them, the causal relationship between wind force and wind direction with the future temperature is smaller. The data can also reflect no causal relationship between rainfall and temperature, for which we obtain a zero SSC.

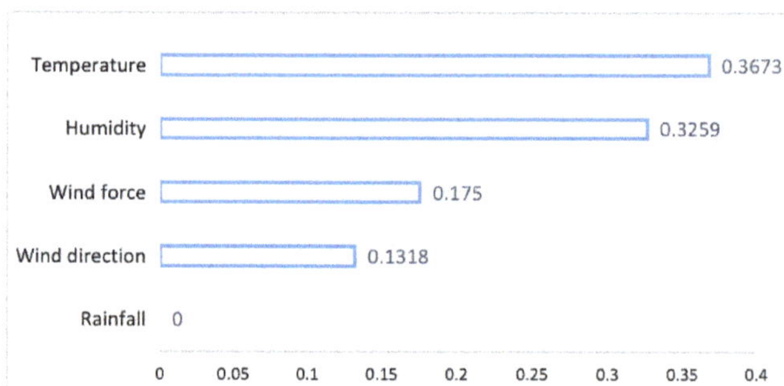

Figure 2. SCC between different measurements and predictions.

Further, we can get the following conclusions. If all the data are used for training, rainfall data can only cause the network to reduce the training's convergence and the temperature prediction performance. Therefore, rainfall data must be eliminated and cannot be used as the input for network training and prediction. Regarding wind force and wind direction data, because of their low causality, even as the network's input data, the performance improvement of the prediction results is limited. It will increase the training time of the network. On the contrary, the humidity data have a high causal correlation with the future temperature. Therefore, using the historical temperature data and humidity data to predict the future temperature may achieve better performance than just using temperature data. The experiments in Section 4 will verify the above points.

3.2. Bayesian LSTM as the Sub-Predictor

The LSTM cell is used in this paper, which is composed of three gating units, i.e., input gate, forget gate, and output gate. The calculation process is the following:

$$\begin{aligned}
f_t &= \sigma\left(W_{fx}x_t + W_{fh}h_{t-1} + b_f\right) \\
i_t &= \sigma(W_{ix}x_t + W_{ih}h_{t-1} + b_i) \\
\bar{c}_t &= \tanh(W_{cx}x_t + W_{ch}h_{t-1} + b_c) \\
c_t &= f_t \cdot c_{t-1} + i_t \cdot \bar{c}_t \\
o_t &= \sigma(W_{ox}x_t + W_{oh}h_{t-1} + b_o) \\
h_t &= o_t \cdot \tanh(c_t)
\end{aligned} \tag{4}$$

where t is the current moment to predict, $w = [W_{fx},\ W_{fh},\ W_{ix},\ W_{cx},\ W_{ox},\ W_{oh}]$ are the weights, and $b = [b_f,\ b_i,\ b_o]$ are the biases. c_t is the hidden state, and h_t is the output of the LSTM cell. The cells can be placed as several layers with different input and output cells depending on the number of input and output steps of the prediction. The structure of the network is shown in Figure 3. The input data x are the given data used to predict the future trend, where $x = [X(1),\ X(2),\ \ldots\ldots,\ X(I)]$ are the input data at each moment with the number of data I, and $x_t = [X_t(1),\ X_t(2),\ \ldots\ldots,\ X_t(I)]$ are the input data at the current moment t. The output of the last layer can be set as the output of the LSTM network, named as y. For the training process, we have $y = [Y(1),\ Y(2),\ \ldots\ldots,\ Y(J)]$, and at the current moment t, we have $y_t = [Y_t(1),\ Y_t(2),\ \ldots\ldots,\ Y_t(J)]$.

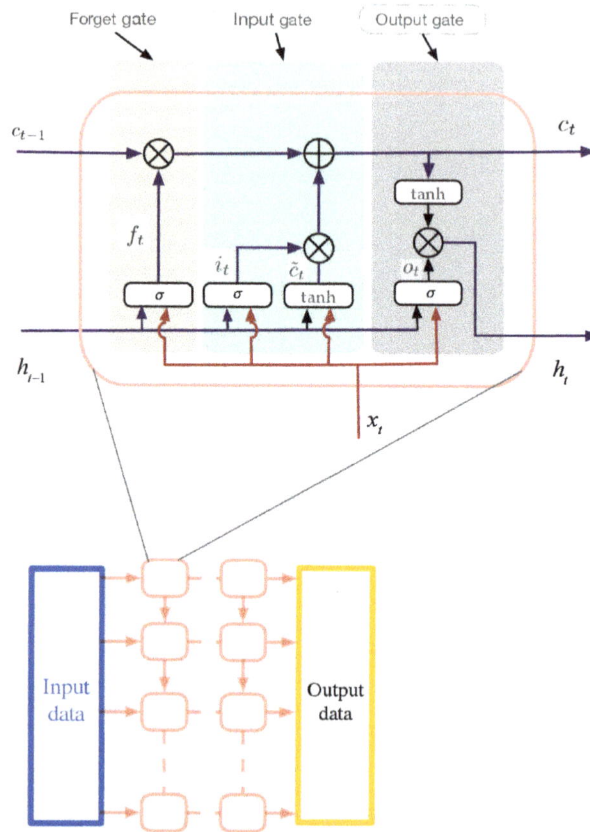

Figure 3. LSTM cell and its networks.

In the normal LSTM network, the parameters, including all the weights and biases, are constants. The Bayesian LSTM can get the weight and bias as a random distribution, not a certain value. Each parameter obtained by the Bayesian LSTM network training is the mean and variance according to the distribution of the weights and biases. The difference between the normal LSTM network and the Bayesian LSTM network is shown in Figure 4.

The LSTM neural network can be seen as a probabilistic model $P(y|x,\theta)$: a probability given an input $x \in \mathbb{R}^p$ to each possible output $y \in Y$, using the set of parameters θ including weights w and biases b, i.e., $\theta = [w,\ b]$. We denote the training data x and y as D, i.e., $D = [x,\ y]$.

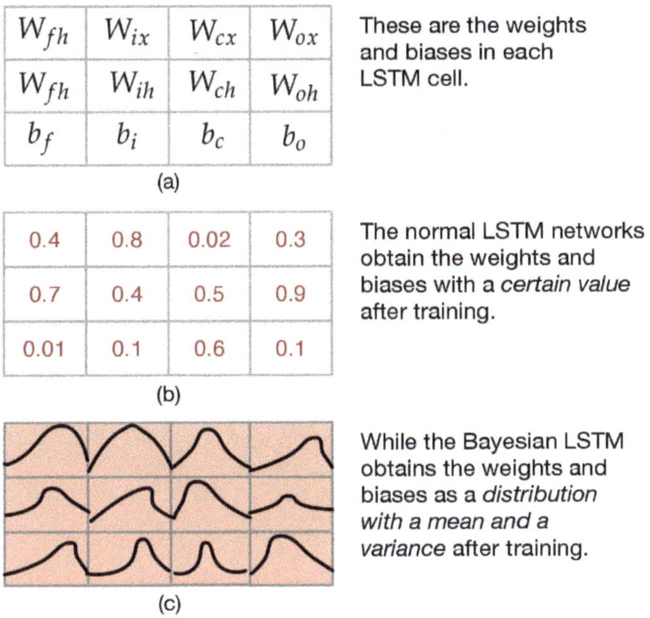

Figure 4. The difference between the normal LSTM network and the Bayesian LSTM network. (**a**) The parameters in the LSTM; (**b**) the example of the parameters in the normal LSTM; (**c**) the example of the parameters in the Bayesian LSTM.

Given the training data D, Bayesian inference can be used to calculate the posterior distribution of weights $P(w \mid D)$ [62]. This distribution answers the predicted distribution of unknown data through the input data value: the predicted distribution of the input data x is given by $P(y \mid x) = E_{P(\theta \mid D)}[P(y \mid x, \theta)]$. Until now, it is still difficult to find $P(w \mid D)$. The variational approximation to the Bayesian posterior distribution on the weights is a feasible method. Variational learning finds the parameters (μ, σ) of a distribution on the weights $q(\theta|\mu, \sigma)$ that minimizes the Kullback–Leibler (KL) divergence [63] with the true Bayesian posterior on the weights:

$$(\mu, \sigma)^* = arg \min_{\mu, \sigma} KL[q(\theta|\mu, \sigma)||P(\theta|D)] \tag{5}$$

According to the Bayesian theory,

$$P(\theta|D) = \frac{P(D|\theta)P(\theta)}{P(D)} \tag{6}$$

and the definition of the Kullback–Leibler (KL) divergence, Equation (5) can be transformed to:

$$(\mu, \sigma)^* = arg \min_{\mu, \sigma} \int q(\theta|\mu, \sigma) \log \frac{q(\theta|\mu, \sigma)}{P(\theta)P(D|\theta)} d\theta \tag{7}$$

Note that we discarded $P(D)$ because it does not affect the optimized parameter solution. Then, the cost function is set as:

$$Loss = \int q(\theta|\mu, \sigma) \log \frac{q(\theta|\mu, \sigma)}{P(\theta)P(D|\theta)} d\theta \tag{8}$$

To keep the variance non-negative, we set it as $\sigma = \log(1 + exp(\rho))$. Set ε as zero mean Gaussian white noise, i.e., $\varepsilon \sim \mathcal{N}(0,1)$. Then, we have $\theta = \mu + \log(1 + exp(\rho)) \otimes \varepsilon$, where $\otimes$ is point-wise multiplication. Further, we can note that $q(\theta|\mu, \rho)\, d\theta = q(\varepsilon)\, d\varepsilon$, then the derivative of Equation (8) can be calculated as the following:

$$\frac{\partial}{\partial\mu}Loss = \frac{\partial}{\partial\mu}\int q(\theta|\mu, \rho) \log\frac{q(\theta|\mu, \rho)}{P(\theta)P(D|\theta)}d\theta \tag{9}$$

$$\frac{\partial}{\partial\rho}Loss = \frac{\partial}{\partial\rho}\int q(\theta|\mu, \rho) \log\frac{q(\theta|\mu, \rho)}{P(\theta)P(D|\theta)}d\theta \tag{10}$$

Then, as for Equation (9), we have:

$$\begin{aligned}\frac{\partial}{\partial\mu}Loss &= \frac{\partial}{\partial\mu}\int q(\theta|\mu, \rho) \log\frac{q(\theta|\mu, \rho)}{P(\theta)P(D|\theta)}d\theta \\ &= \frac{\partial}{\partial\mu}\int \log\frac{q(\theta|\mu, \rho)}{P(\theta)P(D|\theta)}\, q(\theta|\mu, \rho)d\theta \\ &= \frac{\partial}{\partial\mu}\int \log\frac{q(\theta|\mu, \rho)}{P(\theta)P(D|\theta)}\, q(\varepsilon)d\varepsilon \\ &= \frac{\partial}{\partial\mu}\log\frac{q(\theta|\mu, \rho)}{P(\theta)P(D|\theta)}\int q(\varepsilon)d\varepsilon \\ &= \frac{\partial}{\partial\mu}\log\frac{q(\theta|\mu, \rho)}{P(\theta)P(D|\theta)}\end{aligned} \tag{11}$$

Similarly, Equation (10) can be derived further as the following:

$$\frac{\partial}{\partial\rho}Loss = \frac{\partial}{\partial\rho}\log\frac{q(\theta|\mu, \rho)}{P(\theta)P(D|\theta)} \tag{12}$$

Denote that:

$$Loss = \log\frac{q(\theta|\mu, \rho)}{P(\theta)P(D|\theta)} = \log q(\theta|\mu, \rho) - \log P(\theta) - \log P(D|\theta)$$

then we have:

$$\begin{aligned}\frac{\partial}{\partial\mu}Loss &= \frac{\partial Loss}{\partial\theta}\frac{\partial\theta}{\partial\mu} + \frac{\partial Loss}{\partial\mu} \\ &= \frac{\partial Loss}{\partial\theta} + \frac{\partial Loss}{\partial\mu}\end{aligned} \tag{13}$$

$$\begin{aligned}\frac{\partial}{\partial\rho}Loss &= \frac{\partial Loss}{\partial\theta}\frac{\partial\theta}{\partial\rho} + \frac{\partial Loss}{\partial\rho} \\ &= \frac{\partial Loss}{\partial\theta}\frac{\varepsilon}{1+exp(-\rho)} + \frac{\partial Loss}{\partial\rho}\end{aligned} \tag{14}$$

Please note that the standard deviations of the $\frac{\partial Loss}{\partial\theta}$ term of the mean and the gradient are shared, and it happens to be the gradient found by the backpropagation algorithm on the normal LSTM network. Therefore, to learn the mean and standard deviation, we can calculate the gradient by backpropagation and then scale and translate it. We summarize the optimization process as seven steps in Table 2.

Table 2. The optimization process for the Bayesian LSTM networks.

Step	Optimization Process
0	Set the scale parameter α as $\alpha \in (0,\ 1)$.
1	Sample the random variable ε as $\varepsilon \sim \mathcal{N}(0,1)$.
2	Set the initial value of the optimized parameters $(\mu,\ \rho)$.
3	Sample all the parameters as $\theta = \mu\ +\ \log(1\ +\ exp(\rho))\ \otimes\ \varepsilon$.
4	Set the cost function as $Loss = \log q(\theta\|\mu,\ \rho) - \log P(\theta) + \log P(D\|\theta)$.
5	Calculate the gradient by the mean with the training data D as $\triangle\mu = \frac{\partial Loss}{\partial\theta} + \frac{\partial Loss}{\partial\mu}$.
6	Calculate the gradient by the standard deviation with the training data D as $\triangle\rho = \frac{\partial Loss}{\partial\theta}\frac{\varepsilon}{1+exp(-\rho)} + \frac{\partial Loss}{\partial\rho}$.
7	Update the parameters $(\mu,\ \rho)$ as the following: $\mu \leftarrow \mu - \alpha\triangle\mu,\ \rho \leftarrow \rho - \alpha\triangle\rho$.

3.3. Model Framework

We propose a distributed prediction model combining SCC and a deep learning network for the prediction problem. The proposed model framework is shown in Figure 5, and the model consists of three main components: selection nodes, sub-predictors, and fusion nodes.

The selection node calculates the series causality of the data source and selects the variables related to the target data as the network input. For each selected input variable, a Bayesian LSTM sub-predictor is designed. Finally, we use the fusion node to fuse the prediction results of multiple sub-predictors. An artificial neural network MLP is used in the fusion node. MLP is a fully linked combination of artificially designed neurons, which applies a nonlinear activation function to model the relationship between the input and output.

4. Experiments

4.1. Dataset

Our experiments used the meteorological dataset in Shunyi District, Beijing, from 2017 to 2019. The data were measured hourly at meteorological station. The future temperature was chosen to be predicted to test the proposed model. The data set contained 1095 days for a total of 26,280 data samples to ensure sufficient training data. We selected the first 90% of the data for training and the remaining 10% for testing.

4.2. Experimental Setup

A PC with an Intel CORE CPU i5-4200U 1.60 GHz and 6 GB of memory was used for the experiments. In the experiments, the default parameters in Keras and Pytorch were used for deep neural network initialization. We used the ReLU as the activation function of the Bayesian LSTM layer and the linear activation function of the MLP layer.

We set up one Bayesian LSTM layer and one MLP layer, and each layer's size was set to 24. The Adam algorithm was used for the supervised training, and the model was trained by mini-batch sampling. The model hyperparameters, such as learning and batch size, were obtained from experiments and are presented in Table 3.

Table 3. Hyperparameters for the experiments.

Layers	Design Details	Experiment Setup
Bayesian LSTM	Number of layers: 1 Number of neurons: 24 Sampling number: 4	Batch size: 30 Epochs: 100 Learning rate: 0.001
MLP	Number of layers: 1 Number of neurons: 24	

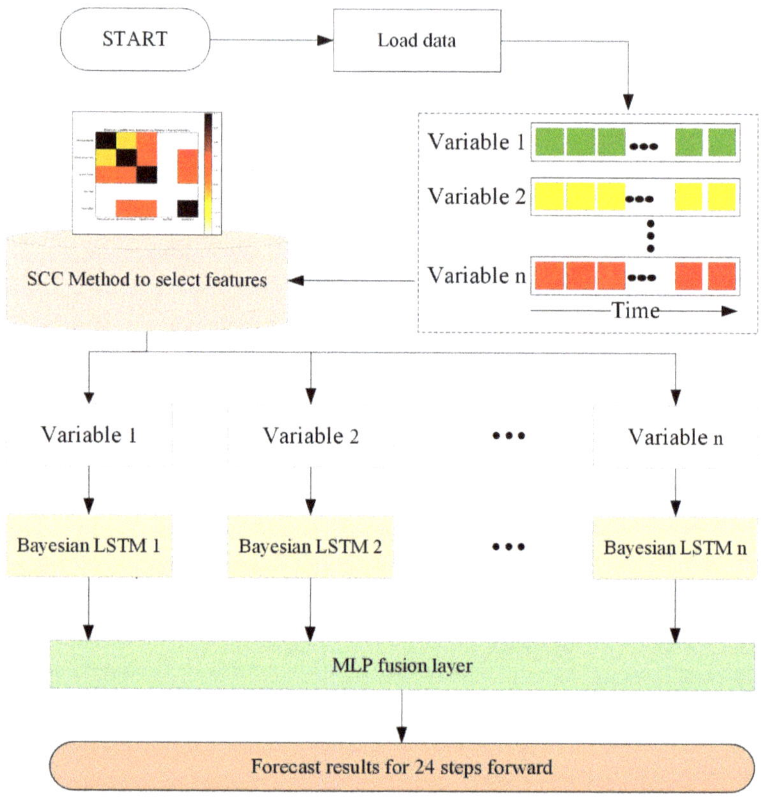

Figure 5. Model framework.

The model's performance was evaluated by the following four factors. The root-mean-squared error (RMSE):

$$RMSE = \sqrt{\frac{1}{n}\sum_{i=1}^{n}(y_i - \hat{y}_i)^2} \tag{15}$$

where $\hat{y}_i$ is the prediction, y_i is the ground truth, and n is the number of data.

The mean-squared error (MSE) can reflect the value of the loss function of network convergence and is defined as:

$$MSE = \frac{1}{n}\sum_{i=1}^{n}(y_i - \hat{y}_i)^2 \tag{16}$$

The mean absolute error (MAE) and Pearson correlation coefficient (R) between the prediction and reference were also explored in the experiments.

$$MAE = \frac{1}{n}\sum_{i=1}^{n}|y_i - \hat{y}_i| \tag{17}$$

$$R = \frac{\sum_{i=1}^{n}(y_i - \overline{y}_i)(\hat{y}_i - \overline{\hat{y}}_i)}{\sqrt{\sum_{i=1}^{n}(y_i - \overline{y}_i)^2 \sum_{i=1}^{n}(\hat{y}_i - \overline{\hat{y}}_i)^2}} \tag{18}$$

4.3. Case 1

In this case, the Bayesian LSTM model's performance is verified and causality evaluated by predicting the further temperature. We used the SCC to compare the correlation between time series variables and selected the temperature and humidity as the distributed deep model's input data. We set the time step to 24 and got a total of 24 prediction steps. The blue and red lines present the ground truth of temperature and the model's predictive results, respectively. The RMSE of the prediction is 3.203.

Figure 6 shows the comparison of the measurement data (the ground truth) and the 24 step forward prediction results. There is a light red band above and below the red line, which is the variance of the Bayesian network's result. It can be seen that the predictive trend is close to the ground truth, and most of the forecast values are within the confidence interval.

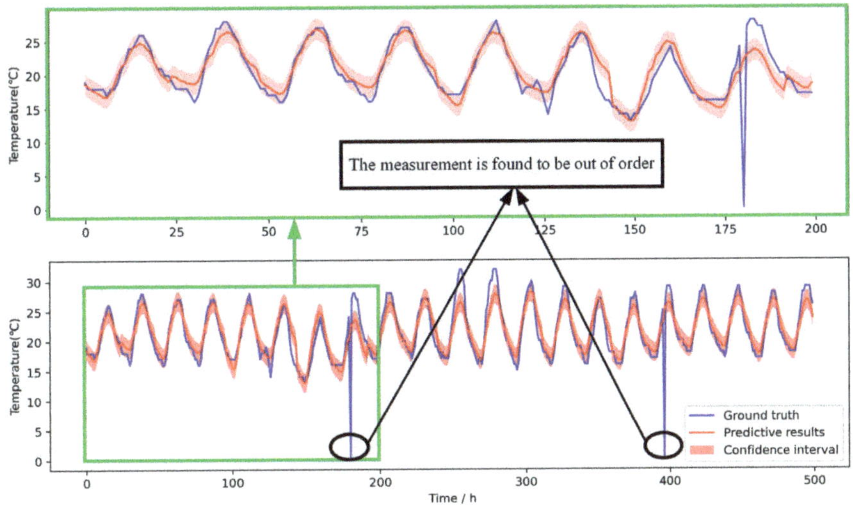

Figure 6. The prediction results of the temperature. The above picture is the prediction for the first 200 hours, which is a part of the bottom picture, in which we draw the results for about 21 days. We can see that in the bottom picture, the sensor is out of order with two hours, in which the sensor measurement data are zero. However, the prediction result effectively overcomes the sensor's failure and gives a daily temperature trend consistent with historical data.

From the actual measurement data, the prediction model's input data caused by sensor failure give the wrong measurement value. We can see that in the bottom picture, the sensor is out of order in two hours, in which the sensor measurement data are zero. However, the prediction result effectively overcomes the sensor's failure and gives a daily temperature trend consistent with historical data. However, the prediction result still maintains the correct trend, which effectively overcomes the sensor failure.

4.4. Case 2

In this case, we calculated the causality of the four meteorological factors in the data set and selected the best data for the network model. Because the SCC is zero between temperature and rainfall, we did not consider the rainfall data in the prediction.

The data set used to predict the temperature is four meteorological elements, i.e., historical temperature, humidity, wind force, and wind direction. We first considered two variables as the input of the network. We found that the predicted performance was different in different combinations. This performance was related to the SCC parameter. In another case, we increased the input signal to three or four. The results show that as the sensor input data increased, the prediction performance would not improve, but would decrease instead.

Table 4 and Figure 7 show the comparison results with two inputs. It can be seen from Table 4 that when historical temperature and humidity are set as the input, the best prediction performance can be obtained, in which the RMSE, MSE, and MAE are 3.203, 10.260, and 2, respectively. Compared with other combinations of input, such as historical temperature and wind force and historical temperature and wind direction, the RMSE, MSE, and MAE decreased.

The larger the SCC, the more it shows that the data have more causality with respect to the target data. As shown in Table 1, the historical temperature data and humidity have the greatest correlation with the future temperature data. Therefore, using these two types of data, compared with historical temperature data as the input, we can significantly improve the prediction performance.

Table 4. Prediction performance with two inputs.

Input Data	RMSE	MSE	MAE	R
Historical temperature and humidity	3.203	10.260	2.000	0.940
Historical temperature and wind force	3.244	10.525	2.108	0.937
Historical temperature and wind direction	3.400	11.559	2.250	0.932

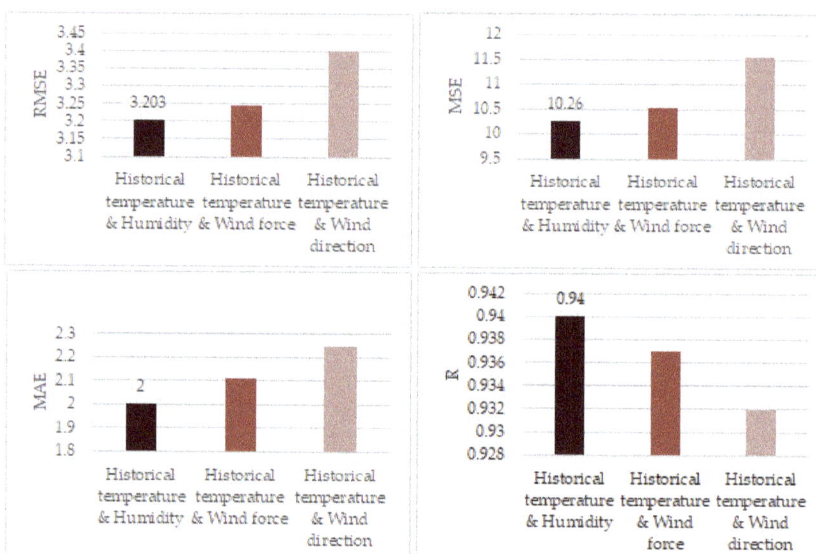

Figure 7. Comparison of prediction performance with two inputs. The input variables are historical temperature and humidity, historical temperature and wind force, and historical temperature and wind direction, respectively. We can find that when the inputs are the historical temperature and humidity, the least RMSE, MSE, and MAE and the largest R can be obtained.

Then, we increased the input variables one-by-one, adding humidity, wind force, and wind direction, separately. The performance of different numbers of inputs are shown in Table 5 and Figure 8. We can see that when there was only historical temperature as the input data, the RMSE, MSE, and MAE were 3.508, 12.305, and 2.331, respectively. Then, when two inputs were used, that is together historical temperature with humidity, the minimum prediction RMSE was 3.203. In addition, the MAE, MSE, and R were the best also. However, when the input data increased and three input data were used, the RMSE increased to 3.235. When four input data were used, the RMSE further increased to 3.230.

Therefore, we can conclude that the experiments show that more input data do not result in better prediction performance.

Table 5. Prediction performance with multiple inputs.

Input Data	RMSE	MSE	MAE	R
Historical temperature	3.508	12.305	2.331	0.930
Historical temperature and humidity	3.203	10.260	2.000	0.940
Historical temperature, humidity, and wind force	3.235	10.465	2.014	0.938
Historical temperature, humidity, wind force, and wind direction	3.230	10.430	2.032	0.938

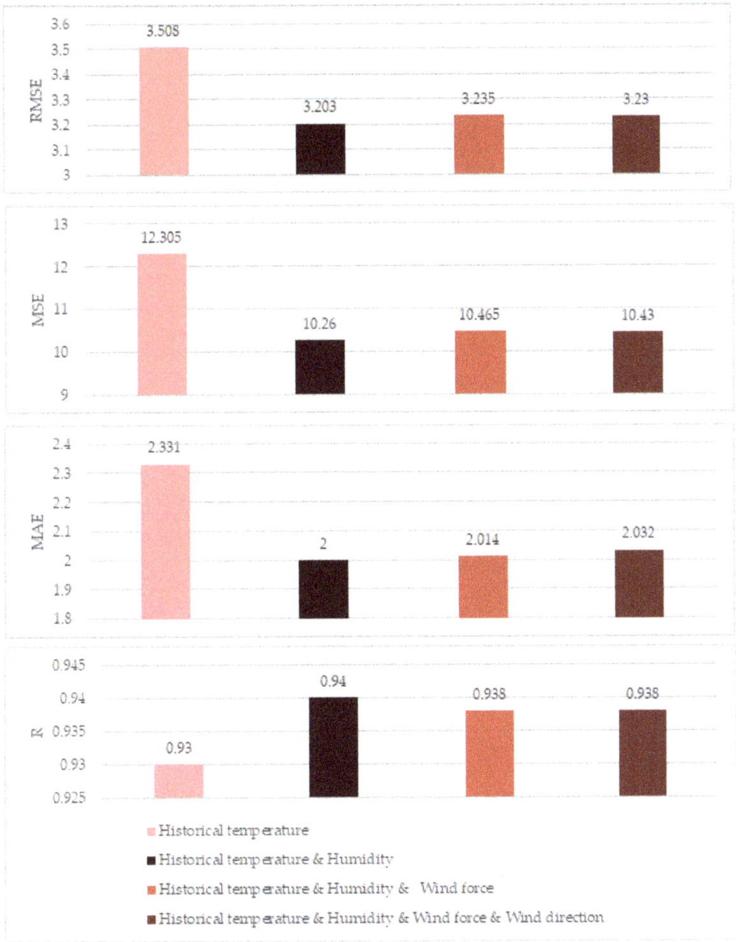

Figure 8. Comparison of prediction performance with multiple inputs. We can see that when two input variables are used, compared with one input variable, the RMSE, MSE, and MAE decrease and R increases, which shows that the performance is getting better. However, as the number of input variables increases, the performance becomes worse. For example, when the input variables are historical temperature, humidity, and wind force, the prediction performance worsens. Further, when we use the four input variables, the performance is the worst.

4.5. Case 3

In this case, we compared other deep network models with the methods proposed in this paper. Among them, no baseline models included a feature selection process and used all features as the network input. As shown in Table 6 and Figure 9, the RMSEs of LSTM [27], GRU [28], CNN-LSTM [11], conv-LSTM [30], and the proposed Bayesian LSTM were 3.714, 3.429, 3.630, 3.594, and 3.203 and the MSEs were 13.797, 11.759, 13.174, 12.915, and 10.260, respectively. The MAEs were 2.467, 2.137, 2.406, 2.344, and 2.000, respectively. Compared with LSTM and GRU, the RMSE of the proposed Bayesian LSTM decreased by 13.76% and 6.59%, and the MSE decreased by 25.64% and 12.75%, while the MAE decreased by 18.93% and 6.41%, respectively. Compared with other hybrid models, such as CNN-LSTM and conv-LSTM, the results show that the Bayesian LSTM was the best, obtaining the minimum RMSE of 3.203 and the least MAE of 2.000. Therefore, the Bayesian LSTM can better fit the data and had the best prediction performance.

Table 6. Prediction performance with different models.

Model	RMSE	MSE	MAE	R
LSTM [27]	3.714	13.797	2.467	0.927
GRU [28]	3.429	11.759	2.137	0.930
CNN-LSTM [11]	3.630	13.174	2.406	0.922
Conv-LSTM [30]	3.594	12.915	2.344	0.931
The proposed Bayesian LSTM	3.203	10.260	2.000	0.940

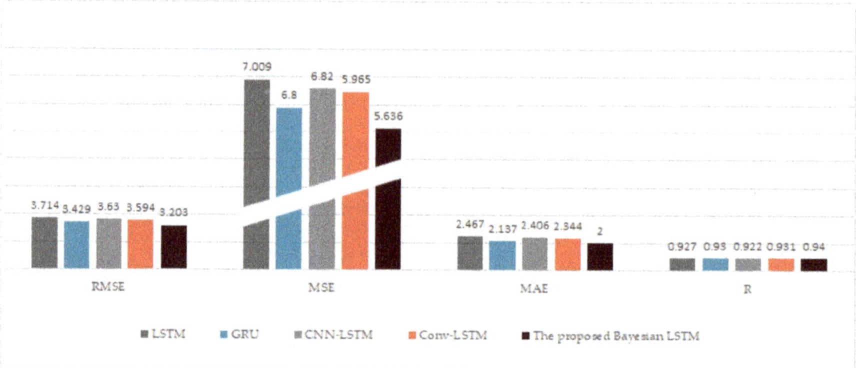

Figure 9. Comparison of the prediction performance with different sub-predictors. We can find that the proposed model with the Bayesian LSTM is the best, obtaining the least RMSE, of 2.374, MSE, and MAE and the largest R.

5. Conclusions

This article focuses on multivariate noisy measurement data modeling and prediction and proposes a distributed deep Bayesian LSTM prediction network based on causality entropy. The performance of the model was verified on real weather data sets.

In a multi-sensor system, the actual data set is usually non-linear and noisy. Therefore, analyzing the correlation between measurement from a multi-sensor system is very important for predicting. We developed the SCC to analyze the original multidimensional variables and then selected the most causal variable for the target variable. The SCC can reduce the total amount of data entered into the network, thereby reducing the computational burden of the network. It also reduces errors caused by unnecessary input.

As we all know, neural networks have a strong ability to fit nonlinearity. However, we found that the measurement data from the multi-sensor system have complex noise. We used the Bayesian LSTM to reduce the influence of noise on the neural network. The model was modeled by weight sampling, and then, the average was taken to obtain a more stable output.

In future research, we can consider other causality analysis methods. We will also replace the MLP with other fusion methods to reduce the network model's parameters for the fusion results. The proposed approaches in the paper can combine other parameter estimation algorithms [32,64–67] to study the parameter identification problems of linear and nonlinear systems with different disturbances [68–72], and to build the soft sensor models and prediction models and can be applied to other fields [73–77] such as signal processing and process control systems.

Author Contributions: Conceptualization, X.-B.J.; data curation, Y.-T.B. and J.-L.K.; formal analysis, T.-L.S. and Y.-T.B.; methodology, X.-H.Y.; software, X.-H.Y.; supervision, L.W.; validation, T.-L.S.; visualization, D.-N.Y.; writing, original draft, X.-H.Y.; writing, review and editing, X.-B.J. All authors have read and agreed to the published version of the manuscript.

Funding: This work was supported in part by the National Key Research and Development Program of China No. 2020YFC1606801, the National Natural Science Foundation of China Nos. 61903009 and 61903008, the Beijing Municipal Education Commission Nos. KM201910011010 and KM201810011005, the Young Teacher Research Foundation Project of BTBU No. QNJJ2020-26, the Defense Industrial Technology Development Program No. 6142006190201, and the Beijing excellent talent training support project for young top-notch team No. 2018000026833TD01.

Data Availability Statement: The data presented in this study are available on request from the corresponding author.

Conflicts of Interest: The authors declare no conflict of interest.

References

1. Rubio, F.; Valero, F.; Llopis-Albert, C. A review of mobile robots: Concepts, methods, theoretical framework, and applications. *Int. J. Adv. Robot. Syst.* **2019**, *16*, 1–22. [CrossRef]
2. Zhao, Z.; Wang, X.; Yao, P.; Bai, Y. A health performance evaluation method of multirotors under wind turbulence. *Nonlinear Dyn.* **2020**, *102*. [CrossRef]
3. Zhang, X; Zhao, Z.; Wang, Z; Wang, X. Fault detection and identification method for quadcopter based on airframe vibration signals. *Sensors* **2021**, 21, 581. [CrossRef]
4. Jin, X.; Yu, X.; Wang, X. Deep Learning Predictor for Sustainable Precision Agriculture Based on Internet of Things System. *Sustainability* **2020**, *12*, 1433. [CrossRef]
5. Jin, X.; Yang, N.; Wang, X. Hybrid Deep Learning Predictor for Smart Agriculture Sensing Based on Empirical Mode Decomposition and Gated Recurrent Unit Group Model. *Sensors* **2020**, *20*, 1334. [CrossRef]
6. Senthilkumar, R.; Venkatakrishnan, P.; Balaji, N. Intelligent based novel embedded system based IoT enabled air pollution monitoring system. *Microprocess. Microsyst.* **2020**, *77*, 103172. [CrossRef]
7. Jin, X.; Yang, N.; Wang, X. Integrated Predictor Based on Decomposition Mechanism for PM2.5 Long-Term Prediction. *Appl. Sci.* **2019**, *9*, 4533. [CrossRef]
8. Jin, X.; Sun, S.; Wei, H.; Yang, F. Advances in Multi-Sensor Information Fusion: Theory and Applications 2017. *Sensors* **2018**, *18*, 1162. [CrossRef] [PubMed]
9. Lee, D.; Kim, K. Recurrent neural network-based hourly prediction of photovoltaic power output using meteorological information. *Energies* **2019**, *12*, 215. [CrossRef]
10. Jin, X.; Yang, N.; Wang, X. Deep hybrid model based on EMD with classification by frequency characteristics for long-term air quality prediction. *Mathematics* **2020**, *8*, 214. [CrossRef]
11. Huang, C.J.; Kuo, P.H. A deep CNN-LSTM model for particulate matter (PM2.5) forecasting in smart cities. *Sensors* **2018**, *18*, 2220. [CrossRef] [PubMed]
12. Bai, Y.; Wang, X.; Sun, Q. Spatio-temporal prediction for the monitoring-blind area of industrial atmosphere based on the fusion network. *Int. J. Environ. Res. Public Health* **2019**, *16*, 3788. [CrossRef] [PubMed]
13. Bai, Y.; Jin, X.; Wang, X.; Wang, X.; Xu, J. Dynamic correlation analysis method of air pollutants in spatio-temporal analysis. *Int. J. Environ. Res. Public Health* **2020**, *17*, 360. [CrossRef] [PubMed]
14. Ziemann, T.; Peri, H.; Singh, A. System and method for enhancing trust for person-related data sources. U.S. Patent 10,542,043, 21 January 2020.

15. Runge, J.; Nowack, P.; Kretschmer, M.; Flaxman, S.; Sejdinovic, D. Detecting and quantifying causal associations in large nonlinear time series datasets. *Sci. Adv.* **2019**, *5*, eaau4996. [CrossRef]
16. Bai, Y.; Jin, X.; Wang, X.; Su, T.; Kong, J.; Lu, Y. Compound autoregressive network for prediction of multivariate time series. *Complexity* **2019**, *2019*, 9107167. [CrossRef]
17. Rendon-Sanchez, J.F.; de Menezes, L.M. Structural combination of seasonal exponential smoothing forecasts applied to load forecasting. *Eur. J. Oper. Res.* **2019**, *275*, 916–924. [CrossRef]
18. Gautam, S. A novel moving average forecasting approach using fuzzy time series data set. *J. Control. Autom. Electr. Syst.* **2019**, *30*, 532–544. [CrossRef]
19. Messner, J.W.; Pinson, P. Online adaptive lasso estimation in vector autoregressive models for high dimensional wind power forecasting. *Int. J. Forecast.* **2019**, *35*, 1485–1498. [CrossRef]
20. Alsharif, M.; Younes, M.; Kim, J. Time series ARIMA model for prediction of daily and monthly average global solar radiation: The case study of Seoul, South Korea. *Symmetry* **2019**, *11*, 240. [CrossRef]
21. Zhang, X.; Wang, J.; Gao, Y. A hybrid short-term electricity price forecasting framework: Cuckoo search-based feature selection with singular spectrum analysis and SVM. *Energy Econ.* **2019**, *81*, 899–913. [CrossRef]
22. Chitsazan, M.A.; Fadali, M.S.; Trzynadlowski, A.M. Wind speed and wind direction forecasting using echo state network with nonlinear functions. *Renew. Energy* **2019**, *131*, 879–889. [CrossRef]
23. Ren, Y.; Mao, J.; Liu, Y. A novel dbn model for time series forecasting. *IAENG Int. J. Comput. Sci.* **2017**, *44*, 79–86.
24. Sulaiman, J.; Wahab, S.H. Heavy rainfall forecasting model using artificial neural network for flood prone area. In *IT Convergence and Security 2017*; Springer: Singapore, 2018; pp. 68–76.
25. Izonin, I.; Tkachenko, R.; Verhun, V.; Zub, K. An approach towards missing data management using improved GRNN-SGTM ensemble method-ScienceDirect. *Eng. Sci. Technol. Int. J.* **2020**, in press.
26. Min, K.; Kim, D.; Park, J.; Huh, K. RNN-based path prediction of obstacle vehicles with deep ensemble. *IEEE Trans. Veh. Technol.* **2019**, *10*, 10252–10256. [CrossRef]
27. Sundermeyer, M.; Schlüter, R.; Ney, H. LSTM neural networks for language modeling. In Proceedings of the Thirteenth Annual Conference of the International Speech Communication Association, Portland, OR, USA, 9–13 September 2012; pp. 194–197.
28. Wang, Y.; Liao, W.; Chang, Y. Gated recurrent unit network-based short-term photovoltaic forecasting. *Energies* **2018**, *11*, 2163. [CrossRef]
29. Tang, X.; Dai, Y.; Wang, T. Short-term power load forecasting based on multi-layer bidirectional recurrent neural network. *IET Gener. Transm. Distrib.* **2019**, *13*, 3847–3854. [CrossRef]
30. Qi, Y.; Li, Q.; Karimian, H.; Liu, D. A hybrid model for spatiotemporal forecasting of PM2.5 based on graph convolutional neural network and long short-term memory. *Sci. Total. Environ.* **2019**, *664*, 1–10. [CrossRef]
31. Tian, C.; Ma, J.; Zhang, C. A deep neural network model for short-term load forecast based on long short-term memory network and convolutional neural network. *Energies* **2018**, *11*, 3493. [CrossRef]
32. Xu, L.; Ding, F. Iterative parameter estimation for signal models based on measured data. *Circuits Syst. Signal Process.* **2018**, *37*, 3046–3069. [CrossRef]
33. Xu, L. The parameter estimation algorithms based on the dynamical response measurement data. *Adv. Mech. Eng.* **2017**, *9*, 1687814017730003. [CrossRef]
34. Ding, F.; Wang, X.H.; Mao, L.; Xu, L. Joint state and multi-innovation parameter estimation for time-delay linear systems and its convergence based on the Kalman filtering. *Dig. Signal Proc.* **2017**, 62, 211–223. [CrossRef]
35. Ding, F.; Xu, L.; Zhu, Q. Performance analysis of the generalised projection identification for time-varying systems. *IET Control. Theory Appl.* **2016**, 10, 2506–2514. [CrossRef]
36. Ding, J.; Cao, Z.X.; Chen, J.Z.; Jiang, G.P. Weighted parameter estimation for Hammerstein nonlinear ARX systems. *Circuits Syst. Signal Proc.* **2020**, *39*, 2178–2192. [CrossRef]
37. Ding, F.; Xu, L.; Meng, D.; Jin, X.B.; Alsaedi, A.; Hayat, T. Gradient estimation algorithms for the parameter identification of bilinear systems using the auxiliary model. *J. Comput. Appl. Math.* **2020**, *369*, 112575. [CrossRef]
38. Pan, J.; Jiang, X.; Wan, X.K.; Ding, W. A filtering based multi-innovation extended stochastic gradient algorithm for multivariable control systems. *Int. J. Control Autom. Syst.* **2017**, *15*, 1189–1197. [CrossRef]
39. Zhang, X. Recursive parameter estimation and its convergence for bilinear systems. *IET Control. Theory Appl.* **2020**, *14*, 677–688. [CrossRef]
40. Li, M.H.; Liu, X.M. The least squares based iterative algorithms for parameter estimation of a bilinear system with autoregressive noise using the data filtering technique. *Signal Process.* **2018**, *147*, 23-34. [CrossRef]
41. Zhang, X.; Liu, Q.Y. Recursive identification of bilinear time-delay systems through the redundant rule. *J. Frankl. Inst.* **2020**, *357*, 726–747. [CrossRef]
42. Xu, L. Parameter estimation algorithms for dynamical response signals based on the multi-innovation theory and the hierarchical principle. *IET Signal Process.* **2017**, *11*, 228–237. [CrossRef]
43. Waldmann, P. On the use of the pearson correlation coefficient for model evaluation in genome-wide prediction. *Front. Genet.* **2019**, *10*, 899. [CrossRef] [PubMed]
44. Amarkhil, Q.; Elwakil, E.; Hubbard, B. A meta-analysis of critical causes of project delay using spearman's rank and relative importance index integrated approach. *Can. J. Civ. Eng.* **2020**, *Just-IN*. [CrossRef]

45. Duan, S.; Yang, W.; Wang, X. Grain pile temperature forecasting from weather factors: A support vector regression approach. In Proceedings of the 2019 IEEE/CIC International Conference on Communications in China (ICCC), Changchun, China, 11–13 August 2019; pp. 255–260. [CrossRef]
46. Ciulla, G.; D'Amico, A. Building energy performance forecasting: A multiple linear regression approach. *Appl. Energy* **2019**, *253*, 113500. [CrossRef]
47. Jing, B.; Qian, Z.; Pei, Y.; Wang, J. Ultra short-term PV power forecasting based on ELM segmentation model. *J. Eng.* **2017**, *2017*, 2564–2568. [CrossRef]
48. Lin, C.Y.; Chang, Y.S.; Chiao, H.T. Design a Hybrid Framework for Air Pollution Forecasting. In Proceedings of the 2019 IEEE International Conference on Systems, Man and Cybernetics (SMC), Bari, Italy, 6–9 October 2019; pp. 2472–2477.
49. Gao, D.; Zhou, Y.; Wang, T.; Wang, Y. A Method for predicting the remaining useful life of lithium-ion batteries based on particle filter using Kendall rank correlation coefficient. *Energies* **2020**, *13*, 4183. [CrossRef]
50. Contreras-Reyes, J.E.; Hernández-Santoro, C. Assessing granger-causality in the southern humboldt current ecosystem using cross-spectral methods. *Entropy* **2020**, *22*, 1071. [CrossRef]
51. Podobnik, B.; Stanley, H.E. Detrended cross-correlation analysis: a new method for analyzing two nonstationary time series. *Phys. Rev. Lett.* **2008**, *100*, 084102. [CrossRef] [PubMed]
52. Jin, X.; Zhang, J.; Su, T.; Bai, Y.; Kong, J.; Wang, X. Wavelet-deep optimized model for nonlinear multi-component data forecasting. *Comput. Intell. Neurosci.* **2021**, accept.
53. Jin, X.; Wang, H.; Wang, X. Deep-learning prediction model with serial two-level decomposition based on bayesian optimization. *Complexity* **2020**, *2020*, 4346803 [CrossRef]
54. Wang, H.; Yeung, D.Y. A survey on Bayesian deep learning. *ACM Comput. Surv. (CSUR)* **2020**, *53*, 1–37. [CrossRef]
55. Mukhopadhyay, P.; Mallick, S. Bayesian deep learning for seismic facies classification and its uncertainty estimation. *Soc. Explor. Geophys.* **2019**, 2488–2492. [CrossRef]
56. Zhang, R.; Li, C.; Zhang, J. Cyclical stochastic gradient MCMC for Bayesian deep learning. *arXiv* **2019**, arXiv:1902.03932.
57. Jin, X.; Lian, X.; Su, T.; Shi, Y.; Miao, B. Closed-loop estimation for randomly sampled measurements in target tracking system. *Math. Probl. Eng.* **2014**, *2014*, 315908.
58. Jin, X.; Du, J.; Bao, J. Target tracking of a linear time invariant system under irregular sampling. *Int. J. Adv. Robot. Syst.* **2012**, *9*, 219–230.
59. Li, G.; Yang, L.; Lee, C.G.; Wang, X.; Rong, M. A Bayesian deep learning RUL framework integrating epistemic and aleatoric uncertainties. *IEEE Trans. Ind. Electron.* **2020**, *1*. [CrossRef]
60. Harper, R.; Southern, J. A Bayesian deep learning framework for end-to-end prediction of emotion from heartbeat. *IEEE Trans. Affect. Comput.* **2020**. [CrossRef]
61. Abdi, H. The Kendall rank correlation coefficient. In *Encyclopedia of Measurement and Statistics*, Salkind, N.J., Ed; SAGE Publications Inc.: Thousand Oaks, CA, USA, 2007; pp. 508–510.
62. Lopez Quintero, F.O.; Contreras-Reyes, J.E.; Wiff, R.; Arellano-Valle, R.B. Flexible Bayesian analysis of the von bertalanffy growth function with the use of a log-skew-t distribution. *Fish. Bull.* **2017**, *115* 13–26. [CrossRef]
63. Bouhlel, N.; Dziri, A. Kullback–Leibler divergence between multivariate generalized gaussian distributions. *IEEE Signal Process. Lett.* **2019**, *26*, 1021–1025. [CrossRef]
64. Xu, L.; Xiong, W.L.; Alsaedi, A.; Hayat, T. Hierarchical parameter estimation for the frequency response based on the dynamical window data. *Int. J. Control Autom. Syst.* **2018**, *16*, 1756–1764. [CrossRef]
65. Gu, Y.; Liu, J.; Li, X.; Chou, Y.; Ji, Y. State space model identification of multirate processes with time-delay using the expectation maximization. *J. Frankl. Inst.* **2019**, *356*, 1623–1639. [CrossRef]
66. Xu, L.; Ding, F.; Zhu, Q.M. Hierarchical Newton and least squares iterative estimation algorithm for dynamic systems by transfer functions based on the impulse responses. *Int. J. Syst. Sci.* **2019**, *50*, 141–151. [CrossRef]
67. Xu, L.; Ding, F.; Lu, X.; Wan, L.J.; Sheng, J. Hierarchical multi-innovation generalised extended stochastic gradient methods for multivariable equation-error autoregressive moving average systems. *IET Control. Theory Appl.* **2020**, *14*, 1276–1286. [CrossRef]
68. Pan, J.; Ma, H.; Zhang, X.; Liu, Q.; Ding, F.; Chang, Y.; Sheng, J. Recursive coupled projection algorithms for multivariable output-error-like systems with coloured noises. *IET Signal Process.* **2020**, *14*, 455–466. [CrossRef]
69. Xu, L.; Ding, F.; Wan, L.J.; Sheng, J. Separable multi-innovation stochastic gradient estimation algorithm for the nonlinear dynamic responses of systems. *Int. J. Adapt. Control Signal Process.* **2020**, *34*, 937–954. [CrossRef]
70. Zhang, X.; Ding, F.; Alsaadi, F.E.; Hayat, T. Recursive parameter identification of the dynamical models for bilinear state space systems. *Nonlinear Dyn.* **2017**, *89*, 2415–2429. [CrossRef]
71. Zhang, X.; Xu, L.; Ding, F.; Hayat, T. Combined state and parameter estimation for a bilinear state space system with moving average noise. *J. Frankl. Inst.* **2018**, *355*, 3079–3103. [CrossRef]
72. Gu, Y.; Zhu, Q.; Nouri, H. Bias compensation-based parameter and state estimation for a class of time-delay nonlinear state-space models. *IET Control. Theory Appl.* **2020**, *14*, 2176–2185. [CrossRef]
73. Zhang, X.; Ding, F.; Xu, L.; Yang, E. State filtering-based least squares parameter estimation for bilinear systems using the hierarchical identification principle. *IET Control. Theory Appl.* **2018**, *12*, 1704–1713. [CrossRef]
74. Wang, L.J.; Ji, Y.; Wan, L.J.; Bu, N. Hierarchical recursive generalized extended least squares estimation algorithms for a class of nonlinear stochastic systems with colored noise. *J. Frankl. Inst.* **2019**, *356*, 10102–10122. [CrossRef]

75. Zhang, X.; Ding, F.; Xu, L.; Yang, E.F. Highly computationally efficient state filter based on the delta operator. *Int. J. Adapt. Control Signal Process.* **2019**, *33*, 875–889. [CrossRef]
76. Fan, Y.M.; Liu, X.M. Two-stage auxiliary model gradient-based iterative algorithm for the input nonlinear controlled autoregressive system with variable-gain nonlinearity. *Int. J. Robust Nonlinear Control* **2020**, *30*, 5492–5509. [CrossRef]
77. Zhang, X.; Ding, F.; Yang, E.F. State estimation for bilinear systems through minimizing the covariance matrix of the state estimation errors. *Int. J. Adapt. Control Signal Process.* **2019**, *33*, 1157–1173. [CrossRef]

entropy

Article

A Modified FlowDroid Based on Chi-Square Test of Permissions

Hongzhaoning Kang, Gang Liu *, Zhengping Wu, Yumin Tian and Lizhi Zhang

School of Computer Science and Technology, Xidian University, Xi'an 710071, China; kanghzn@stu.xidian.edu.cn (H.K.); wuzhenping@stu.xidian.edu.cn (Z.W.); ymtian@mail.xidian.edu.cn (Y.T.); rnzhang@stu.xidian.edu.cn (L.Z.)
* Correspondence: gliu@xidian.edu.cn

Abstract: Android devices are currently widely used in many fields, such as automatic control, embedded systems, the Internet of Things and so on. At the same time, Android applications (apps) always use multiple permissions, and permissions can be abused by malicious apps that disclose users' privacy or breach the secure storage of information. FlowDroid has been extensively studied as a novel and highly precise static taint analysis for Android applications,. Aiming at the problem of complex detection and false alarms in FlowDroid, an improved static detection method based on feature permission and risk rating is proposed. Firstly, the Chi-square test is used to extract correlated permissions related to malicious apps, and mutual information is used to cluster the permissions to generate feature permission clusters. Secondly, risk calculation method based on permissions and combinations of permissions are proposed to identify dangerous data flows. Experiments show that this method can significantly improve detection efficiency while maintaining the accuracy of dangerous data flow detection.

Keywords: automatic control; mutual information; static detection; Chi-square test; permission; FlowDroid

Citation: Kang, H.; Liu, G.; Wu, Z.; Tian, Y.; Zhang, L. A Modified FlowDroid Based on Chi-Square Test of Permissions. *Entropy* **2021**, 23, 174. https://doi.org/10.3390/e23020174

Academic Editor: Quan Min Zhu
Received: 14 December 2020
Accepted: 27 January 2021
Published: 30 January 2021

1. Introduction

Google Android is a mobile operating system that is widely used in many fields [1,2]. With the development of the Internet of Things, Android quickly gained a large proportion of the market share. At the same time, the number of malicious applications (apps) has been increasing [3] and over the last few years, the amount of malware has increased significantly. According to a recent report from McAfee, over 1.6 million new examples of mobile malware were discovered in the first quarter of 2019 [4]. Therefore, the detection of Android malware with a high accuracy rate and high efficiency is an important issue.

Various approaches have been proposed in previous works with the intention of detecting Android malware. These approaches can be categorized into static analysis, dynamic analysis or hybrid analysis [5]. Dynamic analysis means that, in the process of running an application, the flow of privacy information and data is tracked and captured, and the malicious tendency of application behavior is analyzed and judged. Dynamic analysis can monitor and track the flow of private data in real time [6] and is not affected by code obfuscation, encryption and other factors. However, privacy leaks that are not triggered at runtime cannot be detected, and the low code coverage causes a high missing rate. At the same time, real-time operation results in greater resource consumption [7]. In the case of resource shortages on mobile devices, the system efficiency will be seriously affected. In contrast to dynamic analysis, static analysis is done without running an app. In static analysis, features such as permissions and API calls are extracted from the app source code by reverse engineering to analyze and infer suspicious behavior from an app and discover problems in different stages of the entire life cycle, verifying the security of app at the source code level. As a highly influential static analysis tool for Android apps, FlowDroid [8] has the advantage of wide code coverage and it can detect many malicious behaviors that cannot be detected by dynamic analysis. However, source code level analysis will bring a large amount of irrelevant detection while leading to high false positives and low detection efficiency, decreasing the availability of these tools. In our experiments, for apps over 10 MB in size, FlowDroid reports timeouts and insufficient memory. Even for the two apps from the FlowDroid samples, the test takes nearly an hour.

Research and experiments show that app security threats are strongly correlated with some characteristic permissions [9]. When using FlowDroid, only those flow paths that actually cause privacy leakages need to be considered, which significantly reduces the scale of analysis and improves efficiency. This paper proposes a redundancy resolution method for FlowDroid, which can cluster the correlated permissions and calculate the risks of flow paths. This paper provides the following two contributions:

(1) We propose a permission clustering method based on the use of permissions by malware. Compared with all the permissions of the Android system, the proposed permission cluster contains only a few permissions. By monitoring these permissions and the related call paths in FlowDroid, the analysis can be greatly simplified.
(2) We propose a lightweight malicious application detection method based on the permission clusters and FlowDroid in this article. This method is suitable for pre-installed software and user-installed software. We improved FlowDroid so that it can only monitor the call paths that are related to the permissions in the permission cluster and meet certain risk conditions. This greatly reduces the detection time and memory usage.

2. Related Work

The static analysis of Android malware relies on Java bytecode, which is extracted by disassembling an app. The manifest file is also a source of information for static analysis. Kirin [10] is the safety inspection scheme for app installation, which operates by defining security rules to identify dangerous permission combinations; the installation strategy is formulated based on the use of security rules as detection criteria. However, due to the small number of rules and the lack of representation of permission combinations, the detection efficiency and the accuracy cannot be guaranteed. TrustDroid [11] provides two alternative detection modes: real-time detection on the mobile device side and static analysis on the server side, converting the data flow to a tree structure using Jasm in middle code representation to generate a function call graph, preventing untrusted apps from leaking user privacy information. The resource consumption of TrustDroid is also extremely high. LeakMinder [12] analyzes the security of apps from the third-party market and decompiles Android application package (APK) files by reverse engineering. Based on a predefined source and sink, LeakMinder generates a call graph and data flow diagram and finds possible privacy leak paths. However, implicit data leaks cannot be detected. Besides, designed artificially sources and sinks are not particularly representative, which results in contingency and inaccuracy. Cen et al. [13] proposed the use of a probabilistic discriminative model based on regularized logistic regression for Android malware detection. The probabilistic discriminative model works well with permissions and achieves the best detection results by combining both decompiled source code and application permissions. Kang et al. [14] proposed a method that detects and classifies Android malware using static analysis with the combination of the attacker's information. The effectiveness of Android malware detection is improved by integrating the attacker's information as a feature, and the method categorizes illegitimate applications into homogeneous classes. Song et al. [15] proposed an integrated static framework using a filtering technique consisting of four layers to identify and evaluate mobile malware on Android. Sun et al. [16] presented an approach that interfaces static logic-structures and dynamic runtime information to detect Android malware. Behavior similarity is used for the classification of malware. The results showed that the approach is easy to implement and has low computational overheads. Rovelliet al. [17] presented a permission-based malware detection system that uses machine learning classifiers on the behavioral patterns to consequently distinguish inconspicuous applications. DAPASA [18] is an approach used to detect Android piggy-backed apps through sensitive subgraph analysis. DAPASA generates a sensitive subgraph (SSG) to profile the most suspicious behavior of an app. Five features are constructed from SSG to depict the invocation patterns. The five features are fed into the machine learning algorithms to detect whether the app is piggy-backed or benign. Talha et al. [19] presented a permission-based Android malware detection system consisting of three components, namely the central server, Android client and signature database, and static analysis is used to categorize the Android application as normal or harmful. Li et al. [20] raised the issue of considering interaction terms across features for the discovery of malicious behavior patterns in Android applications and proposed a classier for Android malware detection based on a factorization machine architecture.

FlowDroid [8] was proposed by Arztet al. in 2013 and has been widely studied and applied in the field of Android static analysis. FlowDroid is considered as a context, flow, field and object-sensitive and lifecycle-aware static taint analysis tool for Android apps. To increase recall, FlowDroid creates a complete model of Android's app lifecycle. However, a large number of normal paths are also detected while the entire life cycle is analyzed, causing false positives and low efficiency. The main purpose of this paper is to improve the analysis efficiency and applicability of FlowDroid.

3. Preliminaries

3.1. Android Permission

The Android system is an extension based on Linux, which provided the permission mechanism [21]. Operations that apps can perform are specified to limit the software's ability to manipulate systems or other software. The Android permission mechanism requires developers to apply for permissions they need in Android's Manifest.xml and gets user's consent during installation to access system resources and functional components by calling the related API. Android protects sensitive systems and user information by restricting apps from accessing system resources with permissions other than those declared. Android uses a coarse-grained permission management mechanism and no longer reviews the running process after granting permissions; thus, malicious apps exploit users' ignorance of permissions and the coarse-grained permission management of Android's permission mechanism to access or even leak sensitive information. Android 8.0 provides 135 permissions and corresponding APIs to access system resources. In fact, only a small portion of permission usage can lead to sensitive information leakage. If malicious application-independent permission calls are accurately excluded from detection, the data paths that need to be detected in static analysis can be significantly reduced, thus reducing the false alarm and improving detection efficiency.

3.2. FlowDroid

FlowDroid, based on Soot [22], works directly at the bytecode level and does not require access to an app's source code. It parses the APK file of an Android app, converts the Java code into Jimple middle code, simulates the life cycle of an Android app to handle callback functions and generates a call graph (CG) and inter-procedural control-flow graph (ICFG) [23,24] to trace taints (Figure 1). It uses the Interpretural Finite Distributive Subset (IFDS) to model data flow propagation and generates complete, polluted data flow paths by the Heros framework [25]. Therefore, FlowDroid has very high requirements in terms of computing and memory resources. For Enriched1.apk (a sample application of which is shown in [25]), a total of 46 nodes and 78 function call paths were generated (Figure 2 shows a partial call graph of these). The call graph composed of the data paths to be analyzed is very complex, and there is a large number of callbacks and callback relationships between functions, which leads to high time and resource costs.

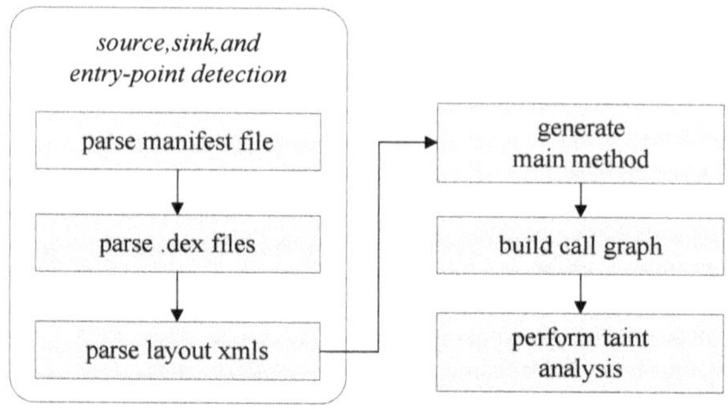

Figure 1. Workflow of FlowDroid.

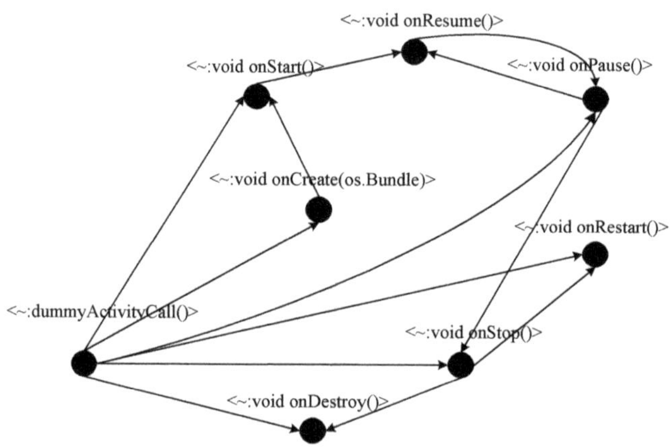

Figure 2. Partial function call graph of Enriched1.apk.

Our experiments show that FlowDroid usually reports timeout or out-of-memory errors for apps with a size larger than 10 MBytes. The reason for this is that an Android app often involves dozens of components at runtime, and the interaction between multiple components leads to hundreds of callback methods.

Although full-scale analysis can ensure high accuracy, it results in an unnecessary amount of analysis. FlowDroid should be improved in two aspects as follows:

(1) There is no further analysis of taint paths, and the large number of false positive paths results in low accuracy.

(2) There is no clear analysis content and there are no taint path identification criteria. This leads to a large number of irrelevant detection paths, resulting in excessive memory and time consumption.

3.3. Mathematical Background

The Chi-square test is a hypothesis testing method used to determine whether two variables are independent. For two discrete variables, it can be concluded whether there is a correlation between them by using the Chi-square test. The larger the Chi-square value, the greater the deviation between the two variables, the smaller the correlation between them and the stronger the independence. When the value of Chi-square reaches 0, that means the factors are exactly the same. The formula of the quaternary Chi-square test is as follows:

$$\chi^2 = \frac{N(AD - BC)^2}{(A+B)(C+D)(A+C)(B+D)} \tag{1}$$

For an abstract random variable, to remove its uncertainty, a certain amount of information needs to be used, and information entropy is a mathematical measure of this. The higher the information entropy is, the larger the amount of information that needs to be introduced and the lower the information entropy is, and the less information is needed. The information entropy of X is defined as:

$$H(X) = -\sum_{n} P(X)log_2 P(X) \tag{2}$$

In order to determine the influence of the information entropy between two variables, the information entropy of X can be obtained when Y appears, as shown in Equations (3) and (4):

$$H(X|Y) = -\sum_{n} P(Y) \sum_{n} P(X|Y)log_2 P(X|Y) \tag{3}$$

$$H(X,Y) = H(X) + H(Y|X) = H(Y) + H(X|Y) \tag{4}$$

where $H(X|Y)$ is the conditional information entropy, $P(X|Y)$ is the conditional probability and $H(X,Y)$ is the joint information entropy. According to the above equations, the mutual information values of X and Y can be obtained as $I(X;Y)$:

$$I(X;Y) = H(X) - H(X\ Y) = -\sum P(X,Y)log\frac{P(X,Y)}{P(X)P(Y)} \tag{5}$$

4. The Improved Detection Method

Flowdroid analyzes all data paths, resulting in high false positives and high resource requirements. This paper presents a redundancy resolution method based on feature permissions and risk. The purpose is to exclude the large number of irrelevant paths (security paths) for static analysis. The lightweight FlowDroid, named Permission-based FlowDroid (PBFlowDroid), is proposed based on above methods. Figure 3 shows the architecture of PBFlowDroid.

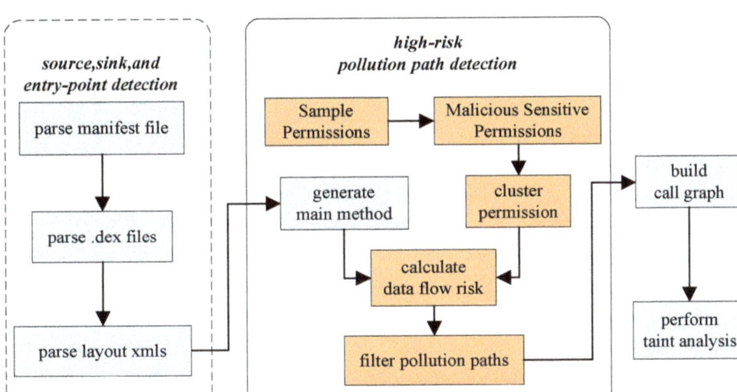

Figure 3. Workflow of PBFlowDroid.

Firstly, the Chi-square test is used to extract permissions related to malicious applications, and these permissions (malicious sensitive permissions) are classified into permission clusters by a clustering algorithm based on mutual information. Thus, the large number of permissions is reduced to a small number of permission clusters which are considered in static analysis. Secondly, different permissions or combinations of permissions bring different risks to user privacy or system security. To improve the accuracy of analysis, a risk assignment and calculation algorithm for single permissions or combinations of permissions is proposed. With these methods, all paths are given a risk value. Using the risk value of each path, the security of the taint data flow propagation path generated by control flow [26] and data flow can be determined, notifying the user whether the taint data flow is a safe path, ensuring the accuracy of static analysis and improving the analysis efficiency.

4.1. Permission Cluster Extraction

In Android, each permission has the two states of "request" and "no request", which are independent of the number of requests. This scenario is suitable for the Chi-square test. In this study, the quaternary Chi-square test is used; the Chi-square of permission p is as follows:

$$\chi^2(p) = \frac{N(A_pD_p - B_pC_p)^2}{(A_p + B_p)(C_p + D_p)(A_p + C_p)(B_p + D_p)} \tag{6}$$

where N denotes the total number of app samples, which consists of X malicious apps and Y normal apps. For permission p, requests by malicious apps and normal apps are counted as A_p and B_p. The numbers that do not apply for p by malicious apps and normal apps are counted as $C_p = (X - A_p)$ and $D_p = (Y - B_p)$, as Table 1 shows.

Table 1. Chi-square test distribution of permission p.

	Number of Malicious Apps (X)	Number of Normal Apps (Y)
Apps with permission p	A_p	B_p
Apps without permission p	$C_p = (X - A_p)$	$D_p = (Y - B_p)$

As for χ^2,the Chi-square test provides a threshold checklist as a criterion of reliability. For each permission, the probability of it relating to a malicious application is obtained by referring to the Chi-square test threshold table [27]. The larger the probability, the more malicious applications tend to have the corresponding permission, while a Chi-square value less than 0.5 indicates that the permission has almost no correlation with malicious applications. In our experiment, the top 20 permissions with a Chi-square value greater than 0.5 are regarded as permissions with a high correlation with malicious applications, as shown in Table 2.

Table 2. Permissions clusters, Chi-square and risk assignment.

Permission Clusters	Permissions	χ^2	Risk Value
	INTERNET	7.693	
c_0	ACCESS_NETWORK_STATE	6.622	7
	CHANGE_WIFI_STATE	6.178	
	CALL_PHONE	6.334	
c_1	READ_CONTACT	4.236	6
	READ_PHONE_STATE	3.855	
	WRITE_EXTERNAL_STORAGE	6.012	
c_2	GET_ACCOUNT	3.821	5
	WRITE_SETTING	2.706	
	SEND_SMS	5.844	
c_3	RECIEVE_SMS	4.689	4
	WRITE_SMS	4.023	
c_4	ACCESS_COARSE_LOCATION	5.672	3
	ACCESS_FINE_LOCATION	2.755	
	RECIEVE_BOOT_COMPLETED	5.522	
c_5	INSTALL_PACKAGE	3.237	2
	WAKE_LOCK	2.468	
	DEVICE_POWER	1.355	
c_6	CAMERA	1.224	1
	FLASHLIGHT	0.698	

The Chi-square test selects the permissions to be investigated and significantly reduces the candidate paths for static analysis. Even with 20 permissions, the number of paths that can be associated with some apps is still quite large. In fact, permissions are not independent of each other. When one app applies for a certain permission, other permissions of the same type that are related to achieve a combined function are also applied, which leads to strong correlation between permissions. For example, "READ_SMS" and "WRITE_SMS" are often applied and used at the same time. In static analysis, if two permissions with high correlation are detected separately, multiple detection results will be generated. This may decrease the accuracy of detection. To solve this problem, we use the clustering algorithm to cluster the selected permissions so that each cluster is representative.

Permission is a discrete kind of feature information, and the similarity between permissions can be measured by mutual information based on information entropy. We use $P_m(X)$ and $P_n(X)$ to represent the probability that permission X will be maliciously applied and normally applied, respectively. Then, the information entropy *H (X)* of permission X is as follows:

$$H(X) = -(P_m(X)log_2P_m(X) + P_n(X)log_2P_n(X)) \tag{7}$$

The mutual information values of permissions X and Y are calculated by formula (5). In order to describe the similarity between permissions X and Y more intuitively, the correlation between the two permissions can be obtained by (8):

$$Cor(X,Y) = 2 \times \left[\frac{I(X,Y)}{H(X)+H(Y)}\right] \tag{8}$$

where the value of $Cor(X,Y)$ is located between [0, 1]. A value of 0 means that permission X and Y are completely unrelated; the larger the value, the greater the correlation between them.

In this paper, a clustering method based on mutual information is proposed to cluster the selected permissions (which in our experiment, the number of permissions is 20 as shown in Table 2) to generate feature permission clusters (FPC) with low similarity between clusters and high similarity within clusters and further remove irrelevant detections in static analysis.

The steps of the clustering algorithm based on mutual information are as follows:

Clustering Algorithm Based on Mutual Information:
Input: Permissions set Obtained by Chi-Square Test: $S = \{p_0, p_1, \cdots, p_n\}$
Output: Cluster Sets: $C = \{c_0, c_1, \cdots, c_m\},\ m \le n$

(1) Take the first element p_0 in S as the first element of c_0, and delete p_0 from S.
(2) For each p_i in S
(3) for each $c_j \neq null$ in C do
(4) if $Cor\left(c_j,\ p_0,\ p_i\right) > T$ then
(5) add p_i to c_j;
(6) else
(7) create a new cin C, and add p_i to c;
(8) end if
(9) end for
(10) end for
(11) for each c_i in C
(12) if $Num(c_i) < 3$ then
(13) cluster the similar Cor in $(c_i$, and put it to a new c in C
(14) end if
(15) end for
(16) output C

After clustering, the permissions with high correlation with malicious apps were clustered into multiple clusters (we have seven clusters in our experiment, from c_0~c_6, as shown in Table 2).

FPC extraction not only eliminates irrelevant paths caused by permissions unrelated to malicious applications but also eliminates the correlation within the permission cluster, further eliminating the redundancy in static detection and improving the detection efficiency.

4.2. Risk Calculation

App security threats are strongly correlated with some characteristic permissions [9]. Operations corresponding to different permissions pose different threats to user privacy and system security [9]. For example, the operation of applying for network permission often transfers privacy settings on a device to other addresses involving the interaction between user information and the outside world, and so the threat degree is greater; the permission of applying for the device's local location only obtains the current user's status and does not interact with the outside world or affect the security of the device, so the threat degree is general.

To visualize the risk level of different permission clusters, we use the risk value to describe it. The risk value is the quantification of the risk of each permission cluster, and the risk value of each permission in the same cluster is the same. The selection of these values is not unique, and only three conditions need to be met:

(1) The risk value decreases as χ^2 of permission cluster decreases;
(2) The risk values of c_0 to c_{k-1} are greater than 1;
(3) The risk values of c_k to c_m are greater than 0;

Where $m + 1$ is the number of cluster sets, and k is used to classify clusters into high risk clusters and low risk clusters. In our experiment, the m is 6 and we choose 4 as the value of k.

These limitations are related to analytical calculation methods. When assigning risk values to each cluster, in this paper, we select a simple assignment that satisfies the above three conditions. It should be noted that the threshold value in 5.1 is related to the risk value; this is an empirical value obtained through experiments. Different risk value assignments will correspond to different threshold values. In our experiment, the risk value for each cluster is shown in Table 2.

When an app performs an operation or acquires a resource, it sometimes applies for more than one permission, which forms a combination of permissions. A combination of permissions may pose a greater threat to the system than a single permission [28]. In our experiment, we refer to [10] to calculate the risk value of an app, and the calculation rules are as follows:

Rule 1: For a single requested permission, the risk value R_S is the sum of the risk value for each permission:

$$R_s = \sum_{i=1}^{M} R(p_i) \tag{9}$$

where $R(p_i)$ is risk value of permission p_i, and p_i is the single permission requested by the app.

Rule 2: For any requested combination of permissions PC_j with where permissions belong to clusters c_0 to c_{k-1}, the risk value $R_C(PC_j)$ is defined as:

$$R_c(PC_j) = \begin{cases} \prod R(p_i),\ p_i \in c_j \ and \ j < k \\ \sum R(p_i),\ p_i \in c_j \ and \ j \geq k \end{cases} \tag{10}$$

and the risk of the combined permissions is defined as the sum of all risk values of each combination of permissions; that is,

$$R_c = \sum_{j=0}^{m} R_c(PC_j) \tag{11}$$

Rule 3: The risk value R of an app is defined as the logarithmic mean of the total risk value:

$$R = \frac{log(R_S + R_C)}{M} \tag{12}$$

where M is the total number of requested single permissions and combinations of permissions.

In general, a normal app provides multiple services to satisfy users' functional needs, and several permissions are used, whereas a malicious app has simpler functions but uses permissions with a higher risk value [29]. Therefore, we use the mean risk rather than the total risk to evaluate the risk of an app. For apps whose R is less than a threshold, it can be preliminarily judged as a secure application. For apps with a higher R than the threshold, the mapping between the permission and corresponding API is constructed and added to the *Source* set, then the security decision of the taint data flow is entered.

4.3. Filtration of Taint Data Flow

The function call graph generated by FlowDroid is very complex. Frequent calls between functions cause a large number of redundant detection paths, which makes the further static analysis cost very high. In this study, data paths from native FlowDroid are further filtered based on the risk value.

On the basis of the definition of native FlowDroid, we define the following extra variables:

(1) *Route*: the taint transmission path set. The element is a pair—<*source, sink*>—and it indicates that there is a taint transmission path from *source* to *sink*, where $source \in Source$ and $sink \in Sink$.

(2) *LeakRoute*: the privacy leak path set. The element is a triple—<*source, sink, risk*>—and it indicates that there is a leak path from *source* to *sink* with a risk value of *risk*. The filtering method is shown in Figure 4.

For the filtering method in PBFlowDroid, the criterion is whether the risk value of the path is higher than the threshold. Based on the risk calculation for the path in the call graph, the filtering method extracts the paths with a higher risk value to *LeakRoute*. In this way, we obtain all the pollution paths and identify the data paths with a high risk of privacy leakage. Because the filtering method reduces the data paths to be analyzed, not only is the analysis time shortened, but also the accuracy of pollution analysis can be improved due to the elimination of false positive paths.

From the above, PBFlowDroid introduces a high-risk pollution path detection method and reduces the scale of pollution paths to be analyzed. Furthermore, PBFlowDroid can solve the challenge of the high false-positive and false-negative rate of analysis in native FlowDroid.

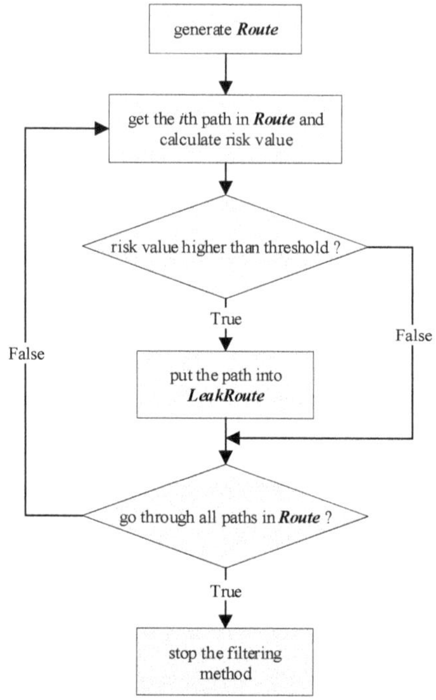

Figure 4. Workflow of filtering method.

5. Experiments

This section tests the accuracy and efficiency of the proposed PBFlowDroid. The computer used was a Z600 WorkStation with an Intel (R) Xeon (R) E5540 @ 2.53 GHz CPU and 4.00 GB of physical memory. All tests were run on Windows 7 with Oracle's Java Runtime version 1.8 (64 bit). Android 6.0 with Android-23 SDK was used in all experiments.

5.1. Accuracy Experiments

In this experiment, 500 normal apps from Google Play and 50 malicious apps from GitHub's Malicious Application Sample Library [5] were used as test samples. The risk value R of each successfully tested app was calculated. An app with an R value less than the threshold was recognized as a normal application; otherwise, it was identified as a malicious application. The risk threshold was an unknown parameter at the beginning of the experiment. We were inspired by machine learning [30] and used one-tenth of the experimental data as a training set to obtain the risk threshold. The training set contained 50 randomly selected normal apps and five malicious apps. By calculating the risk value of all apps in the training set and selecting the threshold to minimize the false positive rate, we obtained the risk value threshold. Other data were used as the validation set. In our experiments, we set 0.12 as the threshold value. Table 3 shows the results.

Table 3. Result of testing.

	Disassembled Successful	$R \geq 0.12$	$R < 0.12$
500 normal apps	413	81	332
50 malicious apps	35	27	8

In our experiments, 413 normal apps and 35 malicious apps were decompiled. Among the 413 normal apps, 332 were correctly identified and the other 81 were identified as malicious apps. The detection rate for the normal sample was 80.4%, and the false alarm rate was 19.6%. In total, 27 of 35 malicious apps were correctly identified. The omission ratio was 22.8% and accuracy rate was 77.2%. Compared with the test results of native FlowDroid in Droid Bench, where 30 malicious applications were detected out of 39 apps, with an accuracy of 76.9% [8], the proposed method guaranteed sufficient detection accuracy.

Table 4 gives the number of permissions in application as M and the taint paths and risk value as R for some apps. We can see that the R values of *FangTianxia* and *MeiPai* were 0.134 and 0.121, respectively; they were reported as malicious apps.

Table 4. Detection results for some apps.

Type	Number	Result			
		App Name	**M**	**Taint Paths**	**R**
News	8	Toutiao	28	53	0.091
		Funinput.Digit	29	43	0.089
		Sina news	36	36	0.071
Social media	8	Zhihu	29	42	0.083
		BaiduTieba	26	61	0.099
		Weibo	27	48	0.096
Services	7	Dianping	27	49	0.097
		FangTianxia	18	38	0.134
		Ganji	24	43	0.103
Tools	8	TencentMobileManager	41	103	0.064
		SougouTypewriting	24	49	0.104
		UC Browser	38	58	0.068
Entertainment	6	MeituXiuXiu	26	47	0.098
		Iqiyi	28	54	0.092
		MeiPai	20	38	0.121
Others	45				
Malicious applications	35	com.estrongs.android.pop. apk	7	20	0.311
		com.evernote. skitch.apk	18	42	0.133
		com.gau.go.launcherex.apk	9	28	0.250
		com.opera.browser.apk	16	46	0.153
		com.outfit7.talkinggina.apk	19	51	0.134

5.2. Efficiency Experiments

In PBFlowDroid, only data paths with a higher risk value are analyzed, which reduces the detection complexity. In this section, a comparative test with native FlowDroid in terms of detection time and memory consumption was performed. When we reproduced FlowDroid in our experimental environment, we found that FlowDroid took more than 10^4 s to analyze some apps, and the magnitude of the results of the completed path analysis was generally between 10^4 to 10^5. Because the time complexity of FlowDroid is orders of magnitude different from the method proposed in this article, we do not compare the apps directly in this article. Thus, 500 test samples were randomly selected from Google Play with sizes ranging from 50 KB to 60 MB. For apps larger than 10 MB, FlowDroid reports a timeout or out-of-memory error. Table 5 shows the time and memory consumption of PBFlowDroid for eight apps.

Table 5. Time and memory consumption comparison (OOM: out of memory).

	Name	Size	Runtime		Memory Consumption	
			FlowDroid	PBFlowDroid	FlowDroid	PBFlowDroid
1	InsecureBank.apk	58.5 KB	29.33 s	14.24 s	82.73 MB	52.26 MB
2	outfit7.talkinggina.apk	109 KB	41.1 s	19.58 s	241.55 MB	130.45 MB
3	com.evernote.skitch.apk	139 KB	51.76 s	18.21 s	272.56 MB	133.66 MB
4	BadNews.apk	1.37 MB	73.19 s	29.01 s	349.33 MB	166.14 MB
5	FakeCallandMessage	3.39 MB	82.01 s	41.73 s	428.16 MB	212.36 MB
6	BaiduNews	23.66 MB	Time Out	48.86 s	OOM	325.88 MB
7	SougouTypewriting	33.96 MB	Time Out	101.14 s	OOM	344.16 MB
8	YoudaoNote	63.23 MB	Time Out	215.84 s	OOM	683.25 MB

With the increase of the app size, the memory consumption increases significantly for both tools. For FlowDroid, memory is quickly exhausted, making the detection fail. For PBFlowDroid, memory consumption is kept within 4 GB and all tests were successful in our experiments.

Table 6 lists the multiple classification algorithms supported by PUMA [31] and their results. However, machine learning algorithms, including the method proposed by PUMA, can only discriminate whether an app is a malicious app through the use of permissions and cannot analyze how apps abuse permissions. In contrast to machine learning algorithms, our method and FlowDroid can not only analyze whether an app is a malicious app but also analyze its usage of permissions. Moreover, the results in Table 5 show that our app analysis consumes less resources than FlowDroid.

Table 6. Android malware detection results for the different algorithms.

Algorithm	Configuration	Accuracy
NaiveBayes	None	67.64%
IBK 10	K = 10	78.94%
RandomForest	I = 10	85.82%
FlowDroid	None	76.9%
PBFlowDroid	None	77.2%

6. Conclusions

FlowDroid is a static taint analysis tool widely used for Android apps. However, the call graph generated by FlowDroid grows exponentially as the size of the app increases, which reduces its availability. Research shows that the security threat of an app mainly comes from its abuse of permissions, and not all permissions will lead to a leakage of sensitive information. This paper proposes a method to identify dangerous data paths, and secure paths are filtered in further analysis. In this way, the call graph is greatly simplified and the resource requirements in the analysis process are significantly reduced. On the other hand, we used the Chi-square test and mutual information values to extract the correlated permissions and proposed risk calculation method considering permission combinations. In this way, a more accurate risk value is taken as a criterion to reduce misjudgment. The experimental results show that our proposed method reduces the complexity of detection significantly, and the detection accuracy is guaranteed.

In our future work, the communication between processes needs to be taken into account, and the assessment of communication risk is worth exploring. This will help to deal with the collusion attack problem. Second, the risk value of the data path should be determined based on API, application components and other features [32] rather than only permissions to improve the accuracy of pollution path identification. Third, the distinction between small malware and large malware should also be considered. We will import some large malware to the test set in our next work and prove the applicability of PBFlowDroid to large malware.

Author Contributions: Conceptualization, G.L. and L.Z.; methodology, H.K.; software, Z.W.; validation, Z.W., L.Z. and H.K.; formal analysis, L.Z.; writing—original draft preparation, H.K.; writing—review and editing, L.Z.; supervision, G.L.; project administration, Y.T. All authors have read and agreed to the published version of the manuscript.

Funding: This work was supported in part by the Shaanxi Key R & D Program (Grant No.2019ZDLGY13-01) and the Science and Technology Projects of Xi'an, China (Grant number:201809170CX11JC12).

Conflicts of Interest: The funders had no role in the design of the study; in the collection, analyses, or interpretation of data; in the writing of the manuscript, or in the decision to publish the results.

References

1. Kumar, P.; Naik, K.K. Android based Wireless Sensor Network application for airborne platforms (AWAAPS). In Proceedings of the 2016 IEEE International Conference on Engineering and Technology (ICETECH), Coimbatore, India, 17–18 March 2016; pp. 776–782. [CrossRef]
2. Serfass, D.; Yoshigoe, K. Wireless Sensor Networks using android virtual devices and Near Field Communication peer-to-peer emulation. In Proceedings of the 2012 Proceedings of IEEE Southeastcon, Orlando, FL, USA, 15–18 March 2012; pp. 1–6. [CrossRef]
3. Grano, G.; Di Sorbo, A.; Mercaldo, F.; Visaggio, C.A.; Canfora, G.; Panichella, S. Android apps and user feedback: A dataset for software evolution and quality improvement. In *Proceedings of the ACM Sigsoft International Workshop on App Market Analytics*; ACM: Chicago, IL, USA, 2017.
4. McAfee Labs Threats Report | McAfee Labs. August 2019. Available online: https://www.mcafee.com/enterprise/en-us/assets/reports/rp-quarterly-threats-aug-2019.pdf (accessed on 20 January 2021).
5. Alzaylaee, M.K.; Yerima, S.Y.; Sezer, S. DL-Droid: Deep learning based android malware detection using real devices. *Comput. Secur.* **2020**, *89*, 101663. [CrossRef]
6. Enck, W.; Gilbert, P.; Han, S.; Tendulkar, V.; Chun, B.-G.; Cox, L.P.; Jung, J.; McDaniel, P.; Sheth, A.N. TaintDroid: An information flow tracking system for real-time privacy monitoring on smartphones. *ACM Trans. Comput. Syst.* **2010**, *32*, 1–29. [CrossRef]
7. Bugiel, S.; Davi, L.; Dmitrienko, A.; Fischer, T.; Sadeghi, A.-R. *Xmandroid: A New Android Evolution to Mitigate Privilege Escalation Attacks*; Technical Report TR-2011-04; Technische Universität Darmstadt: Darmstadt, Germany, 2011.
8. Arzt, S.; Rasthofer, S.; Fritz, C.; Bodden, E.; Bartel, A.; Klein, J.; Traon, Y.L.; Octeau, D.; McDaniel, P. *FlowDroid: Precise Context, Flow, Field, Object-Sensitive and Life Cycle-Aware Taint Analysis for Android Apps*; ACM: Edinburgh, UK, 2014.
9. Han, K.S.; Lee, Y.; Jiang, B.; Im, E.G. How to Violate Android's Permission System without Violating It. In Proceedings of the Third International Conference on Digital Information Processing and Communications (ICDIPC2013), the Society of Digital Information Wireless Communication, Dubai, UAE, 30 January–1 February 2013; pp. 18–25.
10. Enck, W.; Ongtang, M.; Mcdaniel, P. On lightweight mobile phone application certification. In *Proceedings of the ACM Conference on Computer and Communications Security*; ACM: Chicago, IL, USA, 2009.
11. Zhao, Z.; Osono, F.C.C. "TrustDroid™": Preventing the use of SmartPhones for information leaking in corporate networks through the used of static analysis taint tracking. In *Proceedings of the International Conference on Malicious and Unwanted Software, Fajardo, PR, USA, 16–18 October 2012*; IEEE Computer Society: Piscataway, NJ, USA, 2012.
12. Yang, Z.; Yang, M. LeakMiner: Detect Information Leakage on Android with Static Taint Analysis. In *Proceedings of the Software Engineering, Wuhan, China, 6–8 November 2012*; IEEE: Piscataway, NJ, USA, 2013.
13. Cen, L.; Gates, C.S.; Si, L.; Li, N. A probabilistic discriminative model for android malware detection with decompiled source code. *IEEE Trans. Dependable Secur. Comput.* **2014**, *12*, 400–412. [CrossRef]
14. Kang, H.; Jang, J.W.; Mohaisen, A.; Kim, H.K. Detecting and classifying android malware using static analysis along with creator information. *Int. J. Distrib. Sens. Netw.* **2015**, *11*, 479174. [CrossRef]
15. Song, J.; Han, C.; Wang, K.; Zhao, J.; Ranjan, R.; Wang, L. An integrated static detection and analysis framework for Android. *Pervasive Mob. Comput.* **2016**, *32*, 15–25. [CrossRef]
16. Sun, M.; Li, X.; Lui, J.C.; Ma, R.T.; Liang, Z. Monet: A user-oriented behavior-based malware variants detection system for Android. *IEEE Trans. Inf. Forensics Secur.* **2017**, *12*, 1103–1112. [CrossRef]
17. Rovelli, P.; Vigfússon, Ý. PMDS: Permission-based malware detection system. In *ICISS 2014*; LNCS; Prakash, A., Shyamasundar, R., Eds.; Springer: Cham, Germany, 2014; Volume 8880, pp. 338–357. [CrossRef]
18. Fan, M.; Liu, J.; Wang, W.; Li, H.; Tian, Z.; Liu, T. DAPASA: Detecting Android Piggybacked Apps Through Sensitive Subgraph Analysis. *IEEE Trans. Inf. Forensics Secur.* **2017**, *12*, 1772–1785. [CrossRef]
19. Talha, K.A.; Alper, D.I.; Aydin, C. APK Auditor: Permission-based Android malware detection system. *Digit. Investig.* **2015**, *13*, 1–14. [CrossRef]
20. Li, C.; Mills, K.; Niu, D.; Zhu, R.; Zhang, H.; Kinawi, H. Android malware detection based on factorization machine. *IEEE Access* **2019**, *7*, 184008–184019. [CrossRef]
21. Reshetova, E.; Bonazzi, F.; Asokan, N. SELint: An SEAndroid Policy Analysis Tool. In Proceedings of the 3rd International Conference on Information Systems Security and Privacy, Porto, Portugal, 19–21 February 2017.

22. Vallee-Rai, R.; Gagnon, E.; Hendren, L.; Lam, P.; Pominville, P.; Sundaresan, V. Optimizing Java Bytecode Using the Soot Framework: Is It Feasible? In *Proceedings of the International Conference on Compiler Construction, Berlin, Germany, 25 March–2 April 2000*; Springer: Berlin/Heidelberg, Germany, 2000.
23. Reps, T.; Horwitz, S.; Sagiv, M. Precise inter procedural dataflow analysis via graph reachability. In Proceedings of the ACM Sigplan-sigact Symposium on Principles of Programming Languages; ACM: Chicago, IL, USA, 1995.
24. Sagiv, M.; Reps, T.; Horwitz, S. Precise Inter procedural Dataflow Analysis with Applications to Constant Propagation. *Theor. Comput. Sci.* **1996**, *167*, 131–170. [CrossRef]
25. Bodden, E. Inter-procedural data-flow analysis with ifds/ide and soot. In Proceedings of the ACM SIGPLAN International Workshop on State of the Art in Java Program analysis, SOAP '12, Beijing, China, 14 June 2012; pp. 3–8.
26. Allen, F.E. Control flow analysis. *ACM Sigplan Not.* **1970**, *5*, 1–19. [CrossRef]
27. NIST/SEMATECH e-Handbook of Statistical Methods. 2020. Available online: http://www.itl.nist.gov/div898/handbook/ (accessed on 20 January 2021).
28. Sarma, B.P.; Li, N.; Gates, C.; Gates, C.; Potharaju, R.; Nita-Rotaru, C.; Molloy, I. Android Permissions: A Perspective Combining Risks and Benefits. In Proceedings of the ACM Symposium on Access Control Models and Technologies, Barcelona, Spain, 16–18 June 2012; pp. 13–22.
29. Faruki, P.; Bharmal, A.; Laxmi, V.; Ganmoor, V.; Gaur, M.S.; Conti, M.; Rajarajan, M. Android Security: A Survey of Issues, Malware Penetration, and Defenses. *IEEE Commun. Surv. Tutor.* **2015**, *17*, 998–1022. [CrossRef]
30. Alpaydin, E. *Introduction to Machine Learning*; MIT Press: Cambridge, MA, USA, 2020.
31. Sanz, B.; Santos, I.; Laorden, C.; Ugarte-Pedrero, X.; Bringas, P.G.; Álvarez, G. PUMA: Permission Usage to Detect Malware in Android. In *International Joint Conference CISIS'12-ICEUTE'12-SOCO'12 Special Sessions*; Advances in Intelligent Systems and Computing, LNCS; Springer: Berlin/Heidelberg, Germany, 2013; Volume 189.
32. Jiang, X.; Mao, B.; Guan, J.; Huang, X. Android Malware Detection Using Fine-Grained Features. *Sci. Program.* **2020**, *2020*, 1–13. [CrossRef]

MDPI
St. Alban-Anlage 66
4052 Basel
Switzerland
Tel. +41 61 683 77 34
Fax +41 61 302 89 18
www.mdpi.com

Entropy Editorial Office
E-mail: entropy@mdpi.com
www.mdpi.com/journal/entropy

www.ingramcontent.com/pod-product-compliance
Lightning Source LLC
Chambersburg PA
CBHW061604220926
51850CB00005B/420
9783036576619